Environmental Stressors and OxInflammatory Tissue Responses

Environmental risk factors—noise, air pollution, chemical agents, and ultraviolet radiation—impact human health by contributing to the onset and progression of noncommunicable diseases. Accordingly, there is need for preclinical and clinical studies and comprehensive summary of major findings. This book is a state-of-the-art summary of these myriad severe life stressors. The chapters on the different pollutants focus on disease mechanisms (cardiovascular, neurological and metabolic disorders) and on oxidative stress and inflammation. The editors emphasize emerging mechanisms based on dysregulation of the circadian clock, the microbiome, epigenetic pathways, and cognitive function by environmental stressors and introduce the exposome concept while highlighting existing research gaps.

KEY FEATURES

- Links various environmental stressors to the incidence of noncommunicable diseases.
- Includes chapters on airborne toxins, chemical pollutants, noise, and ultraviolet radiation stressors.
- Provides contributions from an international team of leading researchers.
- Summarizes the impacts of stressors on disease mechanisms.

Oxidative Stress and Disease

Series Editors
Helmut Sies, MD
Enrique Cadenas, MD, PhD
University of Southern California School of Pharmacy
Los Angeles, California

Environmental Stressors and OxInflammatory Tissue Responses
edited by Giuseppe Valacchi and Andreas Daiber

Peroxiporins: Redox Signal Mediators In and Between Cells
edited by Iria Medraño-Fernandez, Gerd Patrick Bienert, and Roberto Sitia

Physiological Consequences of Brain Insulin Action
edited by André Kleinridders

Oxidative Eustress in Exercise Physiology
edited by James N. Cobley and Gareth W. Davison

Redox Regulation of Differentiation and De-differentiation
edited by Carsten Berndt and Horst Lillig

Oxidative Stress in Cancer, AIDS, and Neurodegenerative Diseases
edited by Luc Montagnier, René Olivier, and Catherine Pasquier

Understanding the Process of Aging: The Roles of Mitochondria, Free Radicals, and Antioxidants
edited by Enrique Cadenas and Lester Packer

Redox Regulation of Cell Signaling and Its Clinical Application
edited by Lester Packer and Junji Yodoi

Antioxidants in Diabetes Management
edited by Lester Packer, Peter Rösen, Hans J. Tritschler, George L. King, and Angelo Azzi

Free Radicals in Brain Pathophysiology
edited by Giuseppe Poli, Enrique Cadenas, and Lester Packer

Nutraceuticals in Health and Disease Prevention
edited by Klaus Krämer, Peter-Paul Hoppe, and Lester Packer

Environmental Stressors in Health and Disease
edited by Jürgen Fuchs and Lester Packer

Handbook of Antioxidants: Second Edition, Revised and Expanded
edited by Enrique Cadenas and Lester Packer

For more information about this series, please visit: www.crcpress.com/Oxidative-Stress-and-Disease/book-series/CRCOXISTRDIS

Environmental Stressors and OxInflammatory Tissue Responses

Edited by
Giuseppe Valacchi and Andreas Daiber

CRC Press
Taylor & Francis Group
Boca Raton London New York

CRC Press is an imprint of the
Taylor & Francis Group, an **informa** business

Designed cover image: Created with BioRender.com

First edition published 2024
by CRC Press
6000 Broken Sound Parkway NW, Suite 300, Boca Raton, FL 33487-2742

and by CRC Press
4 Park Square, Milton Park, Abingdon, Oxon, OX14 4RN

CRC Press is an imprint of Taylor & Francis Group, LLC

Library of Congress Cataloging-in-Publication Data
Names: Valacchi, Giuseppe, editor. | Daiber, Andreas, 1970– editor.
Title: Environmental stressors and oxinflammatory tissues responses / edited by Giuseppe Valacchi, Andreas Daiber.
Other titles: Oxidative stress and disease (Unnumbered)
Description: First edition. | Boca Raton, FL : CRC Press, 2024. | Series: Oxidative stress and disease | Includes bibliographical references and index.
Identifiers: LCCN 2023019704 (print) | LCCN 2023019705 (ebook) | ISBN 9781032357003 (hardback) | ISBN 9781032357010 (paperback) | ISBN 9781003328100 (ebook)
Subjects: MESH: Oxidative Stress | Environmental Pollutants—adverse effects | Inflammation—physiopathology
Classification: LCC HD75.6 (print) | LCC HD75.6 (ebook) | NLM QZ 160 | DDC 363.7—dc23/eng/20230929
LC record available at https://lccn.loc.gov/2023019704
LC ebook record available at https://lccn.loc.gov/2023019705

ISBN: 978-1-032-35700-3 (hbk)
ISBN: 978-1-032-35701-0 (pbk)
ISBN: 978-1-003-32810-0 (ebk)

DOI: 10.1201/9781003328100

Typeset in Times
by Apex CoVantage, LLC

Contents

Preface

During the last decades, there was a dramatic shift from communicable to noncommunicable diseases (NCDs; mostly of cardiovascular, neurodegenerative and metabolic origin). Environmental risk factors, including noise, air pollution, chemical agents and ultraviolet radiation (UVR), have a considerable impact on human health by contributing significantly to the onset and progression of these NCDs. Recent reports by leading environmental research organizations and consortia reported that all forms of chemical pollution together caused roughly 16–20% of total mortality worldwide. Importantly, these numbers even neglect the contribution of physical stressors such as UV and noise exposure or light pollution at night as well as mental stress to the overall environmental global health burden. In addition, clear evidence of "new pollutants" noxious effects to human health, i.e. nano and microplastic, together with new aspects related to the "old pollutants" need to be discussed and brought to the attention to the public. Accordingly, there is a high societal and scientific need of high-quality preclinical and clinical studies on this topic but also of concise and comprehensive summary of major findings in reviews and books. Since oxidative stress and inflammation represent key players in the molecular pathomechanisms of environmental pollution, we summarized with this book the actual state-of-the-art knowledge on this topic and provided a collection of expert book chapters on air pollution constituents, other chemical pollutants such as nano/microplastics or the behavioral risk factor water-pipe smoking, physical stressors such as noise and UV exposure as well as electromagnetic fields and finally mental stress in the form of social isolation or severe life stress.

In the first section of the book, we address the key concepts of the topic, highlighting in particular how the exposome, the complex mixtures of environmental risk factors deriving from human activities, can influence both human health and disease state. Furthermore, the difficulties in using biomarkers of oxidative stress in disease connected to environmental stressors is also discussed.

The second section examines the pathological mechanisms related to the generation of reactive oxygen species and their negative consequences on living organisms following exposure to airborne toxins, focusing on their effects on different target tissues including primary and distant organs such as lungs, eyes and brain.

In addition to the description of new pathomechanisms from well-known environmental stressors such as particulate matter and heavy metals, this book contains a series of chapters focusing on new aspects related to widely recognized signaling pathways implicated in the stress responses to outdoor stressors such as nuclear factor E2-related factor (NRF2) and aryl hydrocarbon (AHR). Furthermore, this book intends to shed new light on threats and physiopathological processes caused by contaminants of emerging concern including nano/microplastics, waterpipe smoking and e-cigarettes. The concept of exposome encompasses not only the chemical agents but also all environmental hazards components including those of a physical nature such as ultraviolet radiation, noise, and electromagnetic fields. These recognized physical risk factors are examined in the fourth section, emphasizing their involvement in human health through the activation of oxidative and inflammatory pathways able to affect from vascular to brain function, from skin to other organs physiology. Finally, the book outlines how the impact of social stress and childhood adversity can shape not only the brain but also the body via the induction of inflammatory and oxidative stress related processes.

The chapters on the different pollutants will address their impacts on disease mechanisms (cardiovascular, neurological and metabolic disorders) with focus on oxidative stress and inflammation in the underlying pathomechanisms. We will also emphasize emerging mechanisms based on dysregulation of the circadian clock, the microbiome, epigenetic pathways and cognitive function by environmental stressors, introduce the exposome concept and highlight the existing research gaps.

About the Editors

Giuseppe Valacchi, PhD, is David H. Murdock Distinguished Professor and Professor of Regenerative Medicine at the North Carolina State University. In addition, he is Professor of Physiology at the University of Ferrar; Adjunct Professor at Kyung Hee University, Seoul, South Korea; and President Elect for the SFRR-Europe. He earned his PhD in cellular physiology and neuro-immunophysiology from the University of Siena, Italy, followed by postdoctoral training at the Department of Molecular and Cell Biology, University of California-Berkeley and at the Pulmonary Division, University of California-Davis, and as Senior Researcher in the Department of Nutrition, University of California-Davis. Dr. Valacchi's research focuses on the effect of pollution (ozone, ultraviolet radiation, particulate matter and cigarette smoke) on target organs with emphasis on skin, lung and gastrointestinal tract. His research significantly contributed to the mechanism and the possible protection of skin towards ozone exposure and more in general to the OxInflammatory responses, including NLRP1 and NLRP3 inflammasome activation, induced by pollution exposure.

Andreas Daiber, PhD, earned a PhD in biological chemistry from the University of Konstanz and postdoctoral training at University of Konstanz and Experimental Cardiology, University Hospital Hamburg-Eppendorf and University Medical Center, Mainz, Germany, where he now is Professor and Head of Molecular Cardiology. In 2011 he was appointed a guest professorship at University of Grenoble, France, where he worked on protein nitration and redox-activation of sources of oxidative stress (crosstalk of NADPH oxidases and mitochondria) in the group of Prof. Schlattner. From 2014–2016, he was the chair of the European COST Action EU-ROS, a network of excellence with more than 230 participating scientists, dedicated to the mechanistic understanding and therapeutic exploitation of redox biological pathways and antioxidant pharmacological approaches. Dr. Daiber has a longtime research interest in oxidative stress-related cardiovascular disease and redox-regulation of vascular function. More recently, he investigates the effects of environmental stressors such as traffic noise and particulate matter on cardiovascular and metabolic function in humans and model organisms and has identified oxidative stress and inflammation as central players in mediating vascular dysfunction as well as (non-)pharmacological mitigation strategies related to antioxidant response such as NRF2 activation, exercise and fasting.

Contributors

Yumi Abiko
University of Tsukuba
Tsukuba, Japan

Silvia Alvarez
Universidad de Buenos Aires
Argentina

Michael Aschner
Albert Einstein College of Medicine
Bronx, New York, USA

Maria Teresa Bayo Jimenez
University of Granada
Granada, Spain

Barbara Benassi
ENEA-Casaccia Italian National Agency for New Technologies
Rome, Italy

Thomas Berkemeier
Max Planck Institute for Chemistry
Mainz, Germany

Claudia Consales
ENEA-Casaccia Italian National Agency for New Technologies
Italy

Andreas Daiber
University Medical Center
Mainz, Germany

Cristina Espinosa-Diez
Centro de Biología Molecular "Severo Ochoa" (CSIC-UAM)
Madrid, Spain

Pablo A. Evelson
Universidad de Buenos Aires
Argentina

Andre Faria
Federal University of Minas Gerais
Minas Gerais, Brazil

Arthur Faria
Pontifical Catholic University
Minas Gerais, Brazil

Francesca Ferrara
University of Ferrara
Ferrara, Italy

Sandra M. Ferreira
Universidad de Buenos Aires
Argentina

Henry Jay Forman
University of Southern California
Los Angeles, California, USA

Agustina Freire
Universidad de Buenos Aires
Argentina

Katie Frenis
Boston Children's Hospital and Harvard Medical School
Boston, Massachusetts, USA

Janine Fröhlich-Nowoisky
Max Planck Institute for Chemistry
Mainz, Germany

Pietro Ghezzi
University of Urbino
Urbino, Italy

Donya A. Gilan
Leibniz Institute for Resilience Research (LIR)
Mainz, Germany

Filipe Marques Gonçalves
Albert Einstein College of Medicine
Bronx, New York, USA

Florian Gruber
Christian Doppler Laboratory SKINMAGINE, Dermatology
Medical University of Vienna
Austria

Thomas Haarmann-Stemmann
IUF-Leibniz-Research Institute for Environmental Medicine
Düsseldorf, Germany

Omar Hahad
University Medical Center
Mainz, Germany

Moyan Hu
Ludwig-Maximilians-Universität München
Munich, Germany

Ailen Gala Hvozda Arana
Universidad de Buenos Aires
Argentina

Giorgia Innamorati
University of Rome "Tor Vergata"
Rome, Italy

Jasmin Ghaemi Kerahrodi
University Medical Center of the Johannes Gutenberg-University Mainz
Mainz, Germany

Christopher Kremslehner
Christian Doppler Laboratory SKINMAGINE, Dermatology
Medical University of Vienna
Austria

Yoshito Kumagai
University of Tsukuba
Tsukuba, Japan

Marin Kuntic
University Medical Center
Mainz, Germany

Ismail Laher
University of British Columbia
Vancouver, Canada

Romina M. Lasagni Vitar
Universidad de Buenos Aires
Argentina

Huangyuan Li
Fujian Medical University
Fuzhou, Fujian, China

Huige Li
University Medical Center
Mainz, Germany

Xinpei Lin
Fujian Medical University
Fuzhou, Fujian, China

Kurt Lucas
Max Planck Institute for Chemistry
Mainz, Germany

Natalia Magnani
Universidad de Buenos Aires
Argentina

Timoteo O. Marchini
Universidad de Buenos Aires
Argentina

Matthias Michal
University Medical Center of the Johannes Gutenberg-University Mainz
Mainz, Germany

Veronica Miguel
Institute of Experimental Medicine and Systems Biology
Aachen, Germany

Thomas Münzel
Department of Cardiology/Cardiology 1
University Medical Center
Mainz, Germany

Matthias Oelze
University Medical Center
Mainz, Germany

Dušan Palić
Ludwig-Maximilians-Universität München
Munich, Germany

Alessandra Pecorelli
University of Ferrara
Ferrara, Italy

Ulrich Pöschl
Max Planck Institute for Chemistry
Mainz, Germany

Sofía Reynoso
Universidad de Buenos Aires
Argentina

Giuseppe Valacchi
University of Ferrara
Ferrara, Italy

Christoph F. A. Vogel
University of California, Davis
Davis, California, USA

Konrad Wojnarowski
Ludwig-Maximilians-Universität München
Munich, Germany

Ning Xia
University Medical Center
Mainz, Germany

Fuli Zheng
Fujian Medical University
Fuzhou, Fujian, China

Section I

Conceptual

1 Man-Made Environment or Living in the Anthropocene
A Major Health Risk Factor in the 21st Century

Andreas Daiber and Giuseppe Valacchi

1.1 INTRODUCTION

The effect of humans on Earth's ecosystem has been recently identified as Anthropocene, which is a new geological epoch where humankind has caused mass extinctions of plant and animal species, polluted the oceans and altered the atmosphere, among other lasting impacts (Kress and Stine 2017). In addition, in the last century, doubling of the human population from 3 to 6 billion (and probably 9 billion in 2050) has accelerated the impact of humans on Earth "homeostasis" making those changes most likely irreversible (Steffen et al. 2011), probably also as a consequence that more than 50% of the population lives in urban areas. This number will even increase to 70% in the year 2050 according to projections of several public agencies (World Health Organization, United Nations, European Commission, World Bank) (Commission 2020).

The increase in pollution emissions in this new geological epoch, Anthropocene, have transformed the environment for living organisms dramatically to the extent that biogeochemical cycles, air and water quality, in addition to climate, have changed drastically to affect not only the present but also the future of our planet (Lelieveld 2017). Air pollution is probably one or even the leading environmental health risk factor in the Anthropocene and was reported to be associated with almost 9 million premature deaths per year at the global level (Lelieveld et al. 2019; Burnett et al. 2018). However, soil and water pollution also have major impacts on the global burden of disease (Daiber, Lelieveld, et al. 2019), not only by the rise of heavy metals (Neupane et al. 2022; Castro et al. 2021), pesticides (Gerber et al. 2021) and other toxic chemicals (Vane et al. 2011) in soil and water, but also by adverse changes of the soil microbiome that is vital to the primary ecological roles of soil (Guerra et al. 2021) and the most important role for humans, which is food production (Munzel et al. 2022). Also nanoplastic and microplastic pollution of the sea, surface and ground water, as well as soils, becomes a substantial concern as these toxic particles enter the human organism via the food chain and water supply (Munzel et al. 2022). Nanomaterials are already a significant component of ambient particulate matter and represent an emerging health issue (Fadeel et al. 2018; Kik, Bukowska, and Sicinska 2020). They also comprise an appreciable part of household air pollution (Chen, Feng, and Wang 2019; Akdogan and Guven 2019). A number of high-quality studies highlighted the uptake, distribution and pathophysiological effects of nanoplastic and microplastic particles in marine organisms such as plankton, mussels and fish as summarized by meta-analysis (Foley et al. 2018; Bucci, Tulio, and Rochman 2019). Accordingly, in order to achieve a full characterization of the effects of the Anthropocene on our health, a complete bottom-up ("bench to life") or top-down ("life to bench") approach should be applied (Figure 1.1). This would allow a complete understanding of the global (epidemiological) health effects as well as the basic pathomechanisms and toxic impact of the exposures at the cell, organ and organism level.

DOI: 10.1201/9781003328100-2

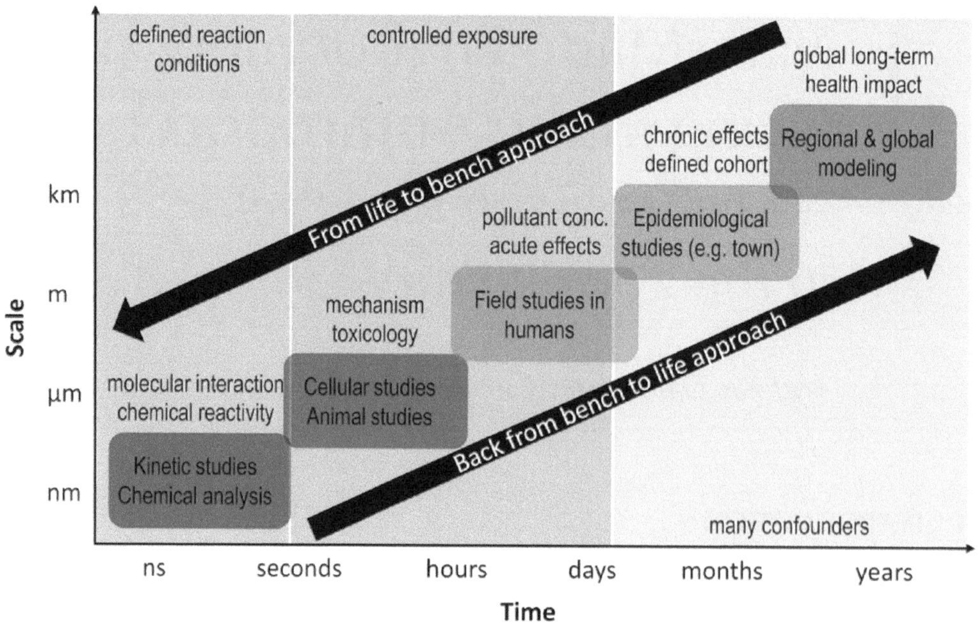

FIGURE 1.1 Studies on exposure-health effects cover a wide range of length and time scales. Different physicochemical and biomedical approaches (tone boxes) are needed to address the multifaceted aspects of exposure-health effects (text above tone boxes). Redrawn and updated from Shiraiwa et al. (2017) with permission.

1.2 HEALTH EFFECTS CAUSED BY THE ANTHROPOCENE

Within the last decades, the global burden of disease (GBD) has shifted from communicable, maternal, perinatal and nutritional causes to noncommunicable diseases such as atherosclerotic complications or metabolic disease (Lim et al. 2012; Murray et al. 2012). Leading scientists postulate a contribution of up to 66% by environmental risk factors to these noncommunicable diseases (Olden and Wilson 2000). There is increasing evidence that environmental stressors such as air pollution, noise and mental stress may facilitate the development of chronic noncommunicable disease (Munzel and Daiber 2018; Daiber, Lelieveld, et al. 2019). Recent reports of the Lancet Commission on Pollution and Health (Landrigan et al. 2018), the World Health Organization (WHO report 2016) and the Global Burden of Disease Study (Cohen et al. 2017; Collaborators 2017) reported that environmental risk factors significantly contribute to the global burden of disease. All forms of pollution together caused 9 to 12.6 million deaths in 2015 and 2012 (Landrigan et al. 2018; WHO report 2016), respectively, reflecting 16–20% of total mortality worldwide and representing a higher number of annual deaths than estimated for smoking.

The Anthropocene via air pollution also affects the potency of allergens, e.g. by modification of biomolecules such as proteins, lipids or DNA/RNA in pollen (or other constituents of the flora) by peroxidation/fragmentation, nucleic acid base oxidation or dimerization, sulfoxidation, tyrosine or tryptophane nitration and amino acid dimerization (Reinmuth-Selzle et al. 2017; Shiraiwa, Selzle, and Poschl 2012). Previously, the redox biology and oxidative stress underlying cardiovascular, neurological and immunological disorders in response to environmental risk factors and the existing research gaps, e.g. dysregulation of the circadian clock, the microbiome, epigenetic pathways and cognitive function by environmental stressors, were highlighted by a special issue in *Redox Biology* (Daiber and Munzel 2020). This also included a definition and discussion of the usefulness of oxidative stress biomarkers in this context (Ghezzi 2020). Focus of this previous special issue was laid on air pollution

(Gangwar et al. 2020; Kunert 2020; Ferrara et al. 2020), nano/microplastic particles (Palic 2020), heavy metals or pesticides (Zheng et al. 2020; Espinosa-Diez 2020; Vogel et al. 2020) and waterpipe smoking (Badran and Laher 2020). In addition, non-chemical environmental stressors such as UV-induced damage (Ferrara et al. 2020; Gruber 2020), traffic noise exposure (Daiber et al. 2020b; Steven et al. 2020) and mental stress (Li 2020; Michal 2020) were evaluated with respect to their impact on oxidative stress pathways and inflammation. With the present book, we attempt to explain the impact of the Anthropocene on redox biological changes and reactive oxygen species formation in full detail.

1.3 CONCLUSIONS

Toxic substances such as heavy (transition) metals and metalloids or pesticides that contaminate soil frequently produce oxidative stress and inflammation that is considered a common initiating event for multiple noncommunicable diseases (Cosselman, Navas-Acien, and Kaufman 2015; Li et al. 2020; Drechsel and Patel 2008). Also for traffic noise exposure there is ample evidence for a key role of oxidative stress and inflammation (Daiber et al. 2020a; Munzel, Sorensen, and Daiber 2021). Air pollution constituents, especially particulate matter, are known to induce oxidative stress and inflammation (Munzel et al. 2018, 2020; Gangwar et al. 2020). Also UV radiation leads to oxinflammation (Ferrara et al. 2020; Nicolaou, Pilkington, and Rhodes 2011) and nano-/microplastics mediate their adverse effects mostly by oxidative stress and inflammation (Hu and Palic 2020). So we can conclude that almost all environmental stressors induce the formation of reactive oxygen species and inflammation (Munzel and Daiber 2018; Daiber and Munzel 2020), either directly or secondary to induction of stress hormone signaling (Daiber, Kroller-Schon, et al. 2019; Ying et al. 2014; Brook and Rajagopalan 2021) or by impaired circadian rhythms (Li et al. 2020; Daiber et al. 2022) (Figure 1.2). Oxidative stress and inflammation will then activate down-stream pathomechanistic

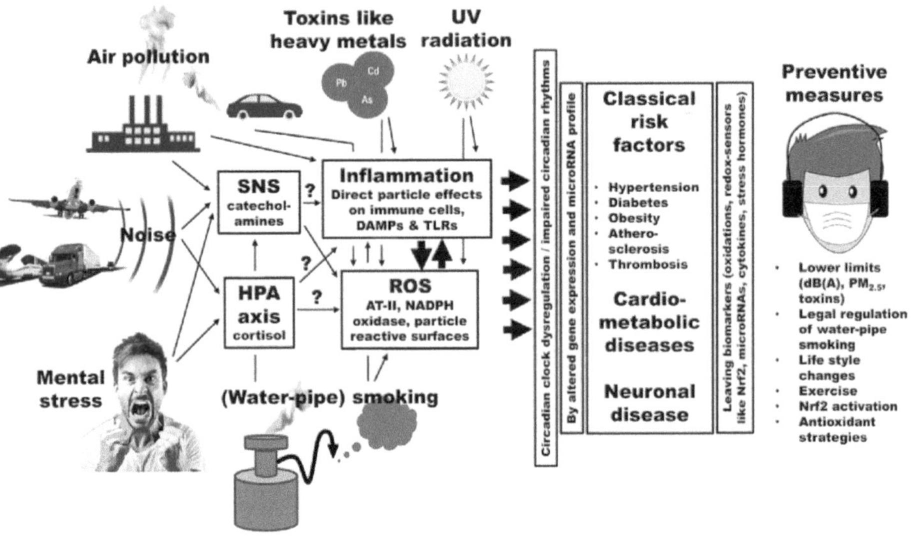

FIGURE 1.2 Environmental risk factors of the Anthropocene and redox-regulated pathomechanisms. Environmental stressors share common pathophysiological pathways centered on stress hormone signaling, oxidative stress and inflammation. An overlap of these central pathways is indicative of additive/synergistic adverse biochemical effects in response to environmental, lifestyle and traditional health risk factors leading to aggravated pathogenesis of noncommunicable diseases in a bonfire fashion (Daiber, Lelieveld, et al. 2019). Therefore, people with existing classical health risk factors or disease (e.g. diabetes or hypertension) may experience additive adverse health effects upon exposure to environmental risk factors. Mitigation measures may be of legislative nature, comprise personal protection strategies such as face masks, lifestyle changes or antioxidant treatments. Reused and updated from data in Munzel and Daiber (2018) with permission.

changes such as blood pressure elevation, platelet aggregation, metabolic dysfunction and atherosclerosis, which will transform upon chronic exposure to manifest cardiometabolic diseases (Daiber, Lelieveld, et al. 2019). Considering the accumulation of environmental risk factors in big cities and large urbanized areas (e.g. noise, air pollution and psychosocial stress), the health problems and disease burden associated with the sum of these environmental stressors may even outperform all previous estimations.

Undoubtedly, contaminating the air is the main risk factor for our health (Lelieveld et al. 2015). Air pollution is a threat to health and the environment as it has also been mentioned by the WHO Director-General Tedro Adhanom Ghebreyesus, "There is nothing more essential to life than air". Therefore, preserving clean air should be considered a human right, and it is among the main goals of the United Nations, including good health, climate action, sustainable cities, clean energy and protecting life on land and in the water (Lelieveld 2017). This was now achieved by a first step by which World Health Organization decreased their annual average exposure limits to 5 $\mu g/m^3$ in their newest air quality recommendations (Weichenthal et al. 2022). Now the national governments and legislations must follow. Of note, the European Union, with its limit of 25 $\mu g/m^3$ is 5-fold higher than recommended by the WHO.

REFERENCES

Akdogan, Z., and B. Guven. 2019. "Microplastics in the environment: A critical review of current understanding and identification of future research needs." *Environ Pollut* 254 (Pt A): 113011. https://doi.org/10.1016/j.envpol.2019.113011. www.ncbi.nlm.nih.gov/pubmed/31404735.

Badran, M., and I. Laher. 2020. "Waterpipe (shisha, hookah) smoking, oxidative stress and hidden disease potential." *Redox Biol*: 101455. https://doi.org/10.1016/j.redox.2020.101455. www.ncbi.nlm.nih.gov/pubmed/32086009.

Brook, R. D., and S. Rajagopalan. 2021. "Getting sympathetic about air pollution exposure." *J Am Heart Assoc* 10 (10): e021675. https://doi.org/10.1161/JAHA.121.021675. www.ncbi.nlm.nih.gov/pubmed/33942623.

Bucci, K., M. Tulio, and C. M. Rochman. 2019. "What is known and unknown about the effects of plastic pollution: A meta-analysis and systematic review." *Ecol Appl*. https://doi.org/10.1002/eap.2044. www.ncbi.nlm.nih.gov/pubmed/31758826.

Burnett, R., H. Chen, M. Szyszkowicz, N. Fann, B. Hubbell, C. A. Pope, 3rd, J. S. Apte, M. Brauer, A. Cohen, S. Weichenthal, et al. 2018. "Global estimates of mortality associated with long-term exposure to outdoor fine particulate matter." *Proc Natl Acad Sci U S A* 115 (38): 9592–9597. https://doi.org/10.1073/pnas.1803222115. www.ncbi.nlm.nih.gov/pubmed/30181279.

Castro, S., W. Luiz-Silva, W. Machado, and E. Valezio. 2021. "Mangrove sediments as long-term mercury sinks: Evidence from millennial to decadal time scales." *Mar Pollut Bull* 173 (Pt B): 113031. https://doi.org/10.1016/j.marpolbul.2021.113031. www.ncbi.nlm.nih.gov/pubmed/34656863.

Chen, G., Q. Feng, and J. Wang. 2019. "Mini-review of microplastics in the atmosphere and their risks to humans." *Sci Total Environ*: 135504. https://doi.org/10.1016/j.scitotenv.2019.135504. www.ncbi.nlm.nih.gov/pubmed/31753503.

Cohen, A. J., M. Brauer, R. Burnett, H. R. Anderson, J. Frostad, K. Estep, K. Balakrishnan, B. Brunekreef, L. Dandona, R. Dandona, et al. 2017. "Estimates and 25-year trends of the global burden of disease attributable to ambient air pollution: An analysis of data from the Global Burden of Diseases Study 2015." *Lancet* 389 (10082): 1907–1918. https://doi.org/10.1016/S0140-6736(17)30505-6. www.ncbi.nlm.nih.gov/pubmed/28408086.

Collaborators, G. B. D. Risk Factors. 2017. "Global, regional, and national comparative risk assessment of 84 behavioural, environmental and occupational, and metabolic risks or clusters of risks, 1990–2016: A systematic analysis for the Global Burden of Disease Study 2016." *Lancet* 390 (10100): 1345–1422. https://doi.org/10.1016/S0140-6736(17)32366-8. www.ncbi.nlm.nih.gov/pubmed/28919119.

Commission, European. 2020. "Competence centre on foresight." Developments and Forecasts on Continuing Urbanisation. https://knowledge4policy.ec.europa.eu/foresight/topic/continuing-urbanisation/developments-and-forecasts-on-continuing-urbanisation_en.

Cosselman, K. E., A. Navas-Acien, and J. D. Kaufman. 2015. "Environmental factors in cardiovascular disease." *Nat Rev Cardiol* 12 (11): 627–642. https://doi.org/10.1038/nrcardio.2015.152. www.ncbi.nlm.nih.gov/pubmed/26461967.

Daiber, A., K. Frenis, M. Kuntic, H. Li, E. Wolf, A. B. Kilgallen, S. Lecour, L. W. Van Laake, R. Schulz, O. Hahad, and T. Munzel. 2022. "Redox regulatory changes of circadian rhythm by the environmental risk factors traffic noise and air pollution." *Antioxid Redox Signal* 37 (10–12): 679–703. https://doi.org/10.1089/ars.2021.0272. www.ncbi.nlm.nih.gov/pubmed/35088601.

Daiber, A., S. Kroller-Schon, K. Frenis, M. Oelze, S. Kalinovic, K. Vujacic-Mirski, M. Kuntic, M. T. Bayo Jimenez, J. Helmstadter, S. Steven, et al. 2019. "Environmental noise induces the release of stress hormones and inflammatory signaling molecules leading to oxidative stress and vascular dysfunction-signatures of the internal exposome." *Biofactors* 45 (4): 495–506. https://doi.org/10.1002/biof.1506. www.ncbi.nlm.nih.gov/pubmed/30937979.

Daiber, A., S. Kroller-Schon, M. Oelze, O. Hahad, H. Li, R. Schulz, S. Steven, and T. Munzel. 2020a. "Oxidative stress and inflammation contribute to traffic noise-induced vascular and cerebral dysfunction via uncoupling of nitric oxide synthases." *Redox Biol* 34: 101506. https://doi.org/10.1016/j.redox.2020.101506. www.ncbi.nlm.nih.gov/pubmed/32371009.

Daiber, A., S. Kroller-Schon, M. Oelze, O. Hahad, H. Li, R. Schulz, S. Steven, and T. Munzel. 2020b. "Oxidative stress and inflammation contribute to traffic noise-induced vascular and cerebral dysfunction via uncoupling of nitric oxide synthases." *Redox Biol*: 101506. https://doi.org/10.1016/j.redox.2020.101506. www.ncbi.nlm.nih.gov/pubmed/32371009.

Daiber, A., J. Lelieveld, S. Steven, M. Oelze, S. Kroller-Schon, M. Sorensen, and T. Munzel. 2019. "The 'exposome' concept: How environmental risk factors influence cardiovascular health." *Acta Biochim Pol* 66 (3): 269–283. https://doi.org/10.18388/abp.2019_2853. www.ncbi.nlm.nih.gov/pubmed/31509369.

Daiber, A., and T. Munzel. 2020. "Special issue 'impact of environmental pollution and stress on redox signaling and oxidative stress pathways'." *Redox Biol* 37: 101621. https://doi.org/10.1016/j.redox.2020.101621. www.ncbi.nlm.nih.gov/pubmed/32863186.

Drechsel, D. A., and M. Patel. 2008. "Role of reactive oxygen species in the neurotoxicity of environmental agents implicated in Parkinson's disease." *Free Radic Biol Med* 44 (11): 1873–1886. https://doi.org/10.1016/j.freeradbiomed.2008.02.008. www.ncbi.nlm.nih.gov/pubmed/18342017.

Espinosa-Diez, C. 2020. "Role of non-coding-RNAs in response to environmental stressors and consequences on human health." *Redox Biol* REDOX_2020_245_R2.

Fadeel, B., L. Farcal, B. Hardy, S. Vazquez-Campos, D. Hristozov, A. Marcomini, I. Lynch, E. Valsami-Jones, H. Alenius, and K. Savolainen. 2018. "Advanced tools for the safety assessment of nanomaterials." *Nat Nanotechnol* 13 (7): 537–543. https://doi.org/10.1038/s41565-018-0185-0. www.ncbi.nlm.nih.gov/pubmed/29980781.

Ferrara, F., B. Woodby, A. Pecorelli, M. L. Schiavone, E. Pambianchi, N. Messano, J. P. Therrien, H. Choudhary, and G. Valacchi. 2020. "Additive effect of combined pollutants to UV induced skin OxInflammation damage: Evaluating the protective topical application of a cosmeceutical mixture formulation." *Redox Biol*: 101481. https://doi.org/10.1016/j.redox.2020.101481. www.ncbi.nlm.nih.gov/pubmed/32336667.

Foley, C. J., Z. S. Feiner, T. D. Malinich, and T. O. Hook. 2018. "A meta-analysis of the effects of exposure to microplastics on fish and aquatic invertebrates." *Sci Total Environ* 631–632: 550–559. https://doi.org/10.1016/j.scitotenv.2018.03.046. www.ncbi.nlm.nih.gov/pubmed/29529442.

Gangwar, R. S., G. H. Bevan, R. Palanivel, L. Das, and S. Rajagopalan. 2020. "Oxidative stress pathways of air pollution mediated toxicity: Recent insights." *Redox Biol*: 101545. https://doi.org/10.1016/j.redox.2020.101545. www.ncbi.nlm.nih.gov/pubmed/32505541.

Gerber, R., H. Bouwman, D. Govender, M. Ishizuka, Y. Ikenaka, Y. B. Yohannes, N. J. Smit, and V. Wepener. 2021. "Levels of DDTs and other organochlorine pesticides in healthy wild Nile crocodiles (Crocodylus niloticus) from a flagship conservation area." *Chemosphere* 264 (Pt 1): 128368. https://doi.org/10.1016/j.chemosphere.2020.128368. www.ncbi.nlm.nih.gov/pubmed/33007566.

Ghezzi, P. 2020. "Environmental risk factors and their footprints in vivo: A proposal for the classification of oxidative stress biomarkers." *Redox Biol*: 101442. https://doi.org/10.1016/j.redox.2020.101442. www.ncbi.nlm.nih.gov/pubmed/32035921.

Gruber, F. 2020. "Imaging of metabolic activity adaptations to UV stress, drugs and differentiation at cellular resolution in skin and skin equivalents: Implications for oxidative UV damage." *Redox Biol* REDOX_2020_255_R2.

Guerra, C. A., M. Delgado-Baquerizo, E. Duarte, O. Marigliano, C. Gorgen, F. T. Maestre, and N. Eisenhauer. 2021. "Global projections of the soil microbiome in the Anthropocene." *Glob Ecol Biogeogr* 30 (5): 987–999. https://doi.org/10.1111/geb.13273. www.ncbi.nlm.nih.gov/pubmed/33867861.

Hu, M., and D. Palic. 2020. "Micro- and nano-plastics activation of oxidative and inflammatory adverse outcome pathways." *Redox Biol* 37: 101620. https://doi.org/10.1016/j.redox.2020.101620. www.ncbi.nlm.nih.gov/pubmed/32863185.

Kik, K., B. Bukowska, and P. Sicinska. 2020. "Polystyrene nanoparticles: Sources, occurrence in the environment, distribution in tissues, accumulation and toxicity to various organisms." *Environ Pollut* 262: 114297. https://doi.org/10.1016/j.envpol.2020.114297. www.ncbi.nlm.nih.gov/pubmed/32155552.

Kress, J. W., and J. K. Stine. 2017. *Living in the Anthropocene: Earth in the Age of Humans*. Smithsonian Books, Washington, DC.

Kunert, A. T. 2020. "Chemical modification of pro-inflammatory proteins by peroxynitrite increases activation of TLR4 and NF-κB: Implications for the health effects of air pollution and oxidative stress." *Redox Biol* REDOX_2020_550_R1.

Landrigan, P. J., R. Fuller, N. J. R. Acosta, O. Adeyi, R. Arnold, N. N. Basu, A. B. Balde, R. Bertollini, S. Bose-O'Reilly, J. I. Boufford, et al. 2018. "The lancet commission on pollution and health." *Lancet* 391 (10119): 462–512. https://doi.org/10.1016/S0140-6736(17)32345-0. www.ncbi.nlm.nih.gov/pubmed/29056410.

Lelieveld, J. 2017. "Clean air in the Anthropocene." *Faraday Discuss* 200: 693–703. https://doi.org/10.1039/c7fd90032e. www.ncbi.nlm.nih.gov/pubmed/28702627.

Lelieveld, J., J. S. Evans, M. Fnais, D. Giannadaki, and A. Pozzer. 2015. "The contribution of outdoor air pollution sources to premature mortality on a global scale." *Nature* 525 (7569): 367–371. https://doi.org/10.1038/nature15371. www.ncbi.nlm.nih.gov/pubmed/26381985.

Lelieveld, J., K. Klingmuller, A. Pozzer, U. Poschl, M. Fnais, A. Daiber, and T. Munzel. 2019. "Cardiovascular disease burden from ambient air pollution in Europe reassessed using novel hazard ratio functions." *Eur Heart J* 40 (20): 1590–1596. https://doi.org/10.1093/eurheartj/ehz135. www.ncbi.nlm.nih.gov/pubmed/30860255.

Li, H. 2020. "The role of oxidative stress in cardiovascular disease caused by social isolation and loneliness." *Redox Biol* REDOX_2020_228_R1.

Li, H., A. B. Kilgallen, T. Munzel, E. Wolf, S. Lecour, R. Schulz, A. Daiber, and L. W. Van Laake. 2020. "Influence of mental stress and environmental toxins on circadian clocks: Implications for redox regulation of the heart and cardioprotection." *Br J Pharmacol* 177 (23): 5393–5412. https://doi.org/10.1111/bph.14949. www.ncbi.nlm.nih.gov/pubmed/31833063.

Lim, S. S., T. Vos, A. D. Flaxman, G. Danaei, K. Shibuya, H. Adair-Rohani, M. Amann, H. R. Anderson, K. G. Andrews, M. Aryee, et al. 2012. "A comparative risk assessment of burden of disease and injury attributable to 67 risk factors and risk factor clusters in 21 regions, 1990–2010: A systematic analysis for the Global Burden of Disease Study 2010." *Lancet* 380 (9859): 2224–2260. https://doi.org/10.1016/S0140-6736(12)61766-8. www.ncbi.nlm.nih.gov/pubmed/23245609.

Michal, M. 2020. "The fear-defense system, emotions, and oxidative stress." *Redox Biol* REDOX_2020_258_R1.

Munzel, T., and A. Daiber. 2018. "Environmental stressors and their impact on health and disease with focus on oxidative stress." *Antioxid Redox Signal* 28 (9): 735–740. https://doi.org/10.1089/ars.2017.7488. www.ncbi.nlm.nih.gov/pubmed/29278923.

Munzel, T., T. Gori, S. Al-Kindi, J. Deanfield, J. Lelieveld, A. Daiber, and S. Rajagopalan. 2018. "Effects of gaseous and solid constituents of air pollution on endothelial function." *Eur Heart J* 39 (38): 3543–3550. https://doi.org/10.1093/eurheartj/ehy481. www.ncbi.nlm.nih.gov/pubmed/30124840.

Munzel, T., O. Hahad, A. Daiber, and P. J. Landrigan. 2022. "Soil and water pollution and human health: What should cardiologists worry about?" *Cardiovasc Res*. https://doi.org/10.1093/cvr/cvac082. www.ncbi.nlm.nih.gov/pubmed/35772469.

Munzel, T., M. Sorensen, and A. Daiber. 2021. "Transportation noise pollution and cardiovascular disease." *Nat Rev Cardiol* 18 (9): 619–636. https://doi.org/10.1038/s41569-021-00532-5. www.ncbi.nlm.nih.gov/pubmed/33790462.

Munzel, T., S. Steven, K. Frenis, J. Lelieveld, O. Hahad, and A. Daiber. 2020. "Environmental factors such as noise and air pollution and vascular disease." *Antioxid Redox Signal* 33 (9): 581–601. https://doi.org/10.1089/ars.2020.8090. www.ncbi.nlm.nih.gov/pubmed/32245334.

Murray, C. J., M. Ezzati, A. D. Flaxman, S. Lim, R. Lozano, C. Michaud, M. Naghavi, J. A. Salomon, K. Shibuya, T. Vos, et al. 2012. "GBD 2010: Design, definitions, and metrics." *Lancet* 380 (9859): 2063–2066. https://doi.org/10.1016/S0140-6736(12)61899-6. www.ncbi.nlm.nih.gov/pubmed/23245602.

Neupane, B., K. Bao, M. Chen, P. Thapa, and M. E. Meadows. 2022. "The timing and magnitude of anthropogenic mercury pollution: A 200-year record from multi-lake sediment cores in northeast China." *Chemosphere* 309 (Pt 2): 136803. https://doi.org/10.1016/j.chemosphere.2022.136803. www.ncbi.nlm.nih.gov/pubmed/36223823.

Nicolaou, A., S. M. Pilkington, and L. E. Rhodes. 2011. "Ultraviolet-radiation induced skin inflammation: Dissecting the role of bioactive lipids." *Chem Phys Lipids* 164 (6): 535–543. https://doi.org/10.1016/j.chemphyslip.2011.04.005. www.ncbi.nlm.nih.gov/pubmed/21524643.

Olden, K., and S. Wilson. 2000. "Environmental health and genomics: Visions and implications." *Nat Rev Genet* 1 (2): 149–153. https://doi.org/10.1038/35038586. www.ncbi.nlm.nih.gov/pubmed/11253655.

Palic, D. 2020. "Micro- and nano-plastics activation of oxidative and inflammatory adverse outcome pathways." *Redox Biol* REDOX_2020_296.

Reinmuth-Selzle, K., C. J. Kampf, K. Lucas, N. Lang-Yona, J. Frohlich-Nowoisky, M. Shiraiwa, P. S. J. Lakey, S. Lai, F. Liu, A. T. Kunert, et al. 2017. "Air pollution and climate change effects on allergies in the Anthropocene: Abundance, interaction, and modification of allergens and adjuvants." *Environ Sci Technol* 51 (8): 4119–4141. https://doi.org/10.1021/acs.est.6b04908. www.ncbi.nlm.nih.gov/pubmed/28326768.

Shiraiwa, M., K. Selzle, and U. Poschl. 2012. "Hazardous components and health effects of atmospheric aerosol particles: Reactive oxygen species, soot, polycyclic aromatic compounds and allergenic proteins." *Free Radic Res* 46 (8): 927–939. https://doi.org/10.3109/10715762.2012.663084. www.ncbi.nlm.nih.gov/pubmed/22300277.

Shiraiwa, M., K. Ueda, A. Pozzer, G. Lammel, C. J. Kampf, A. Fushimi, S. Enami, A. M. Arangio, J. Frohlich-Nowoisky, Y. Fujitani, et al. 2017. "Aerosol health effects from molecular to global scales." *Environ Sci Technol* 51 (23): 13545–13567. https://doi.org/10.1021/acs.est.7b04417. www.ncbi.nlm.nih.gov/pubmed/29111690.

Steffen, W., J. Grinevald, P. Crutzen, and J. McNeill. 2011. "The Anthropocene: Conceptual and historical perspectives." *Philos Trans A Math Phys Eng Sci* 369 (1938): 842–867. https://doi.org/10.1098/rsta.2010.0327. www.ncbi.nlm.nih.gov/pubmed/21282150.

Steven, S., K. Frenis, S. Kalinovic, M. Kvandova, M. Oelze, J. Helmstadter, O. Hahad, K. Filippou, K. Kus, C. Trevisan, et al. 2020. "Exacerbation of adverse cardiovascular effects of aircraft noise in an animal model of arterial hypertension." *Redox Biol*: 101515. https://doi.org/10.1016/j.redox.2020.101515. www.ncbi.nlm.nih.gov/pubmed/32345536.

Vane, C. H., S. R. Chenery, I. Harrison, A. W. Kim, V. Moss-Hayes, and D. G. Jones. 2011. "Chemical signatures of the Anthropocene in the Clyde estuary, UK: Sediment-hosted Pb, (207/206)Pb, total petroleum hydrocarbon, polyaromatic hydrocarbon and polychlorinated biphenyl pollution records." *Philos Trans A Math Phys Eng Sci* 369 (1938): 1085–1111. https://doi.org/10.1098/rsta.2010.0298. www.ncbi.nlm.nih.gov/pubmed/21282161.

Vogel, C. F. A., L. S. Van Winkle, C. Esser, and T. Haarmann-Stemmann. 2020. "The aryl hydrocarbon receptor as a target of environmental stressors: Implications for pollution mediated stress and inflammatory responses." *Redox Biol*: 101530. https://doi.org/10.1016/j.redox.2020.101530. www.ncbi.nlm.nih.gov/pubmed/32354640.

Weichenthal, S., L. Pinault, T. Christidis, R. T. Burnett, J. R. Brook, Y. Chu, D. L. Crouse, A. C. Erickson, P. Hystad, C. Li, et al. 2022. "How low can you go? Air pollution affects mortality at very low levels." *Sci Adv* 8 (39): eabo3381. https://doi.org/10.1126/sciadv.abo3381. www.ncbi.nlm.nih.gov/pubmed/36170354.

WHO report. 2016. "WHO report: Preventing disease through healthy environments." www.who.int/publications/i/item/9789241565196 (last accessed 11/03/23).

Ying, Z., X. Xu, Y. Bai, J. Zhong, M. Chen, Y. Liang, J. Zhao, D. Liu, M. Morishita, Q. Sun, et al. 2014. "Long-term exposure to concentrated ambient PM2.5 increases mouse blood pressure through abnormal activation of the sympathetic nervous system: A role for hypothalamic inflammation." *Environ Health Perspect* 122 (1): 79–86. https://doi.org/10.1289/ehp.1307151. www.ncbi.nlm.nih.gov/pubmed/24240275.

Zheng, F., F. M. Goncalves, Y. Abiko, H. Li, Y. Kumagai, and M. Aschner. 2020. "Redox toxicology of environmental chemicals causing oxidative stress." *Redox Biol* 34: 101475. https://doi.org/10.1016/j.redox.2020.101475. www.ncbi.nlm.nih.gov/pubmed/32336668.

2 The Exposome Concept

Description of Lifelong Environmental Exposure Effects on Metabolism, Health and Disease

Andreas Daiber, Giuseppe Valacchi, Matthias Oelze, Marin Kuntic, and Thomas Münzel

2.1 INTRODUCTION

Results of large global epidemiological and health association studies such as the Global Burden of Disease study suggest that approximately 2/3 of the annual deaths in the world can be attributed to noncommunicable diseases (NCDs) secondary to atherosclerosis or diabetes (Lozano et al. 2012). The decline in the importance of communicable (infectious) diseases can be explained by a general improvement in hygiene (e.g., better sanitation and water quality) and primary medical care, even in low-income countries. Vice versa, the increased contribution of NCDs is most likely due to the aging population in Western societies and the growing importance of environmental risk factors such as air, soil, and water pollution as well as traffic noise and mental stress factors (Landrigan et al. 2018). Whereas scientific efforts during the last decades focused on classical cardiovascular risk factors such as diabetes, hypertension, and being overweight, more recent global investigations show that environmental risk factors, with a leading role of air pollution, induce significant adverse health effects and thereby contribute to a large extent to the global burden of disease (Lozano et al. 2012; Landrigan et al. 2018).

The "Lancet Commission on pollution and health" concluded that chemical pollution is the leading environmental cause of chronic diseases and premature deaths in the world today (Landrigan et al. 2018). All forms of chemical pollution are thought to have caused 9 million premature deaths worldwide in the year 2015, which was 16% of all annual deaths in the world—3-fold more than HIV, tuberculosis, and malaria together—a similar number of deaths as caused by smoking and passive smoking (Figure 2.1a) (updated numbers from Fuller et al. 2022). A comparable number was published by the World Health Organization (WHO), postulating global excess mortality of 12.6 million in 2012 due to people living in unhealthy environments (Webpage 2016b, 2016a). In the most affected countries (those with low income), one in four deaths is caused by environmental risk factors. According to estimations of the Global Burden of Disease study, all forms of environmental pollution are responsible for 268 million disability-adjusted life years (DALYs) in the world, a measure of the population disease burden that includes the loss of life years as well as years lived with severe disease or disability (DALYs and Collaborators 2016). Air pollution represents the most important form of chemical environmental health risk, and it was reported that particulate matter with a diameter ≤ 2.5 µm ($PM_{2.5}$) and ozone are responsible for 8.8 million premature annual deaths in the world (Lelieveld et al. 2019). These numbers will further increase in the future due to improved exposure-response functions and more accurate high-exposure level studies from megacities (Figure 2.1b) (updated from

DOI: 10.1201/9781003328100-3

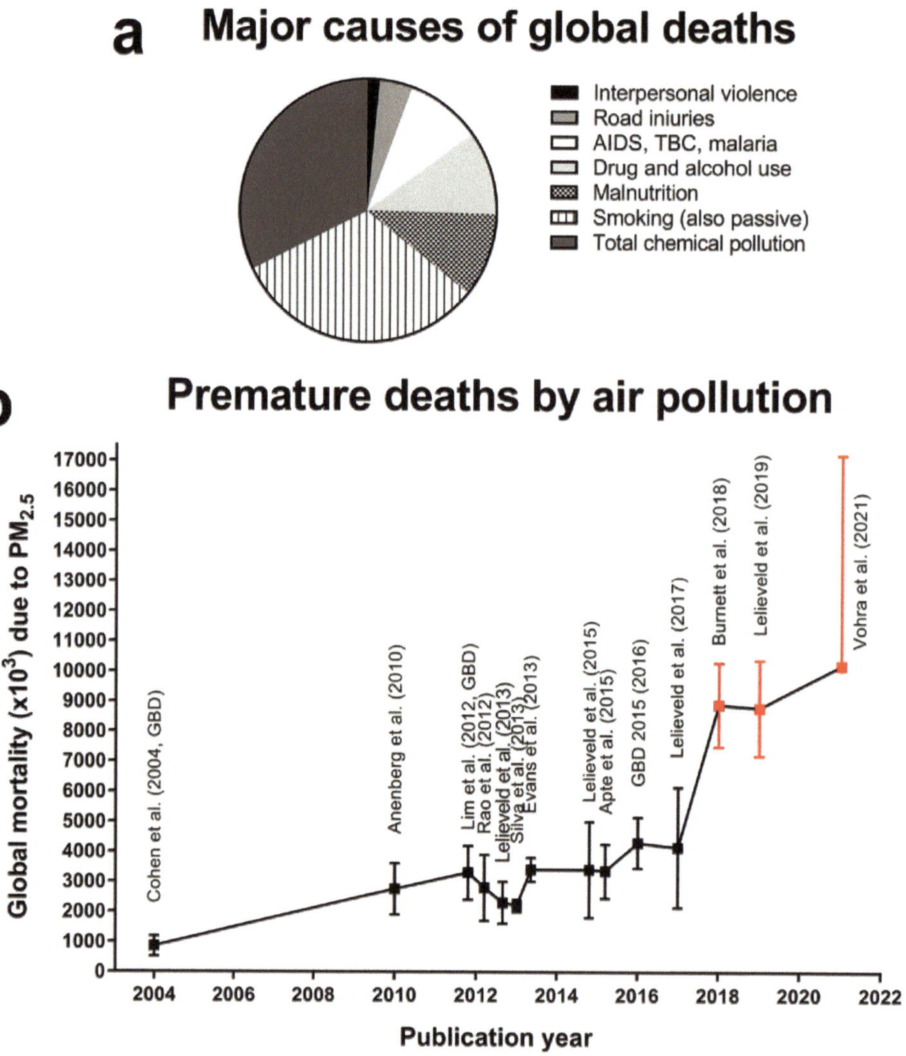

FIGURE 2.1 Environmental factors and the global burden of disease. (a) Relative contribution of the significant causes of death to global mortality. Approximated from data in Fuller et al. (2022) with permission. (b) Literature estimates of global premature mortality attributed to outdoor air pollution by fine particulate matter ($PM_{2.5}$). GBD, Global Burden of Disease study; AIDS, acquired immunodeficiency syndrome; TBC, tuberculosis. Black symbols represent data with mean ± SD; red symbols represent data with mean and 95% confidence intervals. Approximated and updated from data in Daiber, Lelieveld, et al. 2019; Shiraiwa et al. 2017 with permission under the Creative Commons Attribution-ShareAlike 4.0 International License.

Daiber, Lelieveld, et al. 2019; Shiraiwa et al. 2017). Notably, air pollution caused an estimated 3.5% of all healthcare costs in high-income countries and 7.4% in low-income countries in 2013 (Landrigan et al. 2018). Of note, these air pollution-attributable healthcare costs do not reflect this environmental exposure's total socioeconomic burden, since also the loss of productivity due to sick days and invalidity pension of the chronically exposed and disabled people must be considered.

When taking the colocalization of environmental exposures in large cities or metropolitan areas into account, as seen for air pollution, traffic noise, and the light at night as well as psychosocial factors such as mental stress (Munzel and Daiber 2018; Munzel et al. 2021; Munzel, Hahad, and

Daiber 2021), the resulting health impact and disease burden could even exceed the worst scenario estimations (Daiber and Munzel 2020). Accordingly, the socioeconomic savings could be much higher than so far estimated. Therefore, assessment of multi-exposure scenarios is crucial to understanding and estimating the actual disease burden caused by environmental stressors, e.g., as seen for the cumulative diabetes risk by considering the combined exposure of the Danish population to $PM_{2.5}$, ultrafine particles, traffic noise, and lack of green space (Sorensen et al. 2022). This also warrants high-quality mechanistic studies on the combined health effects of two or more environmental stressors in the future. An example is the combined exposure of mice to ambient $PM_{2.5}$ and aircraft noise, revealing dominant biochemical damage by $PM_{2.5}$ in the lung and by noise in the brain. In contrast, additive functional and biochemical damage was observed in the vasculature (Kuntic et al. 2023). Key mechanisms of this observed damage were centered on oxidative stress and inflammation.

Environmental pollution, together with other environmental risk factors of the lifestyle sector (e.g., smoking, nutrition, physical activity), climatic conditions (e.g., high temperature and humidity, UV radiation), and the social capital (e.g., social isolation, work strain), constitute the sum of all environmental exposures, summarized by the term "exposome". All forms of chemical pollution in the air, soil, and water, but also physical exposures such as noise, artificial nocturnal light or climatic parameters, and mental stressors have an enormous impact on the biochemical phenotype, e.g., by influencing metabolic processes, inflammatory cascades, redox homeostasis, and circadian rhythm (Li et al. 2020; Daiber et al. 2022).

2.2 A NEW INITIATIVE: THE "ENVIRONMENT-WIDE ASSOCIATION STUDY" (EWAS)

Despite alarming reports from various international expert commissions and growing scientific evidence for the detrimental impact of environmental risk factors on public health and socioeconomic burden, these environmental exposures were rather insufficiently considered by clinical guidelines (Munzel, Sorensen, and Daiber 2021), global prevention plans (Webpage 2013) or health politics (Landrigan et al. 2018). In addition, the investments made into research on environmental risk factors pale in the light of the astronomic financial resources spent for the characterization of the human genome and the correlation of genetic predisposition with the development of NCDs ("heritable diseases") (Sainani 2016), which defines the research field of Genome-wide Association Studies (GWAS). This becomes even more unintelligible when considering recent estimations by leading experts postulating that almost 2/3 of all chronic NCDs are caused by environmental risk factors, thereby associating a much higher share of the global burden of disease with exposures by unhealthy environment than with genetic predisposition (Rappaport 2016). This made health experts coin the term "genetics loads the gun but environment pulls the trigger" (Olden and Wilson 2000) and call for a new global health project, in analogy to the GWAS initiative, that should investigate primarily all environmentally-triggered diseases by considering various exposures—the Environment-Wide Association Study (EWAS) (Sainani 2016; Zheng et al. 2020).

This kind of a study (EWAS) should measure the sum of all exposures over the entire lifespan in thousands of individuals, correlate the exposures with biochemical changes in the intrinsic systems, and finally assign them to health outcomes and mortality in the whole study population, all of which represents a significant challenge to date (Sainani 2016; Zheng et al. 2020; Daiber, Lelieveld, et al. 2019). The technological progress will enable high-resolution measurement of multiple exposures by satellite-based methods and ground-based detectors as well as mobile sensors and data collection by smartphone apps (Munzel et al. 2017; Munzel et al. 2021). Our social activities and lifestyle are also tracked more and more by mobile devices and social media platforms. In addition, close-mesh determination of toxins, metabolites, proteins, and genes in our organism by so-called OMICs methods (adductome, metabolome, proteome, transcriptome/epigenome) becomes more and more routine due to the rapid biotechnological developments and cheaper prices (Daiber, Lelieveld, et al. 2019; Jones 2016; Vermeulen et al. 2020; Dennis et al. 2017). Also, the analytical approach is challenging since e.g.; the blood exposome requires mass spectrometric analysis that covers a concentration range of the measured biomolecules, metabolites, and toxic compounds from femtomolar to millimolar concentrations (Rappaport et al. 2014). On top of the broad concentration range, differences in the polarity and stability of the measured compounds make exposome analytics a nightmare.

Our health data will soon be archived electronically and could be assessed by study nurses and physicians responsible for exposome studies in which we participate. The handling and analysis of the astronomical data volumes ("big data") will be performed by supercomputers, applying self-learning bioinformatical algorithms (artificial intelligence) and creating interactome and diseasome maps. Combining all exposure data, OMICs datasets, and health outcome knowledge within a holistic, unbiased systems biology approach. Even artificial intelligence-based health prognosis and preventive measure (therapeutic) recommendations for specific environmental exposures may be possible. However, numerous ethical and data protection-related hurdles have to be taken to prevent the misuse of exposome-related data of a "vitreous" participant.

2.3 THE EXPOSOME CONCEPT: A COMPREHENSIVE DESCRIPTION OF THE LIFELONG EXPOSURE HISTORY AND ASSOCIATED BIOCHEMICAL CHANGES AND HEALTH OUTCOMES

The term "exposome" was introduced in 2005 by Christopher P. Wild and described in full detail. The exposome represents the sum of all changes in human physiology (the "internal environment") by environmental factors at a lifespan dimension (Wild 2005). As already discussed in the preceding sections, the exposures not only comprise chemical and physical factors such as air pollution, traffic noise, UV radiation, and climate-related stressors but also socioeconomic determinants related to the social, urban, and work environment (e.g., social isolation, mental stress, economic status, neighborhood) that can all be assigned to the "general external environment" (Figure 2.2) (Vrijheid 2014). Importantly, the availability or lack of green space is also an important environmental factor for our health (Nieuwenhuijsen 2020; Munzel et al. 2021). It is debatable whether exposure to pathogens such as viruses and bacteria is also part of the exposome or not. Our lifestyle, e.g., whether we smoke, drink too much alcohol, eat healthily, do physical exercise, and which consumer products we use, determines the "specific environment" (Vrijheid 2014). Importantly, whereas the individual can easily change the specific environment, the general external environment cannot be altered easily. The exposome can be divided into arbitrary "pollutomes" that are either organized by the periods of life (e.g., pre-birth, lactation, infancy, adolescence, adulthood, seniority) or the place of exposure (e.g., occupational exposome, urbanome) (Landrigan et al. 2018; Faisandier et al. 2011; Morawska et al. 2022). Accordingly, the pollutomes represent spatial and temporal exposures with the associated impact on human physiology and health. Changes in the internal environment by environmental risk factors are mostly related to oxidative stress and inflammation as well as stress hormone signaling (Daiber, Kroller-Schon, et al. 2019; Daiber, Lelieveld, et al. 2019; Munzel and Daiber 2018), with secondary dysregulation of circadian rhythms and metabolic pathways by these primary pathomechanisms (Li et al. 2020; Daiber et al. 2022; Daiber and Munzel 2020).

The influence of environmental stressors is most pronounced during the human development phase (e.g., in the fetus, infants, and children), which represents a largely unexplored field of research since studies among these specific individuals underlie substantial ethical stipulations. However, this vulnerable population is of great interest for exposome studies since even exposures below the legal limits could cause significant impairment of neuronal development and other end-organ damage (Bellinger 2013). Exposures during these time windows of highest

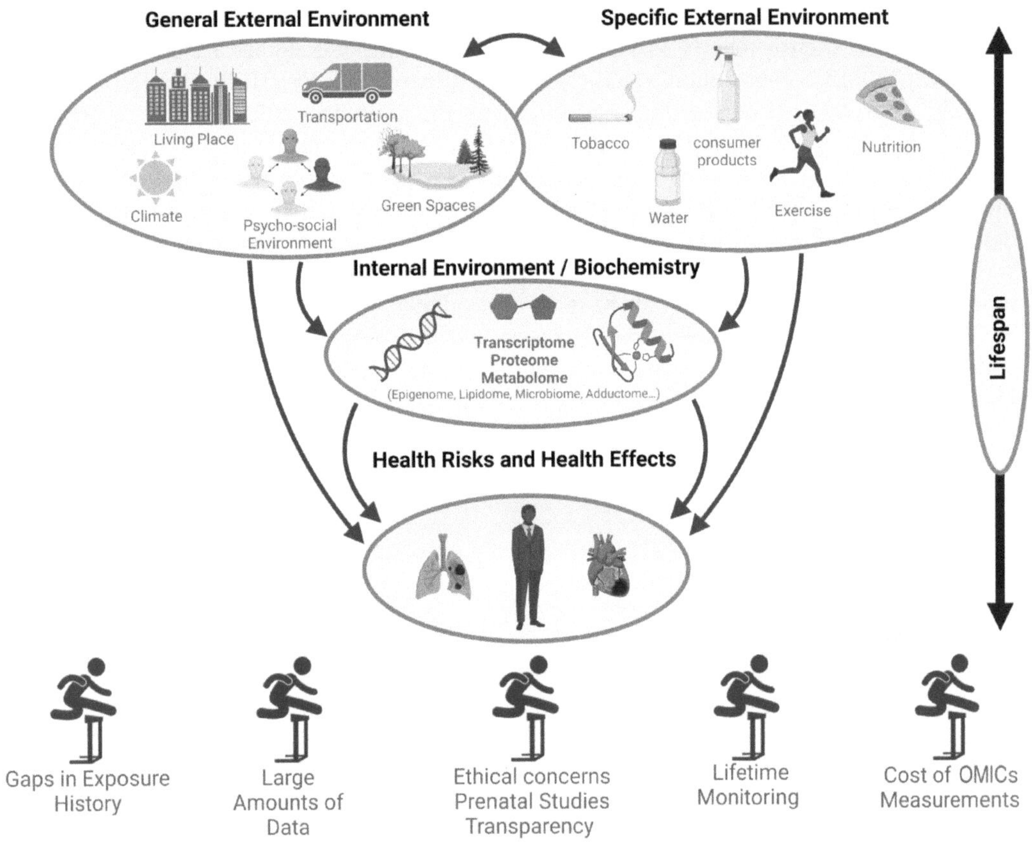

FIGURE 2.2 The exposome concept. The exposome comprises the totality of all lifelong exposures of an individual, from birth (or even in utero) to death. Exposures can be assigned to the general or specific external environment, correlated with biochemical changes in the internal environment, and associated with (adverse) health outcomes, including mortality. At the bottom of the image, the primary "hurdles" of exposome studies (EWAS) are presented. Approximated and modified from Vrijheid et al. *Thorax* 2014 (Vrijheid 2014). Reused and translated from (Daiber et al. 2021) with permission. Copyright © 2021, Rights Managed by Georg Thieme Verlag KG Stuttgart, New York.

susceptibility to environmentally-triggered physical harm, e.g., during early life, could lead to long-term health complications and chronic NCDs, which may also comprise reduced cognitive function, mental disorders, violence during adolescence, or lifelong economic unproductivity (Landrigan et al. 2018). Fetal or infantile reprogramming of specific biological processes by environmental risk factors (e.g., long-term epigenetic imprinting), partially by maternal exposures, represents a major socioeconomic challenge for modern societies. Also, the contribution of environmental exposures to the rapid progression of the prevalence of neurodegenerative diseases such as Alzheimer's or Parkinson's disease is so far not investigated in detail, although the association is supported by large cohort studies on air pollution and traffic noise effects on neurodegeneration (Chen et al. 2017; Cantuaria et al. 2021) and mental health (Iglesias-Vazquez et al. 2022).

Based on the preceding explanations on the importance of exposome studies at the fetal and infantile state, it is not surprising that several large exposome projects and studies focus on this vulnerable population. Examples are the following ongoing projects and studies: HELIX (Maitre et al. 2018; Vrijheid et al. 2014), PACE (Felix et al. 2018; Lozano et al. 2022), UK Biobank (Fu et al. 2022), and several birth cohorts. Promising examples of exposome studies are the following projects: EXPOSOMICS consortium (Vineis et al. 2017; Turner et al. 2018), HERCULES (Niedzwiecki and Miller 2017), HEALS (Steckling et al. 2018), and the EUROPEAN EXPOSOME NETWORK (Webpage 2023). Besides these "real" exposome studies, some established large population classical exposure-health association studies in Europe, such as ESCAPE and ELAPSE, are mostly missing information on biochemical changes of the internal environment, which is required by definition to make-up a complete exposome study. An overview of these and other selected exposome and environment-health-association studies and projects is presented in Figure 2.3. As the exposome studies mostly started 5 or 10 years ago, we may expect the first robust data within the next 5–10 years. A review from 2023 provides an update of ongoing exposome studies, important components of the exposome concept, and the impact of the exposome on cardiovascular disease burden (Münzel et al. 2023).

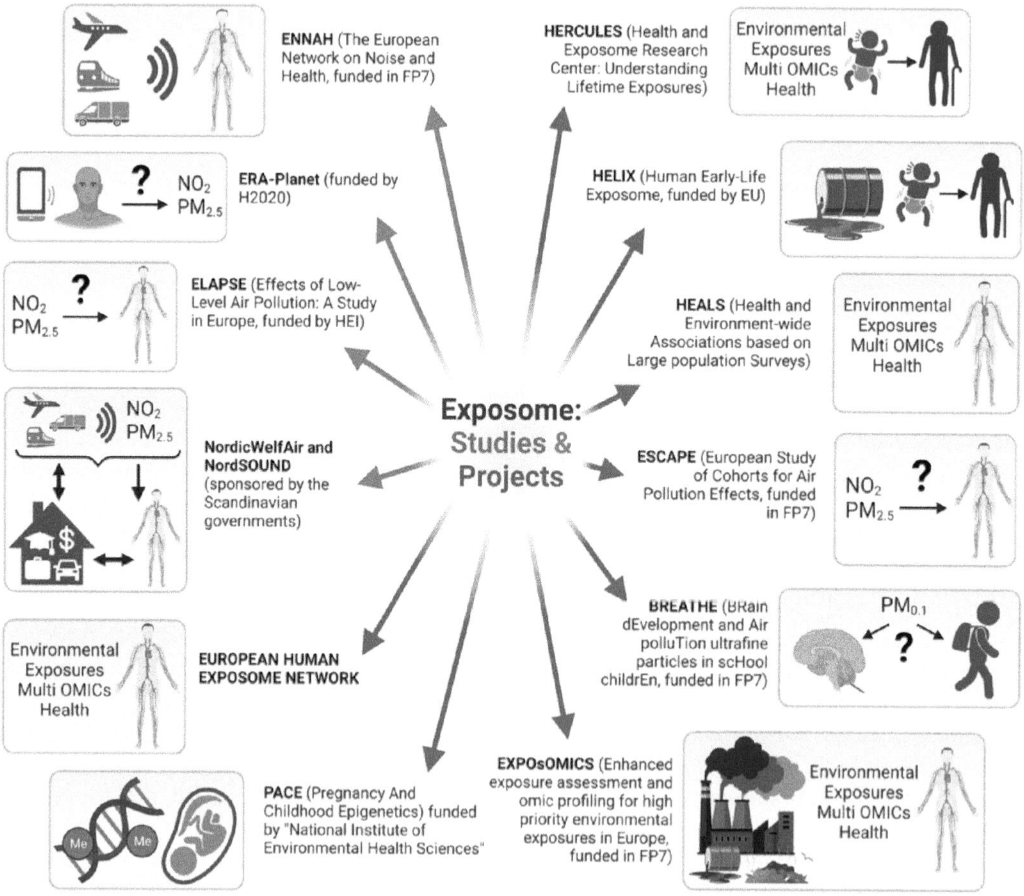

FIGURE 2.3 Selected examples of exposome as well as environment-health-association studies and projects. Summarized from Daiber, Lelieveld, et al. (2019). NO_2, nitrogen dioxide; $PM_{2.5}$, particulate matter with a diameter ≤ 2.5 μm; $PM_{0.1}$, ultra-fine particles with a diameter ≤ 0.1 μm. Reused and translated from (Daiber et al. 2021) with permission. Copyright © 2021, Rights Managed by Georg Thieme Verlag KG Stuttgart, New York.

2.4 HEALTH EFFECTS OF ENVIRONMENTAL RISK FACTORS DURING EARLY LIFE AND IN THE ELDERLY

Environmental risk factors contribute the most significant share to global morbidity and mortality during early and later life. The impact on disease burden (DALYs) by environmental exposures dominates during infancy, but the environmental contribution to premature deaths is most pronounced in the elderly (Landrigan et al. 2018). The high number of DALYs during early life is an impressive reflection of the fact that if only one child dies early because of environmental exposures, this contributes to an unequally high number of lost life years. Likewise, if only one child develops a chronic NCD during early life due to environmental stressors, this leads to many years of severe disability. These results perfectly match the exposome concept, which is intended to monitor all exposure-dependent biochemical changes and adverse health effects over the entire lifespan. The Lancet Commission on pollution and health also highlighted that children (< 4 years) and the elderly (> 60 years) are populations of particular vulnerability concerning effects on DALYs and premature deaths by unhealthy environments, requiring specific protection (Landrigan et al. 2018). Therefore, exposome studies in these population groups are of utmost importance to understand the dynamic exposome changes in the organism, which are most pronounced in these age groups—probably due to the development processes in children and the manifesting cellular decline in the elderly.

Possible explanations for the quite pronounced excess mortality at old age are (Daiber, Lelieveld, et al. 2019): 1) environmental exposures such as chemical pollution represent comorbidities for age-associated classical diseases such as cancer, cardiovascular events, and neurodegeneration, which accelerate the aging process and mechanisms of passing away; 2) lifelong exposures escalate at a particular time point of life (e.g., at older age), when the acceptable threshold of environmentally-triggered biochemical damage is exceeded. Possible explanations for the pronounced effects of unhealthy environments on DALYs in children are (Daiber, Lelieveld, et al. 2019): 1) children have less developed defense systems against environmentally-triggered damage (they are less resilient) since the organism evolves sophisticated stress-response mechanisms during the lifespan by continuous low-grade stress exposure, which represents a kind of "preconditioning"; 2) children are exposed to higher doses of environmental toxins and stressors because of a lower body weight to concentration ratio; 3) unfavorable fetal reprogramming, e.g., due to an unhealthy maternal lifestyle; 4) pre-birth and infantile development processes make cells and organs more vulnerable to environmental toxins and stressors.

2.5 OUTLOOK AND FUTURE MITIGATION STRATEGIES

The sizeable population-based exposome studies yield initial valuable results on the contribution of environmental risk factors (exposures) to the global disease burden (Daiber, Lelieveld, et al. 2019; Vineis et al. 2020). Specific exposome approaches focus on the environmental impact on the onset and progression of diabetes (Beulens et al. 2022), cardiovascular diseases (Juarez et al. 2020), chronic disease development in children (Vrijheid et al. 2020) as well as on accelerated and unhealthy aging (Misra 2020). However, detailed data on the effects of environmental exposures on specific disease clusters or mortality will only become available when the single exposure effects can be clearly distinguished and multi-exposure models are successfully established to assess additive effects of multiple stressors. This requires close-mesh determination of environmental chemical toxins from air, soil, and water pollution; sound pressure levels (e.g., from traffic noise); UV radiation; and temperature as well as questionnaires on mental/social factors (e.g., work strain, economic status) at high temporal and spatial resolution for each participant of an exposome study (Sainani 2016).

In addition, the internal environment of each participant of an exposome study should be assessed at least monthly (better weekly) to correlate functional and biochemical changes with altered exposure, e.g., by assessment of the transcriptome, proteome, and metabolome (OMICs analysis) (Dennis et al. 2016, 2017), as well as selected clinical, functional parameters (Daiber, Kroller-Schon, et al. 2019; Daiber, Lelieveld, et al. 2019). The exposure and biochemical data should be analyzed over the entire lifespan (ideally beginning in utero) to obtain a complete picture of the exposome. They should be correlated with health outcomes and hard clinical endpoints such as event rate of myocardial infarction, stroke, the onset of diabetes, or neurodegeneration. This will also require the development and application of novel bioinformatical and statistical tools (Santos et al. 2020; Sainani 2016), e.g., to solve the problems associated with multi-exposure hazard models (Sorensen et al. 2022). It is quite clear that the measurement of all exposure, biochemical, and health outcome data for only one individual over the entire lifespan is challenging. Extrapolating the data volume and costs for thousands or even millions of people will result in an astronomical number of financial resources. A single country may not easily afford these financial and logistic challenges, calling for a joint global effort as previously done for the human genome project. However, this will also require defining ethical and data protection guidelines that are legally binding at the national and international level for all involved stakeholders (Vlaanderen et al. 2021).

2.6 CONCLUSIONS

The mentioned "hurdles" of exposome research will be completely overcome in the future, enabling large population environment-wide association studies (EWAS). Technological progress will make OMICs analysis cheap enough to apply to large cohorts and measure the changes in the internal environment (primary pathomechanisms are centered on oxidative stress, inflammation, and stress hormones). Advanced computers and software will enable the handling of the resulting big data. Next-generation mobile personal sensors will allow the precise measurement of multiple exposures at high temporal and spatial resolution. When the ethical questions and statistical problems are finally solved, a worldwide exposome study could clarify the contribution of single and multiple exposures on the global disease burden and mortality. This will enable effective mitigation measures that could prevent a significant part of the socioeconomic costs attributable to environmental risk factors.

REFERENCES

Bellinger, D. C. 2013. "Prenatal exposures to environmental chemicals and children's neurodevelopment: An update." *Saf Health Work* 4 (1): 1–11. https://doi.org/10.5491/SHAW.2013.4.1.1. www.ncbi.nlm.nih.gov/pubmed/23515885.

Beulens, J. W. J., M. G. M. Pinho, T. C. Abreu, N. R. den Braver, T. M. Lam, A. Huss, J. Vlaanderen, T. Sonnenschein, N. Z. Siddiqui, Z. Yuan, et al. 2022. "Environmental risk factors of type 2 diabetes-an exposome approach." *Diabetologia* 65 (2): 263–274. https://doi.org/10.1007/s00125-021-05618-w. www.ncbi.nlm.nih.gov/pubmed/34792619.

Cantuaria, M. L., F. B. Waldorff, L. Wermuth, E. R. Pedersen, A. H. Poulsen, J. D. Thacher, O. Raaschou-Nielsen, M. Ketzel, J. Khan, V. H. Valencia, et al. 2021. "Residential exposure to transportation noise in Denmark and incidence of dementia: National cohort study." *BMJ* 374: n1954. https://doi.org/10.1136/bmj.n1954. www.ncbi.nlm.nih.gov/pubmed/34497091.

Chen, H., J. C. Kwong, R. Copes, K. Tu, P. J. Villeneuve, A. van Donkelaar, P. Hystad, R. V. Martin, B. J. Murray, B. Jessiman, et al. 2017. "Living near major roads and the incidence of dementia, Parkinson's disease, and multiple sclerosis: A population-based cohort study." *Lancet* 389 (10070): 718–726. https://doi.org/10.1016/S0140-6736(16)32399-6. www.ncbi.nlm.nih.gov/pubmed/28063597.

Daiber, A., K. Frenis, M. Kuntic, H. Li, E. Wolf, A. B. Kilgallen, S. Lecour, L. W. Van Laake, R. Schulz, O. Hahad, et al. 2022. "Redox regulatory changes of circadian rhythm by the environmental risk factors traffic noise and air pollution." *Antioxid Redox Signal* 37 (10–12): 679–703. https://doi.org/10.1089/ars.2021.0272. www.ncbi.nlm.nih.gov/pubmed/35088601.

Daiber, A., S. Kroller-Schon, K. Frenis, M. Oelze, S. Kalinovic, K. Vujacic-Mirski, M. Kuntic, M. T. Bayo Jimenez, J. Helmstadter, S. Steven, et al. 2019. "Environmental noise induces the release of stress hormones and inflammatory signaling molecules leading to oxidative stress and vascular dysfunction-signatures of the internal exposome." *Biofactors* 45 (4): 495–506. https://doi.org/10.1002/biof.1506. www.ncbi.nlm.nih.gov/pubmed/30937979.

Daiber, A., M. Kuntic, J. Lelieveld, O. Hahad, and T. Münzel. 2021. "Das Exposom charakterisiert die Auswirkungen unserer Umwelt auf Stoffwechsel und Gesundheit." *Aktuel Kardiol* 10: 502–508. https://doi.org/10.1055/a-1546-7401.

Daiber, A., J. Lelieveld, S. Steven, M. Oelze, S. Kroller-Schon, M. Sorensen, and T. Munzel. 2019. "The 'exposome' concept: How environmental risk factors influence cardiovascular health." *Acta Biochim Pol* 66 (3): 269–283. https://doi.org/10.18388/abp.2019_2853. www.ncbi.nlm.nih.gov/pubmed/31509369.

Daiber, A., and T. Munzel. 2020. "Special issue 'impact of environmental pollution and stress on redox signaling and oxidative stress pathways'." *Redox Biol* 37: 101621. https://doi.org/10.1016/j.redox.2020.101621. www.ncbi.nlm.nih.gov/pubmed/32863186.

DALYs, G. B. D., and Hale Collaborators. 2016. "Global, regional, and national disability-adjusted life-years (DALYs) for 315 diseases and injuries and healthy life-expectancy (HALE), 1990–2015: A systematic analysis for the Global Burden of Disease Study 2015." *Lancet* 388 (10053): 1603–1658. https://doi.org/10.1016/S0140-6736(16)31460-X. www.ncbi.nlm.nih.gov/pubmed/27733283.

Dennis, K. K., S. S. Auerbach, D. M. Balshaw, Y. Cui, M. D. Fallin, M. T. Smith, A. Spira, S. Sumner, and G. W. Miller. 2016. "The importance of the biological impact of exposure to the concept of the exposome." *Environ Health Perspect* 124 (10): 1504–1510. https://doi.org/10.1289/EHP140. www.ncbi.nlm.nih.gov/pubmed/27258438.

Dennis, K. K., E. Marder, D. M. Balshaw, Y. Cui, M. A. Lynes, G. J. Patti, S. M. Rappaport, D. T. Shaughnessy, M. Vrijheid, and D. B. Barr. 2017. "Biomonitoring in the era of the exposome." *Environ Health Perspect* 125 (4): 502–510. https://doi.org/10.1289/EHP474. www.ncbi.nlm.nih.gov/pubmed/27385067.

Faisandier, L., V. Bonneterre, R. De Gaudemaris, and D. J. Bicout. 2011. "Occupational exposome: A network-based approach for characterizing occupational health problems." *J Biomed Inform* 44 (4): 545–552. https://doi.org/10.1016/j.jbi.2011.02.010. www.ncbi.nlm.nih.gov/pubmed/21362498.

Felix, J. F., B. R. Joubert, A. A. Baccarelli, G. C. Sharp, C. Almqvist, I. Annesi-Maesano, H. Arshad, N. Baiz, M. J. Bakermans-Kranenburg, K. M. Bakulski, et al. 2018. "Cohort profile: Pregnancy and Childhood Epigenetics (PACE) consortium." *Int J Epidemiol* 47 (1): 22–23u. https://doi.org/10.1093/ije/dyx190. www.ncbi.nlm.nih.gov/pubmed/29025028.

Fu, Z., Q. Liu, J. Liang, T. Huang, G. Liang, Y. Zhou, and A. Gu. 2022. "Association of ambient air pollution exposure with low birth weight." *Environ Res* 215 (Pt 1): 114164. https://doi.org/10.1016/j.envres.2022.114164. www.ncbi.nlm.nih.gov/pubmed/36027958.

Fuller, R., P. J. Landrigan, K. Balakrishnan, G. Bathan, S. Bose-O'Reilly, M. Brauer, J. Caravanos, T. Chiles, A. Cohen, L. Corra, et al. 2022. "Pollution and health: A progress update." *Lancet Planet Health* 6 (6): e535–e547. https://doi.org/10.1016/S2542-5196(22)00090-0. www.ncbi.nlm.nih.gov/pubmed/35594895.

Iglesias-Vazquez, L., A. C. Binter, J. Canals, C. Hernandez-Martinez, N. Voltas, A. Ambros, S. Fernandez-Barres, L. Perez-Crespo, M. Guxens, and V. Arija. 2022. "Maternal exposure to air pollution during pregnancy and child's cognitive, language, and motor function: ECLIPSES study." *Environ Res* 212 (Pt D): 113501. https://doi.org/10.1016/j.envres.2022.113501. www.ncbi.nlm.nih.gov/pubmed/35640710.

Jones, D. P. 2016. "Sequencing the exposome: A call to action." *Toxicol Rep* 3: 29–45. https://doi.org/10.1016/j.toxrep.2015.11.009. www.ncbi.nlm.nih.gov/pubmed/26722641.

Juarez, P. D., D. B. Hood, M. A. Song, and A. Ramesh. 2020. "Use of an exposome approach to understand the effects of exposures from the natural, built, and social environments on cardio-vascular disease onset, progression, and outcomes." *Front Public Health* 8: 379. https://doi.org/10.3389/fpubh.2020.00379. www.ncbi.nlm.nih.gov/pubmed/32903514.

Kuntic, M., I. Kuntic, R. Krishnankutty, A. Gericke, M. Oelze, T. Junglas, M. T. Bayo Jimenez, P. Stamm, M. Nandudu, O. Hahad, et al. 2023. "Co-exposure to urban particulate matter and aircraft noise adversely impacts the cerebro-pulmonary-cardiovascular axis in mice." *Redox Biol* 59: 102580. https://doi.org/10.1016/j.redox.2022.102580. www.ncbi.nlm.nih.gov/pubmed/36566737.

Landrigan, P. J., R. Fuller, N. J. R. Acosta, O. Adeyi, R. Arnold, N. N. Basu, A. B. Balde, R. Bertollini, S. Bose-O'Reilly, J. I. Boufford, et al. 2018. "The lancet commission on pollution and health." *Lancet* 391 (10119): 462–512. https://doi.org/10.1016/S0140-6736(17)32345-0. www.ncbi.nlm.nih.gov/pubmed/29056410.

Lelieveld, J., K. Klingmuller, A. Pozzer, U. Poschl, M. Fnais, A. Daiber, and T. Munzel. 2019. "Cardiovascular disease burden from ambient air pollution in Europe reassessed using novel hazard ratio functions." *Eur Heart J* 40 (20): 1590–1596. https://doi.org/10.1093/eurheartj/ehz135. www.ncbi.nlm.nih.gov/pubmed/30860255.

Li, H., A. B. Kilgallen, T. Munzel, E. Wolf, S. Lecour, R. Schulz, A. Daiber, and L. W. Van Laake. 2020. "Influence of mental stress and environmental toxins on circadian clocks: Implications for redox regulation of the heart and cardioprotection." *Br J Pharmacol* 177 (23): 5393–5412. https://doi.org/10.1111/bph.14949. www.ncbi.nlm.nih.gov/pubmed/31833063.

Lozano, M., P. Yousefi, K. Broberg, R. Soler-Blasco, C. Miyashita, G. Pesce, W. J. Kim, M. Rahman, K. M. Bakulski, L. S. Haug, et al. 2022. "DNA methylation changes associated with prenatal mercury exposure: A meta-analysis of prospective cohort studies from PACE consortium." *Environ Res* 204 (Pt B): 112093. https://doi.org/10.1016/j.envres.2021.112093. www.ncbi.nlm.nih.gov/pubmed/34562483.

Lozano, R., M. Naghavi, K. Foreman, S. Lim, K. Shibuya, V. Aboyans, J. Abraham, T. Adair, R. Aggarwal, S. Y. Ahn, et al. 2012. "Global and regional mortality from 235 causes of death for 20 age groups in 1990 and 2010: A systematic analysis for the Global Burden of Disease Study 2010." *Lancet* 380 (9859): 2095–2128. https://doi.org/10.1016/S0140-6736(12)61728-0. www.ncbi.nlm.nih.gov/pubmed/23245604.

Maitre, L., J. de Bont, M. Casas, O. Robinson, G. M. Aasvang, L. Agier, S. Andrusaityte, F. Ballester, X. Basagana, E. Borras, et al. 2018. "Human early life exposome (HELIX) study: A European population-based exposome cohort." *BMJ Open* 8 (9): e021311. https://doi.org/10.1136/bmjopen-2017-021311. www.ncbi.nlm.nih.gov/pubmed/30206078.

Misra, B. B. 2020. "The chemical exposome of human aging." *Front Genet* 11: 574936. https://doi.org/10.3389/fgene.2020.574936. www.ncbi.nlm.nih.gov/pubmed/33329714.

Morawska, L., W. Miller, M. Riley, S. Vardoulakis, Y. G. Zhu, G. B. Marks, P. Garnawat, P. Kumar, and M. Thynell. 2022. "How to build Urbanome, the genome of the city?" *Sci Total Environ* 810: 152310. https://doi.org/10.1016/j.scitotenv.2021.152310. www.ncbi.nlm.nih.gov/pubmed/34914999.

Munzel, T., and A. Daiber. 2018. "Environmental stressors and their impact on health and disease with focus on oxidative stress." *Antioxid Redox Signal* 28 (9): 735–740. https://doi.org/10.1089/ars.2017.7488. www.ncbi.nlm.nih.gov/pubmed/29278923.

Munzel, T., O. Hahad, and A. Daiber. 2021. "The dark side of nocturnal light pollution: Outdoor light at night increases risk of coronary heart disease." *Eur Heart J* 42 (8): 831–834. https://doi.org/10.1093/eurheartj/ehaa866. www.ncbi.nlm.nih.gov/pubmed/33221876.

Munzel, T., M. Sorensen, and A. Daiber. 2021. "Transportation noise pollution and cardiovascular disease." *Nat Rev Cardiol* 18 (9): 619–636. https://doi.org/10.1038/s41569-021-00532-5. www.ncbi.nlm.nih.gov/pubmed/33790462.

Munzel, T., M. Sorensen, T. Gori, F. P. Schmidt, X. Rao, F. R. Brook, L. C. Chen, R. D. Brook, and S. Rajagopalan. 2017. "Environmental stressors and cardio-metabolic disease: Part II-mechanistic insights." *Eur Heart J* 38 (8): 557–564. https://doi.org/10.1093/eurheartj/ehw294. www.ncbi.nlm.nih.gov/pubmed/27460891.

Münzel, T., M. Sørensen, O. Hahad, M. Nieuwenhuijsen, and A. Daiber. 2023. "The contribution of the exposome to the burden of cardiovascular disease." *Nat Rev Cardiol.* https://doi.org/10.1038/s41569-023-00873-3.

Munzel, T., M. Sorensen, J. Lelieveld, O. Hahad, S. Al-Kindi, M. Nieuwenhuijsen, B. Giles-Corti, A. Daiber, and S. Rajagopalan. 2021. "Heart healthy cities: Genetics loads the gun but the environment pulls the trigger." *Eur Heart J* 42 (25): 2422–2438. https://doi.org/10.1093/eurheartj/ehab235. www.ncbi.nlm.nih.gov/pubmed/34005032.

Niedzwiecki, M. M., and G. W. Miller. 2017. "The exposome paradigm in human health: Lessons from the emory exposome summer course." *Environ Health Perspect* 125 (6): 064502. https://doi.org/10.1289/EHP1712. www.ncbi.nlm.nih.gov/pubmed/28669935.

Nieuwenhuijsen, M. J. 2020. "Urban and transport planning pathways to carbon neutral, liveable and healthy cities: A review of the current evidence." *Environ Int* 140: 105661. https://doi.org/10.1016/j.envint.2020.105661. www.ncbi.nlm.nih.gov/pubmed/32307209.

Olden, K., and S. Wilson. 2000. "Environmental health and genomics: Visions and implications." *Nat Rev Genet* 1 (2): 149–153. https://doi.org/10.1038/35038586. www.ncbi.nlm.nih.gov/pubmed/11253655.

Rappaport, S. M. 2016. "Genetic factors are not the major causes of chronic diseases." *PLoS One* 11 (4): e0154387. https://doi.org/10.1371/journal.pone.0154387. www.ncbi.nlm.nih.gov/pubmed/27105432.

Rappaport, S. M., D. K. Barupal, D. Wishart, P. Vineis, and A. Scalbert. 2014. "The blood exposome and its role in discovering causes of disease." *Environ Health Perspect* 122 (8): 769–774. https://doi.org/10.1289/ehp.1308015. www.ncbi.nlm.nih.gov/pubmed/24659601.

Sainani, K. 2016. "Taking on the exposome: Bringing bioinformatics tools to the environmental side of the health equation." *Biomedical Computation Review* Fall 2016: 14–21.

Santos, S., L. Maitre, C. Warembourg, L. Agier, L. Richiardi, X. Basagana, and M. Vrijheid. 2020. "Applying the exposome concept in birth cohort research: A review of statistical approaches." *Eur J Epidemiol* 35 (3): 193–204. https://doi.org/10.1007/s10654-020-00625-4. www.ncbi.nlm.nih.gov/pubmed/32221742.

Shiraiwa, M., K. Ueda, A. Pozzer, G. Lammel, C. J. Kampf, A. Fushimi, S. Enami, A. M. Arangio, J. Frohlich-Nowoisky, Y. Fujitani, et al. 2017. "Aerosol health effects from molecular to global scales." *Environ Sci Technol* 51 (23): 13545–13567. https://doi.org/10.1021/acs.est.7b04417. www.ncbi.nlm.nih.gov/pubmed/29111690.

Sorensen, M., A. H. Poulsen, U. A. Hvidtfeldt, J. Brandt, L. M. Frohn, M. Ketzel, J. H. Christensen, U. Im, J. Khan, T. Munzel, and O. Raaschou-Nielsen. 2022. "Air pollution, road traffic noise and lack of greenness and risk of type 2 diabetes: A multi-exposure prospective study covering Denmark." *Environ Int* 170: 107570. https://doi.org/10.1016/j.envint.2022.107570. www.ncbi.nlm.nih.gov/pubmed/36334460.

Steckling, N., A. Gotti, S. Bose-O'Reilly, D. Chapizanis, D. Costopoulou, F. De Vocht, M. Gari, J. O. Grimalt, E. Heath, R. Hiscock, et al. 2018. "Biomarkers of exposure in environment-wide association studies: Opportunities to decode the exposome using human biomonitoring data." *Environ Res* 164: 597–624. https://doi.org/10.1016/j.envres.2018.02.041. www.ncbi.nlm.nih.gov/pubmed/29626821.

Turner, M. C., P. Vineis, E. Seleiro, M. Dijmarescu, D. Balshaw, R. Bertollini, M. Chadeau-Hyam, T. Gant, J. Gulliver, A. Jeong, et al. 2018. "EXPOsOMICS: Final policy workshop and stakeholder consultation." *BMC Public Health* 18 (1): 260. https://doi.org/10.1186/s12889-018-5160-z. www.ncbi.nlm.nih.gov/pubmed/29448939.

Vermeulen, R., E. L. Schymanski, A. L. Barabasi, and G. W. Miller. 2020. "The exposome and health: Where chemistry meets biology." *Science* 367 (6476): 392–396. https://doi.org/10.1126/science.aay3164. www.ncbi.nlm.nih.gov/pubmed/31974245.

Vineis, P., M. Chadeau-Hyam, H. Gmuender, J. Gulliver, Z. Herceg, J. Kleinjans, M. Kogevinas, S. Kyrtopoulos, M. Nieuwenhuijsen, D. H. Phillips, et al. 2017. "The exposome in practice: Design of the EXPOsOMICS project." *Int J Hyg Environ Health* 220 (2 Pt A): 142–151. https://doi.org/10.1016/j.ijheh.2016.08.001. www.ncbi.nlm.nih.gov/pubmed/27576363.

Vineis, P., O. Robinson, M. Chadeau-Hyam, A. Dehghan, I. Mudway, and S. Dagnino. 2020. "What is new in the exposome?" *Environ Int* 143: 105887. https://doi.org/10.1016/j.envint.2020.105887. www.ncbi.nlm.nih.gov/pubmed/32619912.

Vlaanderen, J., K. de Hoogh, G. Hoek, A. Peters, N. Probst-Hensch, A. Scalbert, E. Melen, C. Tonne, G. A. de Wit, M. Chadeau-Hyam, et al. 2021. "Developing the building blocks to elucidate the impact of the urban exposome on cardiometabolic-pulmonary disease: The EU EXPANSE project." *Environ Epidemiol* 5 (4): e162. https://doi.org/10.1097/EE9.0000000000000162. www.ncbi.nlm.nih.gov/pubmed/34414346.

Vrijheid, M. 2014. "The exposome: A new paradigm to study the impact of environment on health." *Thorax* 69 (9): 876–878. https://doi.org/10.1136/thoraxjnl-2013-204949. www.ncbi.nlm.nih.gov/pubmed/24906490.

Vrijheid, M., S. Fossati, L. Maitre, S. Marquez, T. Roumeliotaki, L. Agier, S. Andrusaityte, S. Cadiou, M. Casas, M. de Castro, et al. 2020. "Early-life environmental exposures and childhood obesity: An exposome-wide approach." *Environ Health Perspect* 128 (6): 67009. https://doi.org/10.1289/EHP5975. www.ncbi.nlm.nih.gov/pubmed/32579081.

Vrijheid, M., R. Slama, O. Robinson, L. Chatzi, M. Coen, P. van den Hazel, C. Thomsen, J. Wright, T. J. Athersuch, N. Avellana, et al. 2014. "The human early-life exposome (HELIX): Project rationale and design." *Environ Health Perspect* 122 (6): 535–544. https://doi.org/10.1289/ehp.1307204. www.ncbi.nlm.nih.gov/pubmed/24610234.

Webpage. 2013. "World Health Organization: Global action plan for the prevention and control of noncommunicable diseases 2013–2020." https://apps.who.int/iris/bitstream/handle/10665/94384/9789241506236_eng.pdf;jsessionid=71BCEA94B3F85737AB42F3C84216E54A? (last accessed on 23/03/2023).

Webpage. 2016a. "WHO report: Preventing disease through healthy environments." www.who.int/quantifying_ehimpacts/publications/preventingdisease.pdf.

Webpage. 2016b. "WHO: Ambient air pollution: A global assessment of exposure and burden of disease." http://apps.who.int/iris/bitstream/10665/250141/1/9789241511353-eng.pdf?ua=1.

Webpage. 2023. "The European Human Exposome Network (EHEN)." www.humanexposome.eu/(last accessed 11/03/23).

Wild, C. P. 2005. "Complementing the genome with an 'exposome': The outstanding challenge of environmental exposure measurement in molecular epidemiology." *Cancer Epidemiol Biomarkers*

Prev 14 (8): 1847–1850. https://doi.org/10.1158/1055-9965.EPI-05-0456. www.ncbi.nlm.nih.gov/pubmed/16103423.

Zheng, Y., Z. Chen, T. Pearson, J. Zhao, H. Hu, and M. Prosperi. 2020. "Design and methodology challenges of environment-wide association studies: A systematic review." *Environ Res* 183: 109275. https://doi.org/10.1016/j.envres.2020.109275. www.ncbi.nlm.nih.gov/pubmed/32105887.

3 Problems Related to the Use of Biomarkers of Oxidative Stress in Disease

Pietro Ghezzi and Henry Jay Forman

3.1 REACTIVE OXYGEN SPECIES AND THEIR REACTIVITY IN BIOLOGICAL SYSTEMS

The main metabolic pathway for the reduction of molecular oxygen in biological systems is represented by cellular respiration, which generates two molecules of water for each O_2 molecule and requires four electrons. However, there are a number of biochemical pathways that result in the production of intermediate states of oxidation of oxygen, for instance hydrogen peroxide, superoxide radical and hydroxyl radical. Because of their potential to oxidize macromolecules (but only hydroxyl has high rate constants for its reactions), these intermediates are often grouped as reactive oxygen species (ROS), an unfortunate misnomer, as their chemical reactivities and ability to be measured differ markedly (Forman et al. 2015; Murphy et al. 2022).

Superoxide and H_2O_2 can be produced by various enzymes (particularly NADPH oxidases, NOX; xanthine oxidase, XO), but also non-enzymatically (for instance by ionizing radiations via the radiolysis of water or by redox cycling of chemicals such as the herbicide paraquat or drugs in the class of anthracyclines) and by the autooxidation of ubisemiquinone in mitochondria. The most biologically important feature of ROS is their reactivity: they will easily oxidize potential electron donors to result in the formation of the more stable form, water. In a biological system, hydroxyl radical (•OH) and peroxynitrous acid in particular can oxidize various types of cellular molecules, including lipids, nuclei acid/bases and proteins/peptides. Table 3.1 lists some of the oxidation products that oxidants can generate in the cell. These, in some cases, include products generated by so-called reactive nitrogen species such as ONOO-(peroxynitrite) produced by reaction of nitric oxide (•NO) and superoxide. Table 3.1 also reports whether or not the modification is reversible and its biological effect.

3.2 PHYSIOLOGICAL AND PATHOLOGICAL ROLES OF ROS

There are enzymes that specifically evolved to produce O_2^- and/or H_2O_2, such as the NADPH oxidase NOX2 (also known as cytochrome b558 subunit beta), which is an essential part of the bactericidal and fungicidal armamentarium of polymorphonuclear leukocytes (neutrophils), macrophages and other phagocytes (Morel, Doussiere, and Vignais 1991; Segal and Abo 1993). In this case it is appropriate to say that O_2^- and H_2O_2 are produced for a physiological purpose, which is confirmed by the fact that mutations in the NOX2 gene in chronic granulomatous disease lead to increased susceptibility to infections (Kang et al. 2011).

However, most of the literature focuses on the potential pathological role of oxidants. In the 1970s Sies introduced the concept of "oxidative stress" defined as a condition where the production of ROS exceeds the capability of antioxidant enzymes and small-molecular weight antioxidants to eliminate them (reviewed in Sies, Berndt, and Jones [2017]). Under these conditions, one may

DOI: 10.1201/9781003328100-4

TABLE 3.1
Molecules Targeted by ROS

Target molecule	Reaction product	Reversibility	Effects (examples)
Lipids	Lipid peroxides and their breakdown aldehydes	Irreversible	Membrane damage
DNA	Oxidized bases	Irreversible	Single-strand breaks (DNA can then be repaired but the oxidized base cannot be reduced back)
Proteins/peptides	Carbonylation (directly by •OH or via aldehydes breakdown products of lipid peroxidation)	Irreversible	Loss-of function, gain-of function, protein degradation, protein aggregation, generation of autoantigens
	Tyr oxidation to NT via peroxynitrous acid	Irreversible	Above
	Met oxidation to MetSO	Reversible by MetSO reductase	Loss-of-function (inhibition of alpha-1-antitrypsin)
	SH oxidation to SOH (sulfenic)	Reversible by reduction by many compounds	various
	SH oxidation to S-S (disulfide)	Reversible by enzymes and SH reagents	various
	SH oxidation to SO_2H (sulfinic)	Irreversible—can be reduced by action of sulfiredoxin	various
	SH oxidation to SO_3H (sulfonic)	Reversible only in bacteria by sulfonic acid reductases	various

Abbreviations: Met, methionine; MetSO, methionine sulfoxide; NT, nitrotyrosine; Tyr, tyrosine.

observe "oxidative damage". Oxidative damage has been clearly implicated in the toxicity of ionizing radiations, hyperoxia (Gerschman et al. 1954), and some chemicals such as bleomycin (Allawzi et al. 2019) and paraquat (Bus and Gibson 1984). Additionally, thousands of studies have hypothesized a role for an overproduction of oxidants in many diseases, implicitly or explicitly predicting a preventative or therapeutic effect of antioxidants, thus defining an oxidative stress theory of disease (Ghezzi et al. 2016).

More recently, the concept of "redox regulation" emerged. Production of H_2O_2, directly or derived from O_2^- plays a signaling role, since reversible oxidation of specific cysteines can regulate protein activities and functions, similar to the regulation occurring though reversible phosphorylation (Forman, Maiorino, and Ursini 2010; Forman et al. 2017).

Various concepts have been developed to frame the various biological roles of ROS, including "oxidative stress", "oxidative damage", "eustress", "redox regulation" and "redox signaling" (Sies 2017).

It is difficult to precisely define the boundary between oxidative damage vs. redox regulation, as many factors play into this, such as the type and levels of ROS produced, their intracellular location, the levels of ROS-catabolizing enzymes and of low molecular weight antioxidants. Nevertheless, one can assume that reversible forms of oxidation are more likely involved as physiological regulatory mechanisms than irreversible ones. For instance, Table 3.1 shows those forms of cysteine oxidation that, due to their reversibility, could play a role in redox regulation. This highlights the problem of using a generic definition of "ROS" that includes oxidants that potentially induce damage, such as •OH, and those with a regulatory role such as H_2O_2.

For the study of the pathological role of ROS in disease, it is essential to be able to measure individual ROS and/or the products of their reactions with cellular molecules in vitro, in animal models and in patients. However, ROS have extremely short half-lives, from 10^{-9} s for the hydroxyl radical to 10^{-3} s for H_2O_2 (D'Autreaux and Toledano 2007). For this reason, most studies, particularly when measurements are made in biological fluids or tissues, resort to indirectly derive their production by measuring some of the relatively more stable oxidation products, some of which are listed in Table 3.2. Measurements of ROS are meaningless as it is an abbreviation not a molecular entity. The reactions that produce the products are primarily those of $^{\bullet}OH$ and peroxynitrous acid (Murphy et al. 2022).

While the problems related to the reliability and specificity of different biomarkers of oxidative stress have been recently reviewed elsewhere (Murphy et al. 2022; Frijhoff et al. 2015), this review will attempt to define a classification of these biomarkers based on their biological meaning and their place in the causal pathways of diseases.

TABLE 3.2
Molecules Used as "Biomarkers"

Molecule Measured	What Is Measured	Meaning	Problems
ROS	O_2^-, H_2O_2, $^{\bullet}OH$	Actual molecules	Short half-life, specificity
Oxidation products			
MDA	Breakdown product of lipid peroxidation	Indicator of lipid oxidation	Also produced by metabolism
F2α-isoprostanes	Breakdown product of lipid peroxidation	Indicator of lipid oxidation by radical reaction	
8-OH-dG	Oxidized deoxyguanosine	Indicator of DNA oxidation	
Protein carbonyls	Irreversible protein oxidation	Indicator of direct oxidation of proteins or by addiction by aldehydes produced by lipid peroxidation	
Nitrotyrosine	Irreversible oxidation of tyrosine in proteins	Indicator of direct oxidation of proteins by peroxynitrous acid	
Protein Cys SO_2H or SO_3H	Irreversible oxidation of cysteine –SH (except bacteria, see Table 3.1)	Indicator of Cys oxidation	
Reversible modifications			
GSH	Detoxifies H_2O_2	Risk factor (high susceptibility if low) Its decrease may indicate higher H_2O_2	Its levels are induced by many electrophiles via NRF2
GSSG/GSH ratio	Indicator of GSH oxidation	Its increase may indicate higher H_2O_2	Total GSH and GSSG levels are increased later via NRF2
Protein mixed disulfides	Reversible oxidation of cysteine –SH	Indicator of increased GSSG/GSH ratio	May be physiological oxidoreduction with regulatory functions

Abbreviations: 8-OH-dG, 8-hydroxy-2′-deoxyguanosine; GSH, glutathione; GSSG, oxidized GSH.

3.3 BIOMARKERS FOR THE STUDY OF THE ROLE OF ROS PRODUCTION IN DISEASE

In this context, we will discuss biomarkers of oxidative stress in two different theoretical frameworks. The first aspect is specifically related to ROS, outlined in Figure 3.1.

The key players, all of which can be measured, are as follows:

1) The sources of ROS (e.g., levels/activity of NOXs, mitochondrial leak, xanthine oxidase etc.). In toxicology, a chemical or physical agent is at the source of ROS production (e.g., radiolysis of water by ionizing radiations, redox cycling of paraquat); however, these will not be discussed here as this review deals with the role of ROS in pathogenesis of diseases, where ROS are thought to be produced by the cellular metabolism. ROS-generating systems have two components: the enzyme and the substrate used as the electron donor for the reduction of oxygen (e.g., NADPH or xanthine). Thus, increased ROS production can be a result of the activation of an enzyme (as in the case of NOX2 activation by bacterial products in phagocytes [Babior 2000; Nathan and Cunningham-Bussel 2013]) but also of the increased availability of the enzyme substrate. The latter has been hypothesized, for

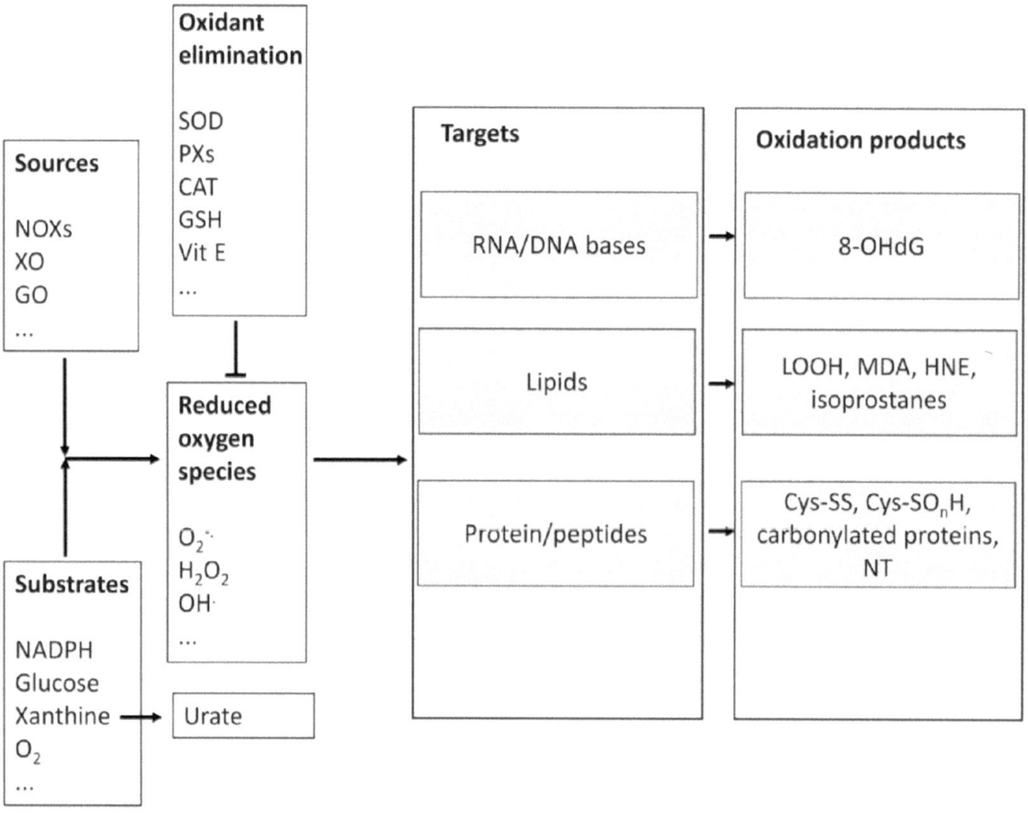

FIGURE 3.1 Biomarkers that can be measured in the context of oxidative stress, their relationships and significance.

Source: Abbreviations: 8-OH-dG, 8-hydroxy-2′-deoxyguanosine; CAT, catalase; GO, glucose oxidase; GPX, GSH peroxidase; GSH, glutathione; HNE, hydroxynonenal; LOOH, lipid hydroperoxide; MDA, malondialdehyde; NT, nitrotyrosine; PX, peroxidase; Vit, vitamin; XO, xanthine oxidase.

instance, for the increased ROS production in post-ischemic reperfusion injury, where the degradation of ATP during the anoxic phase results in the formation of xanthine and hypoxanthine that, upon reoxygenation, will be substrates for ROS-generating XO (Granger 1988). Measuring consumption of O_2 has also been used as an indirect indicator of ROS production, particularly in the earlier studies of the oxidative burst in phagocytes (Sbarra and Karnovsky 1959).

2) The ROS catabolizing systems/molecules, often generically defined as "antioxidants" (e.g., superoxide dismutase, glutathione); lower concentrations of these molecules might result in higher levels of ROS due to decreased elimination rather than increased production.
3) The levels of ROS, as the concept of oxidative stress is that there is an increase in ROS levels, either due to increased production or decreased elimination (points 1 and 2). We use the word "level" here to indicate a relative concentration, as the concentration of any of the ROS varies markedly because there is a very steep gradient between where the species is produced and where it either reacts with a target or is eliminated by an antioxidant enzyme. Average concentrations have been determined, but they are not physically meaningful.
4) The oxidation products of the reaction of ROS with different types of cellular target molecules (nucleic acids bases, lipids or proteins). It could be argued that increased levels of an oxidation product will result from the increased levels of oxidants.

Figure 3.1 also shows how these are related. Ideally, to define oxidative stress we should be able to demonstrate that the levels of ROS are increased. The specific chemical species should be measured because, as mentioned earlier, the species responsible for oxidative damage are mainly ($OH^•$) and peroxynitrous acid, rather than O_2^- and/or H_2O_2. As stated earlier, measuring generically "ROS" is meaningless.

As mentioned earlier, because ROS are short-lived, most studies resort to indirectly measure their production by measuring some of the relatively more stable oxidation products, and, in fact, most studies implying oxidative stress in the pathogenesis of a specific disease (the oxidative stress theory of disease) are based on the measurements of these products in various models.

This is not common in studies of pathogenic mechanisms. For instance, the first of Koch's postulates for the germ theory of disease states that "The bacteria must be present in every case of the disease, but not in healthy subjects"; bacterial cultures are, in fact, a common diagnostic for infectious disease. The cytokine theory of disease, which led to the successful approval of a number of anti-cytokine antibodies (e.g., against TNF, IL-1 or IL-17), was based on the measurement of the relevant inflammatory cytokine in a disease. In all these cases, the agent directly involved in the causal pathway, bacteria or cytokines was measured. There are, of course, surrogate indicators (e.g., fever of neutrophilia for infections; C-reactive protein or the erythrosedimentation rate for inflammatory disease), however these are used as diagnostics rather than to demonstrate the role of bacteria or cytokines in the disease. We will discuss again this problem when discussing the actionability of biomarkers.

3.4 BIOMARKERS RELATED TO OXIDATIVE STRESS: THEIR MEANING

An increase in any of the biomarkers represented by the oxidation products should reflect, when these are specific, an increased production of ROS. Clearly, the biological consequences may be different. Reversible oxidation (for instance formation of disulfides of protein cysteines) may be related to increased production of ROS but not necessarily imply "oxidative damage" and may, in fact, mediate adaptive responses to stress. A typical example is the formation of disulfides in cysteines of the transcription factor NRF2 or of its binding protein KEAP1, where their oxidation leads to transcriptional activation and expression of antioxidant enzymes (e.g., peroxiredoxins, thioredoxin reductase, SOD) or enzymes implicated in the biosynthesis of glutathione (GSH) (Cuadrado et al. 2018).

However, many studies measure ROS-producing enzymes (e.g., XO), their products (e.g., urate), or antioxidants (e.g., SOD, Vit E) and report them as "biomarkers of oxidative stress".

This clearly results from a lack of understanding of basic concepts of epidemiology, one of which is the difference between "causes" and "risk factors", and we point the reader to the model of disease causation represented by the epidemiological triad agent-host-environment (CDC 2012) as well as concepts of causation linked to the exposome, that identified the key steps in the etiogenesis of a disease: exposure, effect, susceptibility, health outcome and mediating factors (Juarez et al. 2020).

In particular, while measuring the molecules on the left side of Figure 3.1 (sources and substrates) is important in getting the full picture of all the determinants of the redox state in a given disease, they cannot be considered "biomarkers of oxidative stress" because a change in their levels is not necessarily reflected in a change in the levels of ROS. Later we expand on the case of antioxidants and ROS-generating enzymes. Figure 3.2 provides a schematic representation of the different

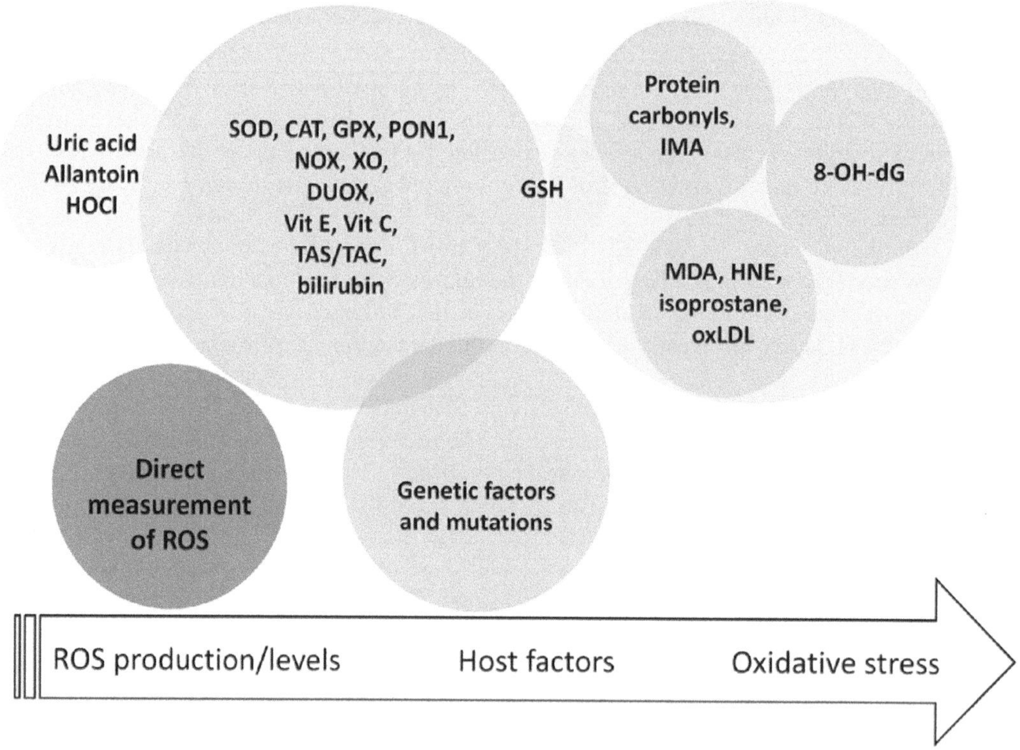

FIGURE 3.2 Biomarkers related to oxidative stress and their meaning.

Source: Abbreviations: 8-OH-dG, 8-hydroxy-2′-deoxyguanosine; CAT, catalase; DUOX, dual oxidase; GPX, GSH peroxidase; GSH, glutathione; HNE, hydroxynonenal; IMA, ischemia-modified albumin; LDL, low-density lipoproteins; MDA, malondialdehyde; MPO, myeloperoxidase; PON1, paraoxonase 1; SOD, superoxide dismutase; TAS, total antioxidant status; TAC, total antioxidant capacity; UA, uric acid; Vit, vitamin; XO, xanthine oxidase. Reproduced from reference (Ghezzi 2020).

meaning of the various biomarkers in the causative pathway of oxidative stress and, potentially, pathological conditions.

1) Antioxidants
   Decreased levels of an antioxidant (enzymes or small molecules) do not necessarily imply "oxidative stress"; as biomarkers, they are ambiguous, i.e., could be interpreted in entirely different ways. For instance
   i) decreased levels of an antioxidant induced by chemicals, such as depletion of GSH by acetaminophen, can cause oxidative stress (Arnaiz et al. 1995).
   ii) decreased levels of an antioxidant can represent a risk factor for oxidative stress when the organism is then exposed to an oxidant; for instance, deficiency of SOD and GSH peroxidase (GPx) increases oxidative stress in response to ionizing radiations or paraquat (Van Remmen et al. 2004).
   iii) higher levels of antioxidants can be due to genetic factors but can also be an adaptive response to oxidative stress mediated by NRF2; in the latter case, increased antioxidants are, in an apparent paradox, an indicator of oxidative stress. For instance, increased ROS production might cause an early GSH depletion (due to its oxidation) but a later increase in its level (due to the transcriptional activation of GSH synthetic enzymes).
2) ROS-producing enzymes, their substrates and their products.
   Changes in the levels of a ROS-generating enzyme do not necessarily cause oxidative stress. For instance, higher levels of XO do not automatically result in increased ROS production unless more substrates (xanthine or hypoxanthine) become available, as in the case, mentioned earlier, of ischemia/reperfusion injury.
   Even the indication that a ROS-producing enzymatic pathway has been activated may be ambiguous. This is, again, the case of XO, whose reaction product, uric acid, is also an antioxidant in extracellular fluids (Ames et al. 1981).

3.5 BIOMARKERS OF OXIDATIVE STRESS: THEIR SPECIFICITY

Table 3.2 lists some of the problems in the interpretation of increased levels of "biomarkers of oxidative stress" used in the literature. A typical example is MDA, probably the most used biomarker of oxidative stress in the past. The same biomarker has also been widely used as an indicator of arachidonic acid metabolism during platelet activation (Smith, Ingerman, and Silver 1976), thus its increase could mean that oxidative stress has occurred but could also be a sign of platelet activation (in the absence of oxidative stress).

One possible solution to this problem, which results from a lack of specificity toward a ROS-mediated reaction, is to disambiguate by measuring more than one biomarker (Daiber et al. 2021; Ghezzi et al. 2018; Turner et al. 2014).

3.6 THE PROBLEM OF ACTIONABILITY

A further problem with biomarkers of oxidative stress is that they are, in most cases, not actionable. As we highlighted elsewhere (Ghezzi et al. 2018), biomarkers can be actionable or not, and this is important when they are measured in the context of theories of disease. We mentioned earlier the difference between biomarkers implicated, or not, in the causal pathway of a disease in the case of the cytokine theory and the germ theory. Many inflammatory diseases involve cytokines such as IL-1, IL-6, IL-17 or TNF in the causal pathway as pathogenic mediators

(Turner et al. 2014). Inflammation, which can be triggered in many ways depending on the etiogenesis of the disease, is associated with production of inflammatory cytokines. These cytokines then cause many of the characteristics of the disease, such as joint damage in the case of rheumatoid arthritis, and this is known as the "cytokine theory of disease" (Czura and Tracey 2005). Cytokines also have other effects; for instance, IL-6 induces the production of the acute-phase protein C-reactive protein (CRP) by the liver (Emery and Luqmani 1993), while IL-1 acts on the thermoregulatory centers in the hypothalamus to cause fever (Dinarello, Cannon, and Wolff 1988).

However, neutralizing CRP with a specific antibody or blocking fever will not improve the disease, while neutralizing IL-1, IL-6, IL-17 or TNF does. Therefore, CRP is a very good biomarker of inflammation, in fact is very valuable diagnostically, but it is not involved in the causal pathway for the disease: it is *not actionable*, while inflammatory cytokines are involved in the causal pathway and are *actionable* (and, if fact, anti-cytokine antibodies are very valuable therapeutically). Likewise, to use the example of the germ theory of disease mentioned earlier, we cannot cure an infection by blocking fever or neutrophilia but only by decreasing the number of bacteria, the agents that are in the causal pathway of infectious diseases.

The concept of non-actionability is not exactly the same as that of a surrogate marker (Katz 2004); it addresses the issue of whether the molecule that can be measured is a pharmacological target, implicated in the causal pathway of the disease.

When we hypothesize a pathogenic role of ROS in a disease, the only actionable molecules are ROS themselves. Decreasing ROS (by scavenging, inhibiting their production or increasing their degradation) ROS should ameliorate a disease if this is due to "oxidative stress", and it is important to measure specific chemical species to distinguish those mainly regulatory to those potentially toxic. However, decreasing markers of protein oxidation or terminal products of lipid peroxidation is not expected to improve a disease. This has, however, a caveat because, in fact, there is evidence that some of the terminal product of oxidation used as biomarkers are not devoid of activity. For instance, aldehydes, such as HNE or MDA, can bind proteins and result in their carbonylation affecting their activity, causing aggregates or the development of neoepitopes (Kurien and Scofield 2008). Therefore, measuring these aldehydes, although they may not be good indicators of the levels of oxidants, can provide information on chemical species that can contribute to a disease. A previous review also provided quality indicators for the different oxidative stress markers (Frijhoff et al. 2015). In addition, leading scientific societies published position papers on the use of oxidative stress markers and the direct measurement of reactive oxygen and nitrogen species (Murphy et al. 2022; Griendling et al. 2016). There is also an ongoing discussion on the quality and meaning of "old" (classical) methods of assessment of oxidative stress representing the majority of publications on oxidative stress for the last decades (Daiber et al. 2017), which may be solved by combining the determination of several biomarkers and at least one advanced method for better understanding of oxidative stress pathologies as discussed earlier.

In conclusion, we have identified some aspects of the use and significance of biomarkers used to study ROS production, oxidative stress and oxidative damage. The main points are 1) taking into account the reversibility of the oxidation product measured is important to distinguish oxidative stress, oxidative damage and redox regulation; 2) it is important to measure specific chemical species (e.g., O_2^-, H_2O_2, $^{\bullet}OH$) rather than generically "ROS"; 3) it is important to distinguish biomarkers of oxidative damage from risk factors, and we should consider the theories of disease causation and the place of each biomarker measured in the theory, usually represented by a directed acyclic graph; 4) while ROS are directly implicated in the causal pathway, other biomarkers may not be and, when possible, an actionable biomarker (such as direct ROS measurement) is preferable; 5) oxidative stress induces adaptive responses that can increase the levels of several antioxidants; 6) some markers are not specific for ROS-mediated reactions and might need disambiguation by measuring other, related biomarkers.

REFERENCES

Allawzi, A., H. Elajaili, E. F. Redente, and E. Nozik-Grayck. 2019. "Oxidative Toxicology of Bleomycin: Role of the Extracellular Redox Environment." *Current Opinion in Toxicology* 13 (February): 68–73. https://doi.org/10.1016/j.cotox.2018.08.001.

Ames, B. N., R. Cathcart, E. Schwiers, and P. Hochstein. 1981. "Uric Acid Provides an Antioxidant Defense in Humans against Oxidant- and Radical-Caused Aging and Cancer: A Hypothesis." *Proc Natl Acad Sci U S A* 78 (11): 6858–6862.

Arnaiz, S. L., S. Llesuy, J. C. Cutrín, and A. Boveris. 1995. "Oxidative Stress by Acute Acetaminophen Administration in Mouse Liver." *Free Radical Biology and Medicine* 19 (3): 303–310. https://doi.org/10.1016/0891-5849(95)00023-Q.

Babior, B. M. 2000. "Phagocytes and Oxidative Stress." *Am J Med* 109 (1): 33–44. https://doi.org/10.1016/s0002-9343(00)00481-2.

Bus, J. S., and J. E. Gibson. 1984. "Paraquat: Model for Oxidant-Initiated Toxicity." *Environmental Health Perspectives* 55 (April): 37–46. https://doi.org/10.1289/ehp.845537.

CDC. 2012. "Lesson 1: Introduction to Epidemiology." www.cdc.gov/csels/dsepd/ss1978/lesson1/section8.html.

Cuadrado, A., G. Manda, A. Hassan, M. J. Alcaraz, C. Barbas, A. Daiber, P. Ghezzi, et al. 2018. "Transcription Factor NRF2 as a Therapeutic Target for Chronic Diseases: A Systems Medicine Approach." *Pharmacol Rev* 70 (2): 348–383. https://doi.org/10.1124/pr.117.014753.

Czura, C. J., and K. J. Tracey. 2005. "Autonomic Neural Regulation of Immunity." *Journal of Internal Medicine* 257 (2): 156–166. https://doi.org/10.1111/j.1365-2796.2004.01442.x.

Daiber, A., O. Hahad, I. Andreadou, S. Steven, S. Daub, and T. Münzel. 2021. "Redox-Related Biomarkers in Human Cardiovascular Disease: Classical Footprints and Beyond." *Redox Biology* 42 (June): 101875. https://doi.org/10.1016/j.redox.2021.101875.

Daiber, A., M. Oelze, S. Steven, S. Kröller-Schön, and T. Münzel. 2017. "Taking Up the Cudgels for the Traditional Reactive Oxygen and Nitrogen Species Detection Assays and Their Use in the Cardiovascular System." *Redox Biology* 12 (August): 35–49. https://doi.org/10.1016/j.redox.2017.02.001.

D'Autreaux, B., and M. B. Toledano. 2007. "ROS as Signalling Molecules: Mechanisms That Generate Specificity in ROS Homeostasis." *Nat Rev Mol Cell Biol* 8 (10): 813–824. https://doi.org/10.1038/nrm2256.

Dinarello, C. A., J. G. Cannon, and S. M. Wolff. 1988. "New Concepts on the Pathogenesis of Fever." *Reviews of Infectious Diseases* 10 (1): 168–189. https://doi.org/10.1093/clinids/10.1.168.

Emery, P., and R. Luqmani. 1993. "The Validity of Surrogate Markers in Rheumatic Disease." *Rheumatology* 32 (suppl_3): 3–8. https://doi.org/10.1093/rheumatology/32.suppl_3.3.

Forman, H. J., O. Augusto, R. Brigelius-Flohe, P. A. Dennery, B. Kalyanaraman, H. Ischiropoulos, G. E. Mann, et al. 2015. "Even Free Radicals Should Follow Some Rules: A Guide to Free Radical Research Terminology and Methodology." *Free Radical Biology and Medicine* 78 (January): 233–235. https://doi.org/10.1016/j.freeradbiomed.2014.10.504.

Forman, H. J., M. J. Davies, A. C. Krämer, G. Miotto, M. Zaccarin, H. Zhang, and F. Ursini. 2017. "Protein Cysteine Oxidation in Redox Signaling: Caveats on Sulfenic Acid Detection and Quantification." *Archives of Biochemistry and Biophysics*, The Chemistry of Redox Signaling 617 (March): 26–37. https://doi.org/10.1016/j.abb.2016.09.013.

Forman, H. J., M. Maiorino, and F. Ursini. 2010. "Signaling Functions of Reactive Oxygen Species." *Biochemistry* 49 (5): 835–842. https://doi.org/10.1021/bi9020378.

Frijhoff, J., P. G. Winyard, N. Zarkovic, S. S. Davies, R. Stocker, D. Cheng, A. R. Knight, et al. 2015. "Clinical Relevance of Biomarkers of Oxidative Stress." *Antioxid Redox Signal* 23 (14): 1144–1170. https://doi.org/10.1089/ars.2015.6317.

Gerschman, R., D. L. Gilbert, S. W. Nye, P. Dwyer, and W. O. Fenn. 1954. "Oxygen Poisoning and X-Irradiation: A Mechanism in Common." *Science* 119 (3097): 623–626. https://doi.org/10.1126/science.119.3097.623.

Ghezzi, P. 2020. "Environmental Risk Factors and Their Footprints in Vivo: A Proposal for the Classification of Oxidative Stress Biomarkers." *Redox Biology* 34 (July): 101442. https://doi.org/10.1016/j.redox.2020.101442.

Ghezzi, P., K. Davies, A. Delaney, and L. Floridi. 2018. "Theory of Signs and Statistical Approach to Big Data in Assessing the Relevance of Clinical Biomarkers of Inflammation and Oxidative Stress." *Proc Natl Acad Sci U S A* 115 (10): 2473–2477. https://doi.org/10.1073/pnas.1719807115.

Ghezzi, P., V. Jaquet, F. Marcucci, and H. H. Schmidt. 2016. "The Oxidative Stress Theory of Disease: Levels of Evidence and Epistemological Aspects." *Br J Pharmacol* (July). https://doi.org/10.1111/bph.13544.

Granger, D. N. 1988. "Role of Xanthine Oxidase and Granulocytes in Ischemia-Reperfusion Injury." *American Journal of Physiology-Heart and Circulatory Physiology* 255 (6): H1269–H1275. https://doi.org/10.1152/ajpheart.1988.255.6.H1269.

Griendling, K. K., R. M. Touyz, J. L. Zweier, S. Dikalov, W. Chilian, Y.-R. Chen, D. G. Harrison, A. Bhatnagar, and American Heart Association Council on Basic Cardiovascular Sciences. 2016. "Measurement of Reactive Oxygen Species, Reactive Nitrogen Species, and Redox-Dependent Signaling in the Cardiovascular System: A Scientific Statement from the American Heart Association." *Circulation Research* 119 (5): e39–e75. https://doi.org/10.1161/RES.0000000000000110.

Juarez, P. D., D. B. Hood, M.-A. Song, and A. Ramesh. 2020. "Use of an Exposome Approach to Understand the Effects of Exposures from the Natural, Built, and Social Environments on Cardio-Vascular Disease Onset, Progression, and Outcomes." *Frontiers in Public Health* 8. www.frontiersin.org/articles/10.3389/fpubh.2020.00379.

Kang, E. M., B. E. Marciano, S. DeRavin, K. A. Zarember, S. M. Holland, and H. L. Malech. 2011. "Chronic Granulomatous Disease: Overview and Hematopoietic Stem Cell Transplantation." *The Journal of Allergy and Clinical Immunology* 127 (6): 1319–1326; quiz 1327–1328. https://doi.org/10.1016/j.jaci.2011.03.028.

Katz, R. 2004. "Biomarkers and Surrogate Markers: An FDA Perspective." *NeuroRx: The Journal of the American Society for Experimental Neuro Therapeutics* 1 (2): 189–195. https://doi.org/10.1602/neurorx.1.2.189.

Kurien, B. T., and R. H. Scofield. 2008. "Autoimmunity and Oxidatively Modified Autoantigens." *Autoimmunity Reviews*, Redox and Autoimmunity 7 (7): 567–573. https://doi.org/10.1016/j.autrev.2008.04.019.

Morel, F., J. Doussiere, and P. V. Vignais. 1991. "The Superoxide-Generating Oxidase of Phagocytic Cells: Physiological, Molecular and Pathological Aspects." *European Journal of Biochemistry* 201 (3): 523–546. https://doi.org/10.1111/j.1432-1033.1991.tb16312.x.

Murphy, M. P., H. Bayir, V. Belousov, C. J. Chang, K. J. A. Davies, M. J. Davies, T. P. Dick, et al. 2022. "Guidelines for Measuring Reactive Oxygen Species and Oxidative Damage in Cells and in Vivo | Nature Metabolism." *Nature Metabolism* 4 (6): 651–662. https://doi.org/10.1038/s42255-022-00591-z.

Nathan, C., and A. Cunningham-Bussel. 2013. "Beyond Oxidative Stress: An Immunologist's Guide to Reactive Oxygen Species." *Nat Rev Immunol* 13 (5): 349–361. https://doi.org/10.1038/nri3423.

Sbarra, A. J., and M. L. Karnovsky. 1959. "The Biochemical Basis of Phagocytosis. I: Metabolic Changes during the Ingestion of Particles by Polymorphonuclear Leukocytes." *The Journal of Biological Chemistry* 234 (6): 1355–1362.

Segal, A. W., and A. Abo. 1993. "The Biochemical Basis of the NADPH Oxidase of Phagocytes." *Trends in Biochemical Sciences* 18 (2): 43–47. https://doi.org/10.1016/0968-0004(93)90051-N.

Sies, H. 2017. "Hydrogen Peroxide as a Central Redox Signaling Molecule in Physiological Oxidative Stress: Oxidative Eustress." *Redox Biol* 11 (April): 613–619. https://doi.org/10.1016/j.redox.2016.12.035.

Sies, H., C. Berndt, and D. P. Jones. 2017. "Oxidative Stress." *Annu Rev Biochem* 86 (June): 715–748. https://doi.org/10.1146/annurev-biochem-061516-045037.

Smith, J. B., C. M. Ingerman, and M. J. Silver. 1976. "Malondialdehyde Formation as an Indicator of Prostaglandin Production by Human Platelets." *The Journal of Laboratory and Clinical Medicine* 88 (1): 167–172.

Turner, M. D., B. Nedjai, T. Hurst, and D. J. Pennington. 2014. "Cytokines and Chemokines: At the Crossroads of Cell Signalling and Inflammatory Disease." *Biochimica et Biophysica Acta (BBA): Molecular Cell Research* 1843 (11): 2563–2582. https://doi.org/10.1016/j.bbamcr.2014.05.014.

Van Remmen, H., W. Qi, M. Sabia, G. Freeman, L. Estlack, H. Yang, Z. M. Guo, et al. 2004. "Multiple Deficiencies in Antioxidant Enzymes in Mice Result in a Compound Increase in Sensitivity to Oxidative Stress." *Free Radical Biology and Medicine* 36 (12): 1625–1634. https://doi.org/10.1016/j.freeradbiomed.2004.03.016.

Section II

Airborne Toxins

4 Formation, Interconversion, and Buffering of Reactive Oxygen Species from Gaseous and Particulate Air Pollutants in Epithelial Lining Fluid

Thomas Berkemeier and Ulrich Pöschl

Anthropogenic air pollution leads to a massive increase of atmospheric aerosol and oxidant concentrations on local, regional, and global scales, posing a major threat to public health (Monks et al., 2009; Pöschl & Shiraiwa, 2015). Although public-health specialists have uncovered connections between air pollution, disease, and mortality, further information is needed to disentangle the causes and effects. For example, the toxicities of different types of particulates, individually and in mixtures, are poorly known (Lelieveld & Pöschl, 2017). The air pollutants that most strongly correlate with negative health outcomes are nitrogen dioxide ($^{\bullet}NO_2$), ozone (O_3), and fine particulate matter with a diameter less than 2.5 µm ($PM_{2.5}$), with the latter likely contributing more than 80% to the total excess mortality (Lelieveld et al., 2021; Turner et al., 2016).

The airways are covered with an epithelial lining fluid (ELF), containing a range of antioxidants and surfactants (van der Vliet et al., 1999). The ELF extends from the nasal cavity to the pulmonary alveoli with a film thickness that decreases from several micrometers in the upper airways to a few hundred nanometers in the distal bronchoalveolar regions (Mudway & Kelly, 2000). The ELF constitutes the first line of defense against air pollutants entering the respiratory tract. Antioxidants and enzymes undergo chemical reactions with dissolved air pollutants in the ELF to suppress the formation of or buffer reactive oxygen species (ROS) (Lelieveld et al., 2021).

$PM_{2.5}$ is a complex mixture that can encompass thousands of different chemical constituents, each having distinct properties. $PM_{2.5}$ originates from both natural and anthropogenic sources, including mineral dust from deserts, gasoline and diesel motor exhausts, tire and brake wear, power generation, residential energy use, agriculture, biomass burning, cooking, and cigarette smoking (Lelieveld et al., 2021). Epidemiological studies show clear correlations between fine particulate matter (PM) in polluted air and adverse health effects including cardiovascular, respiratory, and allergic diseases (Gauderman et al., 2004; Pope et al., 2004). Exposure to high concentrations of aerosol particles can lead to inflammation and damage to cellular proteins, membranes, and DNA (Lewtas, 2007; Nel, 2005). Because of the great heterogeneity in both $PM_{2.5}$ composition and sources, targeted $PM_{2.5}$ pollution control is challenging, and, to date, there is no clear connection between one particular $PM_{2.5}$ constituent and mortality estimates (Baxter et al., 2013; Peng et al., 2009).

$^{\bullet}NO_2$ is an irritant gas that has been linked to mortality in epidemiological studies (Faustini et al., 2014; Mills et al., 2015). However, because $^{\bullet}NO_2$ is often co-emitted with $PM_{2.5}$ and other pollutants in combustion processes, it remains unclear if it poses an independent health risk (Burnett et al., 2004). Experimental studies using bronchoalveolar lavage fluid from healthy volunteers suggest that removal of glutathione and ascorbic acid decreases $^{\bullet}NO_2$ absorption, indicating that $^{\bullet}NO_2$ may react

DOI: 10.1201/9781003328100-6

with and thereby deplete low-molecular-weight antioxidants in the ELF (Velsor & Postlethwait, 1997).

Exposure to O_3 has been shown to exacerbate asthma and increase respiratory and circulatory mortality (McConnell et al., 2002; Mudway & Kelly, 2000). It is known to react with alkenes by addition to the C–C double bond, leading to lipid peroxidation and forming a variety of oxidized reaction products, including Criegee (peroxyl) intermediates and hydroperoxides (Hewitt & Kok, 1991; Zeng & Wilson, 2020). Furthermore, O_3 can contribute to oxidative stress by depleting antioxidants and surfactants in the ELF (Lakey et al., 2016; Mudway & Kelly, 2000).

As illustrated in Figure 4.1, the redox-active pollutants promote and sustain a multitude of radical and redox reaction cycles in the ELF (Lelieveld et al., 2021). Reduced forms of transition metal ions and quinones react with molecular oxygen (O_2), forming superoxide ($O_2^{\bullet-}$) radicals. The enzyme superoxide dismutase (SOD) efficiently shuttles $O_2^{\bullet-}$ radicals into the less reactive H_2O_2, whereas the enzyme catalase is the major natural sink of H_2O_2 in the ELF. H_2O_2 is central to radical reaction cycles and oxidative stress in the respiratory tract (Winterbourn, 2008). The radical reaction cycles in the ELF are maintained by the reaction of antioxidants with the reduced forms of redox-active PM components. $^{\bullet}OH$ radicals, the most reactive form of ROS, can be produced via Fenton-like reactions of H_2O_2 with iron or copper ions (Charrier & Anastasio, 2011) and can also be released upon interaction of secondary organic aerosol (SOA) with water (Tong et al., 2016). $^{\bullet}OH$ radicals cannot be effectively intercepted by antioxidants and are likely the main source of oxidative stress in the ELF due to their unspecific, high reactivity with any biomolecule (e.g., lipids, proteins, DNA) (Cochrane, 1991; Lelieveld et al., 2021; Stadtman, 1993). Such reactivity is not known for $O_2^{\bullet-}$ radicals and H_2O_2, which may result in more specific oxidation and play a role in redox signaling (Pryor, 1986; Sies, 2014).

The kinetic multi-layer model of surface and bulk chemistry in the epithelial lining fluid (KM-SUB-ELF) explicitly treats mass transport and chemical reactions involving air pollutants,

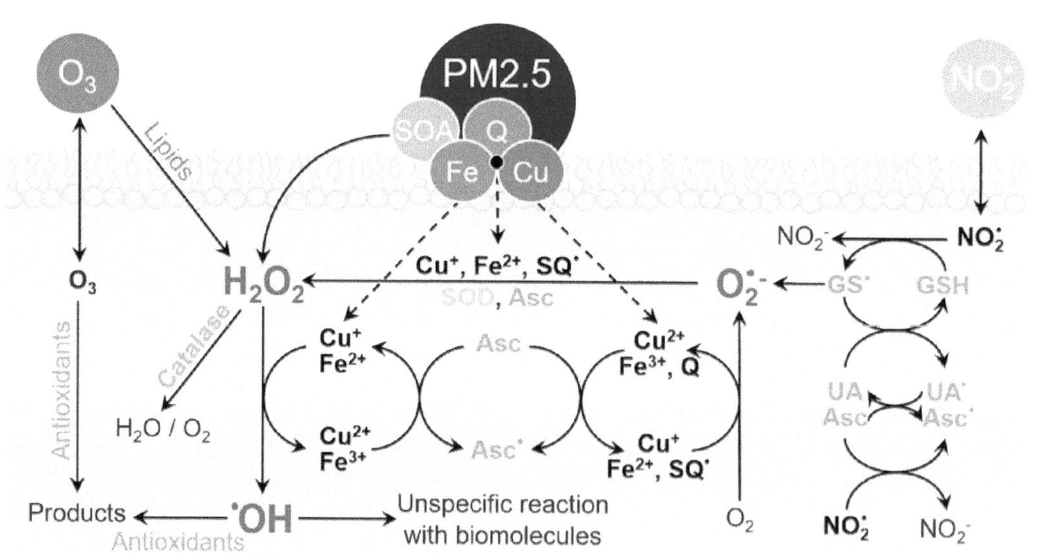

FIGURE 4.1 Production, interconversion, and buffering of reactive oxygen species (ROS) by air pollutants and endogenous molecules in the epithelial lining fluid (ELF). Organic and inorganic constituents of fine particulate matter ($PM_{2.5}$) can produce, convert, and scavenge ROS. Enzymes (catalase; superoxide dismutase, SOD) intercept ROS through the disproportionation of $O_2^{\bullet-}$ and the decomposition of H_2O_2. Antioxidants (ascorbate; glutathione, GSH; uric acid, UA) intercept $^{\bullet}OH$, $O_2^{\bullet-}$, and H_2O_2, but the reaction of antioxidants and surfactant lipids with $^{\bullet}NO_2$ and O_3 can also produce ROS. Figure adapted from Lelieveld et al. (2021).

ROS, antioxidants, and surfactants (Lakey et al., 2016; Lelieveld et al., 2021). Characteristic concentration levels of air pollutants in the ELF can be derived from ambient concentrations, breathing rates, and deposition rates in the respiratory tract. As illustrated in Figure 4.2A, the far most abundant form of ROS in the ELF are H_2O_2 molecules, which are three to four orders of magnitude more abundant than $O_2^{\bullet-}$ radicals, six to seven times more abundant than the hydroperoxy (HO_2) radicals formed by proton transfer to $O_2^{\bullet-}$, and approximately ten orders of magnitude more abundant than $^{\bullet}OH$ radicals. The total concentration of ROS generated by redox-active components of $PM_{2.5}$ deposited in the ELF ranges from ~10 nmol L^{-1} under clean conditions up to almost ~250 nmol L^{-1} under highly polluted conditions as shown in Figure 4.2B. The data points represent explicit $PM_{2.5}$ compositions derived from field data (Lakey et al., 2016).

With KM-SUB-ELF, ROS production, buffering, and conversions can be apportioned to constituents of air pollution, as well as enzymes and antioxidants in the ELF (Lelieveld et al., 2021). Figure 4.3 shows calculation results for three different air pollution scenarios: clean, urban, and megacity.[1] Copper and iron ions are the main sources, i.e., gross producers of ROS in the ELF, largely independent of pollutant concentration. Chemical reactions involving transition-metal ions give rise to ca. 70–90% of all initial ROS formed in the ELF by reduction of O_2 into $O_2^{\bullet-}$ radicals, whereas chemical reactions involving O_3, $^{\bullet}NO_2$, and SOA together are responsible for the remaining ca. 10–30%. O_3 constitutes a significant ROS source in the "clean" and "urban" scenarios but is less important in the "megacity" scenario. Figure 4.3B details the efficacy of all explicit ROS buffers in the ELF. The most potent endogenous ROS sinks include the enzymatic scavenging of $O_2^{\bullet-}$ in the disproportionation by SOD, and the scavenging of H_2O_2 by catalase. Dissolved copper and iron can also scavenge ROS, which becomes increasingly important at higher pollutant concentrations. This signifies the multifaceted role of transition-metal ions in the ELF, i.e., not only inducing formation, but also loss of ROS. To account for such multifaceted roles of pollutants, the net production of ROS from individual sources is presented in Figure 4.3C. In this calculation, all $PM_{2.5}$ constituents taken together are responsible for ~70–80% of all net ROS production. Figure 4.3D shows that SOD and

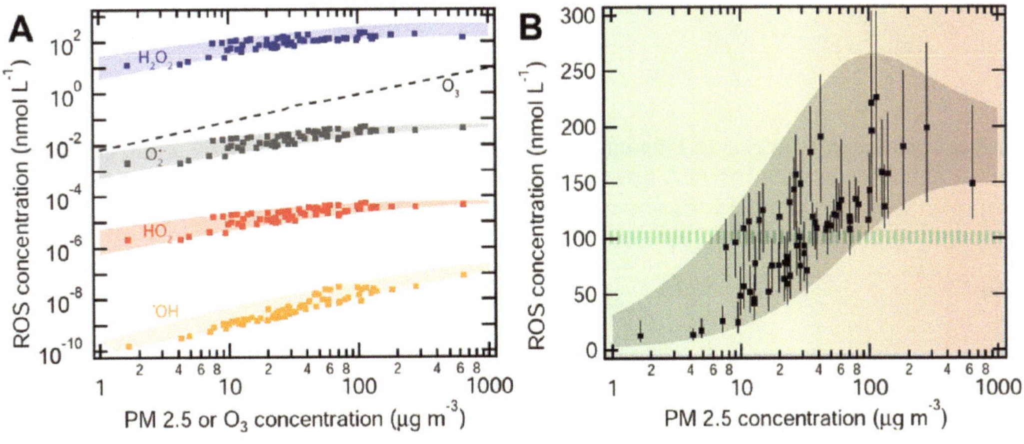

FIGURE 4.2 Chemical exposure-response relations for reactive oxygen species (ROS) produced in the human respiratory tract upon inhalation of fine particulate matter ($PM_{2.5}$) or ozone (O_3). (A) Characteristic concentration levels of different types of ROS in the epithelial lining fluid (ELF) of the respiratory tract. (B) Total ROS concentration in the epithelial lining fluid after 2 hours of inhalation and deposition of ambient $PM_{2.5}$. The green-striped horizontal bar indicates the ROS level characteristic for healthy humans (~100 nmol L^{-1}), and the gray envelope represents the range of aerosol-induced ROS concentrations obtained with the approximate upper and lower limit mass fractions of redox-active components typically observed in ambient $PM_{2.5}$. Figures adapted from Lakey et al. (2016).

antioxidants are the main drivers of H_2O_2 formation from $O_2^{\bullet-}$ in "clean" and "urban" environments, respectively. However, under highly polluted "megacity" conditions, transition-metal ions in $PM_{2.5}$ supersede endogenous molecules in the conversion of $O_2^{\bullet-}$ into H_2O_2. While multiple species in the ELF convert $O_2^{\bullet-}$ into H_2O_2, the model suggests that the conversion of H_2O_2 to $^{\bullet}OH$ almost exclusively involves the $PM_{2.5}$ constituent iron (Figure 4.3E).

At $PM_{2.5}$ concentrations under 10 μg m^{-3}, the buffering of ROS by antioxidants and enzymes is efficient, which leads to low yields of $^{\bullet}OH$ (Figure 4.4A). However, this fraction shows a strong nonlinear increase at typical urban pollution levels and may reach up to 30% at very high pollution according to the model (Lelieveld et al., 2021). The nonlinear increase is due to the competition of catalase and transition metals for reaction with H_2O_2. Catalase scavenges H_2O_2 at a constant rate and when catalase is unable to remove H_2O_2 fast enough, substantial $^{\bullet}OH$ production occurs through Fenton and Fenton-like reactions. Thus, the ELF defense mechanism against oxidative stress acts by antioxidant- and enzyme-driven conversion of the $O_2^{\bullet-}$ radical into the radical reservoir species H_2O_2 and subsequent conversion of H_2O_2 into benign molecules such as O_2 and water (Figure 4.4B, green arrows), with the purpose of maintaining physiological concentration levels of ROS (Sies,

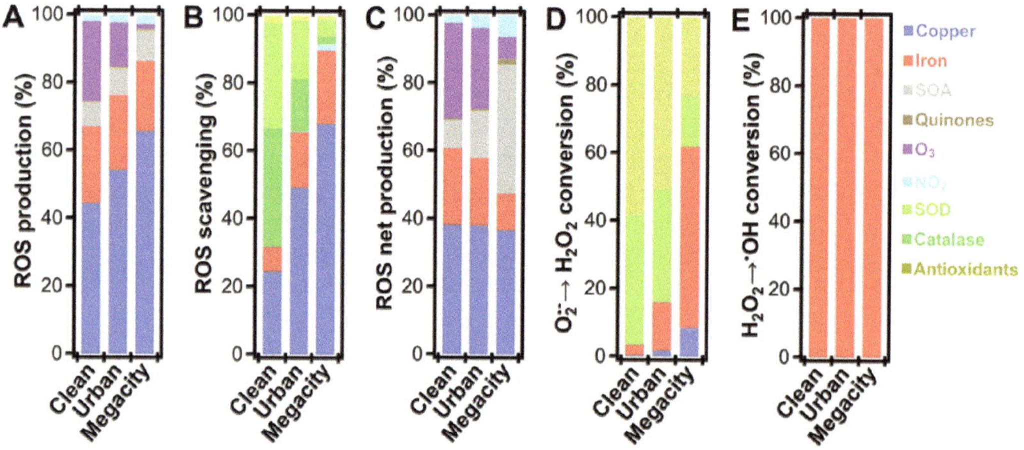

FIGURE 4.3 Relative contributions of pollutants, enzymes, and antioxidants to chemical production, buffering, and conversion of ROS in ELF for three characteristic pollution scenarios (clean, urban, megacity; see note 1 (A) ROS production, (B) ROS scavenging, (C) ROS net production, (D) $O_2^{\bullet-}$-to-H_2O_2 conversion, and (E) H_2O_2-to-$^{\bullet}OH$ conversion. Figure adapted from Lelieveld et al. (2021).

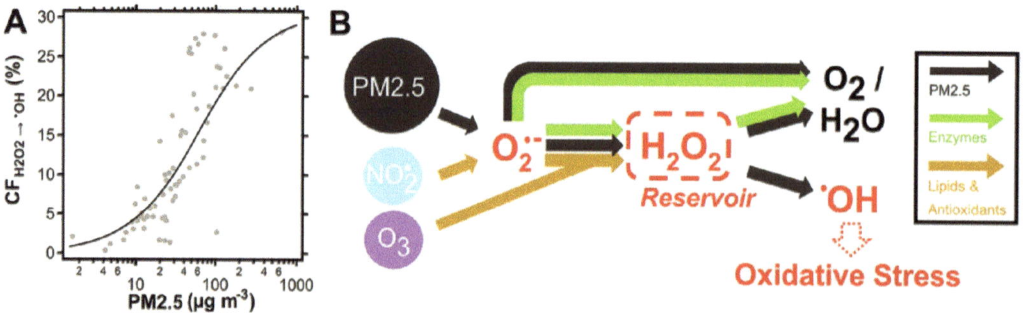

FIGURE 4.4 Shift from the enzyme-controlled ROS buffering regime to the $PM_{2.5}$-controlled $^{\bullet}OH$ radical production regime. (A) The conversion fraction of H_2O_2 to $^{\bullet}OH$ as a function of ambient $PM_{2.5}$ concentration. (B) Schematic summary of the main reaction pathways. Figures adapted from Lelieveld et al. (2021).

1997). Exposure to $PM_{2.5}$ can lead to a shift from the enzyme-controlled ROS buffering regime to the $PM_{2.5}$-controlled $^{\bullet}OH$ radical production regime (black arrows), leading to increased ROS reactivity and oxidative stress.

The exposure-response relations determined with KM-SUB-ELF represent a chemical baseline for the primary chemical production of exogenous ROS and oxidative stress upon inhalation and deposition of redox-active air pollutants in the human respiratory tract (Lakey et al., 2016). On top of this baseline, air pollutants can cause secondary production of endogenous ROS via biological interactions and responses of the human immune system, including the activation of macrophages, mitochondria, and ROS-producing enzymes like NADPH-oxidase ("ROS-induced ROS"; Fang et al., 2022; Zorov et al., 2006) or infections and microbial growth induced by biological and nutrient-rich particles (Borcherding et al., 2014). In certain diseased states, the pH of the ELF may be decreased (Hunt et al., 2000), potentially exacerbating ROS formation through increased transition-metal solubility (Fang et al., 2017) or reduced enzyme activity. Pollution-generated exogenous ROS may also form damage-associated molecular patterns and trigger immune reactions leading to acute or chronic inflammation, e.g., through the toll-like receptor radical cycle (Lucas & Maes, 2013; Ziegler et al., 2020). These and other feedback loops may enhance or dampen the oxidative stress caused by the primary production of ROS from redox-active particulate matter, ozone, and other atmospheric oxidants like nitrogen oxides. Further experimental investigations and model studies will be required to fully unravel and quantify the adverse health effects of air pollution (Lelieveld & Pöschl, 2017). The modeling approach and exposure-response relations presented here provide a basis for such investigations and can help to identify key species and processes, such as the cycling of copper ions, to be addressed in further studies and in the advancement of target-specific regional air quality control strategies (Lakey et al., 2016).

NOTE

1 The three distinct pollutant exposure scenarios that are highlighted have the following characteristics: (1) "clean", with concentrations of 5 $\mu g\ m^{-3}$ PM2.5, 5 $\mu g\ m^{-3}$ $^{\bullet}NO_2$, and 20 $\mu g\ m^{-3}$ O_3; (2) "urban", with 30, 30, and 60 $\mu g\ m^{-3}$; and (3) "megacity", with 300, 300, and 60 $\mu g\ m^{-3}$ for the same pollutants, respectively.

REFERENCES

Baxter, L. K., Duvall, R. M., & Sacks, J. (2013). Examining the effects of air pollution composition on within region differences in PM2.5 mortality risk estimates. *Journal of Exposure Science & Environmental Epidemiology*, *23*(5), 457–465. https://doi.org/10.1038/jes.2012.114

Borcherding, J., Baltrusaitis, J., Chen, H., Stebounova, L., Wu, C. M., Rubasinghege, G., Mudunkotuwa, I. A., Caraballo, J. C., Zabner, J., Grassian, V. H., & Comellas, A. P. (2014). Iron oxide nanoparticles induce Pseudomonas aeruginosa growth, induce biofilm formation, and inhibit antimicrobial peptide function. *Environ Sci Nano*, *1*(2), 123–132. https://doi.org/10.1039/c3en00029j

Burnett, R. T., Stieb, D., Brook, J. R., Cakmak, S., Dales, R., Raizenne, M., Vincent, R., & Dann, T. (2004). Associations between short-term changes in nitrogen dioxide and mortality in Canadian cities. *Archives of Environmental Health: An International Journal*, *59*(5), 228–236. https://doi.org/10.3200/AEOH.59.5.228-236

Charrier, J. G., & Anastasio, C. (2011). Impacts of antioxidants on hydroxyl radical production from individual and mixed transition metals in a surrogate lung fluid. *Atmospheric Environment*, *45*(40), 7555–7562. https://doi.org/10.1016/j.atmosenv.2010.12.021

Cochrane, C. G. (1991). Cellular injury by oxidants. *Am J Med*, *91*(3c), 23s–30s. https://doi.org/10.1016/0002-9343(91)90280-b

Fang, T., Guo, H., Zeng, L., Verma, V., Nenes, A., & Weber, R. J. (2017). Highly acidic ambient particles, soluble metals, and oxidative potential: A link between sulfate and aerosol toxicity. *Environmental Science & Technology*, *51*(5), 2611–2620. https://doi.org/10.1021/acs.est.6b06151

Fang, T., Huang, Y.-K., Wei, J., Monterrosa Mena, J. E., Lakey, P. S. J., Kleinman, M. T., Digman, M. A., & Shiraiwa, M. (2022). Superoxide release by macrophages through NADPH oxidase activation dominating chemistry by isoprene secondary organic aerosols and quinones to cause oxidative damage on membranes. *Environmental Science & Technology*. https://doi.org/10.1021/acs.est.2c03987

Faustini, A., Rapp, R., & Forastiere, F. (2014). Nitrogen dioxide and mortality: Review and meta-analysis of long-term studies. *European Respiratory Journal*, *44*(3), 744–753. https://doi.org/10.1183/09031936.00114713

Gauderman, W. J., Avol, E., Gilliland, F., Vora, H., Thomas, D., Berhane, K., McConnell, R., Kuenzli, N., Lurmann, F., Rappaport, E., Margolis, H., Bates, D., & Peters, J. (2004). The effect of air pollution on lung development from 10 to 18 years of age [Article]. *New England Journal of Medicine*, *351*(11), 1057–1067. https://doi.org/10.1056/NEJMoa040610

Hewitt, C. N., & Kok, G. L. (1991). Formation and occurrence of organic hydroperoxides in the troposphere: Laboratory and field observations. *Journal of Atmospheric Chemistry*, *12*(2), 181–194. https://doi.org/10.1007/BF00115779

Hunt, J. F., Fang, K., Malik, R., Snyder, A., Malhotra, N., Platts-Mills, T. A. E., & Gaston, B. (2000). Endogenous airway acidification. *American Journal of Respiratory and Critical Care Medicine*, *161*(3), 694–699. https://doi.org/10.1164/ajrccm.161.3.9911005

Lakey, P. S. J., Berkemeier, T., Tong, H., Arangio, A. M., Lucas, K., Pöschl, U., & Shiraiwa, M. (2016). Chemical exposure-response relationship between air pollutants and reactive oxygen species in the human respiratory tract [Article]. *Scientific Reports*, *6*, 32916. https://doi.org/10.1038/srep32916

Lelieveld, J., & Pöschl, U. (2017). Chemists can help to solve the air-pollution health crisis. *Nature*, *551*(7680), 291–293. https://doi.org/10.1038/d41586-017-05906-9

Lelieveld, S., Wilson, J., Dovrou, E., Mishra, A., Lakey, P. S. J., Shiraiwa, M., Pöschl, U., & Berkemeier, T. (2021). Hydroxyl radical production by air pollutants in epithelial lining fluid governed by interconversion and scavenging of reactive oxygen species. *Environmental Science & Technology*, *55*(20), 14069–14079. https://doi.org/10.1021/acs.est.1c03875

Lewtas, J. (2007). Air pollution combustion emissions: Characterization of causative agents and mechanisms associated with cancer, reproductive, and cardiovascular effects. *Mutation Research-Reviews in Mutation Research*, *636*(1–3), 95–133. https://doi.org/10.1016/j.mrrev.2007.08.003

Lucas, K., & Maes, M. (2013). Role of the Toll Like Receptor (TLR) radical cycle in chronic inflammation: Possible treatments targeting the TLR4 pathway. *Mol Neurobiol*, *48*(1), 190–204. https://doi.org/10.1007/s12035-013-8425-7

McConnell, R., Berhane, K., Gilliland, F., London, S. J., Islam, T., Gauderman, W. J., Avol, E., Margolis, H. G., & Peters, J. M. (2002). Asthma in exercising children exposed to ozone: A cohort study. *The Lancet*, *359*(9304), 386–391. https://doi.org/10.1016/S0140-6736(02)07597-9

Mills, I. C., Atkinson, R. W., Kang, S., Walton, H., & Anderson, H. R. (2015). Quantitative systematic review of the associations between short-term exposure to nitrogen dioxide and mortality and hospital admissions. *BMJ Open*, *5*(5), e006946. https://doi.org/10.1136/bmjopen-2014-006946

Monks, P. S., Granier, C., Fuzzi, S., Stohl, A., Williams, M. L., Akimoto, H., Amann, M., Baklanov, A., Baltensperger, U., Bey, I., et al. (2009). Atmospheric composition change: Global and regional air quality. *Atmospheric Environment*, *43*(33), 5268–5350. https://doi.org/10.1016/j.atmosenv.2009.08.021

Mudway, I. S., & Kelly, F. J. (2000). Ozone and the lung: A sensitive issue. *Molecular Aspects of Medicine*, *21*(1), 1–48. https://doi.org/10.1016/S0098-2997(00)00003-0

Nel, A. (2005). Air pollution-related illness: Effects of particles. *Science*, *308*(5723), 804–806. https://doi.org/10.1126/science.1108752

Peng, R. D., Bell, M. L., Geyh, A. S., McDermott, A., Zeger, S. L., Samet, J. M., & Dominici, F. (2009). Emergency admissions for cardiovascular and respiratory diseases and the chemical composition of fine particle air pollution. *Environmental Health Perspectives*, *117*(6), 957–963. https://doi.org/doi:10.1289/ehp.0800185

Pope, C. A., Burnett, R. T., Thurston, G. D., Thun, M. J., Calle, E. E., Krewski, D., & Godleski, J. J. (2004). Cardiovascular mortality and long-term exposure to particulate air pollution: Epidemiological evidence of general pathophysiological pathways of disease [Article]. *Circulation*, *109*(1), 71–77. https://doi.org/10.1161/01.cir.0000108927.80044.7f

Pöschl, U., & Shiraiwa, M. (2015). Multiphase chemistry at the atmosphere: Biosphere interface influencing climate and public health in the anthropocene. *Chemical Reviews*, *115*, 4440–4475. https://doi.org/10.1021/cr500487s

Pryor, W. A. (1986). Oxy-radicals and related species: Their formation, lifetimes, and reactions. *Annual Review of Physiology*, *48*(1), 657–667. https://doi.org/10.1146/annurev.ph.48.030186.003301

Sies, H. (1997). Oxidative stress: Oxidants and antioxidants. *Experimental Physiology*, *82*(2), 291–295. https://doi.org/10.1113/expphysiol.1997.sp004024

Sies, H. (2014). Role of metabolic H2O2 generation: Redox signaling and oxidative stress. *J Biol Chem*, *289*(13), 8735–8741. https://doi.org/10.1074/jbc.R113.544635

Stadtman, E. (1993). Oxidation of free amino acids and amino acid residues in proteins by radiolysis and by metal-catalyzed reactions. *Annual Review of Biochemistry*, *62*(1), 797–821.

Tong, H., Arangio, A. M., Lakey, P. S. J., Berkemeier, T., Liu, F., Kampf, C. J., Brune, W. H., Pöschl, U., & Shiraiwa, M. (2016). Hydroxyl radicals from secondary organic aerosol decomposition in water. *Atmos. Chem. Phys.*, *16*(3), 1761–1771. https://doi.org/10.5194/acp-16-1761-2016

Turner, M. C., Jerrett, M., III, C. A. P., Krewski, D., Gapstur, S. M., Diver, W. R., Beckerman, B. S., Marshall, J. D., Su, J., Crouse, D. L., & Burnett, R. T. (2016). Long-term ozone exposure and mortality in a large prospective study. *American Journal of Respiratory and Critical Care Medicine*, *193*(10), 1134–1142. https://doi.org/10.1164/rccm.201508-1633OC

van der Vliet, A., O'Neill, C. A., Cross, C. E., Koostra, J. M., Volz, W. G., Halliwell, B., & Louie, S. (1999). Determination of low-molecular-mass antioxidant concentrations in human respiratory tract lining fluids. *Am J Physiol*, *276*(2), L289–L296. https://doi.org/10.1152/ajplung.1999.276.2.L289

Velsor, L. W., & Postlethwait, E. M. (1997). NO2-induced generation of extracellular reactive oxygen is mediated by epithelial lining layer antioxidants. *Am J Physiol*, *273*(6), L1265–L1275. https://doi.org/10.1152/ajplung.1997.273.6.L1265

Winterbourn, C. C. (2008). Reconciling the chemistry and biology of reactive oxygen species [Review]. *Nature Chemical Biology*, *4*(5), 278–286. https://doi.org/10.1038/nchembio.85

Zeng, M., & Wilson, K. R. (2020). Efficient coupling of reaction pathways of criegee intermediates and free radicals in the heterogeneous ozonolysis of alkenes. *The Journal of Physical Chemistry Letters*, *11*(16), 6580–6585. https://doi.org/10.1021/acs.jpclett.0c01823

Ziegler, K., Kunert, A. T., Reinmuth-Selzle, K., Leifke, A. L., Widera, D., Weller, M. G., Schuppan, D., Frohlich-Nowoisky, J., Lucas, K., & Poschl, U. (2020). Chemical modification of pro-inflammatory proteins by peroxynitrite increases activation of TLR4 and NF-κB: Implications for the health effects of air pollution and oxidative stress. *Redox Biol*, *37*, 101581. https://doi.org/10.1016/j.redox.2020.101581

Zorov, D. B., Juhaszova, M., & Sollott, S. J. (2006). Mitochondrial ROS-induced ROS release: An update and review. *Biochim Biophys Acta*, *1757*(5–6), 509–517. https://doi.org/10.1016/j.bbabio.2006.04.029

5 Chemical Modification of Proteins by Reactive Oxygen and Nitrogen Species under Atmospheric and Physiological Conditions

Janine Fröhlich-Nowoisky, Kurt Lucas, Thomas Berkemeier, and Ulrich Pöschl

5.1 PROTEIN CHEMISTRY AND EFFECTS UNDER ATMOSPHERIC CONDITIONS

In the atmosphere, proteins account for up to ~5 % of fine particulate matter and play a major role as airborne allergens (Franze et al. 2003; Fröhlich-Nowoisky et al. 2016; Miguel et al. 1996; Reinmuth-Selzle et al. 2017). The proteins are released from plant pollen, fungal spores, and other primary biological aerosol particles, for example after rupturing under humid conditions or upon exposure to O_3 and $^{\bullet}NO_2$ (Cecchi et al. 2010; Fröhlich-Nowoisky et al. 2016; Motta et al. 2006; Ouyang et al. 2016; Taylor et al. 2004). In the atmosphere, the proteins are directly exposed to O_3, $^{\bullet}NO_2$, and other photooxidants that can cause chemical modifications and modify the allergenic potential before inhalation and deposition in the respiratory tract (Reinmuth-Selzle et al. 2017). The released proteins can also attach to the surfaces of pollen or other solid particles such as soot and soil or road dust, and they can undergo dissolution and further reactions in aqueous or organic droplets (Backes et al. 2021; Franze et al. 2003; Franze et al. 2005; Miguel et al. 1999; Shiraiwa et al. 2012).

As shown in Figure 5.1, the reactions between $O_3/^{\bullet}NO_2$ and tyrosine (Tyr) residues of proteins involve the formation of tyrosyl radicals resulting from the reaction of Tyr with O_3. These tyrosyl radicals can either react with $^{\bullet}NO_2$ to form nitrotyrosine or cross-link by dityrosine formation. Protein nitration and oligomerization are competing processes, but under atmospheric conditions, Tyr nitration likely dominates over dityrosine formation since the probability of two tyrosyl radical coming close enough for dimerization is very low (Backes et al. 2021; Liu et al. 2017). The extent of tyrosine nitration and cross-linking of proteins is sensitive toward an increase in O_3 concentration and rather insensitive towards changes in ambient $^{\bullet}NO_2$ concentrations (Backes et al. 2021; Liu et al. 2017). As increasing temperature also increases the reaction rate of tyrosyl radical formation by ozonolysis, it is likely that nitrated and cross-linked protein species are important products of exposure to O_3 and $^{\bullet}NO_2$ under tropical or summer smog conditions (Franze et al. 2005; Liu et al. 2017; Sandhiya, Kolandaivel, and Senthilkumar 2014). Furthermore, the increase of the O_3 concentration and temperature in the Anthropocene might thus lead to an increased occurrence of oligomeric proteins in the atmosphere, which in turn may be related to the increased prevalence of allergic diseases that has been observed around the globe (Kampf et al. 2015).

DOI: 10.1201/9781003328100-7

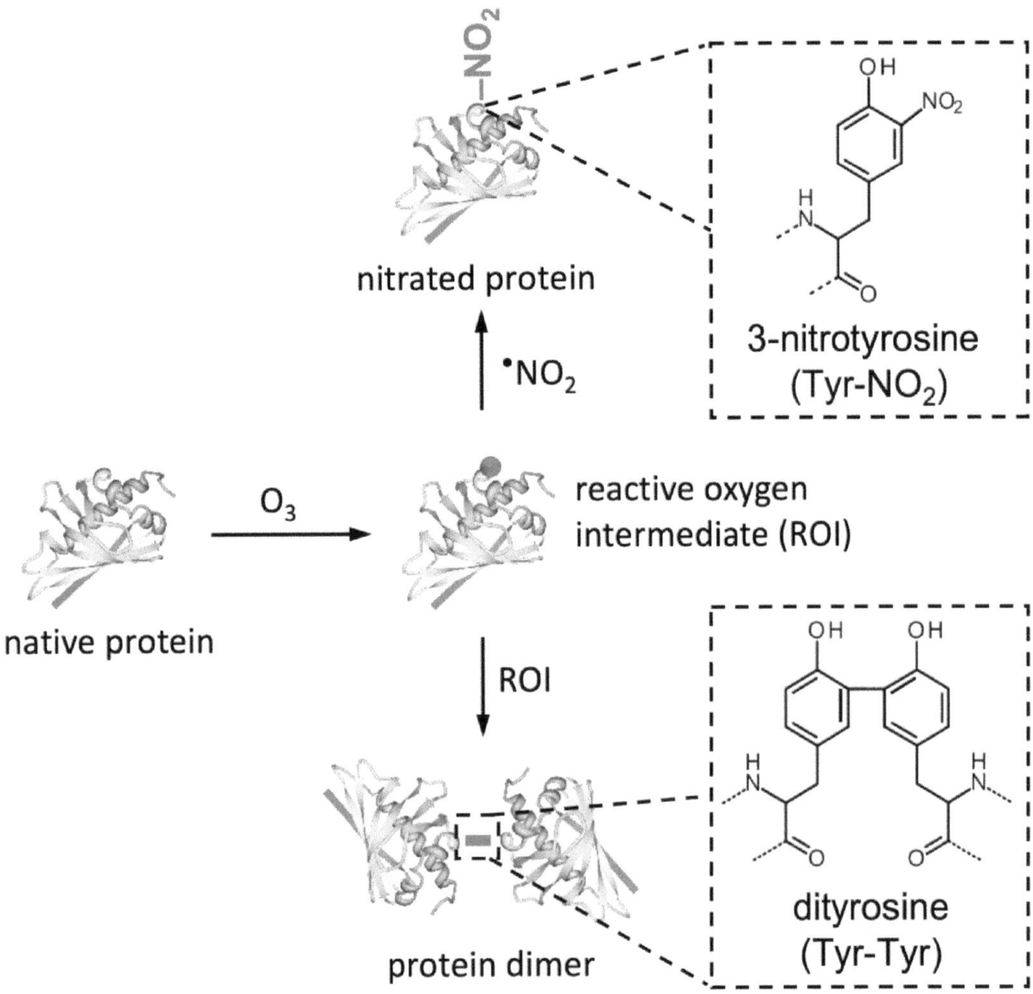

FIGURE 5.1 Protein nitration and oligomerization upon exposure to ozone (O_3) and nitrogen dioxide ($^{\bullet}NO_2$). The initial reaction with O_3 leads to the formation of reactive oxygen intermediates (ROI, tyrosyl radicals), which can further react with $^{\bullet}NO_2$ to form nitrated proteins (nitrotyrosine) or with each other to form cross-linked proteins (dityrosine). The shown protein is the major birch pollen allergen Bet v 1, for which nitration and cross-linking were found to influence the immunogenicity and allergenic potential. Reproduced from Reinmuth-Selzle et al. (2017).

Figure 5.2 illustrates the protein chemistry with O_3 and $^{\bullet}NO_2$ under various atmospheric conditions. The phase state of atmospheric particulates can show a strong dependence on atmospheric conditions by being liquid at high temperature and high humidity, but in a viscous semisolid or even glassy state at low temperature and low humidity (Koop et al. 2011; Shiraiwa et al. 2011; Shiraiwa, Li, et al. 2017). Oxidants such as O_3 and $^{\bullet}NO_2$ diffuse quickly into liquid particles and more slowly in solid or semi-solid phase states (Berkemeier et al. 2016). The yields of protein nitration and oligomerization can thus be strongly influenced by changes in relative humidity (RH). The overall nitration and oligomerization rates are higher in the aqueous phase caused by high RH, indicating an increased relevance of the processes under cloud-processing conditions (Liu et al. 2017;

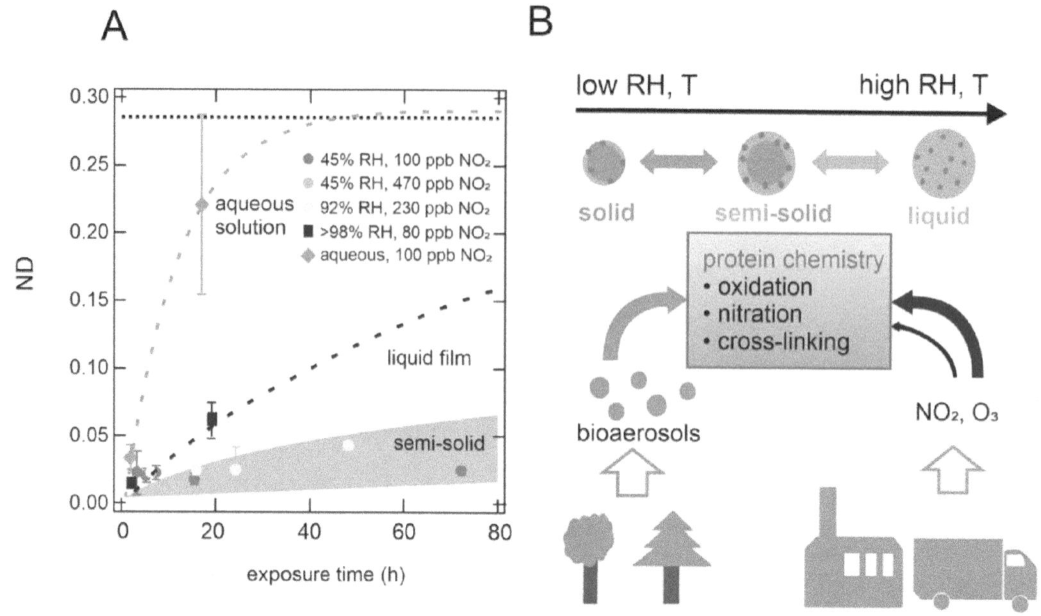

FIGURE 5.2 Atmospheric protein chemistry by ozone (O_3) and nitrogen dioxide ($^{\bullet}NO_2$). (A) Nitration degrees (ND) for the birch pollen allergen Bet v 1 exposed to O_3/$^{\bullet}NO_2$ for different types of reaction conditions; data points and error bars represent the arithmetic mean values and standard errors of four experiments with filter samples and two experiments with liquid samples, respectively. The dashed lines and the gray shaded area show the results of kinetic model simulations. The black dotted line represents the maximum ND for Bet v 1 based on the assumptions used in the model simulations. Reproduced from Reinmuth-Selzle et al. (2014). (B) Reaction rates are limited by the phase state of proteins and hence the diffusivity of oxidants and protein molecules, which changes with relative humidity (RH) and temperature (T). Air pollutants like $^{\bullet}NO_2$ and O_3 can enhance allergen release from bioaerosols (e.g., pollen) with O_3 being more important in triggering the nitration, cross-linking, and oxidation of allergenic proteins in bioaerosols and other protein-containing particles. Reproduced from Liu et al. (2017) with permission from the Royal Society of Chemistry.

Reinmuth-Selzle et al. 2014). At low RH, chemical reactions can be confined to protein films on the surface of bioaerosols or other particles as reaction rates in the bulk material are limited by diffusion (Kampf et al. 2015).

5.2 PROTEIN CHEMISTRY AND EFFECTS UNDER PHYSIOLOGICAL CONDITIONS

The global excess mortality attributed to the chronic impacts of $PM_{2.5}$ and O_3 is estimated to 8.8 (7.11–10.3) million per year (Burnett et al. 2018; Lelieveld et al. 2020). Exposure to $PM_{2.5}$ and other air pollutants can cause oxidant-antioxidant imbalances either through inhalation or through the generation of reactive oxygen and nitrogen species in the epithelial lining fluid upon inhalation of toxic aerosol compounds, leading to inflammatory responses and various acute and chronic respiratory and cardiovascular diseases (Lakey et al. 2016; Münzel et al. 2018; Münzel et al. 2022; Münzel and Daiber 2018; Newby et al. 2015; Rajagopalan, Al-Kindi, and Brook 2018; Xing et al. 2016).

Atmospheric aerosols and oxidants are likely to promote the chemical modification of proteins in the human body, particularly in the respiratory tract. Peroxynitrite especially plays an important role in oxidative stress and adverse health effects upon exposure to environmental pollutants and in the course of inflammatory diseases (Cohen et al. 2017; Kelly 2003; Münzel and Daiber 2018; Shiraiwa, Ueda, et al. 2017). Peroxynitrite can react with a wide range of biomolecules including tyrosine residues of proteins, which may lead to disease and autoimmunity (Ahmad, Hussain, and Ahsan 2019; Li et al. 2019; Strickland et al. 2015; Szabó, Ischiropoulos, and Radi 2007). In *in vitro* studies, peroxynitrite was able to achieve nitration degrees of up to 40% of the grass pollen allergen Phl p 5 and caused oligomerization of up to 50% of protein mass (Backes et al. 2021). Nitrotyrosine and dityrosine are biomarkers for endogenous peroxynitrite activity, inflammation, and oxidative stress (Ahmad, Hussain, and Ahsan 2019; Carmen Baez et al. 2018; DiMarco and Giulivi 2007; Ischiropoulos 2009; Radi 2013). Inhalation of air pollutants further leads to formation of hydroxyl radicals (Lelieveld et al. 2021; Tong et al. 2016), which react quickly, at site of formation, with any biomolecule and may lead to irreversible protein modification.

Animal models have consistently observed inflammatory responses to air pollutants as recently reviewed by Münzel et al. (2022). Among them are the adhesion and infiltration of Ly6Chigh immune cells via CCR2/VCAM interaction, TLR/CD36-mediated activation of NF-κB, NLRP3, and IL1β, associated with MMP-2/9 upregulation, foam cell formation, and plaque instability as well as systemic and neuroinflammation due to activation of the lung-neural reflex arcs (Deiuliis et al. 2012; Goto et al. 2004; Kampfrath et al. 2011; Liu et al. 2013; Rajagopalan and Brook 2012; Simon and Liedtke 2008). Ultrafine particles and toxic contaminations on particle surfaces (e.g., heavy metals, polycyclic aromatic hydrocarbons or fungal pyrogens/bacterial endotoxins) that can directly penetrate the lung tissue into the systemic circulation can lead to immune cell activation or oxidative damage of endothelial cells (Dominici et al. 2007; Kampfrath et al. 2011; Liberda et al. 2010). Moreover, $PM_{2.5}$, ultrafine particles, and in particular diesel exhaust, can cause endothelial dysfunction due to diminished vascular nitric oxide availability and increased production of reactive oxygen species as a consequence of an activation of the vascular and or phagocytic NADPH oxidase and/or endothelial nitric oxide synthase uncoupling (Campen et al. 2005; Knuckles et al. 2008; Sun et al. 2008; Ying, Kampfrath, et al. 2009). In animals, $PM_{2.5}$ activates pathways that are analogous to those of angiotensin II such as Rho/ROCK signaling and heightened calcium sensitivity (Sun et al. 2008; Ying, Yue, et al. 2009). Inflammatory processes can also lead to the formation of oxidized biomolecules such as 7-ketocholesterol (7-KC) or oxidized phospholipid derivatives of 1-palmitoyl-2-arachidonyl-sn-glycero-3-phosphorylcholine (oxPAPC) promoting the migration of bone marrow derived CD11b+Ly6Chigh cells to atherosclerotic plaques (Campen, Lund, and Rosenfeld 2012; Rao et al. 2014; Tsou et al. 2007).

During inflammation, proteinous damage-associated molecular patterns (DAMPs) can be actively secreted by cells (Roh and Sohn 2018). These DAMPs can stimulate pattern recognition receptors (PRR) such as the Toll-like receptor 4 (TLR4) (Vénéreau, Ceriotti, and Bianchi 2015). TLR4 signaling leads to the activation of transcription factors like the nuclear factor 'kappa-light-chain-enhancer' of activated B cells (NF-κB) which is a key activator of inflammatory cascade (Takeda and Akira 2004). NF-κB induces the expression and modulates the secretion of pro-inflammatory cytokines such as TNF-α, IL-1β, and IL-8, which can stimulate the corresponding cytokine receptors leading to further activation of NF-κB and other signaling pathways (Liu et al. 2016, 2014; Varfolomeev and Vucic 2018; Weber, Wasiliew, and Kracht 2010). This positive feedback can amplify and propagate inflammatory processes in autocrine or paracrine fashion (Gan et al. 2014; Lucas and Maes 2013).

As shown in Figure 5.3, chemical modification by peroxynitrite can enhance the TLR4 activity and inflammatory potential of proteinous DAMPs α-synuclein (α-Syn), heat shock protein 60 (HSP60), and high-mobility-group box 1 protein (HMGB1), relevant in neurodegenerative and cardiovascular diseases (Schildknecht et al. 2013; Wenzel et al. 2017). Future *in vivo* studies are needed

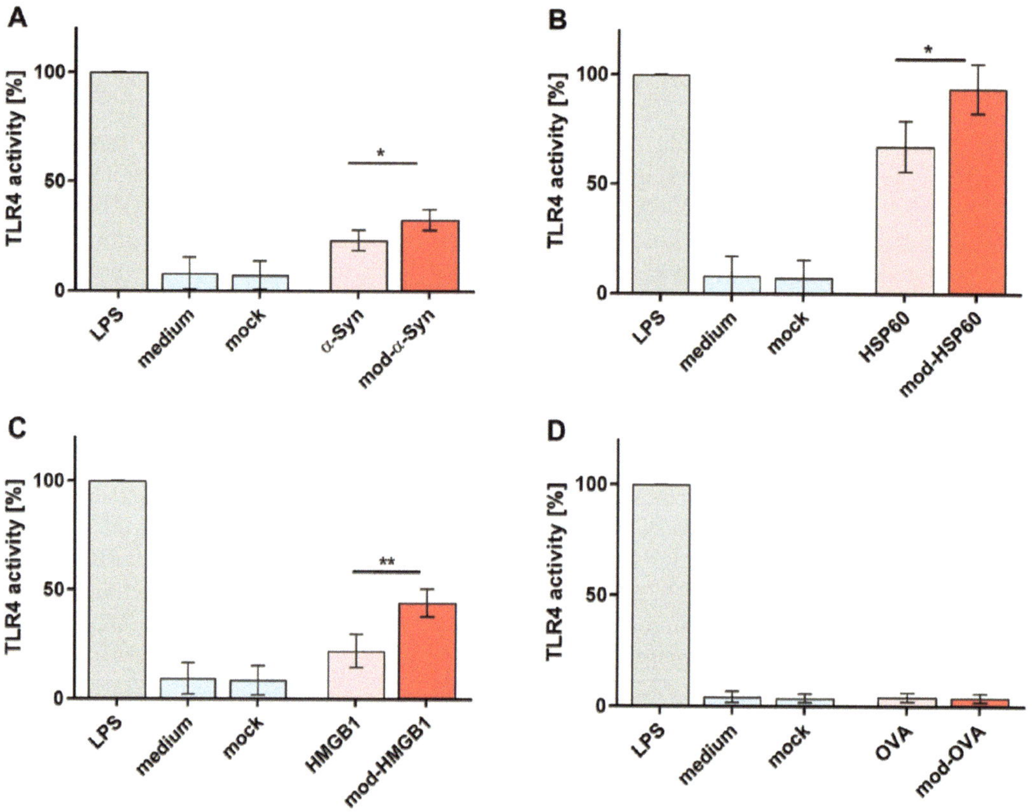

FIGURE 5.3 TLR4 activation by native and peroxynitrite-modified proteins. TLR4 activity in HeLa TLR4 dual luciferase reporter cells determined for (A) α-Syn, (B) HSP60, (C) HMGB1, and (D) ovalbumin (OVA, negative control) normalized to LPS. Arithmetic mean values and standard deviations of three to four independent experiments performed in triplicates (*$p < 0.05$, **$p < 0.01$). Reproduced from Ziegler et al. (2020) with permission.

to clarify whether nitrotyrosine residues and/or dityrosine crosslinks in proteinous DAMPs play a role in amplifying inflammatory processes through enhanced TLR4 signaling. Such insights may also help to advance medical treatment options that involve scavenging of peroxynitrite (Daiber et al. 2013; Das et al. 2015; Pacher, Beckman, and Liaudet 2007) or the inhibition of peroxynitrite formation (Poulos and Li 2017).

Figure 5.4 illustrates how chemically modified DAMPs may amplify oxidative stress and innate immune responses through a positive feedback loop of pro-inflammatory signaling via TLR4. ROS/RNS and oxidative stress, triggering inflammatory processes that lead to the formation of endogenous ROS/RNS and release of DAMPs. The DAMPs can activate PRR (TLR, RAGE, etc.) that induce further pro-inflammatory signaling and responses via transcription factors (NF-κB, IRF3, etc.), cytokines, and cytokine receptors (IL-1, IL-8, etc.). This positive feedback can be additionally enhanced if the DAMPs undergo chemical modification by ROS/RNS and if the modified DAMPs lead to stronger activation of PRR than the native DAMPs (Ziegler et al. 2020). We suggest to consider and elucidate the role of environmental pollutants and potential needs and perspectives for societal action with regard to the steeply increasing and globally pervasive human influence on air quality, climate, and public health in the Anthropocene (Pöschl and Shiraiwa 2015; Reinmuth-Selzle et al. 2017).

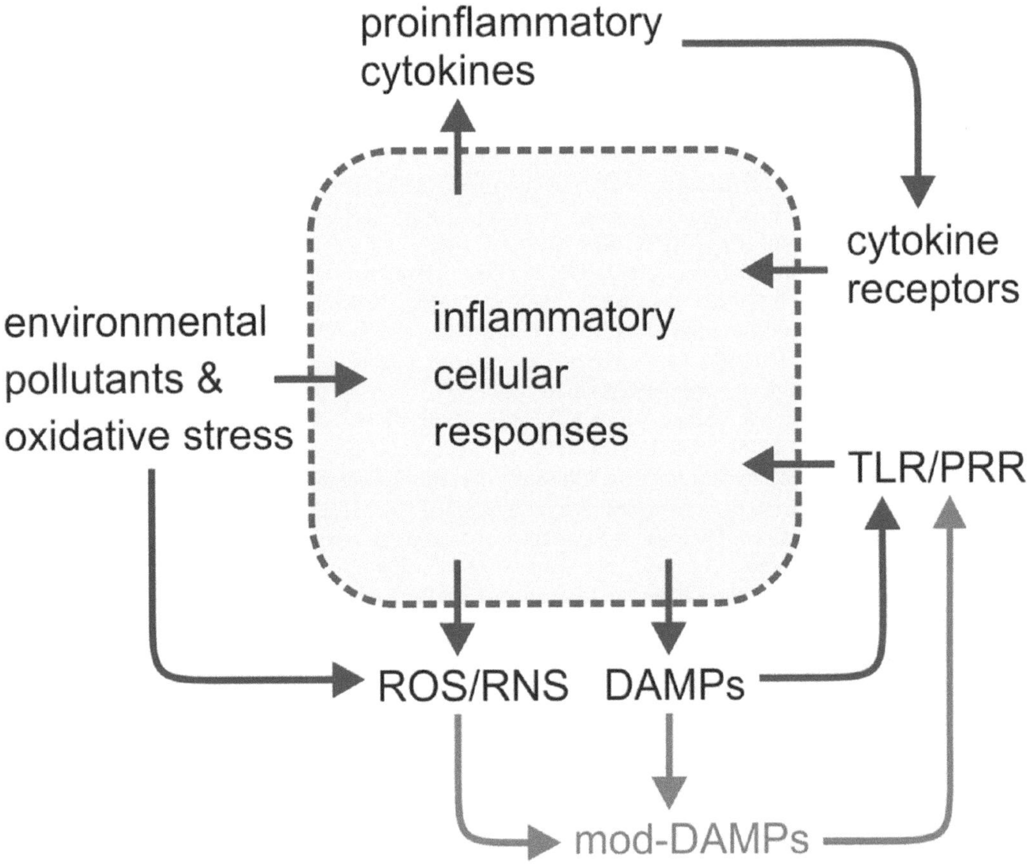

FIGURE 5.4 Amplification of inflammatory processes and innate immune responses through chemically modified DAMPs. Environmental pollutants and oxidative stress can induce an increase of reactive oxygen and nitrogen species (ROS/RNS), the formation of chemically modified damage-associated molecular patterns (mod-DAMPs), an increase of pro-inflammatory signaling via Toll-like receptors and other pattern recognition receptors (TLR/PRR), an increase of proinflammatory cytokines, and further inflammatory cellular responses. Note that the results of this study and the schematic illustration are based on experiments with immortalized cell lines, and that further experiments are required to confirm their applicability to the human immune system. Reproduced from Ziegler et al. (2020) with permission.

5.3 CLINICAL IMPLICATIONS OF ENDOGENOUS PROTEIN NITRATION AND INGESTION OF EXOGENOUS NITRATED PROTEIN ALLERGENS

Various methods such as specific ELISAs and mass spectrometry-based methods allow the immunological detection of nitrotyrosine in patient samples (Knight et al. 2018; Medeiros et al. 2021; Teixeira et al. 2016). Many clinical studies found associations of tyrosine-nitrated proteins with the pathogenesis of various diseases as reviewed by Daiber et al. (2021). For example, 3-nitrotyrosine (3-NT) levels and autoimmune reactions triggered by IgG against 3-NT were found to correlate with coronary artery disease health outcomes (Shishehbor 2003; Thomson et al. 2012). Moreover, tyrosine nitration can lead to enzyme inactivation as found for the important ROS scavenging and vasodilator-producing enzymes Mn-SOD (MacMillan-Crow et al. 1996) and prostacyclin synthase (Davis and Zou 2005) or aggravation of the prothrombotic effects of fibrinogen (Martinez et al. 2012; Vadseth et al. 2004). Note that most other oxidative protein modifications are less specific for peroxynitrite. Thus, regarding the clinical evidence for other protein-based oxidative stress markers, including protein carbonyl groups, we refer to previous overviews (Egea et al. 2017; Frijhoff et al. 2015).

REFERENCES

Ahmad, Rizwan, Ahtesham Hussain, and Haseeb Ahsan. 2019. Peroxynitrite: Cellular Pathology and Implications in Autoimmunity. *Journal of Immunoassay and Immunochemistry* 40 (2). Taylor & Francis:123–138. doi:10.1080/15321819.2019.1583109.

Backes, Anna T., Kathrin Reinmuth-Selzle, Anna Lena Leifke, Kira Ziegler, Carola S. Krevert, Georg Tscheuschner, Kurt Lucas, Michael G. Weller, Thomas Berkemeier, Ulrich Pöschl, et al. 2021. Oligomerization and Nitration of the Grass Pollen Allergen Phl p 5 by Ozone, Nitrogen Dioxide and Peroxynitrite: Reaction Products, Kinetics and Health Effects. *International Journal of Molecular Sciences* 22 (14). doi:10.3390/ijms22147616.

Berkemeier, Thomas, Sarah S. Steimer, Ulrich K. Krieger, Thomas Peter, Ulrich Pöschl, Markus Ammann, and Manabu Shiraiwa. 2016. Ozone Uptake on Glassy, Semi-Solid and Liquid Organic Matter and the Role of Reactive Oxygen Intermediates in Atmospheric Aerosol Chemistry. *Physical Chemistry Chemical Physics* 18 (18). Royal Society of Chemistry:12662–12674. doi:10.1039/c6cp00634e.

Burnett, Richard, Hong Chen, Mieczysław Szyszkowicz, Neal Fann, Bryan Hubbell, C. Arden Pope, Joshua S. Apte, Michael Brauer, Aaron Cohen, Scott Weichenthal, et al. 2018. Global Estimates of Mortality Associated with Long-Term Exposure to Outdoor Fine Particulate Matter. *Proceedings of the National Academy of Sciences* 115 (38):9592–9597. doi:10.1073/pnas.1803222115.

Campen, Matthew J., N. Sathish Babu, G. Andrew Helms, Stuart Pett, Jorge Wernly, Reza Mehran, and Jacob D. McDonald. 2005. Nonparticulate Components of Diesel Exhaust Promote Constriction in Coronary Arteries from ApoE−/− Mice. *Toxicological Sciences* 88 (1):95–102. doi:10.1093/toxsci/kfi283.

Campen, Matthew J., Amie Lund, and Michael Rosenfeld. 2012. Mechanisms Linking Traffic-Related Air Pollution and Atherosclerosis. *Current Opinion in Pulmonary Medicine* 18 (2):155–160. doi:10.1097/MCP.0b013e32834f210a.

Carmen Baez, María del, María de La Paz Scribano, Mariana Tarán, Ismael Fonseca, Ariel Balceda, Sergio Blencio, and Mónica Moya. 2018. Nitrotyrosine as a Biomarker: Recovery of Histopathological Atherogenic Lesions in Rats Treated with Ascorbic Acid. *Trends in Medicine* 18 (4):1–5. doi:10.15761/tim.1000152.

Cecchi, Lorenzo, Gennaro D'Amato, Jon G. Ayres, Carmen Galan, Francesco Forastiere, Bertil Forsberg, Jorrit Gerritsen, Carlos Nunes, Heidrun Behrendt, Cezmi Akdis, et al. 2010. Projections of the Effects of Climate Change on Allergic Asthma: The Contribution of Aerobiology. *Allergy* 65 (9):1073–1081. doi:10.1111/j.1398-9995.2010.02423.x.

Cohen, Aaron J., Michael Brauer, Richard Burnett, H. Ross Anderson, Joseph Frostad, Kara Estep, Kalpana Balakrishnan, Bert Brunekreef, Lalit Dandona, Rakhi Dandona, et al. 2017. Estimates and 25-Year Trends of the Global Burden of Disease Attributable to Ambient Air Pollution: An Analysis of Data from the Global Burden of Diseases Study 2015. *The Lancet* 389 (10082). The Author(s). Published by Elsevier Ltd. This is an Open Access article under the CC BY 4.0 license:1907–1918. doi:10.1016/S0140-6736(17)30505-6.

Daiber, Andreas, Steffen Daub, Markus Bachschmid, Stefan Schildknecht, Matthias Oelze, Sebastian Steven, Patrick Schmidt, Alexandra Megner, Masayuki Wada, Tadashi Tanabe, et al. 2013. Protein Tyrosine Nitration and Thiol Oxidation by Peroxynitrite-Strategies to Prevent These Oxidative Modifications. *International Journal of Molecular Sciences* 14 (4):7542–7570. doi:10.3390/ijms14047542.

Daiber, Andreas, Omar Hahad, Ioanna Andreadou, Sebastian Steven, Steffen Daub, and Thomas Münzel. 2021. Redox-Related Biomarkers in Human Cardiovascular Disease: Classical Footprints and Beyond. *Redox Biology* 42 (June):101875. doi:10.1016/j.redox.2021.101875.

Das, Suvarthi, Firas Alhasson, Diptadip Dattaroy, Sahar Pourhoseini, Ratanesh Kumar Seth, Mitzi Nagarkatti, Prakash S. Nagarkatti, Gregory A. Michelotti, Anna Mae Diehl, Balaraman Kalyanaraman, et al. 2015. NADPH Oxidase-Derived Peroxynitrite Drives Inflammation in Mice and Human Nonalcoholic Steatohepatitis via TLR4-Lipid Raft Recruitment. *American Journal of Pathology* 185 (7). American Society for Investigative Pathology:1944–1957. doi:10.1016/j.ajpath.2015.03.024.

Davis, Bradley, and Ming-Hui Zou. 2005. CD40 Ligand: Dependent Tyrosine Nitration of Prostacyclin Synthase In Vivo. *Circulation* 112 (14):2184–2192. doi:10.1161/CIRCULATIONAHA.105.553206.

Deiuliis, Jeffrey A., Thomas Kampfrath, Jixin Zhong, Steve Oghumu, Andrei Maiseyeu, Lung Chi Chen, Qinghua Sun, Abhay R. Satoskar, and Sanjay Rajagopalan. 2012. Pulmonary T Cell Activation in Response to Chronic Particulate Air Pollution. *American Journal of Physiology-Lung Cellular and Molecular Physiology* 302 (4):L399–L409. doi:10.1152/ajplung.00261.2011.

DiMarco, Theresa, and Cecilia Giulivi. 2007. Current Analytical Methods for the Detection of Dityrosine, a Biomarker of Oxidative Stress, in Biological Samples. *Mass Spectrometry Reviews*. doi:10.1002/mas.20109.

Dominici, Francesca, Roger D. Peng, Keita Ebisu, Scott L. Zeger, Jonathan M. Samet, and Michelle L. Bell. 2007. Does the Effect of PM 10 on Mortality Depend on PM Nickel and Vanadium Content? A Reanalysis of the NMMAPS Data. *Environmental Health Perspectives* 115 (12):1701–1703. doi:10.1289/ehp.10737.

Egea, Javier, Isabel Fabregat, Yves M. Frapart, Pietro Ghezzi, Agnes Görlach, Thomas Kietzmann, Kateryna Kubaichuk, Ulla G. Knaus, Manuela G. Lopez, Gloria Olaso-Gonzalez, et al. 2017. European Contribution to the Study of ROS: A Summary of the Findings and Prospects for the Future from the COST Action BM1203 (EU-ROS). *Redox Biology* 13 (October):94–162. doi:10.1016/j.redox.2017.05.007.

Franze, Thomas, Michael G. Weller, Reinhard Niessner, and Ulrich Pöschl. 2003. Enzyme Immunoassays for the Investigation of Protein Nitration by Air Pollutants. *The Analyst* 128 (7):824–831. doi:10.1039/B303132B.

Franze, Thomas, Michael G. Weller, Reinhard Niessner, and Ulrich Pöschl. 2005. Protein Nitration by Polluted Air. *Environmental Science and Technology* 39 (6):1673–1678. doi:10.1021/es0488737.

Frijhoff, Jeroen, Paul G. Winyard, Neven Zarkovic, Sean S. Davies, Roland Stocker, David Cheng, Annie R. Knight, Emma Louise Taylor, Jeannette Oettrich, Tatjana Ruskovska, et al. 2015. Clinical Relevance of Biomarkers of Oxidative Stress. *Antioxidants & Redox Signaling* 23 (14):1144–1170. doi:10.1089/ars.2015.6317.

Fröhlich-Nowoisky, Janine, Christopher J. Kampf, Bettina Weber, J. Alex Huffman, Christopher Pöhlker, Meinrat O. Andreae, Naama Lang-Yona, Susannah M. Burrows, Sachin S. Gunthe, Wolfgang Elbert, et al. 2016. Bioaerosols in the Earth System: Climate, Health, and Ecosystem Interactions. *Atmospheric Research* 182. The Authors:346–376. doi:10.1016/j.atmosres.2016.07.018.

Gan, Lay T., Derrick M. Van Rooyen, Mark E. Koina, Robert S. McCuskey, Narcissus C. Teoh, and Geoffrey C. Farrell. 2014. Hepatocyte Free Cholesterol Lipotoxicity Results from JNK1-Mediated Mitochondrial Injury and Is HMGB1 and TLR4-Dependent. *Journal of Hepatology* 61 (6):1376–1384. doi:10.1016/j.jhep.2014.07.024.

Goto, Yukinobu, Hiroshi Ishii, James C. Hogg, Chih-Horng Shih, Kazuhiro Yatera, Renaud Vincent, and Stephan F. van Eeden. 2004. Particulate Matter Air Pollution Stimulates Monocyte Release from the Bone Marrow. *American Journal of Respiratory and Critical Care Medicine* 170 (8):891–897. doi:10.1164/rccm.200402-235OC.

Ischiropoulos, Harry. 2009. Protein Tyrosine Nitration-An Update. *Archives of Biochemistry and Biophysics* 484 (2). Elsevier Inc.:117–121. doi:10.1016/j.abb.2008.10.034.

Kampf, Christopher J., Fobang Liu, Kathrin Reinmuth-Selzle, Thomas Berkemeier, Hannah Meusel, Manabu Shiraiwa, and Ulrich Pöschl. 2015. Protein Cross-Linking and Oligomerization through Dityrosine Formation upon Exposure to Ozone. *Environmental Science and Technology* 49 (18):10859–10866. doi:10.1021/acs.est.5b02902.

Kampfrath, Thomas, Andrei Maiseyeu, Zhekang Ying, Zubair Shah, Jeffrey A. Deiuliis, Xiaohua Xu, Nisharahmed Kherada, Robert D. Brook, Kongara M. Reddy, Nitin P. Padture, et al. 2011. Chronic Fine Particulate Matter Exposure Induces Systemic Vascular Dysfunction via NADPH Oxidase and TLR4 Pathways. *Circulation Research* 108 (6):716–726. doi:10.1161/CIRCRESAHA.110.237560.

Kelly, Frank J. 2003. Oxidative Stress: Its Role in Air Pollution and Adverse Health Effects. *Occupational and Environmental Medicine* 60 (8):612–616. doi:10.1136/oem.60.8.612.

Knight, Annie R., Emma L. Taylor, Roman Lukaszewski, Karina Tveen Jensen, Helen E. Jones, Jane E. Carré, Michail N. Isupov, Jennifer A. Littlechild, Stephen J. Bailey, Emily Brewer, et al. 2018. A High-Sensitivity Electrochemiluminescence-Based ELISA for the Measurement of the Oxidative Stress Biomarker, 3-Nitrotyrosine, in Human Blood Serum and Cells. *Free Radical Biology and Medicine* 120 (May):246–254. doi:10.1016/j.freeradbiomed.2018.03.026.

Knuckles, Travis L., Amie K. Lund, Selita N. Lucas, and Matthew J. Campen. 2008. Diesel Exhaust Exposure Enhances Venoconstriction via Uncoupling of ENOS. *Toxicology and Applied Pharmacology* 230 (3):346–351. doi:10.1016/j.taap.2008.03.010.

Koop, Thomas, Johannes Bookhold, Manabu Shiraiwa, and Ulrich Pöschl. 2011. Glass Transition and Phase State of Organic Compounds: Dependency on Molecular Properties and Implications for Secondary Organic Aerosols in the Atmosphere. *Physical Chemistry Chemical Physics* 13 (43):19238–19255. doi:10.1039/c1cp22617g.

Lakey, Pascale S.J., Thomas Berkemeier, Haijie Tong, Andrea M. Arangio, Kurt Lucas, Ulrich Pöschl, and Manabu Shiraiwa. 2016. Chemical Exposure-Response Relationship between Air Pollutants and Reactive Oxygen Species in the Human Respiratory Tract. *Scientific Reports* 6 (1). Nature Publishing Group:32916. doi:10.1038/srep32916.

Lelieveld, Jos, Andrea Pozzer, Ulrich Pöschl, Mohammed Fnais, Andy Haines, and Thomas Münzel. 2020. Loss of Life Expectancy from Air Pollution Compared to Other Risk Factors: A Worldwide Perspective. *Cardiovascular Research* 116 (11):1910–1917. doi:10.1093/cvr/cvaa025.

Lelieveld, Steven, Jake Wilson, Eleni Dovrou, Ashmi Mishra, Pascale S.J. Lakey, Manabu Shiraiwa, Ulrich Pöschl, and Thomas Berkemeier. 2021. Hydroxyl Radical Production by Air Pollutants in Epithelial Lining Fluid Governed by Interconversion and Scavenging of Reactive Oxygen Species. *Environmental Science and Technology* 55 (20):14069–14079. doi:10.1021/acs.est.1c03875.

Li, Wenting, Jinghan Feng, Chong Gao, Meiling Wu, Qiaohui Du, Bun Tsoi, Qi Wang, Dan Yang, and Jiangang Shen. 2019. Nitration of Drp1 Provokes Mitophagy Activation Mediating Neuronal Injury in Experimental Autoimmune Encephalomyelitis. *Free Radical Biology and Medicine* 143 (November):70–83. doi:10.1016/j.freeradbiomed.2019.07.037.

Liberda, Eric N., Azita K. Cuevas, Patricia A. Gillespie, Gabriele Grunig, Qingshan Qu, and Lung Chi Chen. 2010. Exposure to Inhaled Nickel Nanoparticles Causes a Reduction in Number and Function of Bone Marrow Endothelial Progenitor Cells. *Inhalation Toxicology* 22 (sup2):95–99. doi:10.3109/08958378.2010.515269.

Liu, Cuiqing, Zhekang Ying, Jack Harkema, Qinghua Sun, and Sanjay Rajagopalan. 2013. Epidemiological and Experimental Links between Air Pollution and Type 2 Diabetes. *Toxicologic Pathology* 41 (2):361–373. doi:10.1177/0192623312464531.

Liu, Fobang, Pascale S.J. Lakey, Thomas Berkemeier, Haijie Tong, Anna Theresa Kunert, Hannah Meusel, Yafang Cheng, Hang Su, Janine Fröhlich-Nowoisky, Senchao Lai, et al. 2017. Atmospheric Protein Chemistry Influenced by Anthropogenic Air Pollutants: Nitration and Oligomerization upon Exposure to Ozone and Nitrogen Dioxide. *Faraday Discuss.* doi:10.1039/C7FD00005G.

Liu, Qian, Anping Li, Yijun Tian, Jennifer D. Wu, Yu Liu, Tengfei Li, Yuan Chen, Xinwei Han, and Kongming Wu. 2016. Cytokine & Growth Factor Reviews the CXCL8-CXCR1/2 Pathways in Cancer. *Cytokine and Growth Factor Reviews* 31:61–71. doi:10.1016/j.cytogfr.2016.08.002.

Liu, Yu, Heng Yin, Ming Zhao, and Qianjin Lu. 2014. TLR2 and TLR4 in Autoimmune Diseases: A Comprehensive Review. *Clinical Reviews in Allergy and Immunology* 47 (2):136–147. doi:10.1007/s12016-013-8402-y.

Lucas, Kurt, and Michael Maes. 2013. Role of the Toll Like Receptor (TLR) Radical Cycle in Chronic Inflammation: Possible Treatments Targeting the TLR4 Pathway. *Molecular Neurobiology* 48 (1):190–204. doi:10.1007/s12035-013-8425-7.

MacMillan-Crow, L.A., J.P. Crow, J.D. Kerby, J.S. Beckman, and J.A. Thompson. 1996. Nitration and Inactivation of Manganese Superoxide Dismutase in Chronic Rejection of Human Renal Allografts. *Proceedings of the National Academy of Sciences* 93 (21):11853–11858. doi:10.1073/pnas.93.21.11853.

Martinez, Marissa, Adam Cuker, Angela Mills, Richard Lightfoot, Yiying Fan, W.H. Wilson Tang, Stanley L. Hazen, and Harry Ischiropoulos. 2012. Nitrated Fibrinogen Is a Biomarker of Oxidative Stress in Venous Thromboembolism. *Free Radical Biology and Medicine* 53 (2):230–236. doi:10.1016/j.freeradbiomed.2012.05.004.

Medeiros, Romina, Bebiana Sousa, Silvina Rossi, Catarina Afonso, Luis Bonino, Andrew Pitt, Elizabeth López, Corinne Spickett, and Graciela Borthagaray. 2021. Identification and Relative Quantification of 3-Nitrotyrosine Residues in Fibrinogen Nitrated in Vitro and Fibrinogen from Ischemic Stroke Patient Plasma Using LC-MS/MS. *Free Radical Biology and Medicine* 165 (March):334–347. doi:10.1016/j.freeradbiomed.2021.01.049.

Miguel, Ann G., Glen R. Cass, M. Michael Glovsky, and Jay Weiss. 1999. Allergens in Paved Road Dust and Airborne Particles. *Environmental Science and Technology* 33 (23):4159–4168. doi:10.1021/es9904890.

Miguel, Ann G., Glen R. Cass, Jay Weiss, and M. Michael Glovsky. 1996. Latex Allergens in Tire Dust and Airborne Particles. *Environmental Health Perspectives* 104 (11):1180–1186. doi:10.2307/3432911.

Motta, Alexandre C., Maryse Marliere, Gabriel Peltre, Peter A. Sterenberg, and Ghishlaine Lacroix. 2006. Traffic-Related Air Pollutants Induce the Release of Allergen-Containing Cytoplasmic Granules from Grass Pollen. *International Archives of Allergy and Immunology* 139 (4):294–298. doi:10.1159/000091600.

Münzel, Thomas, and Andreas Daiber. 2018. Environmental Stressors and Their Impact on Health and Disease with Focus on Oxidative Stress. *Antioxidants and Redox Signaling* 28 (9):735–740. doi:10.1089/ars.2017.7488.

Münzel, Thomas, Tommaso Gori, Sadeer Al-Kindi, John Deanfield, Jos Lelieveld, Andreas Daiber, and Sanjay Rajagopalan. 2018. Effects of Gaseous and Solid Constituents of Air Pollution on Endothelial Function. *European Heart Journal* 39 (38):3543–3550. doi:10.1093/eurheartj/ehy481.

Münzel, Thomas, Omar Hahad, Mette Sørensen, Jos Lelieveld, Georg Daniel Duerr, Mark Nieuwenhuijsen, and Andreas Daiber. 2022. Environmental Risk Factors and Cardiovascular Diseases: A Comprehensive Expert Review. *Cardiovascular Research* 118 (14):2880–2902. doi:10.1093/cvr/cvab316.

Newby, David E., Pier M. Mannucci, Grethe S. Tell, Andrea A. Baccarelli, Robert D. Brook, Ken Donaldson, Francesco Forastiere, Massimo Franchini, Oscar H. Franco, Ian Graham, et al. 2015. Expert Position Paper on Air Pollution and Cardiovascular Disease. *European Heart Journal* 36 (2):83–93. doi:10.1093/eurheartj/ehu458.

Ouyang, Yuhui, Zhaojun Xu, Erzhong Fan, Ying Li, and Luo Zhang. 2016. Effect of Nitrogen Dioxide and Sulfur Dioxide on Viability and Morphology of Oak Pollen. *International Forum of Allergy and Rhinology* 6 (1):95–100. doi:10.1002/alr.21632.

Pacher, Pál, Joseph S. Beckman, and Lucas Liaudet. 2007. Nitric Oxide and Peroxynitrite in Health and Disease. *Physiological Reviews*. doi:10.1152/physrev.00029.2006.

Pöschl, Ulrich, and Manabu Shiraiwa. 2015. Multiphase Chemistry at the Atmosphere-Biosphere Interface Influencing Climate and Public Health in the Anthropocene. *Chemical Reviews* 115 (10):4440–4475. doi:10.1021/cr500487s.

Poulos, Thomas L., and Huiying Li. 2017. Nitric Oxide Synthase and Structure-Based Inhibitor Design. *Nitric Oxide: Biology and Chemistry* 63:68–77. doi:10.1016/j.niox.2016.11.004.

Radi, Rafael. 2013. Protein Tyrosine Nitration: Biochemical Mechanisms and Structural Basis of Functional Effects. *Accounts of Chemical Research* 46 (2):550–559. doi:10.1021/ar300234c.

Rajagopalan, Sanjay, Sadeer G. Al-Kindi, and Robert D. Brook. 2018. Air Pollution and Cardiovascular Disease: JACC State-of-the-Art Review. *Journal of the American College of Cardiology* 72 (17). Elsevier:2054–2070. doi:10.1016/j.jacc.2018.07.099.

Rajagopalan, Sanjay, and Robert D. Brook. 2012. Air Pollution and Type 2 Diabetes. *Diabetes* 61 (12):3037–3045. doi:10.2337/db12-0190.

Rao, Xiaoquan, Jixin Zhong, Andrei Maiseyeu, Bhavani Gopalakrishnan, Frederick A. Villamena, Lung-Chi Chen, Jack R. Harkema, Qinghua Sun, and Sanjay Rajagopalan. 2014. CD36-Dependent 7-Ketocholesterol Accumulation in Macrophages Mediates Progression of Atherosclerosis in Response to Chronic Air Pollution Exposure. *Circulation Research* 115 (9):770–780. doi:10.1161/CIRCRESAHA.115.304666.

Reinmuth-Selzle, Kathrin, Chloe Chloé Chloe Ackaert, Christopher J. Kampf, Martin Samonig, Manabu Shiraiwa, Stefan Kofler, Hong Yang, Gabriele Gadermaier, Hans Brandstetter, Christian G. Huber, et al. 2014. Nitration of the Birch Pollen Allergen Bet v 1.0101: Efficiency and Site-Selectivity of Liquid and Gaseous Nitrating Agents. *Journal of Proteome Research* 13 (3):1570–1577. doi:10.1021/pr401078h.

Reinmuth-Selzle, Kathrin, Christopher J. Kampf, Kurt Lucas, Naama Lang-Yona, Janine Fröhlich-Nowoisky, Manabu Shiraiwa, Pascale S.J. Lakey, Senchao Lai, Fobang Liu, Anna T. Kunert, et al. 2017. Air Pollution and Climate Change Effects on Allergies in the Anthropocene: Abundance, Interaction, and Modification of Allergens and Adjuvants. *Environmental Science and Technology* 51 (8):4119–4141. doi:10.1021/acs.est.6b04908.

Roh, Jong Seong, and Dong Hyun Sohn. 2018. Damage-Associated Molecular Patterns in Inflammatory Diseases. *Immune Network* 18 (4):1–14. doi:10.4110/in.2018.18.e27.

Sandhiya, Lakshaman, Ponmalai Kolandaivel, and Kittusamy Senthilkumar. 2014. Oxidation and Nitration of Tyrosine by Ozone and Nitrogen Dioxide: Reaction Mechanisms and Biological and Atmospheric Implications. *Journal of Physical Chemistry B* 118 (13):3479–3490. doi: 10.1021/jp4106037.

Schildknecht, Stefan, Hanne R. Gerding, Christiaan Karreman, Malte Drescher, Hilal A. Lashuel, Tiago F. Outeiro, Donato A. Di Monte, and Marcel Leist. 2013. Oxidative and Nitrative Alpha-Synuclein Modifications and Proteostatic Stress: Implications for Disease Mechanisms and Interventions in Synucleinopathies. *Journal of Neurochemistry* 125 (4):491–511. doi:10.1111/jnc.12226.

Shiraiwa, Manabu, Markus Ammann, Thomas Koop, and Ulrich Pöschl. 2011. Gas Uptake and Chemical Aging of Semisolid Organic Aerosol Particles. *Proceedings of the National Academy of Sciences*. doi:10.1073/pnas.1103045108.

Shiraiwa, Manabu, Ying Li, Alexandra P. Tsimpidi, Vlassis A. Karydis, Thomas Berkemeier, Spyros N. Pandis, Jos Lelieveld, Thomas Koop, and Ulrich Pöschl. 2017. Global Distribution of Particle Phase State in Atmospheric Secondary Organic Aerosols. *Nature Communications* 8:1–7. doi:10.1038/ncomms15002.

Shiraiwa, Manabu, Kathrin Selzle, Hong Yang, Yulia Sosedova, Markus Ammann, and Ulrich Pöschl. 2012. Multiphase Chemical Kinetics of the Nitration of Aerosolized Protein by Ozone and Nitrogen Dioxide. *Environmental Science and Technology* 46 (12):6672–6680. doi:10.1021/es300871b.

Shiraiwa, Manabu, Kayo Ueda, Andrea Pozzer, Gerhard Lammel, Christopher J. Kampf, Akihiro Fushimi, Shinichi Enami, Andrea M. Arangio, Janine Fröhlich-Nowoisky, Yuji Fujitani, et al. 2017. Aerosol Health Effects from Molecular to Global Scales. *Environmental Science & Technology* 51 (23):13545–13567. doi:10.1021/acs.est.7b04417.

Shishehbor, Mehdi H. 2003. Association of Nitrotyrosine Levels with Cardiovascular Disease and Modulation by Statin Therapy. *JAMA* 289 (13):1675. doi:10.1001/jama.289.13.1675.

Simon, Sidney A., and Wolfgang Liedtke. 2008. How Irritating: The Role of TRPA1 in Sensing Cigarette Smoke and Aerogenic Oxidants in the Airways. *Journal of Clinical Investigation*, June. doi:10.1172/JCI36111.

Strickland, Faith M., YePeng Li, Kent Johnson, Zhichao Sun, and Bruce C. Richardson. 2015. CD4+ T Cells Epigenetically Modified by Oxidative Stress Cause Lupus-Like Autoimmunity in Mice. *Journal of Autoimmunity* 62 (August):75–80. doi:10.1016/j.jaut.2015.06.004.

Sun, Qinghua, Peibin Yue, Zhekang Ying, Arturo J. Cardounel, Robert D. Brook, Robert Devlin, Jing-Shiang Hwang, Jay L. Zweier, Lung Chi Chen, and Sanjay Rajagopalan. 2008. Air Pollution Exposure Potentiates Hypertension Through Reactive Oxygen Species-Mediated Activation of Rho/ROCK. *Arteriosclerosis, Thrombosis, and Vascular Biology* 28 (10):1760–1766. doi:10.1161/ATVBAHA.108.166967.

Szabó, Csaba, Harry Ischiropoulos, and Rafael Radi. 2007. Peroxynitrite: Biochemistry, Pathophysiology and Development of Therapeutics. *Nature Reviews Drug Discovery* 6 (8):662–680. doi:10.1038/nrd2222.

Takeda, Kiyoshi, and Shizuo Akira. 2004. TLR Signaling Pathways. *Seminars in Immunology* 16 (1):3–9. doi:10.1016/j.smim.2003.10.003.

Taylor, Philip E., Richard C. Flagan, Ann G. Miguel, Rudolf Valenta, and Michael M. Glovsky. 2004. Birch Pollen Rupture and the Release of Aerosols of Respirable Allergens. *Clinical and Experimental Allergy: Journal of the British Society for Allergy and Clinical Immunology* 34 (10):1591–1596. doi:10.1111/j.1365-2222.2004.02078.x.

Teixeira, Dulce, Rúben Fernandes, Cristina Prudêncio, and Mónica Vieira. 2016. 3-Nitrotyrosine Quantification Methods: Current Concepts and Future Challenges. *Biochimie* 125 (June):1–11. doi:10.1016/j.biochi.2016.02.011.

Thomson, Leonor, Margarita Tenopoulou, Richard Lightfoot, Epida Tsika, Ioannis Parastatidis, Marissa Martinez, Todd M. Greco, Paschalis Thomas Doulias, Yuping Wu, W. H. Wilson Tang, et al. 2012. Immunoglobulins against Tyrosine-Nitrated Epitopes in Coronary Artery Disease. *Circulation* 126 (20):2392–2401. doi:10.1161/CIRCULATIONAHA.112.103796.

Tong, Haijie, Andrea M. Arangio, Pascale S.J. Lakey, Thomas Berkemeier, Fobang Liu, Christopher J. Kampf, William H. Brune, Ulrich Poschl, and Manabu Shiraiwa. 2016. Hydroxyl Radicals from Secondary Organic Aerosol Decomposition in Water. *Atmospheric Chemistry and Physics* 16 (3):1761–1771. doi:10.5194/acp-16-1761-2016.

Tsou, Chia-Lin, Wendy Peters, Yue Si, Sarah Slaymaker, Ara M. Aslanian, Stuart P. Weisberg, Matthias Mack, and Israel F. Charo. 2007. Critical Roles for CCR2 and MCP-3 in Monocyte Mobilization from Bone Marrow and Recruitment to Inflammatory Sites. *Journal of Clinical Investigation* 117 (4):902–909. doi:10.1172/JCI29919.

Vadseth, Caryn, Jose M. Souza, Leonor Thomson, Amy Seagraves, Chandrasekaran Nagaswami, Tomas Scheiner, Jim Torbet, Gaston Vilaire, Joel S. Bennett, Juan-Carlos Murciano, et al. 2004. Pro-Thrombotic State Induced by Post-Translational Modification of Fibrinogen by Reactive Nitrogen Species. *Journal of Biological Chemistry* 279 (10):8820–8826. doi:10.1074/jbc.M306101200.

Varfolomeev, Eugene, and Domagoj Vucic. 2018. Intracellular Regulation of TNF Activity in Health and Disease. *Cytokine* 101:26–32. doi:10.1016/j.cyto.2016.08.035.

Vénéreau, Emilie, Chiara Ceriotti, and Marco Emilio Bianchi. 2015. DAMPs from Cell Death to New Life. *Frontiers in Immunology* 6 (AUG):1–11. doi:10.3389/fimmu.2015.00422.

Weber, Axel, Peter Wasiliew, and Michael Kracht. 2010. Interleukin-1 (IL-1) Pathway. *Science Signaling* 3 (105):1–7. doi:10.1126/scisignal.3105cm1.

Wenzel, Philip, Sabine Kossmann, Thomas Münzel, and Andreas Daiber. 2017. Redox Regulation of Cardiovascular Inflammation: Immunomodulatory Function of Mitochondrial and Nox-Derived Reactive Oxygen and Nitrogen Species. *Free Radical Biology and Medicine* 109 (August):48–60. doi:10.1016/j.freeradbiomed.2017.01.027.

Xing, Yu Fei, Yue Hua Xu, Min Hua Shi, and Yi Xin Lian. 2016. The Impact of PM2.5 on the Human Respiratory System. *Journal of Thoracic Disease* 8 (1):E69–E74. doi:10.3978/j.issn.2072-1439.2016.01.19.

Ying, Zhekang, Thomas Kampfrath, George Thurston, Britten Farrar, Mort Lippmann, Aixia Wang, Qinghua Sun, Lung Chi Chen, and Sanjay Rajagopalan. 2009. Ambient Particulates Alter Vascular Function through Induction of Reactive Oxygen and Nitrogen Species. *Toxicological Sciences* 111 (1):80–88. doi:10.1093/toxsci/kfp004.

Ying, Zhekang, Peibin Yue, Xiaohua Xu, Mianhua Zhong, Qinghua Sun, Michael Mikolaj, Aixia Wang, Robert D. Brook, Lung Chi Chen, et al. 2009. Air Pollution and Cardiac Remodeling: A Role for RhoA/Rho-Kinase. *American Journal of Physiology-Heart and Circulatory Physiology* 296 (5):H1540–H1550. doi:10.1152/ajpheart.01270.2008.

Ziegler, Kira, Anna T. Kunert, Kathrin Reinmuth-Selzle, Anna Lena Leifke, Darius Widera, Michael G. Weller, Detlef Schuppan, Janine Fröhlich-Nowoisky, Kurt Lucas, et al. 2020. Chemical Modification of Pro-Inflammatory Proteins by Peroxynitrite Increases Activation of TLR4 and NF-KB: Implications for the Health Effects of Air Pollution and Oxidative Stress. *Redox Biology*, no. May:101581. doi:10.1016/j.redox.2020.101581.

6 Air Pollution and Neuropsychiatric Disease

Omar Hahad, Thomas Münzel, Donya A. Gilan, Maria Teresa Bayo Jimenez, and Jasmin Ghaemi Kerahrodi

6.1 INTRODUCTION

The Global Burden of Disease (GBD) study from 2019 reported that ambient air pollution ranked seventh as a global health risk factor by facilitating the development of chronic noncommunicable diseases (NCD) such as cardiovascular disease (CVD) and by contributing to global mortality (Collaborators 2020). Air pollution caused by particulate matter (PM) is the most significant risk factor among all environmental stressors. There is a strong correlation between ambient PM with an aerodynamic diameter of less than 2.5 μm ($PM_{2.5}$), which is produced by natural sources (primary particles) or formed in the atmosphere through chemical reactions (secondary particles), and adverse health effects (Aghababaeian et al. 2021). In 2015, exposure to those particles ranked fifth among all health risk factors, contributing to 4.2 million annual deaths (Figure 6.1) (Cohen et al. 2017). Additionally, recent World Health Organization (WHO) data confirmed that 90% of people worldwide breathe polluted air ("World Health Organization. Nine out of ten people worldwide breathe polluted air, but more countries are taking action (World Health Organization 2018).

Importantly, besides the fact that air pollution is among the significant environmental contributors to the global burden of NCD, mainly driven by cardiorespiratory diseases, recent evidence also suggested a relationship between air pollution and neuropsychiatric diseases by affecting the brain and central or autonomic nervous system (CNS/ANS) (Landrigan et al. 2018; Costa et al. 2017). For instance, ultrafine particles, whose diameter size is <0.1 μm, can transmigrate through the alveolar cell layers and directly translocate from the pulmonary system to the CNS by crossing the blood-brain barrier (BBB). Then, when these particles enter the brain, they can cause pathophysiological changes in the CNS due to the particle's physical properties (relatively large reactive surface) or to toxic substances that are bound to them, which can lead to the emergence of neuropsychiatric diseases like stroke, dementia, and Parkinson's disease. According to growing evidence, there is a link between air pollution and cerebrovascular and neurological disorders, as well as neuroinflammation and cerebral oxidative stress as main drivers (Wang, Xiong, and Tang 2017). Additionally, in response to the inhalation of these nanometer particles, they can directly cause end-organ damage by immune cell activation, oxidative stress, and structural damage of cells driven by enhanced production of pro-inflammatory mediators and ROS (Lodovici and Bigagli 2011; Anderson, Thundiyil, and Stolbach 2012).

6.2 PATHOPHYSIOLOGY OF AIR POLLUTION-INDUCED DISORDERS

It is postulated that exposure to air pollution causes the vascular endothelium to develop a pathological condition defined by an imbalance between nitric oxide (•NO) production and degradation (Munzel et al. 2018). Furthermore, the redox imbalance brought on by •NO's bioavailability damages endothelial cells physically. It compromises various essential processes that let an intact endothelium retain its anti-inflammatory, antioxidant, antithrombotic, and vasodilatory properties. Accordingly, the ongoing physiological harms from prolonged exposure to air pollution can result in the formation of atherosclerotic plaque and, over time, other CVD phenotypes like stroke,

DOI: 10.1201/9781003328100-8

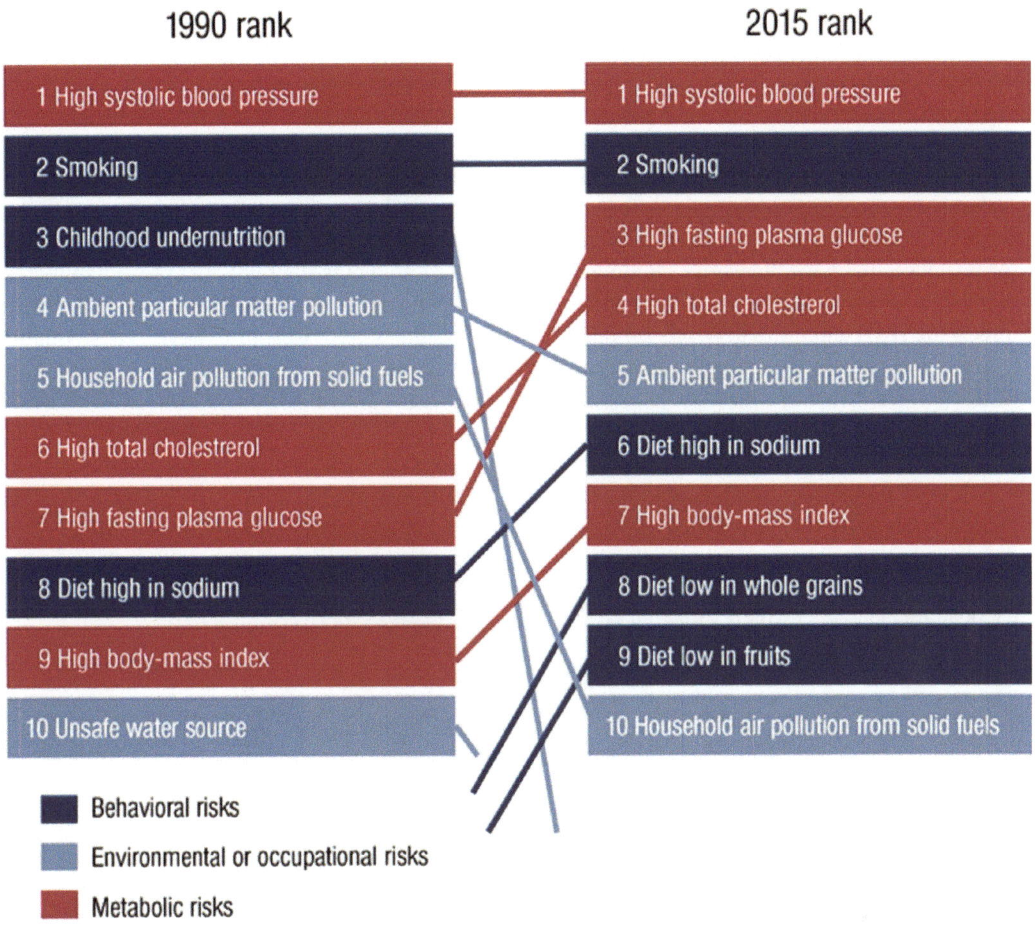

FIGURE 6.1 Global risk factors for global deaths in 1990 compared with 2015 outlined in the Global Burden of Disease (GDB) study. Reused from Hahad et al. (2020) with permission under the terms and conditions of the Creative Commons Attribution (CC BY) license (http://creativecommons.org/licenses/by/4.0/).

arterial hypertension, coronary heart disease, myocardial infarction, heart failure, and arrhythmia (Munzel et al. 2008), as well as neuropsychiatric conditions like cognitive decline, dementia, anxiety, depression, schizophrenia, and attention-deficit disorder (Kim et al. 2020). Increasing evidence from animal and human studies suggests that long-term exposure to air pollution impacts the CNS through various cellular, molecular, inflammatory, and oxidative stress mechanisms (Hahad et al. 2021). However, the consequences of air pollution on the CNS are poorly understood, the underlying processes are still not fully understood, and complicated interactions with other risk and lifestyle variables may occur. Neuropsychiatric diseases (including stroke) are among the leading causes of DALYs and global mortality. Since air pollution is a factor in 30% of all strokes, a greater understanding of these relationships is crucial and warrants further attention (The Lancet 2018). Two pathways are possible by which air pollutants enter the CNS: via nasal inhalation as a unique direct route to transport particles into the brain or via an indirect route through systemic inflammation upon initial recruitment of immune cells in the lung tissue (Cipriani et al. 2018). The direct route may allow smaller particles to pass through the nose-brain barrier and enter the brain through olfactory receptor neurons or the trigeminal nerve, which can cross the central nervous system to enter other brain regions.

On the other hand, particles can travel to the alveolar region through the lungs and enter the bloodstream. Small particles may now enter the systemic circulation through a transition process, depending on their size. For instance, nanoparticles can move through a direct pathway. In contrast, microparticles most likely do so by attracting immune cells, being taken up by phagocytic cells, and moving from the lung tissue to the bloodstream, where they cause localized inflammation and oxidative stress that, after prolonged exposure, spread to the bloodstream (Wilson, Miller, and Newby 2018) and then proceed to either simple diffusion or energy-dependent active transport across the BBB to the brain parenchyma. Once inside the organism, these small particles can potentially harm the brain through three main mechanisms (Moulton and Yang 2012). Inducing the release of pro-inflammatory mediators first results in long-term respiratory and systemic inflammation (Rao et al. 2018), which affects the BBB and ultimately sets off a neural-immune interaction. Resulting increased production of ROS and chronic oxidative stress causes an Alzheimer phenotype in children who are exposed (Calderon-Garciduenas et al. 2008). Second, the particles can change the permeability of the BBB by directly producing ROS, which can harm the barrier (Heidari Nejad et al. 2015; Hajipour et al. 2020). Finally, particles can stimulate the lung arc reflex via mechanical stimulation of specific mechanoreceptors in pulmonary tissue (Robinson et al. 2018; Perez, Hazari, and Farraj 2015) leading to sympathetic activation with the release of vasoconstrictors such as catecholamine (Hajat et al. 2019). If these mechanisms are considered collectively, they play a key role in promoting brain inflammation, neuronal dysfunction, and neuropathology (reviewed in Jayaraj et al. [2017]; D'Angiulli [2018]; Chen and Nakagawa [2018]; Block and Calderon-Garciduenas [2009]) (Figure 6.2).

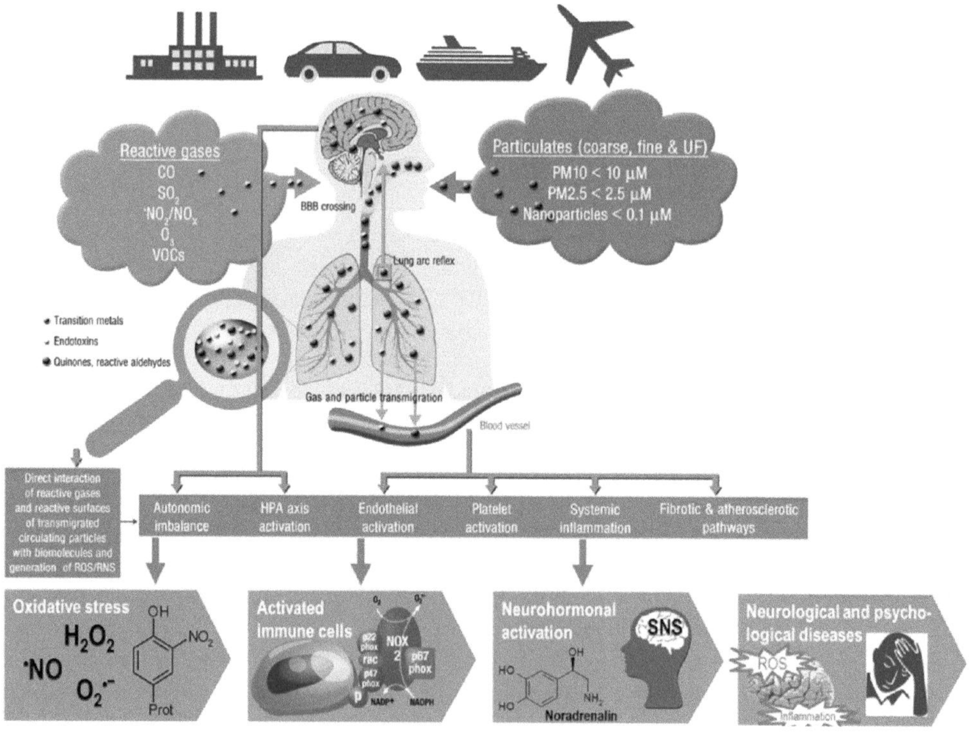

FIGURE 6.2 Summary of pathophysiological mechanisms by which air pollutants cause increased oxidative stress and inflammation, contributing to cerebrovascular, neurological, mental, and cardiorespiratory disorders. (A) Uptake and cardiorespiratory health effects triggered by air pollution constituents. (B) Key events that contribute to neurological and mental disorders by air pollution constituents. Reused from Hahad et al. (2020) with permission under the terms and conditions of the Creative Commons Attribution (CC BY) license (http://creativecommons.org/licenses/by/4.0/).

6.3 OXIDATIVE STRESS AND INFLAMMATION INDUCED BY AIR POLLUTION

Human studies using body scans (positron emission tomography) showed that PM induces coronary inflammation, associated with higher cardiovascular risk (Abohashem et al. 2021). Moreover, in the last decades, more human studies demonstrated that diesel exhaust inhalation causes vascular dysfunction, impaired endogenous fibrinolysis, and ischemic and thrombotic effects in men with coronary heart disease (Mills et al. 2005, 2007). The most predominant hypothesis for the mechanisms that inhaled PM causes cardiovascular effects is through immune cell activation in the lung. Activated immune cells release cytokines into circulation, causing an inflammatory cascade generating ROS and reactive nitrogen species, such as $^{\bullet}NO$ and $O_2^{\bullet-}$ through nicotinamide adenine dinucleotide phosphate (NADPH) oxidase in activated immune cells and the vascular wall, consequently inducing direct scavenging of $^{\bullet}NO$ and endothelial $^{\bullet}NO$ synthase (eNOS) uncoupling through oxidation of tetrahydrobiopterin (BH_4) and eNOS S-glutathionylation. $O_2^{\bullet-}$ and $^{\bullet}NO$ produced by infiltrating immune cells (neutrophils, natural killer cells, and monocytes/macrophages) promote peroxynitrite formation and subsequent oxidative damage.

Moreover, diesel exhaust particles possess active surfaces capable of directly stimulating the formation of $O_2^{\bullet-}$, representing a second mechanism by which ROS could be produced (as demonstrated by 2,2,6,6-tetramethylpiperidinyl-1-oxy [TEMPO] spin trapping) (Miller et al. 2009). Also, studies demonstrate two possible mechanisms by which inhaled particles can negatively affect the cardiovascular system, e.g., via activation of the autonomic nervous system through the alveolar sensory receptors or direct translocation of (nano)particles along the olfactory nerve into the olfactory bulb. These two mechanisms could work synergistically to exert the neuropsychiatric and cerebrovascular effects (Kan, Pan, and Castranova 2018). An overview focusing on the cardiovascular oxidative stress pathways and their interaction with PM is provided by Rao et al. (2018).

On the other hand, there is an apparent central role of inflammatory processes in PM-induced cardiovascular consequences. A potential role of CD36-mediated abnormal accumulation of oxidized lipids was reported, such as 7-KCh, in PM-induced atherosclerosis progression. It was concluded that PM could release or generate damage-associated molecular patterns (DAMP) that activate CD44, TLR4, and CD36 to trigger activation of nuclear factor 'kappa-light-chain-enhancer' of activated B cells (NFκB), kinase signaling, and cytokine synthesis (Rao et al. 2014). In addition, chronic fine PM exposure induces systemic vascular dysfunction via NADPH oxidase and TLR4 pathways, as demonstrated by a study describing TLR4-dependent activation of NADPH oxidase 2, an essential source of $O_2^{\bullet-}$ in phagocytes (Kampfrath et al. 2011). Finally, a cohort study of 18 highway maintenance workers showed that exposure to $PM_{2.5}$ and traffic noise was associated with higher C-reactive protein, serum amyloid A, and systolic and diastolic blood pressure. (Meier et al. 2014). Overall, exposure to PM is likely to act through various oxidative and inflammatory mechanisms to cause serious cardiovascular complications, which seem to be exacerbated in the setting of established disease and other risk factors. It is important to acknowledge that these described mechanisms likely contribute not only to the risk of cardiovascular but also neuropsychiatric diseases (Figure 6.3).

6.4 EVIDENCE LINKING AIR POLLUTION AND NEUROPSYCHIATRIC DISEASES FROM HUMAN OBSERVATIONAL/EPIDEMIOLOGICAL RESEARCH

6.4.1 Stroke

Encompassed by ambient air pollution-induced diseases of the cardiovascular system, stroke is one of the most prominent neurological diseases reported to be a consequence of both short- and long-term exposure to air pollution. Early studies reported this relationship after the Great Fog of London of 1952, a notorious pollution episode of the 20th century, showing an increased risk of stroke deaths in response to an extreme short-term rise in air pollution (Mateen and Brook 2011).

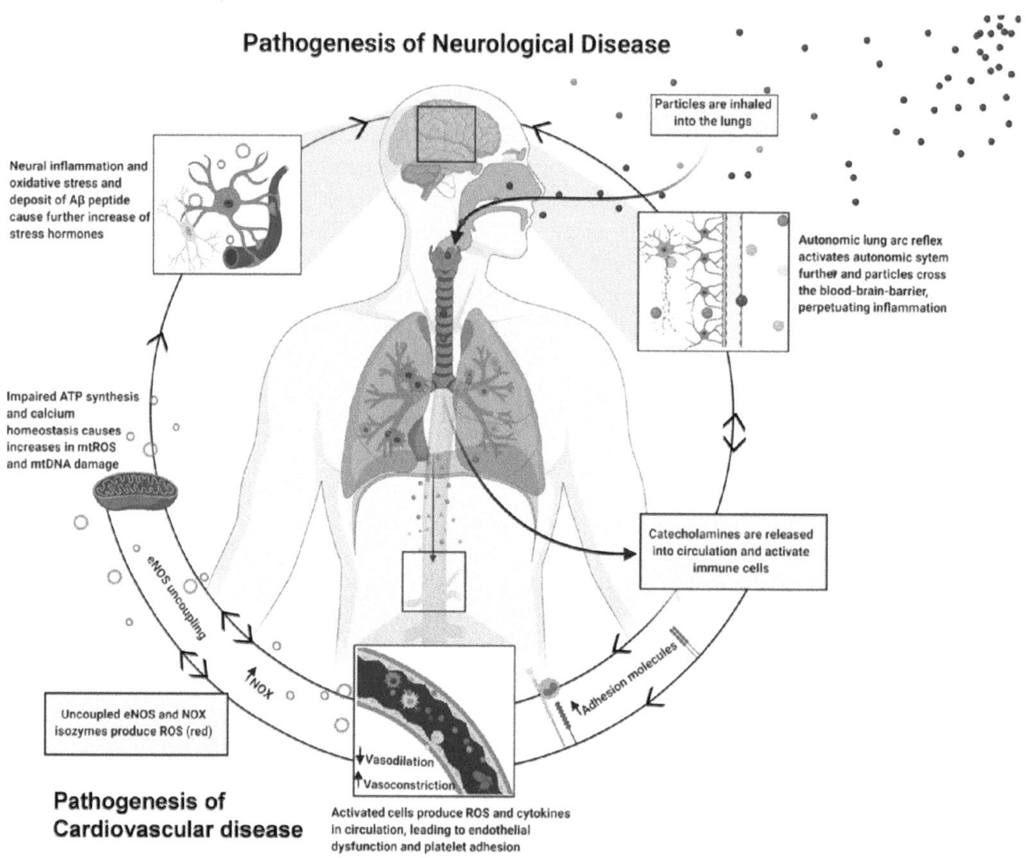

FIGURE 6.3 Pathophysiological mechanisms of neurological and cardiovascular disease induced by air pollution. Arrows indicate directions of pathways and pathomechanisms. Red circles indicate the release adverse signaling reactive oxygen species (ROS). Reused from Hahad et al. (2021) with permission under the terms and conditions of the Creative Commons Attribution (CC BY) license (http://creativecommons.org/licenses/by/4.0/).

Recent evidence also supports this relationship. A short-term rise in $PM_{2.5}$, PM_{10}, carbon monoxide (CO), sulphur dioxide (SO_2), and nitrogen dioxide (NO_2) levels was linked to stroke hospitalization or stroke mortality (6.2 million events), according to a meta-analysis of 94 studies from 28 countries (Shah et al. 2015). In contrast, a meta-analysis including 45 studies on the impact of short-term changes in levels of PM did not suggest an association between $PM_{2.5-10}$ and cerebrovascular mortality or morbidity, whereas $PM_{2.5}$ and PM_{10} were associated with a 1.4% (95% CI 0.9–1.9) and 0.5% (95% CI 0.3–0.7) elevated mortality due cerebrovascular disease, respectively (Wang, Eliot, and Wellenius 2014). Similar results were found in two prospective studies from Sweden and Spain, and they concluded that there are no associations between long-term exposure to $PM_{2.5}$ and/or PM_{10} and the incidence of stroke, which could have been affected by the relatively low exposure levels (Ljungman et al. 2019; Vivanco-Hidalgo et al. 2019).

6.4.2 Dementia

Air pollution has been considered a possible dementia risk factor. However, evidence from epidemiological studies is still limited. The role of PM as a potential determinant of dementia was analyzed in two recent meta-analyses. A more extensive meta-analysis that analyzed four cohorts from the United States, Canada, the United Kingdom, and Taiwan with more than 12 million elderly subjects aged ≥50 years reported a more than 3-fold increased dementia risk (HR 3.26, 95% CI 1.20–5.31 per 10 μg/m³) with long-term exposure to $PM_{2.5}$ (Tsai et al. 2019). A second comprehensive meta-analysis that included 80 studies from 26 countries examined the effects of $PM_{2.5}$ on several cerebrovascular and neurological conditions, including dementia, Parkinson's disease, stroke, autism spectrum disorder, and Alzheimer's disease (Fu et al. 2019). The authors found short- and long-term $PM_{2.5}$ exposure associated with increased stroke risks and stroke mortality. Moreover, the authors revealed that long-term $PM_{2.5}$ exposure was associated with increased risks of dementia, Alzheimer's disease, autism spectrum disorder, and Parkinson's disease. They concluded that there are strong associations between $PM_{2.5}$ and neurological disorders.

6.4.3 Parkinson's Disease

Results appear to be usually ambiguous, and little evidence links air pollution to Parkinson's disease. In 2020, a meta-analysis analyzed the association between long-term exposure to air pollutants such as PM and second-hand smoke and the onset of Parkinson's disease, including 21 studies with 222,051 patients with Parkinson's disease. Marginally increased risks were revealed in response to increased concentrations in $PM_{2.5}$, NO_2, O_3, and CO, suggesting that exposure to $PM_{2.5}$, NO_2, and O_3 may confer to a higher risk of Parkinson's disease (Han et al. 2020). Accordingly, a further meta-analysis found low-risk increases in the incidence of Parkinson's disease based on 15 studies. In contrast, only 13 reports were identified investigating associations of PD with long-term exposure to $PM_{2.5}$ (Kasdagli et al. 2019).

6.4.4 Cognitive Dysfunction

Cognitive dysfunction is considered an important intermediate event in the pathogenesis of dementia. For this reason, some studies suggest that chronic air pollution exposure has neurotoxic effects that culminate over time, leading to neuronal damage and loss and further cognitive dysfunction. This is predicted by cumulative exposure across the lifetime (Cipriani et al. 2018).

A recent study from the United States in 5,330 participants has demonstrated that long-term exposure to higher levels of air pollution, i.e., $PM_{2.5}$, PM_{10}, and NO_2, were cross-sectionally and longitudinally related to pronounced cognitive decline among older adults (Kulick et al. 2020). Among elderly Taiwanese, another study that investigated the cognitive consequences of short-term and long-term exposure to ambient air pollution found that long-term exposure to PM_{10} (OR 1.094, 95% CI 1.020–1.174) and O_3 (OR 1.878, 95% CI 1.363–2.560) was significantly associated with cognitive impairment with evidence of a joint effect of both exposures (Lo et al. 2019). After adjusting for various individual factors, Zhang et al. used the air pollution index to show that continuous decreased air quality hinders cognitive function in verbal and math examinations. This effect was most apparent in older adults, particularly in men and the less educated (Zhang, Chen, and Zhang 2018). Interestingly, a study using data from 779 U.S. adults aged 55 and older demonstrated the effects of long-term exposure to $PM_{2.5}$ on cognitive decline. The effect was particularly strong among respondents who resided in high-stress areas, suggesting that residents of underprivileged areas, where social stressors and environmental risks are more prevalent, may be especially vulnerable to the adverse health effects of air pollution (Ailshire, Karraker, and Clarke 2017).

6.4.5 Neurodevelopmental Disorders

Several studies suggest that young individuals may be particularly susceptible to air pollution-induced neurotoxicity and that perinatal exposure may cause or contribute to developmental disabilities and behavioral abnormalities (Costa, Chang, and Cole 2017). In the last decade, associations between air pollution and autism spectrum disorders (ASD) have been revealed. Any difficulty in socialization and communication, as well as the presence of repetitive and odd activities, are the main characteristics of ASD. Indeed, evidence from three meta-analyses suggests exposure to air pollution may raise the risk of ASD (Lam et al. 2016; Flores-Pajot et al. 2016; Chun et al. 2020). Maternal exposure to NO was linked to an elevated risk of ASD development in children in a large Canadian cohort of 132,256 births, but correlations with $PM_{2.5}$ and NO_2 were weaker (Pagalan et al. 2019). Interestingly, a recent cohort study in the United States, which included 246,420 singleton children born in Kaiser Permanente Southern California hospitals in 1999–2009, has shown that gestational diabetes mellitus onset early in pregnancy may increase children's susceptibility to prenatal O_3-associated autism risk, likely due to shared pathways that include inflammation and oxidative stress (Jo et al. 2019). A recent study in Sweden found an increased risk of ASD and attention-deficit/hyperactivity disorder in response to maternal exposure to air pollution, with an OR of 1.40 (95% CI 1.02–1.93) with increased NO_x exposure (top quartile). Still, no significant associations were found for the risk of developing attention-deficit/hyperactivity disorder (Oudin et al. 2019). Additionally, data from Denmark indicates that exposure to air pollution during early infancy—but not during pregnancy—increases the likelihood of being diagnosed with autism and Asperger's syndrome (Ritz et al. 2018). This was supported by an Israeli study that found that postnatal exposure to NO_2 and the risk of autism were more likely to be increased than prenatal exposure (Raz et al. 2018).

6.4.6 Mental Illness

Given that inflammation and oxidative stress are common aspects of many mental and psychiatric disorders and given the potential for air pollution to cause such processes, the relationship between air pollution and mental and psychiatric disorders is particularly interesting. Indeed, mounting evidence suggests that some mental and psychological problems, such as depression, anxiety disorders, suicidal ideation, and psychoses, may be influenced by air pollution. Recently, a meta-analysis showed that chronic exposure to fine PM was associated with elevated odds of depression (OR 1.102, 95% CI 1.023–1.189 per 10 μg/m³ increase in $PM_{2.5}$), anxiety disorder (assessed by two primary study results Power et al. [2015]; Pun, Manjourides, and Suh [2017]), and suicide (relative risk [RR] 1.02, 95% CI 1.00–1.03 per 10 μg/m³ increase in PM_{10}) (Braithwaite et al. 2019). In line, a meta-analysis reported higher odds of depression (OR 1.19, 95% CI 1.07–1.33) and suicide (OR 1.05, 95% CI 0.99–1.11) per 10 μg/m³ increase in $PM_{2.5}$, while for PM_{10} no association was found (Gu et al. 2019). A meta-analysis of observational studies investigating the link between outdoor air pollution and depression was carried out by Zeng et al. They demonstrated that several air pollutants, including chronic exposure to $PM_{2.5}$ and short-term exposure to PM_{10}, NO_2, SO_2, and CO, increased the risk of depression, as determined by a meta-analysis of 15 studies. At the same time, there was no evidence of an association between exposure to O_3 and depression. But further research using standardized techniques is still needed to back up the findings (Zeng et al. 2019). In contrast, a recent meta-analysis displayed weaker associations of air pollutants and risk of depression, including both long- and short-term exposure to $PM_{2.5}$, PM_{10}, and NO_2, among others, with difficulties in interpreting the results due to high heterogeneity between studies (Fan et al. 2020). Two investigations found a possible connection between air pollution and schizophrenia and psychotic experiences. First, following thorough adjustment for co-variables, a prospective study from the United Kingdom was able to show an elevated risk of psychotic events, with the risk being highest in those with the strongest (top quartile) long-term exposure to $PM_{2.5}$ (OR 1.45, 95% CI 1.11–1.90), NO_2 (OR 1.71, 95% CI 1.28–2.28), and NO_x (OR 1.72, 95% CI 1.30–2.29) (Newbury et al. 2019). The authors concluded that a combination of biological causes, such as neuroinflammation, and psychosocial factors, such as mental stress, may explain the higher risk of psychotic experiences associated with air pollution. Second, a Chinese study published in 2019 found that greater short-term exposure to NO_2 was associated with an increased risk of hospitalizations for schizophrenia (RR 1.10, 95% CI 1.01–1.18 per interquartile range increase) (Bai et al. 2019).

6.5 CONCLUSIONS

It is crucial to conclude that, in contrast to other personal or lifestyle risk factors like smoking, inactivity, excessive alcohol use, and poor diets, air pollution as a systemic health risk factor is scarcely anything that can be avoided or addressed by lifestyle decisions. This is why it is difficult for patients and doctors to control environmental quality changes; instead, policymakers must control air pollution to safeguard the public from harmful health effects. Reviewing the health implications of air pollution, as done by the current review, is crucial to bolstering these efforts. This should make it easier to recognize and include air pollution as a major health risk factor in official preventative guidelines.

When evaluating the current literature, it can be assumed that air pollution can raise the risk of neuropsychiatric diseases. It is suggested that air pollution causes brain damage through two main pathophysiological mechanisms: inflammation and oxidative stress. The idea that air pollution is not just a cardiorespiratory risk factor but also a neuropsychiatric risk factor has to be strengthened. However, since this study area is still relatively young, more credible data is required. Disparities in determining the most appropriate time of exposure for the outcome of interest (acute vs. chronic), accurate assessment of air pollutants, confounding (individual- and area-variables), and adjustment for pertinent environmental co-exposures are possible sources of bias and error among studies. To comprehend why certain people are more resistant to the detrimental impacts of air pollution on their health, resilience should also be discussed in greater detail. To combat the harmful health impacts of air pollution, this may also help to identify novel targets for pharmaceutical or lifestyle therapies.

CONFLICT OF INTEREST

None.

REFERENCES

Abohashem, S., M. T. Osborne, T. Dar, N. Naddaf, T. Abbasi, A. Ghoneem, A. Radfar, T. Patrich, B. Oberfeld, B. Tung, et al. 2021. "A leucopoietic-arterial axis underlying the link between ambient air pollution and cardiovascular disease in humans." *Eur Heart J*. https://doi.org/10.1093/eurheartj/ehaa982. www.ncbi.nlm.nih.gov/pubmed/33428721.

Aghababaeian, H., A. Ostadtaghizadeh, A. Ardalan, A. Asgary, M. Akbary, M. S. Yekaninejad, and C. Stephens. 2021. "Global health impacts of dust storms: A systematic review." *Environ Health Insights* 15: 11786302211018390. https://doi.org/10.1177/11786302211018390. www.ncbi.nlm.nih.gov/pubmed/34103932.

Ailshire, J., A. Karraker, and P. Clarke. 2017. "Neighborhood social stressors, fine particulate matter air pollution, and cognitive function among older U.S. adults." *Soc Sci Med* 172: 56–63. https://doi.org/10.1016/j.socscimed.2016.11.019. www.ncbi.nlm.nih.gov/pubmed/27886528.

Anderson, J. O., J. G. Thundiyil, and A. Stolbach. 2012. "Clearing the air: A review of the effects of particulate matter air pollution on human health." *J Med Toxicol* 8 (2): 166–175. https://doi.org/10.1007/s13181-011-0203-1. www.ncbi.nlm.nih.gov/pubmed/22194192.

Bai, L., X. Zhang, Y. Zhang, Q. Cheng, J. Duan, J. Gao, Z. Xu, H. Zhang, S. Wang, and H. Su. 2019. "Ambient concentrations of NO2 and hospital admissions for schizophrenia." *Occup Environ Med* 76 (2): 125–131. https://doi.org/10.1136/oemed-2018-105162. www.ncbi.nlm.nih.gov/pubmed/30366962.

Block, M. L., and L. Calderon-Garciduenas. 2009. "Air pollution: Mechanisms of neuroinflammation and CNS disease." *Trends Neurosci* 32 (9): 506–516. https://doi.org/10.1016/j.tins.2009.05.009. www.ncbi.nlm.nih.gov/pubmed/19716187.

Braithwaite, I., S. Zhang, J. B. Kirkbride, D. P. J. Osborn, and J. F. Hayes. 2019. "Air pollution (particulate matter) exposure and associations with depression, anxiety, bipolar, psychosis and suicide risk: A systematic review and meta-analysis." *Environ Health Perspect* 127 (12): 126002. https://doi.org/10.1289/EHP4595. www.ncbi.nlm.nih.gov/pubmed/31850801.

Calderon-Garciduenas, L., A. C. Solt, C. Henriquez-Roldan, R. Torres-Jardon, B. Nuse, L. Herritt, R. Villarreal-Calderon, N. Osnaya, I. Stone, R. Garcia, et al. 2008. "Long-term air pollution exposure is associated with neuroinflammation, an altered innate immune response, disruption of the blood-brain barrier, ultrafine particulate deposition, and accumulation of amyloid beta-42 and alpha-synuclein in children and young adults." *Toxicol Pathol* 36 (2): 289–310. https://doi.org/10.1177/0192623307313011. www.ncbi.nlm.nih.gov/pubmed/18349428.

Chen, C., and S. Nakagawa. 2018. "Planetary health and the future of human capacity: The increasing impact of planetary distress on the human brain." *Challenges* 9 (41): 1–10. https://doi.org/10.3390/challe9020041.

Chun, H., C. Leung, S. W. Wen, J. McDonald, and H. H. Shin. 2020. "Maternal exposure to air pollution and risk of autism in children: A systematic review and meta-analysis." *Environ Pollut* 256: 113307. https://doi.org/10.1016/j.envpol.2019.113307. www.ncbi.nlm.nih.gov/pubmed/31733973.

Cipriani, G., S. Danti, C. Carlesi, and G. Borin. 2018. "Danger in the air: Air pollution and cognitive dysfunction." *Am J Alzheimers Dis Other Demen* 33 (6): 333–341. https://doi.org/10.1177/1533317518777859. www.ncbi.nlm.nih.gov/pubmed/29874918.

Cohen, A. J., M. Brauer, R. Burnett, H. R. Anderson, J. Frostad, K. Estep, K. Balakrishnan, B. Brunekreef, L. Dandona, R. Dandona, et al. 2017. "Estimates and 25-year trends of the global burden of disease attributable to ambient air pollution: An analysis of data from the Global Burden of Diseases Study 2015." *Lancet* 389 (10082): 1907–1918. https://doi.org/10.1016/S0140-6736(17)30505-6. www.ncbi.nlm.nih.gov/pubmed/28408086.

Collaborators, G. B. D. Risk Factors. 2020. "Global burden of 87 risk factors in 204 countries and territories, 1990–2019: A systematic analysis for the Global Burden of Disease Study 2019." *Lancet* 396 (10258): 1223–1249. https://doi.org/10.1016/S0140-6736(20)30752-2. www.ncbi.nlm.nih.gov/pubmed/33069327.

Costa, L. G., Y. C. Chang, and T. B. Cole. 2017. "Developmental neurotoxicity of traffic-related air pollution: Focus on autism." *Curr Environ Health Rep* 4 (2): 156–165. https://doi.org/10.1007/s40572-017-0135-2. www.ncbi.nlm.nih.gov/pubmed/28417440.

Costa, L. G., T. B. Cole, J. Coburn, Y. C. Chang, K. Dao, and P. J. Roque. 2017. "Neurotoxicity of traffic-related air pollution." *Neurotoxicology* 59: 133–139. https://doi.org/10.1016/j.neuro.2015.11.008. www.ncbi.nlm.nih.gov/pubmed/26610921.

D'Angiulli, A. 2018. "Severe urban outdoor air pollution and children's structural and functional brain development, from evidence to precautionary strategic action." *Front Public Health* 6: 95. https://doi.org/10.3389/fpubh.2018.00095. www.ncbi.nlm.nih.gov/pubmed/29670873.

Fan, S. J., J. Heinrich, M. S. Bloom, T. Y. Zhao, T. X. Shi, W. R. Feng, Y. Sun, J. C. Shen, Z. C. Yang, B. Y. Yang, et al. 2020. "Ambient air pollution and depression: A systematic review with meta-analysis up to 2019." *Sci Total Environ* 701: 134721. https://doi.org/10.1016/j.scitotenv.2019.134721. www.ncbi.nlm.nih.gov/pubmed/31715478.

Flores-Pajot, M. C., M. Ofner, M. T. Do, E. Lavigne, and P. J. Villeneuve. 2016. "Childhood autism spectrum disorders and exposure to nitrogen dioxide, and particulate matter air pollution: A review and meta-analysis." *Environ Res* 151: 763–776. https://doi.org/10.1016/j.envres.2016.07.030. www.ncbi.nlm.nih.gov/pubmed/27609410.

Fu, P., X. Guo, F. M. H. Cheung, and K. K. L. Yung. 2019. "The association between PM2.5 exposure and neurological disorders: A systematic review and meta-analysis." *Sci Total Environ* 655: 1240–1248. https://doi.org/10.1016/j.scitotenv.2018.11.218. www.ncbi.nlm.nih.gov/pubmed/30577116.

Gu, X., Q. Liu, F. Deng, X. Wang, H. Lin, X. Guo, and S. Wu. 2019. "Association between particulate matter air pollution and risk of depression and suicide: Systematic review and meta-analysis." *Br J Psychiatry* 215 (2): 456–467. https://doi.org/10.1192/bjp.2018.295. www.ncbi.nlm.nih.gov/pubmed/30719959.

Hahad, O., K. Frenis, M. Kuntic, A. Daiber, and T. Munzel. 2021. "Accelerated aging and age-related diseases (CVD and neurological) due to air pollution and traffic noise exposure." *Int J Mol Sci* 22 (5). https://doi.org/10.3390/ijms22052419. www.ncbi.nlm.nih.gov/pubmed/33670865.

Hahad, O., J. Lelieveld, F. Birklein, K. Lieb, A. Daiber, and T. Munzel. 2020. "Ambient air pollution increases the risk of cerebrovascular and neuropsychiatric disorders through induction of inflammation and oxidative stress." *Int J Mol Sci* 21 (12). https://doi.org/10.3390/ijms21124306. www.ncbi.nlm.nih.gov/pubmed/32560306.

Hajat, A., A. V. Diez Roux, C. Castro-Diehl, K. Cosselman, S. H. Golden, M. F. Hazlehurst, A. Szpiro, S. Vedal, and J. D. Kaufman. 2019. "The association between long-term air pollution and urinary catecholamines: Evidence from the multi-ethnic study of atherosclerosis." *Environ Health Perspect* 127 (5): 57007. https://doi.org/10.1289/EHP3286. www.ncbi.nlm.nih.gov/pubmed/31095432.

Hajipour, S., Y. Farbood, M. K. Gharib-Naseri, G. Goudarzi, M. Rashno, H. Maleki, N. Bakhtiari, A. Nesari, S. E. Khoshnam, M. Dianat, et al. 2020. "Exposure to ambient dusty particulate matter impairs spatial memory and hippocampal LTP by increasing brain inflammation and oxidative stress in rats." *Life Sci* 242: 117210. https://doi.org/10.1016/j.lfs.2019.117210. www.ncbi.nlm.nih.gov/pubmed/31874166.

Han, C., Y. Lu, H. Cheng, C. Wang, and P. Chan. 2020. "The impact of long-term exposure to ambient air pollution and second-hand smoke on the onset of Parkinson disease: A review and meta-analysis." *Public Health* 179: 100–110. https://doi.org/10.1016/j.puhe.2019.09.020. www.ncbi.nlm.nih.gov/pubmed/31770719.

Heidari Nejad, S., R. Takechi, B. J. Mullins, C. Giles, A. N. Larcombe, D. Bertolatti, K. Rumchev, S. Dhaliwal, and J. Mamo. 2015. "The effect of diesel exhaust exposure on blood-brain barrier integrity and function in a murine model." *J Appl Toxicol* 35 (1): 41–7. https://doi.org/10.1002/jat.2985. www.ncbi.nlm.nih.gov/pubmed/24477878.

Jayaraj, R. L., E. A. Rodriguez, Y. Wang, and M. L. Block. 2017. "Outdoor ambient air pollution and neurodegenerative diseases: The neuroinflammation hypothesis." *Curr Environ Health Rep* 4 (2): 166–179. https://doi.org/10.1007/s40572-017-0142-3. www.ncbi.nlm.nih.gov/pubmed/28444645.

Jo, H., S. P. Eckel, J. C. Chen, M. Cockburn, M. P. Martinez, T. Chow, F. W. Lurmann, W. E. Funk, A. H. Xiang, and R. McConnell. 2019. "Gestational diabetes mellitus, prenatal air pollution exposure, and autism spectrum disorder." *Environ Int* 133 (Pt A): 105110. https://doi.org/10.1016/j.envint.2019.105110. www.ncbi.nlm.nih.gov/pubmed/31610366.

Kampfrath, T., A. Maiseyeu, Z. Ying, Z. Shah, J. A. Deiuliis, X. Xu, N. Kherada, R. D. Brook, K. M. Reddy, N. P. Padture, et al. 2011. "Chronic fine particulate matter exposure induces systemic vascular dysfunction via NADPH oxidase and TLR4 pathways." *Circ Res* 108 (6): 716–726. https://doi.org/10.1161/CIRCRESAHA.110.237560. www.ncbi.nlm.nih.gov/pubmed/21273555.

Kan, H., D. Pan, and V. Castranova. 2018. "Engineered nanoparticle exposure and cardiovascular effects: The role of a neuronal-regulated pathway." *Inhal Toxicol* 30 (9–10): 335–342. https://doi.org/10.1080/08958378.2018.1535634. www.ncbi.nlm.nih.gov/pubmed/30604639.

Kasdagli, M. I., K. Katsouyanni, K. Dimakopoulou, and E. Samoli. 2019. "Air pollution and Parkinson's disease: A systematic review and meta-analysis up to 2018." *Int J Hyg Environ Health* 222 (3): 402–409. https://doi.org/10.1016/j.ijheh.2018.12.006. www.ncbi.nlm.nih.gov/pubmed/30606679.

Kim, H., W. H. Kim, Y. Y. Kim, and H. Y. Park. 2020. "Air pollution and central nervous system disease: A review of the impact of fine particulate matter on neurological disorders." *Front Public Health* 8: 575330. https://doi.org/10.3389/fpubh.2020.575330. www.ncbi.nlm.nih.gov/pubmed/33392129.

Kulick, E. R., G. A. Wellenius, A. K. Boehme, N. R. Joyce, N. Schupf, J. D. Kaufman, R. Mayeux, R. L. Sacco, J. J. Manly, and M. S. V. Elkind. 2020. "Long-term exposure to air pollution and trajectories of cognitive decline among older adults." *Neurology* 94 (17): e1782–e1792. https://doi.org/10.1212/WNL.0000000000009314. www.ncbi.nlm.nih.gov/pubmed/32269113.

Lam, J., P. Sutton, A. Kalkbrenner, G. Windham, A. Halladay, E. Koustas, C. Lawler, L. Davidson, N. Daniels, C. Newschaffer, and T. Woodruff. 2016. "A systematic review and meta-analysis of multiple airborne pollutants and autism spectrum disorder." *PLoS One* 11 (9): e0161851. https://doi.org/10.1371/journal.pone.0161851. www.ncbi.nlm.nih.gov/pubmed/27653281.

The Lancet, Neurology. 2018. "Air pollution and brain health: An emerging issue." *Lancet Neurol* 17 (2): 103. https://doi.org/10.1016/S1474-4422(17)30462-3. www.ncbi.nlm.nih.gov/pubmed/29413304.

Landrigan, P. J., R. Fuller, N. J. R. Acosta, O. Adeyi, R. Arnold, N. N. Basu, A. B. Balde, R. Bertollini, S. Bose-O'Reilly, J. I. Boufford, et al. 2018. "The lancet commission on pollution and health." *Lancet* 391 (10119): 462–512. https://doi.org/10.1016/S0140-6736(17)32345-0. www.ncbi.nlm.nih.gov/pubmed/29056410.

Ljungman, P. L. S., N. Andersson, L. Stockfelt, E. M. Andersson, J. Nilsson Sommar, K. Eneroth, L. Gidhagen, C. Johansson, A. Lager, K. Leander, et al. 2019. "Long-term exposure to particulate air pollution, black carbon, and their source components in relation to ischemic heart disease and stroke." *Environ Health Perspect* 127 (10): 107012. https://doi.org/10.1289/EHP4757. www.ncbi.nlm.nih.gov/pubmed/31663781.

Lo, Y. C., Y. C. Lu, Y. H. Chang, S. Kao, and H. B. Huang. 2019. "Air pollution exposure and cognitive function in taiwanese older adults: A repeated measurement study." *Int J Environ Res Public Health* 16 (16). https://doi.org/10.3390/ijerph16162976. www.ncbi.nlm.nih.gov/pubmed/31430886.

Lodovici, M., and E. Bigagli. 2011. "Oxidative stress and air pollution exposure." *J Toxicol* 2011: 487074. https://doi.org/10.1155/2011/487074. www.ncbi.nlm.nih.gov/pubmed/21860622.

Mateen, F. J., and R. D. Brook. 2011. "Air pollution as an emerging global risk factor for stroke." *JAMA* 305 (12): 1240–1241. https://doi.org/10.1001/jama.2011.352. www.ncbi.nlm.nih.gov/pubmed/21427378.

Meier, R., W. E. Cascio, A. J. Ghio, P. Wild, B. Danuser, and M. Riediker. 2014. "Associations of short-term particle and noise exposures with markers of cardiovascular and respiratory health among highway maintenance workers." *Environ Health Perspect* 122 (7): 726–732. https://doi.org/10.1289/ehp.1307100. www.ncbi.nlm.nih.gov/pubmed/24647077.

Miller, M. R., S. J. Borthwick, C. A. Shaw, S. G. McLean, D. McClure, N. L. Mills, R. Duffin, K. Donaldson, I. L. Megson, P. W. Hadoke, and D. E. Newby. 2009. "Direct impairment of vascular function by diesel exhaust particulate through reduced bioavailability of endothelium-derived nitric oxide induced by superoxide free radicals." *Environ Health Perspect* 117 (4): 611–616. https://doi.org/10.1289/ehp.0800235. www.ncbi.nlm.nih.gov/pubmed/19440501.

Mills, N. L., H. Tornqvist, M. C. Gonzalez, E. Vink, S. D. Robinson, S. Soderberg, N. A. Boon, K. Donaldson, T. Sandstrom, A. Blomberg, and D. E. Newby. 2007. "Ischemic and thrombotic effects of dilute diesel-exhaust inhalation in men with coronary heart disease." *N Engl J Med* 357 (11): 1075–1082. https://doi.org/10.1056/NEJMoa066314. www.ncbi.nlm.nih.gov/pubmed/17855668.

Mills, N. L., H. Tornqvist, S. D. Robinson, M. Gonzalez, K. Darnley, W. MacNee, N. A. Boon, K. Donaldson, A. Blomberg, T. Sandstrom, and D. E. Newby. 2005. "Diesel exhaust inhalation causes vascular dysfunction and impaired endogenous fibrinolysis." *Circulation* 112 (25): 3930–3936. https://doi.org/10.1161/CIRCULATIONAHA.105.588962. www.ncbi.nlm.nih.gov/pubmed/16365212.

Moulton, P. V., and W. Yang. 2012. "Air pollution, oxidative stress, and Alzheimer's disease." *J Environ Public Health* 2012: 472751. https://doi.org/10.1155/2012/472751. www.ncbi.nlm.nih.gov/pubmed/22523504.

Munzel, T., T. Gori, S. Al-Kindi, J. Deanfield, J. Lelieveld, A. Daiber, and S. Rajagopalan. 2018. "Effects of gaseous and solid constituents of air pollution on endothelial function." *Eur Heart J* 39 (38): 3543–3550. https://doi.org/10.1093/eurheartj/ehy481. www.ncbi.nlm.nih.gov/pubmed/30124840.

Munzel, T., C. Sinning, F. Post, A. Warnholtz, and E. Schulz. 2008. "Pathophysiology, diagnosis and prognostic implications of endothelial dysfunction." *Ann Med* 40 (3): 180–196. https://doi.org/10.1080/07853890701854702. www.ncbi.nlm.nih.gov/pubmed/18382884.

Newbury, J. B., L. Arseneault, S. Beevers, N. Kitwiroon, S. Roberts, C. M. Pariante, F. J. Kelly, and H. L. Fisher. 2019. "Association of air pollution exposure with psychotic experiences during adolescence." *JAMA Psychiatry* 76 (6): 614–623. https://doi.org/10.1001/jamapsychiatry.2019.0056. www.ncbi.nlm.nih.gov/pubmed/30916743.

Oudin, A., K. Frondelius, N. Haglund, K. Kallen, B. Forsberg, P. Gustafsson, and E. Malmqvist. 2019. "Prenatal exposure to air pollution as a potential risk factor for autism and ADHD." *Environ Int* 133 (Pt A): 105149. https://doi.org/10.1016/j.envint.2019.105149. www.ncbi.nlm.nih.gov/pubmed/31629172.

Pagalan, L., C. Bickford, W. Weikum, B. Lanphear, M. Brauer, N. Lanphear, G. E. Hanley, T. F. Oberlander, and M. Winters. 2019. "Association of prenatal exposure to air pollution with autism spectrum disorder." *JAMA Pediatr* 173 (1): 86–92. https://doi.org/10.1001/jamapediatrics.2018.3101. www.ncbi.nlm.nih.gov/pubmed/30452514.

Perez, C. M., M. S. Hazari, and A. K. Farraj. 2015. "Role of autonomic reflex arcs in cardiovascular responses to air pollution exposure." *Cardiovasc Toxicol* 15 (1): 69–78. https://doi.org/10.1007/s12012-014-9272-0. www.ncbi.nlm.nih.gov/pubmed/25123706.

Power, M. C., M. A. Kioumourtzoglou, J. E. Hart, O. I. Okereke, F. Laden, and M. G. Weisskopf. 2015. "The relation between past exposure to fine particulate air pollution and prevalent anxiety: Observational cohort study." *BMJ* 350: h1111. https://doi.org/10.1136/bmj.h1111. www.ncbi.nlm.nih.gov/pubmed/25810495.

Pun, V. C., J. Manjourides, and H. Suh. 2017. "Association of ambient air pollution with depressive and anxiety symptoms in older adults: Results from the NSHAP study." *Environ Health Perspect* 125 (3): 342–348. https://doi.org/10.1289/EHP494. www.ncbi.nlm.nih.gov/pubmed/27517877.

Rao, X., J. Zhong, R. D. Brook, and S. Rajagopalan. 2018. "Effect of particulate matter air pollution on cardiovascular oxidative stress pathways." *Antioxid Redox Signal* 28 (9): 797–818. https://doi.org/10.1089/ars.2017.7394. www.ncbi.nlm.nih.gov/pubmed/29084451.

Rao, X., J. Zhong, A. Maiseyeu, B. Gopalakrishnan, F. A. Villamena, L. C. Chen, J. R. Harkema, Q. Sun, and S. Rajagopalan. 2014. "CD36-dependent 7-ketocholesterol accumulation in macrophages mediates progression of atherosclerosis in response to chronic air pollution exposure." *Circ Res* 115 (9): 770–780. https://doi.org/10.1161/CIRCRESAHA.115.304666. www.ncbi.nlm.nih.gov/pubmed/25186795.

Raz, R., H. Levine, O. Pinto, D. M. Broday, Yuval, and M. G. Weisskopf. 2018. "Traffic-related air pollution and autism spectrum disorder: A population-based nested case-control study in Israel." *Am J Epidemiol* 187 (4): 717–725. https://doi.org/10.1093/aje/kwx294. www.ncbi.nlm.nih.gov/pubmed/29020136.

Ritz, B., Z. Liew, Q. Yan, X. Cui, J. Virk, M. Ketzel, and O. Raaschou-Nielsen. 2018. "Air pollution and Autism in Denmark." *Environ Epidemiol* 2 (4). https://doi.org/10.1097/EE9.0000000000000028. www.ncbi.nlm.nih.gov/pubmed/31008439.

Robinson, R. K., M. A. Birrell, J. J. Adcock, M. A. Wortley, E. D. Dubuis, S. Chen, C. M. McGilvery, S. Hu, M. S. P. Shaffer, S. J. Bonvini, et al. 2018. "Mechanistic link between diesel exhaust particles and respiratory reflexes." *J Allergy Clin Immunol* 141 (3): 1074–1084 e9. https://doi.org/10.1016/j.jaci.2017.04.038. www.ncbi.nlm.nih.gov/pubmed/28532657.

Shah, A. S., K. K. Lee, D. A. McAllister, A. Hunter, H. Nair, W. Whiteley, J. P. Langrish, D. E. Newby, and N. L. Mills. 2015. "Short term exposure to air pollution and stroke: Systematic review and meta-analysis." *BMJ* 350: h1295. https://doi.org/10.1136/bmj.h1295. www.ncbi.nlm.nih.gov/pubmed/25810496.

Tsai, T. L., Y. T. Lin, B. F. Hwang, S. F. Nakayama, C. H. Tsai, X. L. Sun, C. Ma, and C. R. Jung. 2019. "Fine particulate matter is a potential determinant of Alzheimer's disease: A systemic review and meta-analysis." *Environ Res* 177: 108638. https://doi.org/10.1016/j.envres.2019.108638. www.ncbi.nlm.nih.gov/pubmed/31421449.

Vivanco-Hidalgo, R. M., C. Avellaneda-Gomez, P. Dadvand, M. Cirach, A. Ois, A. Gomez Gonzalez, A. Rodriguez-Campello, P. de Ceballos, X. Basagana, A. Zabalza, E. Cuadrado-Godia, J. Sunyer, J. Roquer, and G. A. Wellenius. 2019. "Association of residential air pollution, noise, and greenspace with initial ischemic stroke severity." *Environ Res* 179 (Pt A): 108725. https://doi.org/10.1016/j.envres.2019.108725. www.ncbi.nlm.nih.gov/pubmed/31539843.

Wang, Y., M. N. Eliot, and G. A. Wellenius. 2014. "Short-term changes in ambient particulate matter and risk of stroke: A systematic review and meta-analysis." *J Am Heart Assoc* 3 (4). https://doi.org/10.1161/JAHA.114.000983. www.ncbi.nlm.nih.gov/pubmed/25103204.

Wang, Y., L. Xiong, and M. Tang. 2017. "Toxicity of inhaled particulate matter on the central nervous system: Neuroinflammation, neuropsychological effects and neurodegenerative disease." *J Appl Toxicol* 37 (6): 644–667. https://doi.org/10.1002/jat.3451. www.ncbi.nlm.nih.gov/pubmed/28299803.

Wilson, S. J., M. R. Miller, and D. E. Newby. 2018. "Effects of diesel exhaust on cardiovascular function and oxidative stress." *Antioxid Redox Signal* 28 (9): 819–836. https://doi.org/10.1089/ars.2017.7174. www.ncbi.nlm.nih.gov/pubmed/28540736.

World Health Organization. 2018. "9 out of 10 people worldwide breathe polluted air, but more countries are taking action." Available online: www.who.int/news-room/detail/02-05-2018-9-out-of-10-people-worldwide-breathe-polluted-air-but-more-countries-are-taking-action (accessed on 29 April 2020).

Zeng, Y., R. Lin, L. Liu, Y. Liu, and Y. Li. 2019. "Ambient air pollution exposure and risk of depression: A systematic review and meta-analysis of observational studies." *Psychiatry Res* 276: 69–78. https://doi.org/10.1016/j.psychres.2019.04.019. www.ncbi.nlm.nih.gov/pubmed/31029037.

Zhang, X., X. Chen, and X. Zhang. 2018. "The impact of exposure to air pollution on cognitive performance." *Proc Natl Acad Sci U S A* 115 (37): 9193–9197. https://doi.org/10.1073/pnas.1809474115. www.ncbi.nlm.nih.gov/pubmed/30150383.

7 Underlying Mechanisms of the Effects of Particulate Matter in Primary and Distant Organs

Natalia Magnani, Timoteo O. Marchini, Agustina Freire, Sofía Reynoso, Silvia Alvarez, and Pablo A. Evelson

7.1 INTRODUCTION

Humans are continuously and involuntarily exposed to air pollutants, mainly through tissue or organs that have a direct interface with the outside environment such as the respiratory tract by inhalation, penetration through the skin or eyes, and ingestion via the gastrointestinal tract (Brook and Rajagopalan, 2020). In its last report, the World Health Organization (WHO) estimated that 91% of the world's population breathe polluted air leading to almost 7 million premature deaths. Increased morbidity and mortality rates have been pointed out by several epidemiological studies to be associated not only with chronic air pollution exposures, but also with short-term daily exposures as well (Marchini et al., 2020). It was estimates that 37% of outdoor air pollution-related premature deaths were due to cardiovascular diseases, such as ischemic heart disease and strokes, while 18% of deaths were due to chronic obstructive pulmonary disease and 23% to acute lower respiratory infections; 11% of deaths were due to lung cancer (WHO, 2018; Landrigan et al., 2018). These findings confirm that air pollution is now the world's largest single environmental health risk (Lelieveld et al., 2020). Impaired cellular redox metabolism followed by oxidative damage are central mechanisms for the adverse effects of airborne particulate matter (PM). Reactive oxygen species (ROS) are generated during phagocytosis of the particles, leading to enhancement of oxidative stress and an inflammatory response. PM exposure triggered these mechanisms first on primary organs that directly interact with air pollution (Valacchi et al., 2020; Magnani et al., 2020). Both after acute and chronic pollutants exposure, PM uptake has been described to induce local cell activation and recruitment, cytokine release, and inflammation. Afterwards, inflammatory mediators have been shown to be able to reach the bloodstream and induce a systemic response that affects metabolism in distant organs such as heart and brain (Marchini et al., 2020; Fuller et al., 2018).

Here we will describe the mechanisms by which PM exposure results in adverse health effects, in terms of the relationship between the redox responses triggered by the presence of the particles and the inflammation observed in the different organs.

7.2 AIR POLLUTION—PARTICULATE MATTER

Air pollution is a complex heterogeneous mix whose complexity is increased due to the variation in its components and concentrations between places and over time (Valacchi et al., 2020). Air pollutants can be classified by their source, chemical composition, size, and mode of release into indoor or outdoor environments (Chen et al., 2022).

Atmospheric pollutants can be differentiated into two large groups according to the chemical characteristics of their constituents: (a) gaseous components, such as nitrogen dioxide (NO_2), nitric oxide (NO), sulfur dioxide (SO_2), ozone (O_3), and carbon monoxide (CO); and (b) particulate matter

DOI: 10.1201/9781003328100-9

(PM), which is composed of solid and liquid particles suspended in air. PM includes not only vehicle exhaust and fossil combustion-derived carbon-centered particles, nitrates and sulfates, coming mainly from anthropogenic sources, but also particles emitted from natural sources such as geological wind-blown dust, potentially containing endotoxin, and biological particles (spores, pollen) with their associated allergens (Loxham et al., 2019; Zhang et al., 2019) (Figure 7.1).

There is an established association between the toxicity effects of air pollution and specific PM components PM (Wu et al., 2018). Among the numerous compounds present in PM, the most frequently found are non-metallic ions (mostly ammonium $-NH_4^+$, nitrate $-NO_3^-$, and sulfate $-SO_4^{2-}$) (Losacco and Perillo, 2018), organic compounds such as polycyclic aromatic hydrocarbons (PAHs) and polychlorinated biphenyls (PBCs), endotoxins, and transition metals (Cáceres et al., 2020). Moreover, environmental parameters, such as temperature and humidity, may interact with PM altering its composition and distribution (Mudway et al., 2020). Besides the complex nature of air pollution, epidemiological studies indicate that PM is primarily responsible for the adverse outcomes resulting in a major concern from a health perspective (Fuller et al., 2018). It is now well established that PM causes respiratory diseases and increased lung cancer, as well as death due to cardiovascular and neurological diseases (Pope et al., 2019; Rao et al., 2018; Rajagopalan et al., 2018).

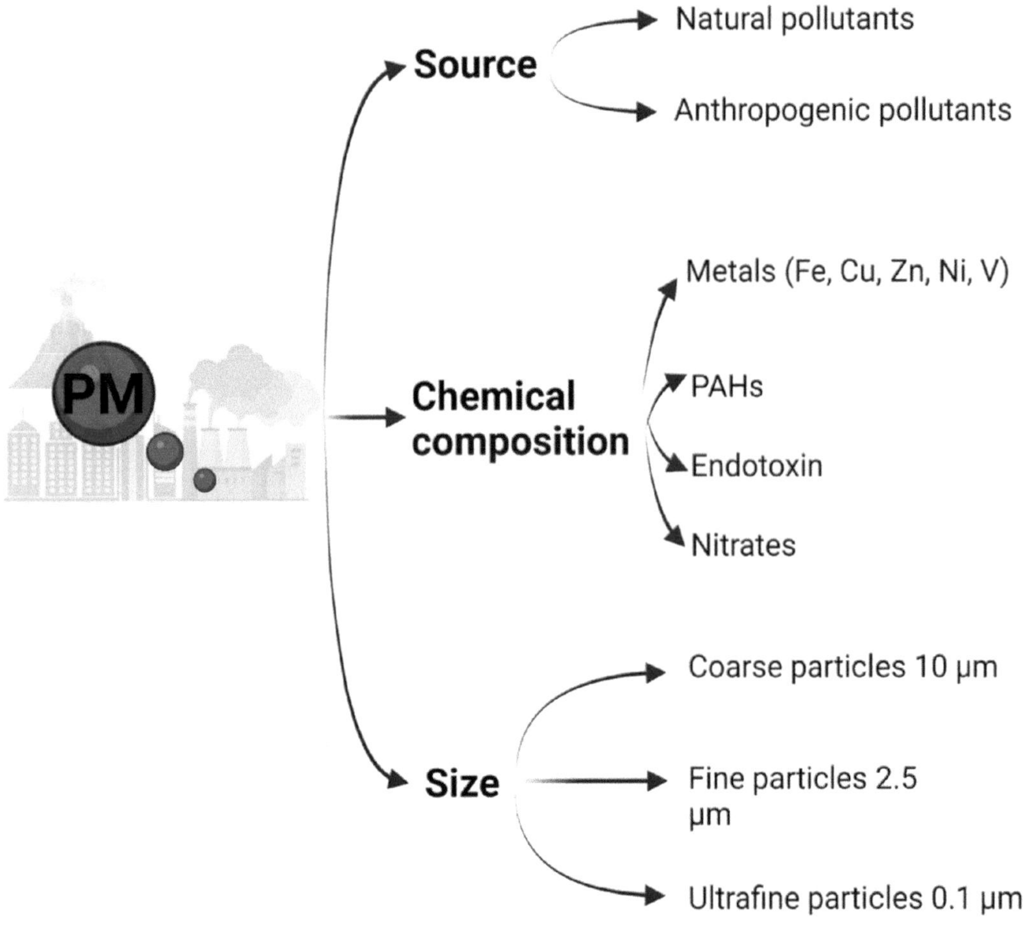

FIGURE 7.1 Particulate matter (PM) classification. Created with schematics from BioRender.com.

Due to PM relative composition, which is different in each microenvironment, the most widely classification used is based on their aerodynamic size into three main groups: coarse, also known as "thoracic" particles (2.5–10 microns, PM_{10}), fine (0.1–2.5 microns, $PM_{2.5}$), and ultrafine (less than 0.1 micron, $PM_{0.1}$) (Kim et al., 2015). The smaller the particle, the deeper they can penetrate the tissues in organs directly exposed to the atmospheric surroundings.

PM characterization regarding its composition is relevant to establish health outcomes as potential toxic effects may be linked with specific components rather than the whole mixture (Ghio et al., 2020). For example, transition metals typically found within the fine PM fraction, including iron (Fe), vanadium (V), nickel (Ni), chromium (Cr), copper (Cu), cadmium (Cd), and zinc (Zn), have the ability of participate in Fenton-like reactions contributing to an increased oxidant species release, altering the redox metabolism (Magnani et al., 2016). Organic compounds, like PAHs that can be found coating particles are also associated with cytotoxic mechanisms that involve O_2-derived free radicals mainly from mitochondria (Manisalidis et al., 2020).

PM-increased health risks have been associated with characteristics of susceptibility namely age (i.e., children and older adults), preexisting diseases (asthma, chronic obstructive pulmonary disease [COPD], cystic fibrosis, lung transplant patients, heart failure, obesity), obesity, genetic polymorphisms, age, sex, ethnicity, and low-socioeconomic status (Kaufman et al., 2016; Valacchi et al., 2020). In addition, socioeconomic position and nutritional status are dynamic elements that may vary throughout lifetime, modifying individuals' exposure (Hooper and Kaufman, 2018).

Considering the nature of PM, a deeper understanding of the PM-triggered toxic mechanisms that impact human health are required in order to prevent the deleterious effects of exposure to this global environmental hazard.

7.3 PM DETRIMENTAL EFFECTS ON PRIMARY ORGANS

Humans are continually exposed to pollutants mainly through tissue or organs that directly interact with PM present in the atmosphere (Valacchi et al., 2020). The consequences of PM exposure on different organs results in oxidative damage, decreased function, and inflammation, which can lead to the exacerbation of proinflammatory disorders (Magnani et al., 2020).

7.3.1 Toxic Mechanisms of PM Exposure on the Respiratory System

After breathing $PM_{2.5}$, both following acute or chronic exposures, impairment of the redox metabolism and inflammatory pathways largely account for the lung biological effects (Cho et al., 2018; Daellenbach et al., 2020; Marchini et al., 2022). The mechanisms by which PM inhalation increases ROS production in the lung leading to a imbalanced redox status may be attributed to different causes: (a) generation of oxidants by the compounds adsorbed at the particle surface and release of transition metals, (b) an interplay between PM and cellular ROS production sources, and (c) recruitment of inflammatory cells (Evelson et al., 2016).

The direct generation of ROS at the particles is supported by the concept that its surface offers a unique physicochemical interface to catalyze reactions resulting in oxidant production. Moreover, the smaller the particles the higher the surface/mass ratio, where more potentially toxic substances, such as carcinogenic hydrocarbons and heavy metals, can be adsorbed outside and cause health complications (Singh and Singh, 2021; Chan et al., 2016). Transition metals mainly found within the fine PM fraction are of great toxicological interest given that they can increase the production of free radicals in biological tissues through their ability to participate in Fenton-like chemical reactions and generate hydroxyl radicals, a potent oxidant responsible for the oxidative damage to macromolecules (Magnani et al., 2016). These toxicological mechanisms were confirmed by the inhibition through the presence of antioxidants, such as dimethylthiourea or iron chelator deferoxamine (Sørensen et al., 2005). This data have led to the suggestion that the dose of bioavailable transition metal, rather than particulate mass, may be the principal determinant of the acute inflammatory response (Cáceres et al., 2020).

Lung injury after PM inhalations has been suggested to be triggered by local ROS production, which in turn could be generated not only from the particles themselves but from the interaction of pollutants components with cellular enzymes and organelles (Magnani et al., 2013). It has been shown that an altered mitochondrial function is an important event in PM-induced cytotoxicity (Magnani et al., 2020). The first response to PM observed in isolated mitochondria was an augmented production of superoxide anion (O_2^-) and hydrogen peroxide (H_2O_2), and a decreased mitochondrial membrane potential, which in turn affects the bioenergetic function of lung mitochondria as ATP production is impaired (Magnani et al., 2013). However, an aberrant ROS production may overwhelm the antioxidant system, shifting the cellular redox status toward an oxidant environment and the loss of redox homeostasis. It was well described as a ROS production enhancement for mitochondria-to-mitochondria communication, where ROS released from one organelle triggers ROS production by another organelle (Zandalinas et al., 2018). An increase of the mitochondrial respiratory chain-ROS release activates the mitochondrial permeability transition pore (mPTP) causing depolarization of the inner and outer membranes, which in turn yields a burst of ROS released to the cytosol (Zandalinas et al., 2018). From this point of view, mitochondria are not only a relevant PM-initiated ROS production source but an oxidative damage target as well; all main ROS sources are, at the same time, regulated by oxidation. For example, PM-induced oxidative damage of the mitochondrial respiratory chain constituents leads to mild uncoupling resulting in an augmented ROS production (Magnani et al., 2020). The mitochondrial ROS-induced ROS release concept was then widened to communication between different ROS sources such as NADPH oxidase (Nox). This crosstalk between mitochondria and Nox, both main sources of lung tissue O_2 consumption, might result in a vicious feed-forward cycle of ROS production, which might be relevant under conditions of oxidative stress. The interplay can either be triggered by Nox-released ROS to mitochondria or vice versa from mitochondrial site to Nox (Kröller-Schön et al., 2014). In this scenario, various ROS sources interaction stimulates each other in a positive feedback fashion, resulting in a complex oxidative stress and redox signaling network (Schulz et al., 2014).

Mitochondrial mediators, especially H_2O_2, are also engaged in intracellular signaling. The ROS burden elicited by several environmental agents leads to an airway inflammatory response, which includes leukocyte recruitment, activation, and increased alveolar macrophages count (Marchini et al., 2020). In our model of chronic exposure to Buenos Aires City urban air altered lung redox metabolism indicated by the decreased GSH/GSSG ratio, a more oxidant environment, represents an earlier event for the development of local and systemic inflammation following PM exposure (Marchini et al., 2022). In accordance, exposure to polluted air has been associated with increased levels of circulating oxidative damage markers to macromolecules and inflammatory mediators in humans (Pope et al., 2016).

Activation of inflammatory cells capable of generating ROS and reactive nitrogen species (RNS) is another proposed mechanism as it has been shown that in chronic inflammatory lung diseases, PM inhalation induced the innate immune system, initiating the release of proinflammatory mediators, including interleukin (IL)-1β, tumor necrosis factor (TNF)-α, IL-6, and monocyte chemoattractant protein (MCP)-1, from alveolar macrophages, that are the major population of immune cells in the airways able to respond rapidly to inhaled pathogens (Glencross et al., 2020; Miyata and van Eeden, 2011; Mitschik et al., 2008). Alveolar macrophages are responsible for the pathogen's clearance from the respiratory surface by phagocytosis to maintain lung homeostasis (Cáceres et al., 2020). Therefore, tissue-resident macrophages, through the PM internalization, initiate the onset of lung inflammation, inflammatory cells (neutrophils, eosinophils, monocytes, and lymphocytes) recruitment, both in humans (Losacco and Perillo, 2018) and in different animal models (Marchini et al., 2016, 2021; Pope et al., 2016). The central role of alveolar macrophages in the cellular response after pollutants exposure was showed *in vivo* through a therapeutic depletion of lung macrophages, which impaired the classical PM-initiated adverse effects (Marchini et al., 2016; Hamanaka and

Mutlu, 2018; Pope et al., 2016). Moreover, the proinflammatory mediators secreted from alveolar macrophages after particles uptake, are able to stimulate other cell types in the surroundings, such as lung epithelial cells, fibroblasts, and dendritic cells, to release cytokines enhancing and perpetuating the inflammatory response in the lung (Arantes-Costa et al., 2014; Marchini et al., 2016; Orona et al., 2014). The inflammatory genes induced after exposure to PM (TNF-α and IL-8, for example) are regulated by redox-sensitive transcription factors such as NF-κB, one of the key transcription factors that is involved in the inflammatory responses to PM in the lungs (Kampfrath et al., 2011; Magnani et al., 2013).

Taking all this data into consideration, the probable sequence of events for PM-induced lung inflammation involves the following: (a) injury to epithelial cells by ROS, possibly enhanced in the presence of metals via Haber-Weiss and Fenton chemistry; (b) increased production of ROS/RNS and their related chemical species in the airways, which are efficient secondary messengers able to act as a link between the redox metabolism and the nuclear translocation of transcription factors involved in the activation of pro-inflammatory events; (c) activation of vascular endothelium and circulating leukocytes and emigration of inflammatory cells from blood to tissue sites, which involves up-regulation of adhesion molecules and other markers on vascular endothelium and on circulating leukocytes.

The association between inflammation and oxidative stress and the redox signaling involved in inflammatory-associated diseases has been termed "OxInflammation". OxInflammation embodies the outcome of the crosstalk between oxidative stress and inflammation that can aggravate or promote the development of health effects triggered by exposure to PM (Valacchi et al., 2018). The amplified oxidative processes represent a key event of PM lung injury (Magnani et al., 2011), commonly associated with increased damage-associated molecular patterns (DAMPs) levels, which can be recognized by toll-like receptor (TLR)-4 (Kampfrath et al., 2011; Shoenfelt et al., 2009), a membrane receptor involved in the NF-κB translocation pathway, establishing a positive feedback loop between a ROS generation and inflammation (Marchini et al., 2020; Shukla et al., 2000).

7.4 TOXIC MECHANISMS OF PM EXPOSURE ON SECONDARY ORGANS

PM exposure on primary organs produce inflammation leading to systemic complications resulting in distant organs defects, such as heart and brain (Danesh Yazdi et al., 2021; Marchini et al., 2020; Clifford et al., 2016; Shih et al., 2018). Cardiovascular diseases largely account for most of the increase in morbidity and mortality rates (WHO, 2018). In fact, according to the Global Burden of Disease study, air pollution is responsible for one-fourth of the total death count from ischemic heart disease and stroke (Fuller et al., 2018). Regarding PM-associated neurodegenerative diseases, the way in which the particles reach the brain and cause damage has not been clarified yet.

7.4.1 Cardiorespiratory Diseases Associated to PM Exposure

Exposure to air pollution leads to increased cardiometabolic diseases. As mentioned earlier, particles can be uptaken by tissue-resident macrophages exerting a pro-inflammatory response in the lung that can be extended to circulation. The systemic inflammatory response, characterized by alterations in circulating factors such as IL-1 and IL-6 and TNF-α. Secreted cytokines may reach blood circulation to remotely act on peripheral tissues (Kampfrath et al., 2011). Accordingly, it has been shown that therapeutic depletion of alveolar macrophages significantly reduced some of the detrimental effects observed in heart function after PM inhalation (Marchini et al., 2016), reinforcing the central role of these inflammatory cells in the initiation and progression of the oxinflammation that will lead to cardiovascular alterations.

Given the small size of fine and ultrafine particles that can go through the alveolar-capillary barrier, therefore, PM deposits have been found in several non-pulmonary tissues, including the blood, ventricular microvascular walls, liver, spleen, heart, and brain. PM by several direct and indirect effects may promote endothelial dysfunction, impair vasomotor function, induce platelet activation and aggregation pathways, and increase the expression of pro-thrombotic mediators on the endothelium (Chen et al., 2022; Marchini et al., 2020).

Impaired cardiac mitochondrial function also arises as a central feature of air pollution PM toxicology. Mice exposed to PM from urban air showed diminished cardiac mitochondrial function evidenced by decreased oxygen consumption resulting in a deficient ATP production, increased H_2O_2 release, pathogenic ultrastructural modifications and altered dynamics (Marchini et al., 2022) (Figure 7.2). In line with these findings, impaired cardiac mitochondrial function and contractility have been reported after an acute exposure to $PM_{2.5}$ (Li et al., 2015; Marchini et al., 2013; Nichols et al., 2015). Interestingly, this cardiac mitochondrial bioenergetic dysfunction seems to be partially mediated by an inflammatory response triggered by PM exposure, since impaired mitochondrial respiration and cardiac contractility is attenuated by pretreatment with a chimeric anti-TNF-α antibody (Infliximab) in PM-exposed mice (Marchini et al., 2015).

Taken together, impaired mitochondrial respiration, enhanced ROS release, and deficient ATP supply play a central role in the adverse health effects reported after PM exposure. In this context, redox metabolism through mitochondrial function modulation arises as a potential therapeutic target to prevent not only an excessive lung inflammatory response, but an impaired cardiac contractility function in PM-exposed individuals at particularly high risk (Wang et al., 2018; Marchini et al., 2020; Magnani et al., 2020).

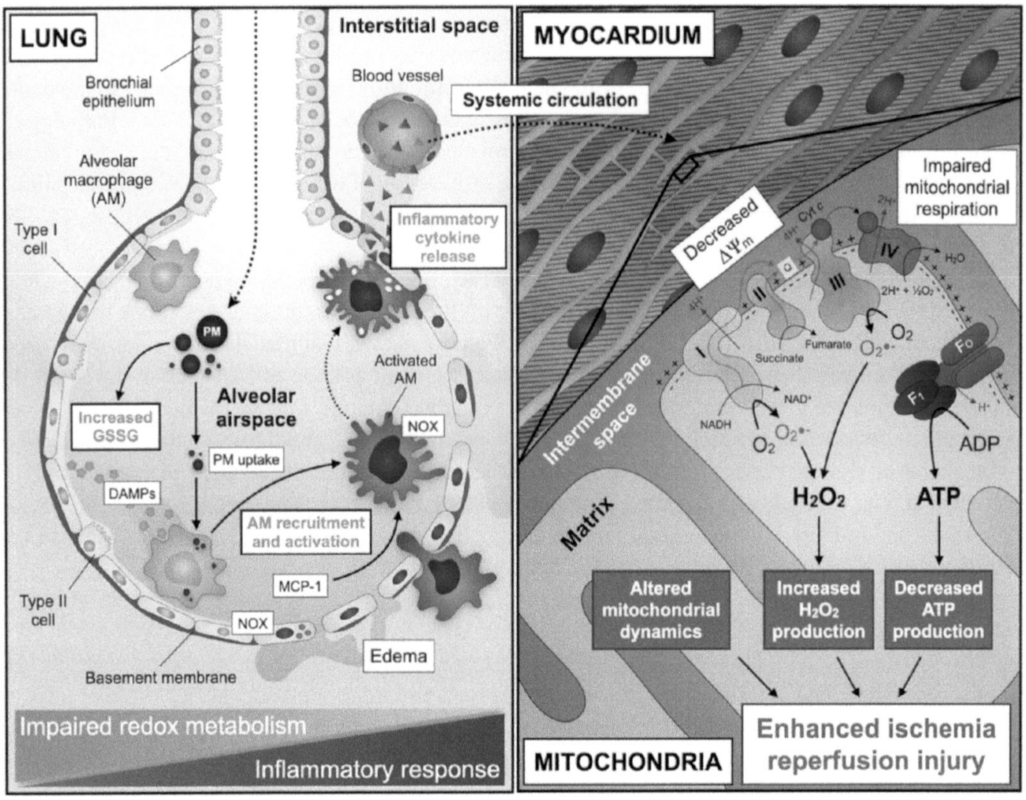

FIGURE 7.2 Proposed model of heart function aggravation by a chronic exposure particulate matter present in polluted urban air. Taken from Marchini et al. (2022) with permission from Elsevier [License Number 5462580450175].

7.4.2 PM Toxicity within the Brain

While the association between polluted air inhalation and the morbidity and mortality caused by cardiorespiratory diseases has been clearly established, current studies regarding PM neurotoxicity are not as numerous. However, growing evidence points to the central nervous system (CNS) as another highly relevant target of environmental pollutants. Epidemiological studies in humans have shown that high levels of PM are associated with cognitive function changes in children, adults, and the elderly (Clifford et al., 2016). Alterations at the olfactory level, hearing deficits, symptoms of depression, and other neuropsychological effects have also been reported (Costa et al., 2019). The way in which the PM reaches the brain and how the particles cause the damage has not been clarified yet. It has been proposed that ultrafine particulate matter (UFPM < 0.1 μm), given its smaller size, can reach the circulation easily, and be distributed to different tissues, including the brain (Oberdörster et al., 2002). PM that circulates through the nasal compartment can cross the epithelial barrier into the bloodstream or translocate along the axon of the olfactory nerve and reach the central nervous system. Other studies suggest that ultrafine PM can directly translocate to the brain through the olfactory mucosa and trigger different pathological responses (Peters et al., 2006). Taking all these findings into account, the deleterious effects observed after PM inhalation could be the result of two different mechanisms: (a) a direct effect on the brain via an olfactory route (from the nasal mucosa to the axon of the olfactory nerve and from there to the olfactory bulb) or (b) an indirect effect through an systemic inflammatory reaction and oxidative stress initiated by increased levels of proinflammatory mediators released into the bloodstream as a consequence of a local response in the lung. The combination of both mechanisms taking place should not be discarded (Costa et al., 2017). Hence, like the gastrointestinal tract, brain toxicity might begin through direct as well as indirect PM-initiated mechanism.

It is postulated that the main mechanism of neurotoxicity produced by PM would involve different alterations including oxidative stress and neuroinflammation, which are also related to the pathophysiology of various neurodegenerative diseases (Jankowska-Kieltyka et al., 2021). Pathological manifestations of neurological diseases are often related to the impairment of cellular redox homeostasis and, subsequently, oxidative and nitrosative stress (Angelova et al., 2018). Numerous studies reported dysfunction and damage to mitochondria elicited by ROS in the pathogenesis of CNS disorders (Chew et al., 2020).

In a previous work from our group, we evaluated the effects of air pollution using an *in vivo* whole-body urban air chronic exposure protocol, focusing on the changes in brain cortex redox homeostasis and mitochondrial function (Calabró et al., 2021). We have shown that that chronic exposure to urban air induces an early shift towards a more oxidative environment as soon as 1 week after the beginning of the exposure, leading to brain lipid and protein oxidative damage with increased Nox activity and impaired mitochondrial function that increases ROS production and overwhelms antioxidant defenses. Mitochondrial dysfunction was mainly characterized by ATP mismanagement and an increase in H_2O_2 production that, together with the contribution of NOX-derived superoxide anion, suggests that oxidative stress and impaired mitochondrial function might have a central role in CNS damage mechanisms triggered by air pollution. A summary of the main findings is described in Figure 7.3.

7.5 CONCLUSIONS

Epidemiological data and studies performed in different animal models suggest a variety of possible mechanisms being an altered redox metabolism and inflammation, the most relevant mechanism following PM exposure, that may contribute to the increased morbidity and mortality associated with polluted areas. A better understanding of these mechanisms underlying PM-induced health problems would allow a more targeted approach to face the toxic effects of pollutants and could possibly provide different ways to decrease individual sensitivity to PM, including a cardiac and

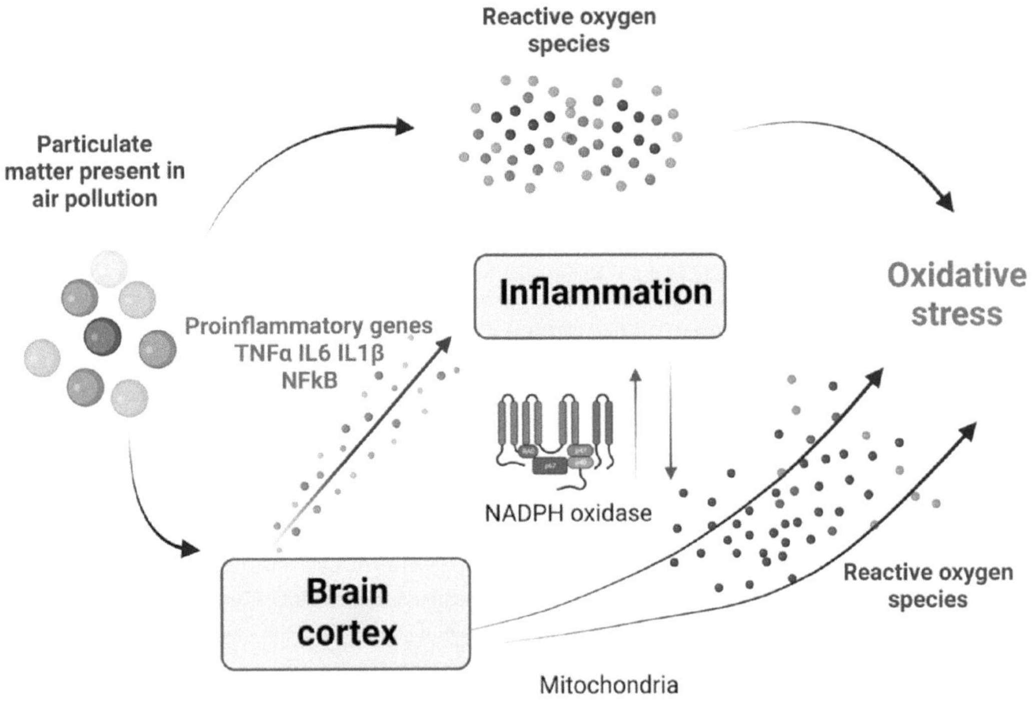

FIGURE 7.3 Proposed mechanisms triggered by PM inhalation in the brain cortex. Created with schematics from BioRender.com.

neurological effects induced, at least in part, by proinflammatory mediators that affects oxygen metabolism and mitochondrial function.

FUNDING

This study was supported by research grants from the Universidad de Buenos Aires (20020170100441BA and PIDAE 24/2020 to PE, 20020190200323BA to NM) and from Agencia Nacional de Promoción Científica y Tecnológica (ANPCyT) (PICT 2018-03250 to NM, and PICT 2019-0150 to PE).

REFERENCES

Angelova PR, Abramov AY. Role of mitochondrial ROS in the brain: From physiology to neurodegeneration. *FEBS Lett.* 2018 592(5):692–702. doi: 10.1002/1873-3468.12964.

Arantes-Costa FM, Grund LZ, Martins MA, Lima C. Airborne pollutant ROFA enhances the allergic airway inflammation through direct modulation of dendritic cells in an uptake-dependent mechanism. *Int Immunopharmacol.* 2014 22(1):9–20. doi: 10.1016/j.intimp.2014.06.020.

Brook RD, Rajagopalan S. Inhaling hypertension: Clearing the air to prevent cardiometabolic diseases. *Hypertension.* 2020 76(1):32–34. doi: 10.1161/HYPERTENSIONAHA.120.14613.

Cáceres L, Paz ML, Garcés M, Calabró V, Magnani ND, Martinefski M, Martino Adami PV, Caltana L, Tasat D, Morelli L, et al. NADPH oxidase and mitochondria are relevant sources of superoxide anion in the OxInflammatory response of macrophages exposed to airborne particulate matter. *Ecotoxicol Environ Saf.* 2020 205:111186. doi: 10.1016/j.ecoenv.2020.111186. Epub 2020 Aug 24. PMID: 32853868.

Calabró V, Garcés M, Cáceres L, Magnani ND, Marchini T, Freire A, Vico T, Martinefski M, Vanasco V, Tripodi V, et al. Urban air pollution induces alterations in redox metabolism and mitochondrial dysfunction in mice brain cortex. *Arch Biochem Biophys.* 2021 704:108875. doi: 10.1016/j.abb.2021.108875.

Chan EA, Buckley B, Farraj AK, Thompson LC. The heart as an extravascular target of endothelin-1 in particulate matter-induced cardiac dysfunction. *Pharmacol Ther.* 2016 165:63–78. doi: 10.1016/j.pharmthera.2016.05.006.

Chen H, Oliver BG, Pant A, Olivera A, Poronnik P, Pollock CA, Saad S. Effects of air pollution on human health: Mechanistic evidence suggested by in vitro and in vivo modelling. *Environ Res.* 2022 212(Pt C):113378. doi: 10.1016/j.envres.2022.113378.

Chew S, Kolosowska N, Saveleva L, Malm T, Kanninen KM. Impairment of mitochondrial function by particulate matter: Implications for the brain. *Neurochem Int.* 2020 135:104694. doi: 10.1016/j.neuint.2020.104694.

Cho CC, Hsieh WY, Tsai CH, Chen CY, Chang HF, Lin CS. In vitro and in vivo experimental studies of PM2.5 on disease progression. *Int J Environ Res Public Health.* 2018 15(7):1380. doi: 10.3390/ijerph15071380.

Clifford A, Lang L, Chen R, Anstey KJ, Seaton A. Exposure to air pollution and cognitive functioning across the life course: A systematic literature review. *Environ Res.* 2016 147:383–398. doi: 10.1016/j.envres.2016.01.018.

Costa LG, Cole TB, Coburn J, Chang YC, Dao K, Roqué PJ. Neurotoxicity of traffic-related air pollution. *Neurotoxicology.* 2017 59:133–139. doi: 10.1016/j.neuro.2015.11.008.

Costa LG, Cole TB, Dao K, Chang YC, Garrick JM. Developmental impact of air pollution on brain function. *Neurochem Int.* 2019 131:104580. doi: 10.1016/j.neuint.2019.104580.

Daellenbach KR, Uzu G, Jiang J, Cassagnes LE, Leni Z, Vlachou A, Stefenelli G, Canonaco F, Weber S, Segers A, et al. Sources of particulate-matter air pollution and its oxidative potential in Europe. *Nature.* 2020 587(7834):414–419. doi: 10.1038/s41586-020-2902-8.

Danesh Yazdi M, Wang Y, Di Q, Wei Y, Requia WJ, Shi L, Sabath MB, Dominici F, Coull BA, Evans JS, et al. Long-term association of air pollution and hospital admissions among medicare participants using a doubly robust additive model. *Circulation.* 2021 143(16):1584–1596. doi: 10.1161/CIRCULATIONAHA.120.050252.

Evelson P, Marchini T, Garces M, Cáceres L, Magnani N, Alvarez S. Inhaled particulate matter leads to myocardial dysfunction. In: Gelpi R, Boveris A, Poderoso J (eds) Biochemistry of Oxidative Stress. *Advances in Biochemistry in Health and Disease*, vol 16. Springer, Cham, 2016. https://doi.org/10.1007/978-3-319-45865-6_18.

Fuller R, Rahona E, Fisher S, Caravanos J, Webb D, Kass D, Matte T, Landrigan PJ. Pollution and noncommunicable disease: Time to end the neglect. *Lancet Planet Health.* 2018 2(3):e96–e98. doi: 10.1016/S2542-5196(18)30020-2.

Ghio AJ, Soukup JM, Dailey LA, Madden MC. Air pollutants disrupt iron homeostasis to impact oxidant generation, biological effects, and tissue injury. *Free Radic Biol Med.* 2020 151:38–55. doi: 10.1016/j.freeradbiomed.2020.02.007.

Glencross DA, Ho TR, Camiña N, Hawrylowicz CM, Pfeffer PE. Air pollution and its effects on the immune system. *Free Radic Biol Med.* 2020 151:56–68. doi: 10.1016/j.freeradbiomed.2020.01.179

Hamanaka RB, Mutlu GM. Particulate matter air pollution: Effects on the cardiovascular system. *Front Endocrinol (Lausanne).* 2018 9:680. doi: 10.3389/fendo.2018.00680.

Hooper LG, Kaufman JD. Ambient air pollution and clinical implications for susceptible populations. *Ann Am Thorac Soc.* 2018 15(Suppl 2): S64–S68. doi: 10.1513/AnnalsATS.201707-574MG.

Jankowska-Kieltyka M, Roman A, Nalepa I. The air we breathe: Air pollution as a prevalent proinflammatory stimulus contributing to neurodegeneration. *Front Cell Neurosci.* 2021 15:647643. doi: 10.3389/fncel.2021.647643.

Kampfrath T, Maiseyeu A, Ying Z, Shah Z, Deiuliis JA, Xu X, Kherada N, Brook RD, Reddy KM, Padture NP, et al. Chronic fine particulate matter exposure induces systemic vascular dysfunction via NADPH oxidase and TLR4 pathways. *Circ Res.* 2011 108(6):716–726. doi: 10.1161/CIRCRESAHA.110.237560.

Kaufman JD, Spalt EW, Curl CL, Hajat A, Jones MR, Kim SY, Vedal S, Szpiro AA, Gassett A, Sheppard L, et al. Advances in understanding air pollution and CVD. *Glob Heart.* 2016 11(3):343–352. doi: 10.1016/j.gheart.2016.07.004. PMID: 27741981; PMCID: PMC5082281.

Kim KH, Kabir E, Kabir S. A review on the human health impact of airborne particulate matter. Environ Int. 2015 74:136–143. doi: 10.1016/j.envint.2014.10.005.

Kröller-Schön S, Steven S, Kossmann S, Scholz A, Daub S, Oelze M, Xia N, Hausding M, Mikhed Y, Zinssius E, et al. Molecular mechanisms of the crosstalk between mitochondria and NADPH oxidase through reactive oxygen species-studies in white blood cells and in animal models. *Antioxid Redox Signal.* 2014 20(2):247–266. doi: 10.1089/ars.2012.4953.

Landrigan PJ, Fuller R, Acosta NJR, Adeyi O, Arnold R, Basu NN, Baldé AB, Bertollini R, Bose-O'Reilly S, Boufford JI, et al. The lancet commission on pollution and health. *Lancet.* 2018 391(10119):462–512. doi: 10.1016/S0140-6736(17)32345-0.

Lelieveld J, Pozzer A, Pöschl U, Fnais M, Haines A, Münzel T. Loss of life expectancy from air pollution compared to other risk factors: A worldwide perspective. Cardiovasc Res. 2020 116(11):1910–1917. doi: 10.1093/cvr/cvaa025. Erratum in: *Cardiovasc Res.* 2020 116(7):1334.

Li R, Kou X, Geng H, Xie J, Tian J, Cai Z, Dong C. Mitochondrial damage: An important mechanism of ambient PM2.5 exposure-induced acute heart injury in rats. *J Hazard Mater.* 2015 287:392–401. doi: 10.1016/j.jhazmat.2015.02.006.

Losacco C, Perillo A. Particulate matter air pollution and respiratory impact on humans and animals. *Environ Sci Pollut Res Int.* 2018 25(34):33901–33910. doi: 10.1007/s11356-018-3344-9.

Loxham M, Nieuwenhuijsen MJ. Health effects of particulate matter air pollution in underground railway systems: A critical review of the evidence. *Part Fibre Toxicol.* 2019 16(1):12. doi: 10.1186/s12989-019-0296-2.

Magnani ND, Marchini T, Calabró V, Alvarez S, Evelson P. Role of mitochondria in the redox signaling network and its outcomes in high impact inflammatory syndromes. *Front Endocrinol (Lausanne).* 2020 11:568305. doi: 10.3389/fendo.2020.568305.

Magnani ND, Marchini T, Garcés M, Mebert A, Cáceres L, Diaz L, Desimone M, Evelson PA. Role of transition metals present in air particulate matter on lung oxygen metabolism. Int J Biochem Cell Biol. 2016 81(Pt B):419–426. doi: 10.1016/j.biocel.2016.10.009.

Magnani ND, Marchini T, Tasat DR, Alvarez S, Evelson PA. Lung oxidative metabolism after exposure to ambient particles. *Biochem Biophys Res Commun.* 2011 412(4):667–672. doi: 10.1016/j.bbrc.2011.08.021.

Magnani ND, Marchini T, Vanasco V, Tasat DR, Alvarez S, Evelson P. Reactive oxygen species produced by NADPH oxidase and mitochondrial dysfunction in lung after an acute exposure to residual oil fly ashes. *Toxicol Appl Pharmacol.* 2013 270(1):31–38. doi: 10.1016/j.taap.2013.04.002.

Manisalidis I, Stavropoulou E, Stavropoulos A, Bezirtzoglou E. Environmental and health impacts of air pollution: A review. *Front Public Health.* 2020 8:14. doi: 10.3389/fpubh.2020.00014.

Marchini T, D'Annunzio V, Paz ML, Cáceres L, Garcés M, Perez V, Tasat D, Vanasco V, Magnani N, Gonzalez Maglio D, et al. Selective TNF-α targeting with infliximab attenuates impaired oxygen metabolism and contractile function induced by an acute exposure to air particulate matter. *Am J Physiol Heart Circ Physiol.* 2015 309(10):H1621–H1628. doi: 10.1152/ajpheart.00359.2015.

Marchini T, Magnani ND, D'Annunzio V, Tasat D, Gelpi RJ, Alvarez S, Evelson P. Impaired cardiac mitochondrial function and contractile reserve following an acute exposure to environmental particulate matter. *Biochim Biophys Acta.* 2013 830(3):2545–2552. doi: 10.1016/j.bbagen.2012.11.012.

Marchini T, Magnani ND, Garces M, Kelly J, Paz M, Caceres L, Calabro V, Lasagni Vitar R, Caltana L, Contin M, et al. Chronic exposure to polluted urban air aggravates myocardial infarction by impaired cardiac mitochondrial function and dynamics. *Environ Pollut.* 2022 295:118677. doi: 10.1016/j.envpol.2021.118677.

Marchini T, Wolf D, Michel NA, Mauler M, Dufner B, Hoppe N, Beckert J, Jäckel M, Magnani N, Duerschmied D, et al. Acute exposure to air pollution particulate matter aggravates experimental myocardial infarction in mice by potentiating cytokine secretion from lung macrophages. *Basic Res Cardiol.* 2016 111(4):44. doi: 10.1007/s00395-016-0562-5.

Marchini T, Zirlik A, Wolf D. Pathogenic role of air pollution particulate matter in cardiometabolic disease: Evidence from mice and humans. *Antioxid Redox Signal.* 2020 33(4):263–279. doi: 10.1089/ars.2020.8096.

Mitschik S, Schierl R, Nowak D, Jörres RA. Effects of particulate matter on cytokine production in vitro: A comparative analysis of published studies. Inhal Toxicol. 2008 20(4):399–414. doi: 10.1080/08958370801903784.

Miyata R, van Eeden SF. The innate and adaptive immune response induced by alveolar macrophages exposed to ambient particulate matter. *Toxicol Appl Pharmacol.* 2011 257(2):209–226. doi: 10.1016/j.taap.2011.09.007.

Mudway IS, Kelly FJ, Holgate ST. Oxidative stress in air pollution research. *Free Radic Biol Med.* 2020 151:2–6. doi: 10.1016/j.freeradbiomed.2020.04.031.

Nichols CE, Shepherd DL, Knuckles TL, Thapa D, Stricker JC, Stapleton PA, Minarchick VC, Erdely A, Zeidler-Erdely PC, Alway SE, et al. Cardiac and mitochondrial dysfunction following acute pulmonary exposure to mountaintop removal mining particulate matter. *Am J Physiol Heart Circ Physiol.* 2015 309(12):H2017–H2030. doi: 10.1152/ajpheart.00353.2015.

Oberdörster G, Sharp Z, Atudorei V, Elder A, Gelein R, Lunts A, Kreyling W, Cox C. Extrapulmonary translocation of ultrafine carbon particles following whole-body inhalation exposure of rats. *J Toxicol Environ Health A.* 2002 65(20):1531–1543. doi: 10.1080/00984100290071658.

Orona NS, Astort F, Maglione GA, Saldiva PH, Yakisich JS, Tasat DR. Direct and indirect air particle cytotoxicity in human alveolar epithelial cells. *Toxicol In Vitro.* 2014 28(5):796–802. doi: 10.1016/j.tiv.2014.02.011.

Peters A, Veronesi B, Calderón-Garcidueñas L, Gehr P, Chen LC, Geiser M, Reed W, Rothen-Rutishauser B, Schürch S, Schulz H. Translocation and potential neurological effects of fine and ultrafine particles a critical update. *Part Fibre Toxicol.* 2006 3:13. doi: 10.1186/1743-8977-3-13.

Pope CA, 3rd, Bhatnagar A, McCracken JP, Abplanalp W, Conklin DJ, O'Toole T. Exposure to fine particulate air pollution is associated with endothelial injury and systemic inflammation. *Circ Res.* 2016 119(11):1204–1214. doi: 10.1161/CIRCRESAHA.116.309279.

Pope CA, 3rd, Lefler JS, Ezzati M, Higbee JD, Marshall JD, Kim SY, Bechle M, Gilliat KS, Vernon SE, Robinson AL, Burnett RT. Mortality risk and fine particulate air pollution in a large, representative cohort of U.S. adults. *Environ Health Perspect.* 2019 127(7):77007. doi: 10.1289/EHP4438.

Rajagopalan S, Al-Kindi SG, Brook RD. Air pollution and cardiovascular disease: JACC state-of-the-art review. *J Am Coll Cardiol.* 2018 72(17):2054–2070. doi: 10.1016/j.jacc.2018.07.099.

Rao X, Zhong J, Brook RD, Rajagopalan S. Effect of particulate matter air pollution on cardiovascular oxidative stress pathways. *Antioxid Redox Signal.* 2018 28(9):797–818. doi: 10.1089/ars.2017.7394.

Schulz E, Wenzel P, Münzel T, Daiber A. Mitochondrial redox signaling: Interaction of mitochondrial reactive oxygen species with other sources of oxidative stress. *Antioxid Redox Signal.* 2014 20(2):308–324. doi: 10.1089/ars.2012.4609.

Shih CH, Chen JK, Kuo LW, Cho KH, Hsiao TC, Lin ZW, Lin YS, Kang JH, Lo YC, Chuang KJ, et al. Chronic pulmonary exposure to traffic-related fine particulate matter causes brain impairment in adult rats. *Part Fibre Toxicol.* 2018 15(1):44. doi: 10.1186/s12989-018-0281-1.

Shoenfelt J, Mitkus RJ, Zeisler R, Spatz RO, Powell J, Fenton MJ, Squibb KA, Medvedev AE. Involvement of TLR2 and TLR4 in inflammatory immune responses induced by fine and coarse ambient air particulate matter. *J Leukoc Biol.* 2009 86(2):303–312. doi: 10.1189/jlb.1008587.

Shukla A, Timblin C, BeruBe K, Gordon T, McKinney W, Driscoll K, Vacek P, Mossman BT. Inhaled particulate matter causes expression of nuclear factor (NF)-kappaB-related genes and oxidant-dependent NF-kappaB activation in vitro. *Am J Respir Cell Mol Biol.* 2000 23(2):182–187. doi: 10.1165/ajrcmb.23.2.4035.

Singh N, Singh S. Interstitial lung diseases and air pollution: Narrative review of literature. *Pulm Ther.* 2021 10:1–12. doi: 10.1007/s41030-021-00148-7.

Sørensen M, Schins RP, Hertel O, Loft S. Transition metals in personal samples of PM2.5 and oxidative stress in human volunteers. *Cancer Epidemiol Biomarkers Prev.* 2005 14(5):1340–1343. doi: 10.1158/1055-9965.EPI-04-0899.

Valacchi G, Magnani N, Woodby B, Ferreira SM, Evelson P. Particulate matter induces tissue OxInflammation: From mechanism to damage. *Antioxid Redox Signal.* 2020 33(4):308–326. doi: 10.1089/ars.2019.8015.

Valacchi G, Virgili F, Cervellati C, Pecorelli A. OxInflammation: From subclinical condition to pathological biomarker. *Front Physiol.* 2018 9:858. doi: 10.3389/fphys.2018.00858.

Wang H, Shen X, Tian G, Shi X, Huang W, Wu Y, Sun L, Peng C, Liu S, Huang Y, et al. AMPKα2 deficiency exacerbates long-term PM2.5 exposure-induced lung injury and cardiac dysfunction. *Free Radic Biol Med.* 2018 121:202–214. doi: 10.1016/j.freeradbiomed.2018.05.008.

World Health Organization (WHO). Burden of disease from the joint effects of household and ambient Air pollution for 2016. *Public Health, Social and Environmental Determinants of Health Department*, WHO report, Geneva, Switzerland, 2018.

Wu W, Jin Y, Carlsten C. Inflammatory health effects of indoor and outdoor particulate matter. *J Allergy Clin Immunol.* 2018 141(3):833–844. doi: 10.1016/j.jaci.2017.12.981.

Zandalinas SI, Mittler R. ROS-induced ROS release in plant and animal cells. *Free Radic Biol Med.* 2018 122:21–27. doi: 10.1016/j.freeradbiomed.2017.11.028.

Zhang H, Haghani A, Mousavi AH, Cacciottolo M, D'Agostino C, Safi N, Sowlat MH, Sioutas C, Morgan TE, Finch CE, et al. Cell-based assays that predict in vivo neurotoxicity of urban ambient nano-sized particulate matter. *Free Radic Biol Med.* 2019 145:33–41. doi: 10.1016/j.freeradbiomed.2019.09.016.

8 The Ocular Surface as a Target of Air Pollution

Romina M. Lasagni Vitar, Ailen Gala Hvozda Arana, Timoteo O. Marchini, Pablo A. Evelson, and Sandra M. Ferreira

8.1 INTRODUCTION

The effects of air pollution on human health increasingly generate great concern worldwide (Al-Kindi et al., 2020; Brook et al., 2010). Particulate matter (PM) is a critical component of polluted air and has been pointed out as the main responsible for the adverse health outcomes associated with air pollution exposure (Marchini et al., 2020). PM is classified by size according to their aerodynamic diameter in coarse (<10 μm, PM_{10}), fine (<2.5 μm, $PM_{2.5}$), and ultrafine (<0.1 μm) particles. The fine and ultrafine particle fractions are especially toxic since they have the largest surface area to volume ratio, containing the greatest amount of adsorbed toxic compounds (Nel, 2005; World Health Organization (WHO), 2016). In urban polluted air, the combustion-derived particles present in traffic emissions and fossil fuel combustion during power generation and industrial processes are the main contributors to the environmental PM burden (Zhang et al., 2015). These particles are in the fine and ultrafine range and contain polycyclic aromatic hydrocarbons (PAHs) and metal ions adsorbed to the surface (Maglione et al., 2019).

The toxic mechanism triggered by PM exposure has been well-studied in the lungs (Magnani et al., 2013; Magnani, Marchini, et al., 2016; Marchini et al., 2014), the heart (Brook et al., 2010; Marchini et al., 2013, 2015; Sundar et al., 2018), and the skin (Kim et al., 2016; Lefebvre et al., 2015; Magnani, et al., 2016; Marrot, 2018). It has been suggested that redox-active chemical compounds adsorbed to the particle surface interact with different cell targets inducing redox imbalance and inflammation. In this context, oxidative damage to macromolecules occurs, leading to pathogenic changes that can affect the tissue function (Valacchi et al., 2020).

The eyes are constantly exposed to the environment, only separated from the outdoor stressors by the tear film. Therefore, they are particularly vulnerable to the effects of air pollutants. In fact, the ocular surface plays a fundamental role in the defence against the oxidant species generated by environmental stressors, which makes it highly susceptible to oxidative damage. In this chapter, we will review the effects in the ocular surface triggered by PM exposure, highlighting the role of oxidative stress and inflammation in the cornea and the conjunctiva.

8.2 OXIDATIVE STRESS AND INFLAMMATION ARE THE MAIN TOXIC MECHANISMS TRIGGERED BY PM

Redox imbalance and the subsequent oxidative stress have been suggested to play a critical role in the toxic effects triggered by PM. The mechanisms by which PM induces oxidative stress have not yet been fully elucidated. Different events have been proposed: (i) oxidant species generation at the particle surface, (ii) metals or organic compound release from the particle, and/or (iii) activation of pro-inflammatory pathways (Xia et al., 2006). In addition, PM can directly activate inflammatory cells by interacting with membrane receptors (Mazzoli-Rocha et al., 2010), as well as during phagocytosis (Becker et al., 2002). This scenario contributes to impaired tissue redox metabolism since

 DOI: 10.1201/9781003328100-10

inflammatory cells can produce reactive oxygen and nitrogen species (ROS and RNS, respectively) (Iles & Forman, 2002). Therefore, generating a positive feedback between inflammation and oxidative stress would perpetuate PM toxic effects in the tissue. In fact, ROS and RNS can also initiate and maintain the pro-inflammatory state by oxidizing macromolecules such as lipids, proteins, and DNA and by inducing several inflammatory signalling pathways (i.e., nuclear factor kappa-light-chain-enhancer of activated B cells [NF-kB], mitogen-activated protein kinases [MAPK]) (Michael et al., 2013). The activation of these pathways results in the release of pro-inflammatory cytokines and chemokines including tumor necrosis factor-alpha (TNF-α), interleukin-6 (IL-6), interleukin-8 (IL-8), interleukin-1-beta (IL-1β), monocyte chemoattractant protein-1 (MCP-1), which in turn further promote tissue inflammation and stimulate the production of ROS and RNS (Pope et al., 2016).

In this context, a well-established antioxidant defence system is crucial to maintain the redox balance and to avoid cell oxidative damage. In this regard, the nuclear factor erythroid 2-related factor-2 (NRF2) is central in the modulation of the adaptive cellular response to oxidative stress. The NRF2 pathway maintains the cellular antioxidant capacity by up-regulating the expression of phase II detoxifying and antioxidant enzymes such as superoxide dismutase (SOD), glutathione peroxidase (GPx), glutathione S-transferase (GST), and glutathione reductase (GR) (Rubio et al., 2010). Additionally, the homeostasis of glutathione (GSH) levels, the main non-protein thiol, together with its oxidized form (GSSG), is essential to maintain the intracellular redox status. NRF2 up-regulates glutamate cysteine ligase (GCL), a key enzyme in GSH synthesis (Ma, 2013). Several cellular processes, such as redox signalling, detoxification of xenobiotics, regulation of cell proliferation, apoptosis, immune function, and fibrogenesis, are modulated by the GSH/GSSG balance (Lu, 2014).

8.3 THE OCULAR SURFACE AS A TARGET OF PM

The ocular surface, and, in particular, the corneal and the conjunctival epithelia, represents the first barrier of the eye that protects the underlying tissues from air pollutants. The capacity of the cornea and the conjunctiva to efficiently respond to infections and trauma is important to protect the eye from the environment. In fact, the defence mechanisms of the ocular surface have evolved to repair the injury by minimally affecting the transparency of the cornea. However, in the presence of severe or chronic damage (e.g., constant exposure to air pollutants), an overwhelming response may occur, resulting in permanent alterations of the ocular surface morphology and function.

Several studies have reported that inhabitants of highly polluted urban centers frequently present symptoms of eye discomfort such as irritation, burning, foreign body sensation, redness, and itching (Berra et al., 2015; Fu et al., 2017; Hwang et al., 2016; Nucci et al., 2017; Torricelli et al., 2014). Moreover, it has been suggested that air pollutants aggravate preexisting chronic eye diseases (e.g., dry eye) (Gupta et al., 2007; Novaes et al., 2010), highlighting the deleterious effects of air pollution on the ocular surface.

In the recent years, only a few studies aimed to elucidate the toxic mechanism underlying the effects of PM on the cornea and the conjunctiva, suggesting that oxidative stress and inflammation are at play (Gao et al., 2016; Kang et al., 2020; Kashiwagi & Iizuka, 2020; Lasagni Vitar et al., 2015, 2018; Lee et al., 2020; Niu et al., 2021a; Park et al., 2017; Tau et al., 2013; Tesone et al., 2018). In this vein, it has been reported that the cornea, the conjunctiva, and the aqueous humor have an efficient antioxidant defense system (Ferreira et al., 2009, 2010), which is essential for the regulation of redox homeostasis in the eye and protection against oxidative damage. Moreover, several studies suggested that changes in GSH homeostasis is associated with various human diseases affecting the ocular surface epithelia (Chen et al., 2009).

8.4 PM EFFECTS ON THE CORNEA

The cornea is a transparent and avascular tissue that serves as an impermeable barrier between the eye and the environment. It is the main refractive surface of the eye and essentially consists of five layers: epithelium, the Bowman's layer, stroma, Descemet's membrane, and endothelium. Both the anterior and the posterior surfaces contribute to its optical function. The visual information is processed in the retina by two types of photoreceptors, rods and cones. The visual impulse is conducted through the optic nerve to vision-related areas of the brain. Therefore, it is important that corneal transparency and structural order are maintained to allow adequate transmission and refraction of light (DelMonte & Kim, 2011).

The deleterious effects of PM on the cornea include inflammation and oxidative stress. In this regard, *in vitro* studies have demonstrated that corneal epithelial cells respond to different types of PM by releasing pro-inflammatory cytokines such as IL-6, IL-8, TNF-α, IL-1β, and MCP-1 (Park et al., 2017; Tau et al., 2013; Tesone et al., 2018; Xiang et al., 2016; Yoon et al., 2018). *In vivo* models involving topical application of PM on the eye confirmed the release of these pro-inflammatory mediators (Kang et al., 2020). In contrast, using a continuous whole-body-exposure model, a decrease in TNF-α and IL-6 after a short exposure time was found in the cornea, with a concomitant increase of interleukin-10 (IL-10) (Lasagni Vitar et al., 2019) (Figure 8.1A). The involvement of IL-10 could be important at early stages after the exposure to urban air to regulate the inflammatory response triggered by air pollutants. In line with these results, it has been reported a decrease in the Th2 profile cytokines on the tear of individuals chronically exposed to high PM levels in the metropolitan area of San Pablo, Brazil (Matsuda et al., 2015), suggesting an adaptive strategy of the ocular surface to the continuous exposure to PM.

With the cornea being highly susceptible to environmental stressors, and in particular, to the continuous PM exposure, it is not surprising that the antioxidant system has evolved to protect it from oxidative injury. In this vein, primary antioxidant enzymes such SOD, catalase (CAT), and GPx, as well as non-enzymatic antioxidants, are found in relatively high levels among the corneal tissue (Cabrera & Chihuailaf, 2011). Changes in the corneal redox homeostasis have been also described after PM exposure. An increase in ROS and RNS production has been reported (Gao et al., 2016; Lasagni Vitar et al., 2019; Niu et al., 2021a; Park et al., 2017; Yang et al., 2019). The

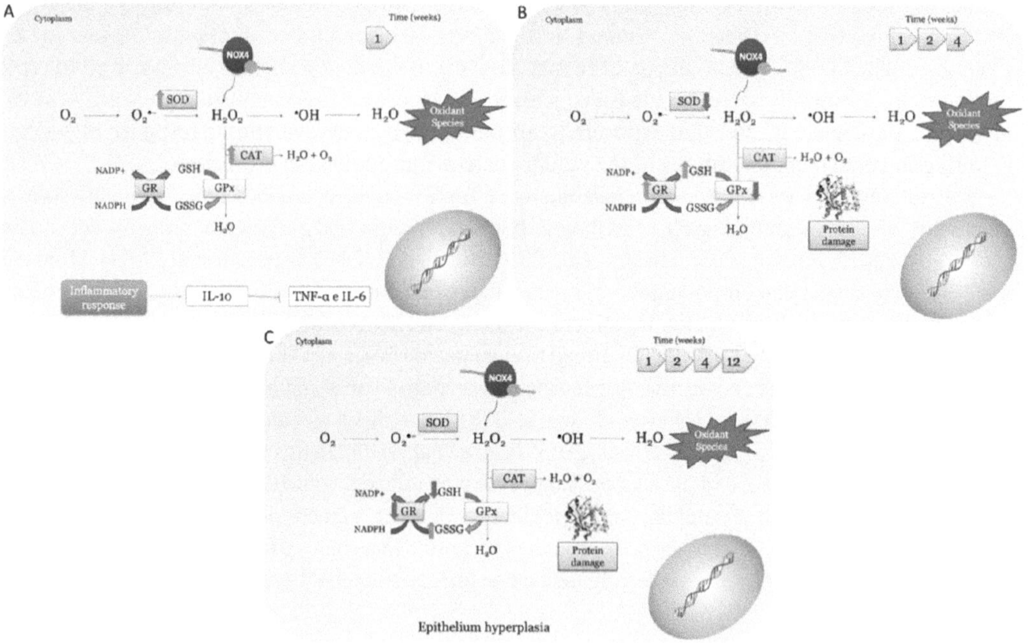

FIGURE 8.1 Time course of oxidative stress markers in the cornea after exposure to air pollutants. The expression of NADPH oxidase-4 (NOX), carbonyl content as a marker of protein oxidation, the activities of superoxide dismutase (SOD), catalase (CAT), glutathione peroxidase (GPx), and glutathione reductase (GR), as well as the levels of reduced and oxidized glutathione (GSH and GSSG, respectively) were measured in the cornea of mice exposed to urban air for (A) 1, (B) 4, and (C) 12 weeks.

increased oxidant species production has been attributed to an overactivation of NADPH-oxidase-4 (NOX4) (Lasagni Vitar et al., 2019) and alterations in mitochondrial function (Choi et al., 2018; Park et al., 2017; Xiang et al., 2016). In response, an early activation of the antioxidant defences has been reported after acute PM exposure, which could be interpreted as an adaptive mechanism to the rapid increase in ROS production (Figure 8.1A). However, after a long-term exposure, a decreased antioxidant response with an increased production of oxidant species was described, leading to redox imbalance (Figure 8.1B–C) and making the cornea even more susceptible to oxidative damage as evidenced by the high levels of oxidized proteins (Lasagni Vitar et al., 2019) and DNA damage (Gao et al., 2016). The increase in ROS and RNS production may activate different intracellular pathways such as the ones modulated by NF-kB (Tan et al., 2018; Yang et al., 2018) and the nod-like receptor family pyrin domain containing three (NLRP3) (Long et al., 2020; Niu et al., 2021b) that will initiate and/or maintain the oxidative and the inflammatory response.

Interestingly, the continuous exposure to urban air alters the histoarchitecture of the cornea, characterized by epithelium hyperplasia. In fact, the increase in the corneal epithelial thickness and disorganization has been suggested to be a consequence of PM exposure (Kang et al., 2020; Li et al., 2017; Yang et al., 2018). If these morphological changes triggered by PM remain unattended, it might result in an irreversible damage that could be vision-threatened (Del Monte & Kim, 2011).

8.5 PM EFFECTS ON THE CONJUNCTIVA

The conjunctiva is a multi-layered epithelium that extends from the mucocutaneous junction of the internal margins of the eyelid to the limbus area. The epithelium also contains goblet cells, which produce mucin, the glycoproteins in the tear film. The main functions of this epithelium are lubrication (by synthesizing mucins) and the protection of the eye (by phagocytosis of particles and bacteria and secretion of inflammatory mediators). The ability of the conjunctival epithelium to rapidly respond to infections and trauma is crucial to protect the eye from the environment, but it also makes it particularly susceptible to mechanical, toxic, and immunological injuries.

It has been reported that air pollution impacts on the conjunctiva function and morphology. In this regard, inhabitants from a highly polluted urban centre show an increase in mucin 5AC (MUC5AC) expression, which is related to conjunctival goblet cell hyperplasia (Torricelli et al., 2014). Increased expression of the conjunctival mucins MUC1, MUC5AC, and MUC16 was also described *in vitro* as a response to PM incubation (Lee et al., 2020; Tau et al., 2013), probably as an attempt to promote particle clearance. Interestingly, conjunctival epithelial hyperplasia has been only observed after long-term exposure to PM (Lasagni Vitar et al., 2019); however, a decrease in the number of goblet cells was described (Kang et al., 2020; Yang et al., 2019). These observations suggest that the morphological changes of the conjunctiva that occur in response to long-term exposure to air pollutants can affect tissue function and make it prone to develop pathological conditions such as dry eye.

Oxidative stress and inflammation are also at play in the toxic effects that follow PM exposure in the conjunctiva (Tau et al. 2013; Fujishima et al. 2013; Lasagni Vitar et al. 2015, 2018; Qian Yang et al. 2019). *In vitro* and *in vivo* studies have demonstrated that conjunctival epithelial cells respond to PM by releasing pro-inflammatory cytokines such as IL-6, IL-8, TNF-α, IL-1β, interleukin-18 (IL-18), interleukin-22 (IL-22), interleukin-23 (IL-23), and MCP-1 (Fujishima et al., 2013; Lasagni Vitar et al., 2018; Lee et al., 2020; Maglione et al., 2019; Tau et al., 2013; Yang et al., 2019). Interestingly, ROS production seems to be an earlier event triggered by exposure to PM compared to the pro-inflammatory response (Lasagni Vitar et al., 2018). In this context, increased NOX4 expression and altered mitochondrial function have been proposed as the intracellular sources of ROS in the conjunctiva (Lasagni Vitar et al., 2018).

As a response to increased ROS production, the NRF2 pathway is rapidly activated, enhancing the cellular antioxidant capacity by stimulating the expression of enzymatic and non-enzymatic antioxidants. However, after long-term exposure to PM, the epithelial cells lose the capacity to maintain a homeostatic redox balance. In this scenario, the antioxidant defences are overwhelmed, leading to oxidative damage (Lasagni Vitar et al., 2015, 2018) (Figure 8.2).

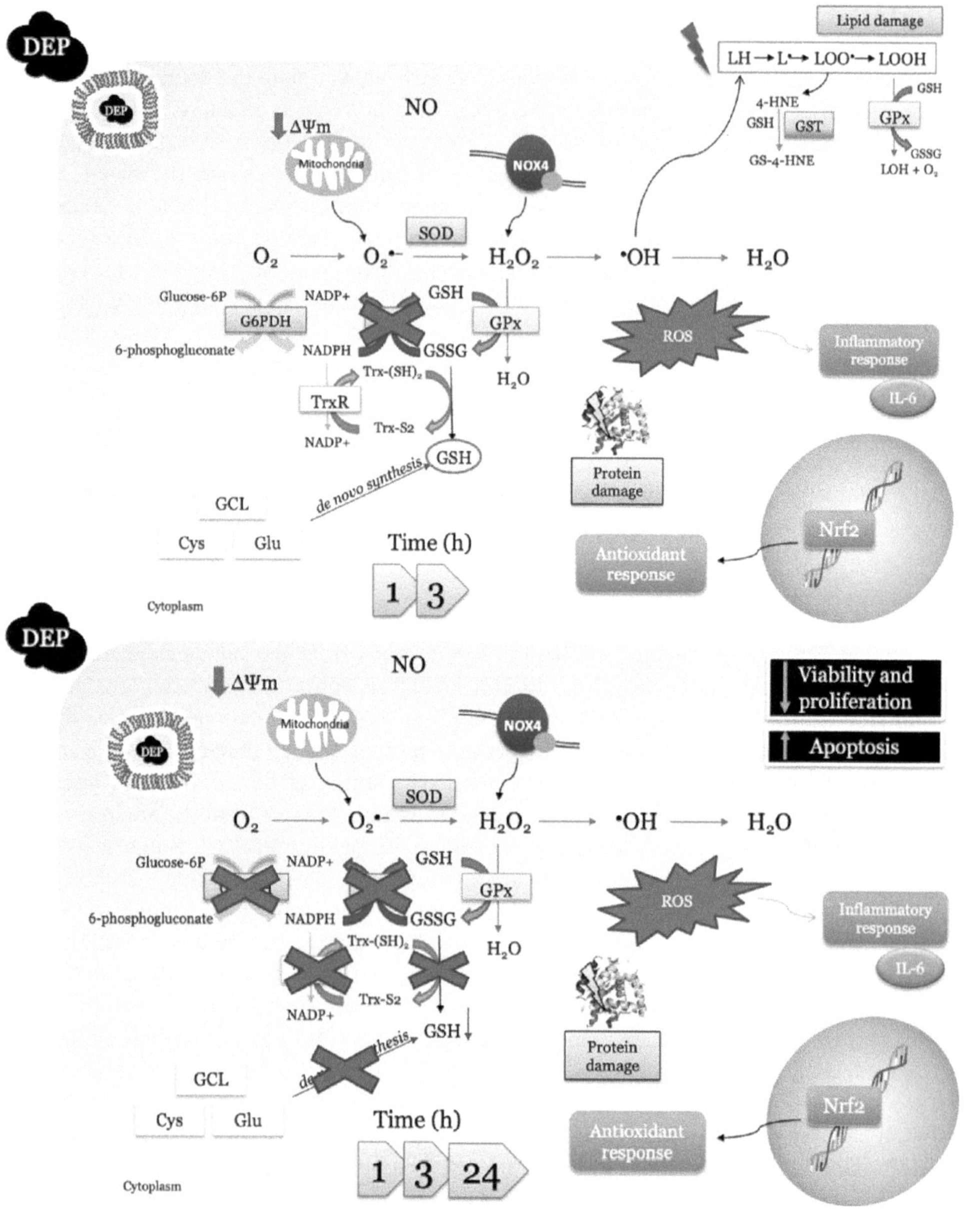

FIGURE 8.2 Time course of oxidative stress markers in human conjunctival epithelial cells after PM exposure. (A) Diesel exhaust particles (DEP, 100 μg/mL) were used as representative PM present in the urban air. At first, the oxidative environment generated principally by mitochondrial superoxide anion produced a cellular adaptive response consisting of an increase of both enzymatic and non-enzymatic antioxidants. After 3 h, NADPH oxidase-4 is presented as the main source of ROS, when the cell also starts to experiment with a proinflammatory response mediated by IL-6. At this time point, the nucleus translocation of NRF2 is stimulated, attempting to enhance the cellular antioxidant capacity. Despite the fact that under short periods of exposure to DEP lipids and then proteins are targets of oxidative damage, the viability of the cells is not affected at early stages, since cell hyperplasia was detected as compensatory mechanism. (B) Although after 24 h NRF2 pathway is still enhanced, the epithelial cell capacity to maintain redox balance is exceeded, as the antioxidant enzymes activation and the depleted GSH pool are not capable of counteracting the increased ROS production. Under these conditions, significant increase of oxidative damage, a decrease in cell proliferation viability, and apoptosis take place. Modified from "Particulate Matter Induces Tissue OxInflammation: From Mechanism to Damage". *Antioxid Redox Signal.* 2020 doi: 10.1089/ars.2019.8015

Interestingly, it has been suggested that these toxic effects can result from PM accumulation inside the cells, as conjunctival epithelia are capable to uptake PM by phagocytosis (Lasagni Vitar et al., 2015). In fact, the interaction between the toxic chemical compounds with intracellular targets could be one of the leading causes promoting oxidative stress and inflammation.

In summary, long-term exposure to air pollutants would affect the conjunctival epithelial integrity and function, which may increase the susceptibility to ocular surface affections.

8.6 CONCLUSION

Continuous exposure to air pollutants affects the ocular surface integrity and function, becoming a novel and important target when considering the health effects of air pollution. The elucidation of the cellular mechanism triggered by PM might help to understand the ocular symptoms frequently presented by the urban population. In this sense, it would be essential to consider air pollutants, especially PM, as a risk factor for the development and progression of ocular diseases that usually affect the inhabitants of highly polluted urban areas, especially those in which oxidative stress and inflammation play a key role in their pathophysiology (e.g., conjunctivitis, dry eye, keratitis).

REFERENCES

Al-Kindi, S. G., Brook, R. D., Biswal, S., & Rajagopalan, S. (2020). Environmental determinants of cardiovascular disease: Lessons learned from air pollution. *Nature Reviews Cardiology*, *17*(10), 656–672. https://doi.org/10.1038/s41569-020-0371-2

Becker, S., Soukup, J. M., & Gallagher, J. E. (2002). Differential particulate air pollution induced oxidant stress in human granulocytes, monocytes and alveolar macrophages. *Toxicology in Vitro*, *16*(3), 209–218. https://doi.org/10.1016/S0887-2333(02)00015-2

Berra, M., Galperín, G., Dawidowski, L., Tau, J., Márquez, I., & Berra, A. (2015). Impact of wildfire smoke in Buenos Aires, Argentina, on ocular surface. *Arquivos Brasileiros de Oftalmologia*, *78*(2), 110–114. https://doi.org/10.5935/0004-2749.20150028

Brook, R. D., Rajagopalan, S., Pope, C. A., Brook, J. R., Bhatnagar, A., Diez-Roux, A. V., Holguin, F., Hong, Y., Luepker, R. V., Mittleman, M. A., et al. (2010). Particulate matter air pollution and cardiovascular disease: An update to the scientific statement from the American Heart Association. *Circulation*, *121*(21), 2331–2378. https://doi.org/10.1161/CIR.0b013e3181dbece1

Cabrera, M. P., & Chihuailaf, R. H. (2011). Antioxidants and the integrity of ocular tissues. In *Veterinary Medicine International* (Vol. 2011, p. 905153). Hindawi Limited. https://doi.org/10.4061/2011/905153

Chen, Y., Mehta, G., & Vasiliou, V. (2009). Antioxidant defences in the ocular surface. *The Ocular Surface*, *7*(4), 176–185. https://doi.org/10.1016/S1542-0124(12)70185-4

Choi, Y. H., Park, S. J., Paik, H. J., Kim, M. K., Wee, W. R., & Kim, D. H. (2018). Unexpected potential protective associations between outdoor air pollution and cataracts. *Environmental Science and Pollution Research*, *25*(11), 10636–10643. https://doi.org/10.1007/s11356-018-1266-1

Del Monte, D. W., & Kim, T. (2011). Anatomy and physiology of the cornea. *Journal of Cataract and Refractive Surgery*, *37*(3), 588–598. https://doi.org/10.1016/j.jcrs.2010.12.037

Ferreira, S. M., Lerner, S. F., Brunzini, R., Evelson, P. M., & Llesuy, S. F. (2009). Antioxidant status in the aqueous humour of patients with glaucoma associated with exfoliation syndrome. *Eye*, *23*(8), 1691–1697. https://doi.org/10.1038/eye.2008.352

Ferreira, S. M., Lerner, S. F., Brunzini, R., Reides, C. G., Evelson, P. A., & Llesuy, S. F. (2010). Time course changes of oxidative stress markers in a rat experimental glaucoma model. *Investigative Opthalmology & Visual Science*, *51*(9), 4635. https://doi.org/10.1167/iovs.09-5044

Fu, Q., Mo, Z., Lyu, D., Zhang, L., Qin, Z., Tang, Q., Yin, H., Xu, P., Wu, L., Lou, X., et al. (2017, December 1). Air pollution and outpatient visits for conjunctivitis: A case-crossover study in Hangzhou, China. *Environmental Pollution*, *231*, 1344–1350. https://doi.org/10.1016/j.envpol.2017.08.109

Fujishima, H., Satake, Y., Okada, N., Kawashima, S., Matsumoto, K., & Saito, H. (2013). Effects of diesel exhaust particles on primary cultured healthy human conjunctival epithelium. *Annals of Allergy, Asthma and Immunology*, *110*(1), 39–43. https://doi.org/10.1016/j.anai.2012.10.017

Gao, Z. X., Song, X. L., Li, S. S., Lai, X. R., Yang, Y. L., Yang, G., Li, Z. J., Cui, Y. H., & Pan, H. W. (2016). Assessment of DNA damage and cell senescence in corneal epithelial cells exposed to airborne particulate matter (PM2.5) collected in Guangzhou, China. *Investigative Ophthalmology and Visual Science*, *57*(7), 3093–3102. https://doi.org/10.1167/iovs.15-18839

Gupta, S. K., Gupta, S. C., Agarwal, R., Sushma, S., Agrawal, S. S., & Saxena, R. (2007). A multicentric case-control study on the impact of air pollution on eyes in a metropolitan city of India. *Indian Journal of Occupational and Environmental Medicine*, *11*(1), 37–40. https://doi.org/10.4103/0019-5278.32463

Hwang, S. H., Choi, Y.-H., Paik, H. J., Wee, W. R., Kim, M. K., & Kim, D. H. (2016). Potential importance of ozone in the association between outdoor air pollution and dry eye disease in South Korea. *JAMA Ophthalmology*, *134*(5), 503. https://doi.org/10.1001/jamaophthalmol.2016.0139

Iles, K. E., & Forman, H. J. (2002). Macrophage signaling and respiratory burst. *Immunologic Research*, *26*(1–3), 95–105. https://doi.org/10.1385/ir:26:1-3:095

Kang, W. S., Choi, H., Jang, G., Lee, K. H., Kim, E., Kim, K. J., Jeong, G. Y., Kim, J. S., Na, C. S., & Kim, S. (2020). Long-term exposure to urban particulate matter on the ocular surface and the incidence of deleterious changes in the cornea, conjunctiva and retina in rats. *International Journal of Molecular Sciences*, *21*(14), 1–18. https://doi.org/10.3390/ijms21144976

Kashiwagi, K., & Iizuka, Y. (2020). Effect and underlying mechanisms of airborne particulate matter 2.5 (PM2.5) on cultured human corneal epithelial cells. *Scientific Reports*, *10*(1), 1–10. https://doi.org/10.1038/s41598-020-76651-9

Kim, K. E., Cho, D., & Park, H. J. (2016). Air pollution and skin diseases: Adverse effects of airborne particulate matter on various skin diseases. *Life Sciences*, *152*, 126–134. https://doi.org/10.1016/J.LFS.2016.03.039

Lasagni Vitar, R. M., Hvozda Arana, A. G., Janezic, N. S., Marchini, T., Tau, J., Martinefski, M., Tesone, A. I., Racca, L., Reides, C. G., Tripodi, V., et al. (2019). Urban air pollution induces redox imbalance and epithelium hyperplasia in mice cornea. *Toxicology and Applied Pharmacology*, *384*. https://doi.org/10.1016/j.taap.2019.114770

Lasagni Vitar, R. M., Tau, J., Janezic, N. S., Tesone, A. I., Hvozda Arana, A. G., Reides, C. G., Berra, A., Ferreira, S. M., & Llesuy, S. F. (2018). Diesel exhaust particles (DEP) induce an early redox imbalance followed by an IL-6 mediated inflammatory response on human conjunctival epithelial cells. *Experimental Eye Research*, *171*, 37–47. https://doi.org/10.1016/j.exer.2018.03.005

Lasagni Vitar, R. M., Tau, J., Reides, C. G., Berra, A., Ferreira, S. M., & Llesuy, S. F. (2015). Evaluation of oxidative stress markers in human conjunctival epithelial cells exposed to diesel exhaust particles (DEP). *Investigative Ophthalmology & Visual Science*, *56*(12), 7058–7066. https://doi.org/10.1167/iovs.15-16864

Lee, H. S., Han, S., Seo, J. W., & Jeon, K. J. (2020). Exposure to traffic-related particulate matter 2.5 triggers th2-dominant ocular immune response in a murine model. *International Journal of Environmental Research and Public Health*, *17*(8), 1–12. https://doi.org/10.3390/ijerph17082965

Lefebvre, M. A., Pham, D. M., Boussouira, B., Bernard, D., Camus, C., & Nguyen, Q. L. (2015). Evaluation of the impact of urban pollution on the quality of skin: A multicentre study in Mexico. *International Journal of Cosmetic Science*, *37*(3), 329–338. https://doi.org/10.1111/ics.12203

Li, J., Tan, G., Ding, X., Wang, Y., Wu, A., Yang, Q., Ye, L., & Shao, Y. (2017). A mouse dry eye model induced by topical administration of the air pollutant particulate matter 10. *Biomedicine and Pharmacotherapy*, *96*(October), 524–534. https://doi.org/10.1016/j.biopha.2017.10.032

Long, Q., Huang, Y., Shang, J., Li, Y., & Chen, C. (2020). Black carbon induces cytotoxicity and NLRP3 inflammasome activation in human corneal epithelial cells. *Current Eye Research*, *45*(6), 680–685. https://doi.org/10.1080/02713683.2019.1698051

Lu, S. C. (2014). Glutathione synthesis. *Biochim Biophys Acta*, *1830*(5), 3143–3153. https://doi.org/10.1016/j.bbagen.2012.09.008.GLUTATHIONE

Ma, Q. (2013). Role of NRF2 in oxidative stress and toxicity. *Annual Review of Pharmacology and Toxicology*, *53*, 401–426. https://doi.org/10.1146/annurev-pharmtox-011112-140320

Maglione, G. A., Kurtz, M. L., Orona, N. S., Astort, F., Busso, I. T., Mandalunis, P. M., Berra, A., & Tasat, D. R. (2019). Chronic exposure to urban air pollution from Buenos Aires: The ocular mucosa as an early biomarker. *Environmental Science and Pollution Research*. https://doi.org/10.1007/s11356-019-05966-6

Magnani, N. D., Marchini, T., Garcés, M., Mebert, A., Cáceres, L., Diaz, L., Desimone, M., & Evelson, P. A. (2016). Role of transition metals present in air particulate matter on lung oxygen metabolism. *International Journal of Biochemistry and Cell Biology*, *81*, 419–426. https://doi.org/10.1016/j.biocel.2016.10.009

Magnani, N. D., Marchini, T., Vanasco, V., Tasat, D. R., Alvarez, S., & Evelson, P. (2013). Reactive oxygen species produced by NADPH oxidase and mitochondrial dysfunction in lung after an acute exposure to Residual Oil Fly Ashes. *Toxicology and Applied Pharmacology*, *270*(1), 31–38. https://doi.org/10.1016/j.taap.2013.04.002

Magnani, N. D., Muresan, X. M., Belmonte, G., Cervellati, F., Sticozzi, C., Pecorelli, A., Miracco, C., Marchini, T., Evelson, P., & Valacchi, G. (2016). Skin damage mechanisms related to airborne particulate matter exposure. *Toxicological Sciences*, *149*(1), 227–236. https://doi.org/10.1093/toxsci/kfv230

Marchini, T., D'Annunzio, V., Paz, M. L., Cáceres, L., Garcés, M., Perez, V., Tasat, D., Vanasco, V., Magnani, N., Maglio, D. G., Gelpi, R. J., Alvarez, S., & Evelson, P. (2015). Selective TNF-α targeting with infliximab attenuates impaired oxygen metabolism and contractile function induced by an acute exposure to air particulate matter. *American Journal of Physiology: Heart and Circulatory Physiology*, *309*(10), H1621–H1628. https://doi.org/10.1152/ajpheart.00359.2015

Marchini, T., Magnani, N. D., D'Annunzio, V., Tasat, D., Gelpi, R. J., Alvarez, S., & Evelson, P. (2013). Impaired cardiac mitochondrial function and contractile reserve following an acute exposure to environmental particulate matter. *Biochimica et Biophysica Acta: General Subjects*, *1830*(3), 2545–2552. https://doi.org/10.1016/j.bbagen.2012.11.012

Marchini, T., Magnani, N. D., Paz, M. L., Vanasco, V., Tasat, D., González Maglio, D. H., Alvarez, S., & Evelson, P. A. (2014). Time course of systemic oxidative stress and inflammatory response induced by an acute exposure to Residual Oil Fly Ash. *Toxicology and Applied Pharmacology*, *274*(2), 274–282. https://doi.org/10.1016/j.taap.2013.11.013

Marchini, T., Zirlik, A., & Wolf, D. (2020). Pathogenic role of air pollution particulate matter in cardiometabolic disease: Evidence from mice and humans. *Antioxidants & Redox Signaling, 33*(4), 263–279. https://doi.org/10.1089/ars.2020.8096

Marrot, L. (2018). Pollution and sun exposure: A deleterious synergy: Mechanisms and opportunities for skin protection. *Current Medicinal Chemistry*, *25*(40), 5469–5486. https://doi.org/10.2174/0929867324666170918123907

Matsuda, M., Bonatti, R., Marquezini, M. V., Garcia, M. L. B., Santos, U. P., Braga, A. L. F., Alves, M. R., Saldiva, P. H. N., & Monteiro, M. L. R. (2015). Lacrimal cytokines assessment in subjects exposed to different levels of ambient air pollution in a large metropolitan area. *PLoS One*, *10*(11), e0143131. https://doi.org/10.1371/journal.pone.0143131

Mazzoli-Rocha, F., Fernandes, S., Einicker-Lamas, M., & Zin, W. A. (2010). Roles of oxidative stress in signaling and inflammation induced by particulate matter. *Cell Biology and Toxicology*, *26*(5), 481–498. https://doi.org/10.1007/s10565-010-9158-2

Michael, S., Montag, M., & Dott, W. (2013). Pro-inflammatory effects and oxidative stress in lung macrophages and epithelial cells induced by ambient particulate matter. *Environmental Pollution*, *183*, 19–29. https://doi.org/10.1016/j.envpol.2013.01.026

Nel, A. (2005). Air pollution: Related illness: Effects of particles. *Science*, *308*(5723), 804–806. https://doi.org/10.1126/science.1108752

Niu, L., Li, L., Xing, C., Luo, B., Hu, C., Song, M., Niu, J., Ruan, Y., Sun, X., & Lei, Y. (2021a). Airborne particulate matter (PM2.5) triggers cornea inflammation and pyroptosis via NLRP3 activation. *Ecotoxicology and Environmental Safety*, *207*(August 2020), 111306. https://doi.org/10.1016/j.ecoenv.2020.111306

Niu, L., Li, L., Xing, C., Luo, B., Hu, C., Song, M., Niu, J., Ruan, Y., Sun, X., & Lei, Y. (2021b). Airborne particulate matter (PM2.5) triggers cornea inflammation and pyroptosis via NLRP3 activation. *Ecotoxicology and Environmental Safety*, *207*(August 2020), 111306. https://doi.org/10.1016/j.ecoenv.2020.111306

Novaes, P., Hilário do Nascimento Saldiva, P., Matsuda, M., Macchione, M., Peres Rangel, M., Kara-José, N., & Berra, A. (2010). The effects of chronic exposure to traffic derived air pollution on the ocular surface. *Environmental Research*, *110*(4), 372–374. https://doi.org/10.1016/j.envres.2010.03.003

Nucci, P., Sacchi, M., Pichi, F., Allegri, P., Serafino, M., Dello Strologo, M., De Cillà, S., & Villani, E. (2017). Pediatric conjunctivitis and air pollution exposure: A prospective observational study. *Seminars in Ophthalmology*, *32*(4), 407–411. https://doi.org/10.3109/08820538.2015.1115088

Park, E. J., Chae, J. B., Lyu, J., Yoon, C., Kim, S., Yeom, C., Kim, Y., & Chang, J. (2017). Ambient fine particulate matters induce cell death and inflammatory response by influencing mitochondria function in human corneal epithelial cells. *Environmental Research*, *159*(August), 595–605. https://doi.org/10.1016/j.envres.2017.08.044

Pope, C. A. 3rd, Bhatnagar, A., McCracken, J. P., Abplanalp, W., Conklin, D. J., & O'Toole, T. (2016). Exposure to fine particulate air pollution is associated with endothelial injury and systemic inflammation. *Circulation Research, 119*(11), 1204–1214. https://doi.org/10.1161/CIRCRESAHA.116.309279.

Rubio, V., Valverde, M., & Rojas, E. (2010). Effects of atmospheric pollutants on the NRF2 survival pathway. *Environmental Science and Pollution Research, 17*(2), 369–382. https://doi.org/10.1007/s11356-009-0140-6

Sundar, I. K., Sellix, M. T., & Rahman, I. (2018). Redox regulation of circadian molecular clock in chronic airway diseases. *Free Radical Biology & Medicine, 119*, 121. https://doi.org/10.1016/J.FREERADBIOMED.2017.10.383

Tan, G., Li, J., Yang, Q., Wu, A., Qu, D. Y., Wang, Y., Ye, L., Bao, J., & Shao, Y. (2018). Air pollutant particulate matter 2.5 induces dry eye syndrome in mice. *Scientific Reports, 8*(1), 17828. https://doi.org/10.1038/s41598-018-36181-x

Tau, J., Novaes, P., Matsuda, M., Tasat, D. R., Saldiva, P. H., & Berra, A. (2013). Diesel exhaust particles selectively induce both proinflammatory cytokines and mucin production in cornea and conjunctiva human cell lines. *Investigative Opthalmology & Visual Science, 54*(7), 4759. https://doi.org/10.1167/iovs.12-10541

Tesone, A. I., Lasagni Vitar, R. M., Tau, J., Maglione, G. A., Llesuy, S., Tasat, D. R., & Berra, A. (2018). Volcanic ash from Puyehue-Cordón Caulle Volcanic Complex and Calbuco promote a differential response of pro-inflammatory and oxidative stress mediators on human conjunctival epithelial cells. *Environmental Research, 167*, 87–97. https://doi.org/10.1016/j.envres.2018.07.013

Torricelli, A. A. M., Matsuda, M., Novaes, P., Braga, A. L. F., Saldiva, P. H. N., Alves, M. R., & Monteiro, M. L. R. (2014). Effects of ambient levels of traffic-derived air pollution on the ocular surface: Analysis of symptoms, conjunctival goblet cell count and mucin 5AC gene expression. *Environmental Research, 131*, 59–63. https://doi.org/10.1016/j.envres.2014.02.014

Valacchi, G., Magnani, N. D., Woodby, B., Ferreira, S. M., & Evelson, P. A. (2020). Particulate matter induces tissues oxinflammation: From mechanism to damage. *Antioxidants & Redox Signaling*. https://doi.org/10.1089/ars.2019.8015

World Health Organization (WHO). (2016). Ambient air pollution: A global assessment of exposure and burden of disease. In World Health Organization (Ed.), *World Health Organization*. https://doi.org/9789241511353

Xia, T., Kovochich, M., & Nel, A. (2006). The role of reactive oxygen species and oxidative stress in mediating particulate matter injury. *Clinics in Occupational and Environmental Medicine, 5*(4), 817–836. https://doi.org/10.1016/j.coem.2006.07.005

Xiang, P., He, R. W., Han, Y. H., Sun, H. J., Cui, X. Y., & Ma, L. Q. (2016). Mechanisms of housedust-induced toxicity in primary human corneal epithelial cells: Oxidative stress, proinflammatory response and mitochondrial dysfunction. *Environment International, 89–90*, 30–37. https://doi.org/10.1016/j.envint.2016.01.008

Yang, Q., Li, K., Li, D., Zhang, Y., Liu, X., & Wu, K. (2019). Effects of fine particulate matter on the ocular surface: An in vitro and in vivo study. *Biomedicine and Pharmacotherapy, 117*, 109177. https://doi.org/10.1016/j.biopha.2019.109177

Yang, Q., Tang, L., Shen, M., Wang, Y., Wei, Y., Jeyalatha, V., Chen, P., Dong, F., Wang, G., Wu, S., et al. (2018). Effects of diesel exhaust particles on the condition of mouse ocular surface. In *Ecotoxicology and Environmental Safety* (Vol. 163, pp. 585–593). Academic Press. https://doi.org/10.1016/j.ecoenv.2018.07.101

Yoon, S., Han, S., Jeon, K. J., & Kwon, S. (2018). Effects of collected road dusts on cell viability, inflammatory response, and oxidative stress in cultured human corneal epithelial cells. *Toxicology Letters, 284*, 152–160. https://doi.org/10.1016/j.toxlet.2017.12.012

Zhang, R., Wang, G., Guo, S., Zamora, M. L., Ying, Q., Lin, Y., Wang, W., Hu, M., & Wang, Y. (2015). Formation of urban fine particulate matter. In *Chemical Reviews* (Vol. 115, Issue 10, pp. 3803–3855). *American Chemical Society*. https://doi.org/10.1021/acs.chemrev.5b00067

Section III

Other Chemical Pollutants

9 Environmental Heavy Metals, Oxidative Stress and Disease Potential

NRF2 Centered Genetic and Epigenetic Mechanisms

Xinpei Lin, Fuli Zheng, Filipe Marques Gonçalves, Yumi Abiko, Huangyuan Li, Yoshito Kumagai, and Michael Aschner

9.1 INTRODUCTION

Heavy metals are commonly termed as metallic elements with high atomic weight, which may impair living organisms at low concentrations and tend to accumulate in the food chain. A growing body of evidence has indicated strong association between heavy metals and the onset of a plethora of pathological diseases, such as neurodegenerative/psychiatric disorders and cardiovascular diseases (Cuadrado et al. 2018). Although in-depth mechanisms associated with the toxic effects of heavy metals need to be further elucidated, the generation of reactive oxygen species (ROS) and imbalance in the antioxidant defense system is one of the most predominant mechanisms. At homeostasis, free radicals/ROS are generated in a controlled manner to regulate several signaling pathways including cell division, inflammation, autophagy and stress response. However, an excess of oxidants production can damage a wide range of biochemical pathways, causing pathological consequences. Evidence from literature implies that heavy metals could induce ROS causing disruption of circadian rhythms with ensuring pathogenesis (Dai et al. 2019).

ROS are closely related to redox signaling, but can also cause oxidative stress playing, therefore, both physiological and pathophysiological roles in biology (Sies et al. 2022). ROS is a collective term for species derived from O_2 but are more reactive than O_2 itself. ROS contains a series of species, including superoxide anion radical ($O_2^{\bullet-}$), hydrogen peroxide (H_2O_2), hypochlorous acid (HOCl) and peroxynitrite/peroxynitrous acid ($ONOO^-$/ONOOH). The measurement and characterization of various types of ROS have been thoroughly summarized by Murphy et al. (2022). It is noteworthy that not only oxidative stress, but electrophilic stress is also classified as "oxidative stress" because of the modification of protein thiols in either case (Figure 9.1) (Zheng et al. 2020).

The nuclear transcription factor erythroid 2-related factor 2 (NRF2 or NFE2L2) is considered as a critical regulator of oxidative stress and a target of various intracellular signaling pathways involved in cellular responses to various toxicants (Silva-Islas and Maldonado 2018). Therefore, this chapter provides an up-to-date overview of the mechanisms involved in NRF2 pathway in response to environmental heavy metals—with considerable attention focused on NRF2 activation and its epigenetic modifications.

DOI: 10.1201/9781003328100-12

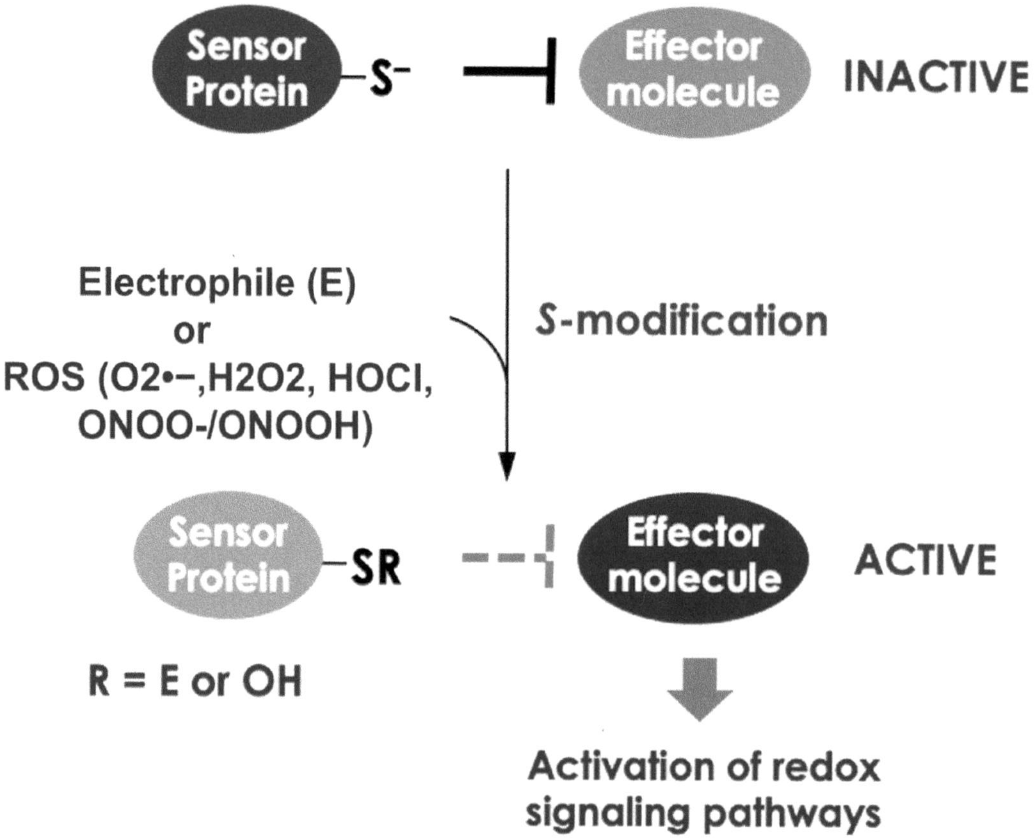

FIGURE 9.1 Scheme illustration of the activation of redox signaling pathways. Redox signaling pathways consist of sensor proteins with deprotonated thiol groups (S) and their targeted effector molecules. Under physiological condition, the effector molecules are inhibited by the sensor proteins. When electrophiles (E) or ROS occurs, the sensor proteins are suppressed by S-modification, resulting in activation of the effector molecules. This activation of effector molecules such as transcription factors or kinases leads to the activation of redox signaling pathways. Reused from Zheng et al. (2020) with permission.

9.2 ACTIVATION OF NRF2 SIGNALING PATHWAYS

NRF2 is a cytoplasmic transcription factor encoded by the *nfe2l2* gene in humans, which belongs to the cap'n'collar family for its basic leucine zipper structure. Once translocated to the nucleus, NRF2 binds to the antioxidant response element (ARE) in the DNA thereby regulating the expression of phase-II detoxifying enzymes and antioxidant proteins/enzymes such as glutamate cysteine ligase and heme oxygenase-1 (HO-1) (Wild, Moinova, and Mulcahy 1999). NRF2-knockout mice show increased vulnerability to toxicants and deteriorated pathological conditions induced by oxidative stress (Tarantini et al. 2018; Bouvier et al. 2017), while upregulated NRF2 protects against oxidative damage in both *in vitro* and *in vivo* models (Dai et al. 2019; Toyama et al. 2011). NRF2 not only acts as a master regulator in antioxidant response, but also plays an important role in regulating inflammation, autophagy and cell proliferation (Hou et al. 2018; Jiang, Harder, et al. 2015; Kahroba et al. 2019; Rojo de la Vega, Chapman, and Zhang 2018).

The sophisticated mechanism regulating NRF2-associated gene expression needs to be further elucidated. Under physiological conditions, NRF2 combines with the Kelch-like ECH-associated protein 1 (KEAP1), promoting NRF2 ubiquitination and degradation. Two KEAP1 molecules bind to one molecule of NRF2. Each of the Kelch domains in the KEAP1 binds to one of two motifs in the NRF2 allowing the attachment of ubiquitin-ligase enzymes, leading to NRF2 polyubiquitination and subsequent degradation. Upon cellular stress, the KEAP1/NRF2 association is disrupted, promoting NRF2 translocation into the nucleus where it heterodimerizes with other nuclear factors (Jun and small MAF protein) and binds to the ARE located in the regulatory regions of antioxidant response genes. After its nuclear localization, NRF2 recruits a complex transcriptional machinery to transactivate the ARE-driven genes, such as the receptor associated co-activator (RAC3) and co-regulators (CREB binding protein and protein arginine methyl-transferase etc.).

9.3 REGULATION OF NRF2 ACTIVITY

In addition to the regulatory role of KEAP1 on NRF2 activity, several intracellular signaling pathways may induce post-translational modifications in NRF2, in turn, regulating its function. The following sections will describe the main mechanisms associated with NRF2 activation.

9.3.1 KEAP1-Dependent Mechanism

The negative regulation of KEAP1 on NRF2 activation has been extensively described. The KEAP1 structure consists of an *N*-terminal region, a BTB domain, and a *C*-terminal Kelch domain as well as an intervening region (IVR) (Dinkova-Kostova, Kostov, and Canning 2017). The BTB/POZ domain maintains the basal NRF2 levels, while the IVR domain controls the interaction between KEAP1 and NRF2, and consequently NRF2 ubiquitination (Krajka-Kuzniak, Paluszczak, and Baer-Dubowska 2017).

Various activators could activate NRF2 via divergent mechanisms. KEAP1 interacts with Cul-3 through the BTB and IVR domain to negatively regulate NRF2 (Taguchi, Motohashi, and Yamamoto 2011). The reactive Cys residues of KEAP1 can sense ROS and electrophiles to alter its conformation, thereby avoiding NRF2 degradation by KEAP1-dependent pathways. Various inducers can modify specific combinations of cysteines to fine-tune the control of the KEAP1/NRF2 association and stress-sensing response.

Other intracellular processes also regulate the stability of the KEAP1/NRF2 complex. P21 completes with KEAP1 for NRF2 binding, decreasing KEAP1-dependent NRF2 ubiquitination, resulting in increased NRF2 transcriptional activity (Shan et al. 2016). P62 possesses a KEAP1-interacting region that binds to the KEAP1 domain. Phosphorylated P62 increases KEAP1 affinity and deranges the KEAP1/NRF2 complex, ultimately reducing KEAP1 levels by autophagy (Sanchez-Martin and Komatsu 2018). Moreover, NRF2 phosphorylated by protein kinase C promotes the dissociation of NRF2 from the KEAP1, increasing the transcription of ARE-driven genes (Numazawa et al. 2003). Mitogen-activated protein kinases (MAPKs) may mediate the NRF2 phosphorylation, but the specific mechanism remains to be investigated (Huang et al. 2015).

NRF2 becomes also activated in the absence of oxidative-stress. The endoplasmic reticulum stress response can induce NRF2 phosphorylation through ER-mediated protein kinases such as PKR-like endoplasmic reticulum kinase (PERK). The PERK mediated phosphorylation appears to destabilize the KEAP1/NRF2 complex to promote NRF2 activation (Cullinan et al. 2003).

9.3.2 KEAP1-Independent Mechanism

Besides the fact that NRF2 activation is negatively regulated by KEAP1, the stability and nuclear localization of NRF2 is also controlled by glycogen synthase kinase 3 (GSK3) signaling pathway. In addition, there is a positive correlation between AKT and NRF2 activation. When AKT was activated by phosphorylation, it would inhibit GSK3 to increase NRF2 activation (Culbreth, Zhang, and Aschner 2017; Wang et al. 2016).

On the one hand, GSK3 modulated the degradation of NRF2. The Neh6 domain of NRF2 contains a KEAP1-independent degradation region with a putative sequence for GSK3 phosphorylation. Therefore, GSK3 phosphorylates the Nhe6 domain at S355 and S338 creating a degradation domain that is recognized by the ubiquitin ligase adapter β-TrCP and labeled for proteasome degradation by a Cullin1/Rbx1 complex (Cuadrado 2015). On the other hand, GSK3 has been shown to indirectly modulate the nuclear localization of NRF2. GSK3 kinase activates Fyn by phosphorylation at tyrosine 216 residue or dephosphorylation at serine 9 residue (Grimes and Jope 2001). In turn, Fyn can be translocated to the nucleus and phosphorylates NRF2, thus promoting NRF2 nuclear export (Jain and Jaiswal 2007). Taken together, these findings indicate that the modulation of GSK-3 and/or Fyn might connect the activation of NRF2 after the upregulation of the AKT pathway, but additional evidence is still necessary to confirm this hypothesis.

9.4 HEAVY METALS ACTIVATE NRF2 SIGNALING PATHWAY AND ITS ASSOCIATED DISEASES

As mentioned earlier, NRF2 is a master regulator of cellular responses to fight against environmental stressors. In this section, we discuss the NRF2 pathway as it relates to the effects of heavy metals. At the same time, we summarized the diseases related to heavy metals (See **Table 9.1**).

TABLE 9.1
Environmental Heavy Metal Exposure and Associated Diseases

Exposure source	Disease	Refs
Methylmercury	Motor coordination disorders, learning and memory processing disturbances, and depression-like behaviors	(Ke et al. 2019; Albores-Garcia et al. 2016).
Manganese	Parkinsonism-like symptoms	(Jiang et al. 2017)
Cadmium	Bone disease, emphysema and hypertension	(Thevenod and Lee 2013)
Arsenic	Reproductive defects	(Li et al. 2016).
Cobalt and cobalt nanoparticles	Metallosis	(Clark et al. 2014).

9.4.1 METHYLMERCURY (MEHG)

Mercury (Hg) is a heavy metal with omnipresent occurrence in the environment. Hg released in the environment can deposit into aquatic systems and then is biomethylated to form methylmercury where it undergoes biomagnification, accumulating across the food chain (Hintelmann 2010). Thus, aquatic products contaminated with MeHg are the most relevant exposure source in humans.

MeHg is a potent electrophilic substance, possessing high affinity for thiol and selenol groups. MeHg could be readily absorbed by the gastrointestinal tract and accumulates preferably in the central nervous system (CNS). Therefore, MeHg exposure can induce neurotoxicity, such as impairment in motor coordination, disturbances in learning and memory processing and depressive-like behavior (Ke et al. 2019; Albores-Garcia et al. 2016). MeHg cytotoxicity was mainly related to glutathione (GSH) depletion, oxidative stress and mitochondrial dysfunction. Nuclear translocation of NRF2 has been hiked after exposure to MeHg both *in vivo* (Feng et al. 2017) and *in vitro* (Culbreth, Zhang, and Aschner 2017). NRF2-associated gene expressions (such as *hmox-1*, *nqo-1*, *cglc* and *nfe2l2*) are also augmented upon metal exposure. Thus, this data indicates that MeHg activates NRF2.

MeHg can increase NRF2 activation in both KEAP1-dependent and KEAP1-independent manners. The cysteine residues on KEAP1 are a crucial target for MeHg for its electrophilic properties. MeHg can directly bind to the Cys151 and Cys319 on KEAP1, in which the residues are important for the ubiquitin E3 ligase activity and consequent NRF2 degradation (Unoki et al. 2018).

On the other hand, Aschner et al. have shown that MeHg augments NRF2 activation by the downregulation of Src family kinase Fyn in primary cortical astrocytes. In detail, MeHg activates Akt, which inhibits GSK3β. Then, GSK3β could not promote the expression and nuclear import of Fyn (Culbreth, Zhang, and Aschner 2017). These results corroborate previous reports, demonstrating that MeHg-induced NRF2 activation is linked to PI3K/Akt activation (Wang et al. 2009).

9.4.2 Manganese (Mn)

Mn is an important trace element for human normal developments and functions (Underwood 1981). The enzyme, Mn-superoxide dismutase (Mn-SOD), one of the most important antioxidant defense enzymes (its genetic deletion is lethal), requires Mn (Stephenson et al. 2013). However, excess of Mn may lead to neurotoxicity, which is characterized by Parkinsonism-like symptoms with degeneration of dopaminergic neurons in the basal ganglia. The similarities and differences between idiopathic Parkinson's disease and manganism (=manganese poisoning)-induced parkinsonism have been extensively reviewed elsewhere (Guilarte and Gonzales 2015). Li et al. have demonstrated that $MnCl_2$ upregulates the level of NRF2 protein and augments the binding activity of NRF2 with ARE. This upregulation is accompanied with the promotion of HO-1 expression and the reduction in levels of reduced GSH (Li, Wu, Shi, Lian, et al. 2011). Upregulation of NRF2 has been shown in aquatic organisms and the gill of grass carp after exposure to Mn. In contrast, Mn deficiency causes inhibition of intestinal immunity, induction of inflammation and dysfunction of the intestinal physical barrier secondary to NRF2 activation in grass carp (Jiang, Tang, et al. 2015). Reproductive toxicology studies have indicated that Mn exposure suppresses NRF2 expression, unexpectedly, disrupting gonadotropin-releasing hormone (GnRH) synthesis both *in vitro* and *in vivo* (Qi et al. 2019).

Mn-induced damage can be antagonized by many chemicals. For example, pretreatment with *tert*-butylhydroquinone (tBHQ) activates the NRF2/ARE pathway and attenuates the cytotoxicity and apoptosis of $MnCl_2$ on PC12 cells (Li, Wu, Shi, Lin, et al. 2011). Similarly, melatonin, quercetin and 15-deoxy-Δ-12,14-prostaglandin J2 (15d-PGJ2) play a protective role in astrocytes against Mn-induced toxicity (inflammation and oxidative stress) by modulating the activation of NRF2/ARE signaling pathway in various organisms (Bahar, Kim, and Yoon 2017).

9.4.3 Cadmium (Cd) and Arsenic (As)

Cadmium and arsenic are widespread pollutants of great concern. Cd and As derived from cigarette smoke, rice, water or other environmental/occupational conditions may cause harm to multiple organs (Thevenod and Lee 2013; Hettick et al. 2015). Cd and As can activate NRF2 as well (Fu et al. 2010). It has been demonstrated that KEAP1 is covalently modified by Cys249, Cys368 and/or Cys613 to participate in Cd-mediated NRF2 activation. NRF2 is also related to lung injury by Cd stemming from cigarettes (Tuder and Petrache 2012). For example, the knockdown of NRF2 enhanced Cd cytotoxicity, while KEAP1 knockdown reduced the toxicity of heavy metals in vascular endothelial cells. Pretreatment with the NRF2 activator sulforaphane reduces Cd toxicity. Pretreatment with sulforaphane, a NRF2 activator, represses As cytotoxicity by increasing the expression of NRF2-related proteins, such as glycine cleavage system protein H (GCSH) and protein L (GCSL), glutathione S-transferases (GSTs) and multidrug resistance protein 1 (MRP1) in primary mouse hepatocytes (Shinkai et al. 2006). Human keratinocytes overexpressing NRF2 are hyper-resistant to arsenic toxicity, and this effect could be mitigated by stable deletion of NRF2 (Wu et al. 2017).

Curcumin and lycopene also activate the NRF2/KEAP1/ARE pathway to attenuate Cd-induced lipid peroxidation, GSH depletion and alternations in antioxidant enzymes (Chen, Huang, and Chen 2019). When cardiomyocytes were pretreated with L-ascorbic acid and α-tocopherol the toxicity induced by As_2O_3 was mostly prevented through activation of NRF2 and BCL-2 (Vineetha et al. 2018). In male mice, lutein activates the NRF2 signaling pathway, alleviating As induced reproductive defects (Li et al. 2016).

9.4.4 Cobalt (Co) and Cobalt Nanoparticles (CoNPs)

As a component of vitamin B12, cobalt (Co) is an essential metal element for humans. However, a variety of sources (including diet, occupational exposure who take part in hard metal manufacturing or from alloys used in tooth and hip joint replacements) could induce the accumulation of cobalt in the serum (Leyssens et al. 2017). In addition, excess of cobalt results in metallosis, which is characterized by sensorineural hearing loss, visual and cognitive impairment and peripheral neuropathy (Clark et al. 2014).

Co is also known as a hypoxia inducer, which is directly linked with oxidative stress and cytotoxicity in varying cell types. Previously, Zheng et al. have demonstrated that an excess of cobalt and its nanoparticles leads to defects in rat neuronal cells, including lethality, hypoxia and the loss of learning, memory and spatial-exploration abilities (Zheng et al. 2019; Akhtar et al. 2017). Tungsten carbide-cobalt particles activate NRF2 and its downstream targets such as GST and NQO-1 initiated by ROS and contribute to adverse effects in lymphoma cells JB6. Compared with CoO NPs and Co_3O_4 NPs, CoNPs (cobalt elements at nanoparticle size) have higher cytotoxicity and induce NRF2 signaling associated with oxidative stress and DNA strand breaks formation (Cappellini et al. 2018; Zheng and Li 2019). Microarray and mass spectrometry analysis have shown that cobalt exposure causes transcriptomic and proteomic alteration in rat liver cells, such as HIF-1a signaling, NRF2-mediated response, protein degradation and GSH production. In human keratinocytes, NRF2 has been shown to be involved in the antioxidant response induced by cobalt chloride. Yuan et al. have demonstrated that NRF2 overexpression in mesenchymal stem cells prevents apoptosis induced by Co, maintaining the stemness. Both thymoquinone and *mfat-1* gene expression (which cause over-production of omega-3 polyunsaturated fatty acids) showed neuroprotective effects on cobalt chloride by upregulating NRF2. Andrographolide, the main bioactive component generated from plant *Andrographis paniculate*, inhibits cobalt induced inflammation by activating NRF2/HO-1 pathway and suppressing p38 MAPK signaling (Lin et al. 2017).

In summary, heavy metals induce oxidative stress and upregulate the ROS sensor NRF2. Metal-induced defects may be rescued by further induction of NRF2 and its downstream antioxidant compounds or by the elimination of excess ROS.

9.5 EPIGENETIC REGULATIONS OF NRF2 LINKED TO CANCER AND OTHER DISEASES BY ENVIRONMENTAL CHEMICALS

Epigenetics is the process of causing heritable changes in gene expression or cell phenotype without altering the DNA sequence. Epigenetic regulations such as DNA methylation, histone modifications, microRNA (miRNA), long noncoding RNA (lncRNA) degradation and N6-methyladenosine modification (m^6A) have been emergingly reported as NRF2 modulations.

9.5.1 DNA Methylation

DNA methylation is observed in a wide range of organisms. In mammalian cells, DNA methylation is composed of DNA methyltransferases (DNMTs), which includes DNMT3a, DNMT3b and DNMT1. The ten 11 translocation (TET) enzymes can reverse DNA methylation. CpG islands are

GC-rich DNA stretches mainly found in promoter regions of genes and are predominantly non-methylated. The methylation state of CpG islands is related to the regulation in chromatin structure, genomic imprinting, transposon and chromosome X inactivation, differentiation and cancer. The function of DNA methylation is granted by proteins containing methyl-CpG binding domains (MBDs). These MBD-containing proteins appear to act as structural proteins that recruit a variety of histone deacetylase complexes and chromatin remodeling factors, leading to chromatin condensation that results in transcription inhibition.

The relationship between DNA methylation and environmental heavy metals exposure has been summarized (Martin and Fry 2018). For instance, MeHg has been shown to increase DNA methylation and DNMT1 levels both *in vivo* and *in vitro* (Go et al. 2021). Arsenic has been shown to trigger the degradation of DNMT3A by directly binding DNMT3A, leading to genome-wide DNA hypomethylation and causing deficits of recognition and spatial memory in the offspring of mice exposed to arsenic (Yan et al. 2022).

DNA methylation on both KEAP1 and NRF2 can regulate the NRF2 signaling pathway. The hypermethylation of KEAP1 promoter leads to NRF2 nuclear accumulation in solid tumors and cancers (such as lung, breast, renal cell carcinoma, colorectal cancer, head and neck cancer tissues) (Fabrizio et al. 2018). On the other hand, the hypermethylation of KEAP1 promoter is primarily playing protective roles in oxidative stress associated diseases, such as diabetes and aging. Yet, epigenetic modifications on NRF2 are investigated to a lesser extent. Yu and his colleagues have found CpG islands in the promoter region of NRF2. Moreover, the repression of NRF2 in prostate tumors is due to the hypermethylation of those CpG islands. A DNMT inhibitor could restore NRF2 expression and dissociation of MBD2 and methylated histones (Yu et al. 2010).

Higher NRF2 and NQO-1 protein levels seem to be associated with the suppression of NRF2 hypermethylation induced by γ-tocopherol—rich mixture of tocopherols (γ-TmT) in mice. Furthermore, ROS produced by 5-fluorouracil (5-FU) could activate DNA demethylases and promote the hypomethylation of NRF2 promoter, which activates the NRF2 translocation and enhances the resistance of cancer cells to 5-FU (Kang et al. 2014).

9.5.2 Histone Modifications

Four core histone proteins (which include H2A, H2B, H3 and H4) wrap up eukaryotic DNA and fold it into chromatin fibers. This highly organized protein-DNA complex consists of two configurations: concentrated heterochromatin and easily transcribed euchromatin (Huisinga, Brower-Toland, and Elgin 2006). The post-translational modifications of histone residues modify chromatin structure and further control the transcriptional activity of different genes on the chromatin. Histone modifications include histone acetylation, methylation and modification on readers.

Histone acetylation destabilizes the nucleosome structure and promotes the access of transcription factors to genetic locus, thereby activating gene transcription. Histone acetyltransferases (HATs) regulate the histone acetylation, whereas the histone deacetylases (HDACs) remove the acetyl groups. The mammalian sirtuins (SIRT1–7) belong to the group of NAD^+ dependent HDACs. SIRT1, as the most studied sirtuin, can deacetylate histone and non-histone targets, and it is therefore deeply involved in cellular processes such as apoptosis, autophagy, senescence and tumorigenesis. NRF2 deacetylated by SIRT1 can enhance the activity of KEAP1/NRF2/ARE pathway in glomerular mesangial cells under oxidative stress. Huang et al. have further shown that NRF2 positively manipulates the expression of SIRT1 protein and the activity of deacetylase, forming a positive feedback loop to suppress the development of diabetic nephropathy (DN). SRT2104, a potent SIRT1 activator, also conferred a protective effect thereby supporting the beneficial role of SIRT1. In addition to diabetic nephropathy, NRF2 activation induced by SIRT1 and its protective effects have also been found in other diseases, such as acute lung injury, optic neuritis, oocytes aging, and vascular calcification, etc. Extensive reviews have been conducted to explore the major roles of SIRT1 in response to environmental stress and DNA damage (Ong and Ramasamy 2018).

With respect to histone modifications, arsenic and cobalt were shown to upregulate H3Me2K9, H3Me3K4 and H3Me3K36, while suppressing H3MeK4 (Dik, Scheepers, and Godderis 2012). Both MeHg and Mn have been shown to increase the level of HDAC, finally decreasing histone modifications. Mn markedly inhibits the acetylation of histone H3 and H4 in SHSY5Y and PC12 cells by hiking the activity of HDAC3 and HDAC4 and inactivating HATs (Guo et al. 2018). Down-regulated histone acetylation by HAT inhibitor anacardic acid can inhibit NRF2 nuclear translocation and suppress Mn activated NRF2/HO-1 pathway. This histone deacetylation increases ROS production and decreases GSH in neurons as well. Yet, HDAC inhibition does not always contribute to the activation of NRF2, especially in airway epithelial or lung inflammatory conditions. NRF2-mediated oxidative stress responses can have epigenetic effects by conversely regulating HDAC activity (Lam et al. 2013).

EZH2, a methyltransferase involved in histone methylation, can trimethylate lysine 27 on histone H3 (H3K27Me3) in the NFR2 promoter region and promote transcriptional repression (Li et al. 2014). In addition, histone-lysine N-methyltransferase (SetD7) monomethylates histone H3 lysine 4 (H3K4me1) and increases the binding of transcription factors such as TAF10 and p53 (Ruthenburg, Allis, and Wysocka 2007). These findings indicate that histone methylation can contribute to either gene activation or suppression depending on which residue is methylated and the methylation degree.

ROS production induced by H_2O_2 can be ameliorated by the inhibition of BET (bromodomain and extraterminal) proteins, known as acetyl-lysine readers, which are negative regulators of the NRF2 signaling pathway (Michaeloudes et al. 2014). However, not much research has been done in the field of toxicology.

9.5.3 Noncoding RNAs

miRNAs are single stranded non-coding RNA molecules of approximately 22 nucleotides. miRNAs have the ability to regulate gene expression by complementarily binding to the 3′-untranslated regions (3′-UTRs) of target mRNAs. The alteration in miRNA levels can be induced by exposure to persistent chemicals, endocrine disrupting chemicals, heavy metals and pesticides. In addition to redox stress, toxicity of these environmental chemicals can also be mediated by changes in miRNAs, which participate in cellular processes by regulating the expression of target genes. With respect to noncoding RNAs, MeHg has been shown to generally downregulate miRNA expression (Culbreth and Aschner 2019), while Cd was reported to downregulate miRNA in liver and carcinogenesis and upregulate miRNA in kidney and reproductive system. Cd and As have also been shown to dysregulate miRNA expression, contributing to disease development (Ferragut Cardoso, Udoh, and States 2020), (Gu et al. 2020).

miR-28 is the first documented miRNA that negatively controls the expression of NRF2 in MCF-7 breast cancer cells (Yang et al. 2011). Subsequently other miRNAs that modulate NRF2 were found, such as miR155, miR144, miR93 miR153, miR27a, miR-142-5p miR340, miR-507, miR-634, miR-450a, miR-129-5p and miR-140-5p, alleviate oxidative stress in cisplatin-induced-acute kidney injury by activating NRF2/ARE pathway through a KEAP1-independent mechanism. miR-140-5p directly targets the 3′-UTR of NRF2 mRNA and plays a positive role in the regulation of NRF2 expression.

The direct binding of miRNAs to NRF2 mRNA not only changes its activity, but also regulates NRF2 activation through the binding of miRNAs to KEAP1 mRNA. miR-141 was the first miRNA that has the ability to promote the nuclear translocation of NFR2 in ovarian carcinoma cell lines (Van Jaarsveld et al. 2013).

Other than miRNAs, lncRNAs characterized as more than 200 nucleotides are also involved in manipulating the KEAP1/NRF2 pathway. The smoke and cancer-associated lncRNA1 (SCAL1) is the first lncRNA activated by NRF2, and it functions as a downstream mediator against oxidative stress (White et al. 2014).

9.5.4 N6-Methyladenosine Modification (m^6A)

The most prevalent RNA modification is RNA methylation, which accounts for about 60%, and four ribonucleic acid bases have been identified. Of the four ribonucleic acid bases, methylation on the sixth nitrogen of adenosine, called N6-methyladenosine (m^6A), is the most universal modification on mammalian messenger RNA and long non-coding RNA (lncRNA). m^6A plays a role in a variety of biological processes by regulating the fate of m^6A-modified RNAs through splicing, export, translation and degradation. Transcriptome analysis showed that m^6A sites were preferentially distributed in long exons, three untranslated regions (3′ UTR) and near the stop codons of mRNA and non-coding RNA in eukaryotes and some nuclear replications viruses. The dynamic modification of m^6A is controlled by "writers", "erasers" and "readers". "Writers", which include methyltransferase-like 14 (METTL14), methyltransferase-like 3 (METTL3) and Wilm's tumor 1-associated protein (WTAP), are responsible for methylating m^6A. Whereas "erasers", which contain fat mass and obesity-associated protein (FTO) and ALKB homolog 5 (ALKBH5), are able to demethylate m^6A. In addition, YT521-B homology domain-containing proteins 1 and 2 (YTHDC1 and YTHDC2) and the YT521-B homology domain family proteins 1, 2 and 3 (YTHDF1, YTHDF2, and YTHDF3) are recognized as "readers".

Exposure to environmental heavy metals can cause changes in m^6A levels. Cadmium could significantly decrease the level of m^6A in mRNA via upregulating ALKBH5 in BEAS-2B cells (Li et al. 2021). Cobalt can increase m^6A level via decreasing the activity of demethylases (Tang et al. 2020). Moreover, upregulated m^6A levels of KEAP1 and NRF2 decreases the expression of KEAP1 and increases the expression of NRF2 (Arumugam, Ghazi, and Chuturgoon 2021). Nevertheless, the interplay between NRF2 and epigenetic regulations in response to heavy metals is still at the early stage and future investigations are needed.

9.6 DISCUSSION

In summary, KEAP1 binds to NRF2 under homeostatic conditions and subsequently promotes NRF2 ubiquitination and degradation. When exposed to environmental heavy metals, oxidative stress is induced including ROS production, leading to dissociation of KEAP1 and NRF2. NRF2 then translocates from cytoplasm to nucleus and binds to ARE, followed by interaction with the partner protein small MAF (sMAF), activating the transcription of downstream detoxifying enzymes and antioxidants such as HO-1 and NQO-1. NRF2 signaling can be regulated by p21 and p62 induced KEAP1 inhibition or GSK-3 related NRF2 activation. Moreover, epigenetic modifications of NRF2 have attracted increased attention, including DNA methylation, histone acetylation, noncoding RNAs and N6-methyladenosine modification (Figure 9.2).

When placing the regulation of NRF2 to heavy metals in the aspect of clinical disorders, genes targeted by NRF2 protect cells from age-related pathologies by either neutralizing free radicals to prevent damages from occurring or fighting against the damages. The loss of NRF2 leads to neurodegenerative disorders such as PD, AD and age-related cancers (Cuadrado et al. 2018). Furthermore, the activators of NRF2 have been proposed to be a therapeutic target in AD. In diabetic cardiomyopathy and hypertension, the protective role of NRF2 is also suggested. However, NRF2 activation must occur in conjunction with caution. Researchers have also shown that hyperactivation of NRF2 and its target gene HO-1 has an adverse effect in atherosclerosis and vascular calcification owing to NRF2-mediated inflammation, which may have contributed to the adverse cardiovascular outcome by bardoxolone methyl in the BEACAN trial. Moreover, sustained activation of NRF2 and its subsequent upregulation of iron storage protein can contribute to increase cancer proliferation, resulting in therapeutic resistance (Kerins and Ooi 2018). Besides NRF2, other redox cycling activated pathways play an important role in the development and progress of diseases, such as PTP1B/EGFR, HSP90/HSF1 and the aryl hydrocarbon receptor (AHR) (Vogeley et al. 2019).

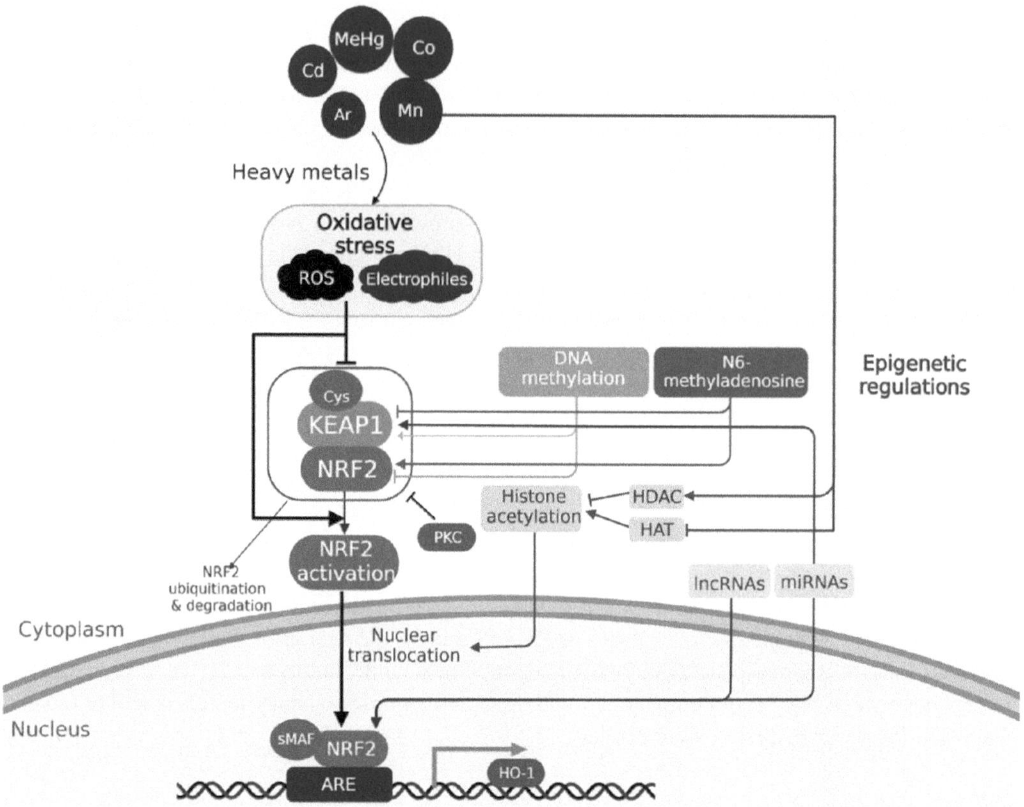

FIGURE 9.2 NRF2 signaling pathway in the context of environmental heavy metals induced redox toxicity and its regulation. Exposure of environmental heavy metals could cause oxidative stress, including ROS and electrophiles, followed by activated NRF2. This NRF2 activation can be induced by the negative regulation of KEAP1 through Cys residues or by the inhibition of KEAP1/NRF2 complex stability via p21, p62 and PKC. Upon the dissociation of KEAP1, NRF2 no longer proceeds to the ubiquitination and degradation, resulting in the translocation to the nucleus. The translocated NRF2 heterodimerizes with other nuclear factors such as sMAF and binds to AREs, driving the activation of detoxifying enzymes and antioxidants. This process is also epigenetically regulated. For instance, the hypermethylation of KEAP1 promoter results in NRF2 nuclear translocation, while the hypermethylation of NRF2 promoter represses NRF2. Similarly, upregulation of the m^6A level of KEAP1 and NRF2 decreases the expression of KEAP1 and increases the expression of NRF2. Environmental pollutants could modulate histone acetylation through histone acetyltransferases (HATs) or histone deacetylases (HDACs). Finally, miRNAs and lncRNAs also engage in this regulation process. Modified from Zheng et al. (2020) with permission.

Heavy metals induce ferroptosis via different pathways ranging from GPX4 inhibition, mitochondria dysfunction and generation of mtROS, SLC7A11 inhibition and GSH depletion to TfR-dependent iron uptake and NCOA4-dependent ferritinophagy (Figure 9.3) (Aschner et al. 2022). The interference with transition and heavy metal-induced ferroptosis pathways by NRF2-confered antioxidant protection may be a relevant feature of this defense mechanism. On one hand, the antioxidant activity of NRF2 is closely involved in ROS deactivation and GPX4 reduction (Sun et al. 2016). Vitamin D inhibits ferroptosis by increasing NRF2, and subsequent HO-1, FTH1 and GPX4 expression, so as to enhance total glutathione peroxidase (GPx) activity and reduce ROS, lipid peroxidation, MDA and iron ion level (Cheng, Huang, and Wang 2021). On the other hand, NRF2 can directly bind to the ARE sequence of the SLC7A11 promoter, increasing the expression of SLC7A11

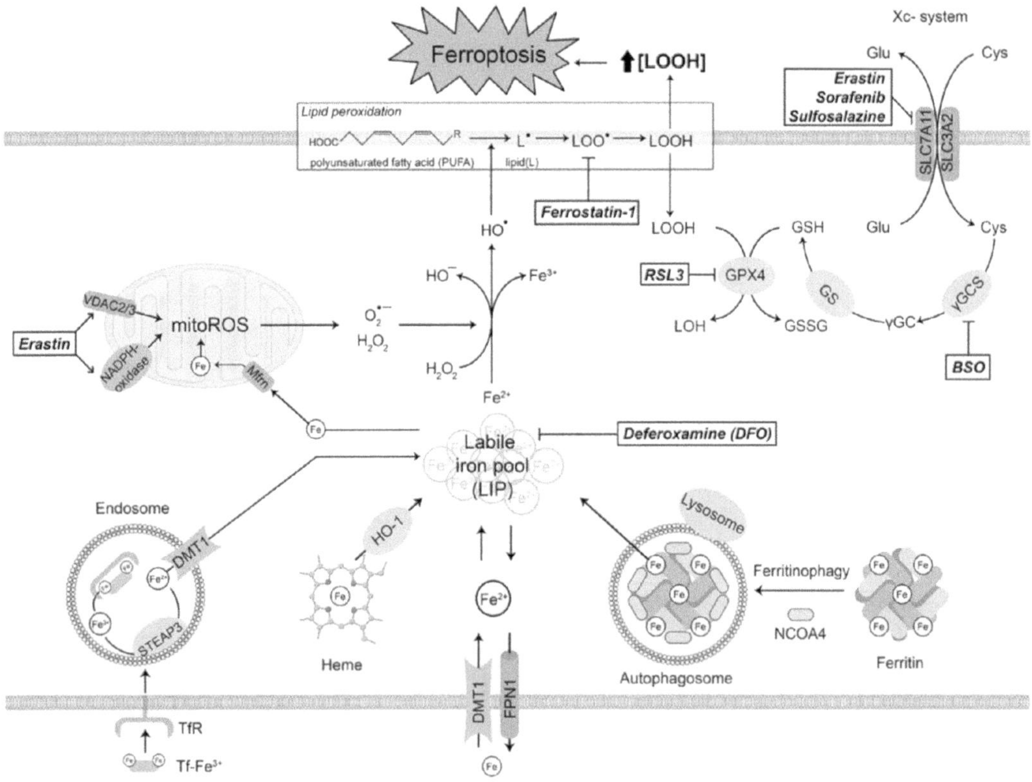

FIGURE 9.3 Impact of metal exposure on the mechanisms involved in ferroptosis. The existing data demonstrate that metal overload results in inhibition (T) of Xc-system, thus limiting GSH precursors and leading to GSH pool depletion. Dysfunction of glutathione system is also aggravated by metal-induced inhibition of GPX activity with subsequent alteration of hydroperoxide removal. At the same time, the discussed metals promote iron accumulation through up-regulation (up arrow) of ferroptosis, heme degradation and TfR-dependent iron uptake. Along with an increase in labile iron pool, metals may also possess prooxidant activity through alteration of mitochondrial function with subsequent mitoROS generation. Reused from Aschner et al. (2022) with permission.

and thus inhibiting ferroptosis (Zeng et al. 2022). Furthermore, NRF2 exports intracellular free iron by activating iron metabolism-related genes (SLC40A1) (Zeng et al. 2022).

Although NRF2 has been widely discussed and studied, further investigations will be required as additional environmental issues emerge and (epi) genetic mechanisms are characterized. For instance, the mechanism of activation of NRF2 pathway independent of oxidative stress and KEAP1 needs to be further investigated. Moreover, in-depth understanding of heavy metals associated NRF2 regulations should be addressed on both upstream and downstream of NRF2-centered signaling pathways. Finally, given that several studies have shown correlative effects between toxicants and epigenetic regulation, more research is needed to map the connected networks of cellular responses to environmental chemicals.

In conclusion, our review summarizes the activation mechanisms of NRF2 and associated signaling pathways after exposure to environmental heavy metals. This regulatory paradigm of NRF2 can be realized from both genetic and epigenetic aspects. The detrimental redox signaling activated by environmental chemicals impacts multiple clinical disorders, such as cancer, cardiovascular and neurodegenerative diseases. Nonetheless, additional research on this timely topic is needed to better characterize up-stream and down-stream regulators of the NRF2 pathway.

ACKNOWLEDGMENTS

This work is supported by the National Institutes of Health [grant numbers R01ES07331, R01ES10563], the National Natural Science Foundation of China [grant number 81903352, 81973083], the Joint Funds for the Innovation of Science and Technology, Fujian province [grant number 2019Y9020] and grant-in-aids for scientific research from the Ministry of Education, Culture, Sports, Science, and Technology of Japan [grant number 18H05293, 17K15489].

REFERENCES

Akhtar, M. J., M. Ahamed, H. A. Alhadlaq, and A. Alshamsan. 2017. "Nanotoxicity of cobalt induced by oxidant generation and glutathione depletion in MCF-7 cells." *Toxicology in Vitro* 40:94–101. doi: 10.1016/j.tiv.2016.12.012.

Albores-Garcia, D., L. C. Acosta-Saavedra, A. J. Hernandez, M. J. Loera, and E. S. Calderon-Aranda. 2016. "Early developmental low-dose methylmercury exposure alters learning and memory in periadolescent but not young adult rats." *Biomed Res Int* 2016:6532108. doi: 10.1155/2016/6532108.

Arumugam, T., T. Ghazi, and A. A. Chuturgoon. 2021. "Fumonisin B-1 alters global m6A RNA methylation and epigenetically regulates Keap1-NRF2 signaling in human hepatoma (HepG2) cells." *Archives of Toxicology* 95 (4):1367–1378. doi: 10.1007/s00204-021-02986-5.

Aschner, M., A. V. Skalny, A. C. Martins, A. I. Sinitskii, M. Farina, R. Lu, F. Barbosa, Jr., Y. G. Gluhcheva, A. Santamaria, and A. A. Tinkov. 2022. "Ferroptosis as a mechanism of non-ferrous metal toxicity." *Arch Toxicol* 96 (9):2391–2417. doi: 10.1007/s00204-022-03317-y.

Bahar, E., J.-Y. Kim, and H. Yoon. 2017. "Quercetin attenuates manganese-induced neuroinflammation by alleviating oxidative stress through regulation of apoptosis, iNOS/NF-κB and HO-1/NRF2 pathways." *International Journal of Molecular Sciences* 18 (9):1989.

Bouvier, E., F. Brouillard, J. Molet, D. Claverie, J. H. Cabungcal, N. Cresto, N. Doligez, C. Rivat, K. Q. Do, C. Bernard, J. J. Benoliel, and C. Becker. 2017. "NRF2-dependent persistent oxidative stress results in stress-induced vulnerability to depression." *Mol Psychiatry* 22 (12):1701–1713. doi: 10.1038/mp.2016.144.

Cappellini, F., Y. Hedberg, S. McCarrick, J. Hedberg, R. Derr, G. Hendriks, I. O. Wallinder, and H. L. Karlsson. 2018. "Mechanistic insight into reactivity and (geno) toxicity of well-characterized nanoparticles of cobalt metal and oxides." *Nanotoxicology* 12 (6):602–620. doi: 10.1080/17435390.2018.1470694.

Chen, D., C. Huang, and Z. Chen. 2019. "A review for the pharmacological effect of lycopene in central nervous system disorders." *Biomed Pharmacother* 111:791–801. doi: 10.1016/j.biopha.2018.12.151.

Cheng, K., Y. Huang, and C. Wang. 2021. "1,25(OH)(2)D(3) inhibited ferroptosis in Zebrafish Liver Cells (ZFL) by regulating keap1-NRF2-GPx4 and NF-κB-hepcidin axis." *Int J Mol Sci* 22 (21). doi: 10.3390/ijms222111334.

Clark, M. J., J. R. Prentice, N. Hoggard, M. N. Paley, M. Hadjivassiliou, and J. M. Wilkinson. 2014. "Brain structure and function in patients after metal-on-metal hip resurfacing." *American Journal of Neuroradiology* 35 (9):1753–1758. doi: 10.3174/ajnr.A3922.

Cuadrado, A. 2015. "Structural and functional characterization of NRF2 degradation by glycogen synthase kinase 3/beta-TrCP." *Free Radic Biol Med* 88 (Pt B):147–157. doi: 10.1016/j.freeradbiomed.2015.04.029.

Cuadrado, A., G. Manda, A. Hassan, M. J. Alcaraz, C. Barbas, A. Daiber, P. Ghezzi, R. León, M. G. López, B. Oliva, et al. 2018. "Transcription factor NRF2 as a therapeutic target for chronic diseases: A systems medicine approach." *Pharmacol Rev* 70 (2):348–383. doi: 10.1124/pr.117.014753.

Culbreth, M., and M. Aschner. 2019. "Methylmercury epigenetics." *Toxics* 7 (4). doi: 10.3390/toxics7040056.

Culbreth, M., Z. Zhang, and M. Aschner. 2017. "Methylmercury augments NRF2 activity by downregulation of the Src family kinase Fyn." *Neurotoxicology* 62:200–206. doi: 10.1016/j.neuro.2017.07.028.

Cullinan, S. B., D. Zhang, M. Hannink, E. Arvisais, R. J. Kaufman, and J. A. Diehl. 2003. "NRF2 is a direct PERK substrate and effector of PERK-dependent cell survival." *Mol Cell Biol* 23 (20):7198–7209. doi: 10.1128/mcb.23.20.7198-7209.2003.

Dai, P., D. Shen, J. Shen, Q. Tang, M. Xi, Y. Li, and C. Li. 2019. "The roles of NRF2 and autophagy in modulating inflammation mediated by TLR4: NFkappaB in A549cell exposed to layer house particulate matter 2.5 (PM2.5)." *Chemosphere* 235:1134–1145. doi: 10.1016/j.chemosphere.2019.07.002.

Dik, S., P. T. Scheepers, and L. Godderis. 2012. "Effects of environmental stressors on histone modifications and their relevance to carcinogenesis: A systematic review." *Crit Rev Toxicol* 42 (6):491–500. doi: 10.3109/10408444.2012.684146.

Dinkova-Kostova, A. T., R. V. Kostov, and P. Canning. 2017. "Keap1, the cysteine-based mammalian intracellular sensor for electrophiles and oxidants." *Arch Biochem Biophys* 617:84–93. doi: 10.1016/j.abb.2016.08.005.

Fabrizio, F. P., A. Sparaneo, D. Trombetta, and L. A. Muscarella. 2018. "Epigenetic versus genetic deregulation of the KEAP1/NRF2 axis in solid tumors: Focus on methylation and noncoding RNAs." *Oxid Med Cell Longev* 2018:2492063. doi: 10.1155/2018/2492063.

Feng, S., Z. Xu, F. Wang, T. Yang, W. Liu, Y. Deng, and B. Xu. 2017. "Sulforaphane prevents methylmercury-induced oxidative damage and excitotoxicity through activation of the NRF2-ARE pathway." *Mol Neurobiol* 54 (1):375–391. doi: 10.1007/s12035-015-9643-y.

Ferragut Cardoso, A. P., K. T. Udoh, and J. C. States. 2020. "Arsenic-induced changes in miRNA expression in cancer and other diseases." *Toxicol Appl Pharmacol* 409:115306. doi: 10.1016/j.taap.2020.115306.

Fu, J., C. G. Woods, E. Yehuda-Shnaidman, Q. Zhang, V. Wong, S. Collins, G. Sun, M. E. Andersen, and J. Pi. 2010. "Low-level arsenic impairs glucose-stimulated insulin secretion in pancreatic beta cells: Involvement of cellular adaptive response to oxidative stress." *Environmental Health Perspectives* 118 (6):864–870. doi: 10.1289/ehp.0901608.

Go, S., H. Kurita, M. Hatano, K. Matsumoto, H. Nogawa, M. Fujimura, M. Inden, and I. Hozumi. 2021. "DNA methyltransferase- and histone deacetylase-mediated epigenetic alterations induced by low-level methylmercury exposure disrupt neuronal development." *Arch Toxicol* 95 (4):1227–1239. doi: 10.1007/s00204-021-02984-7.

Grimes, C. A., and R. S. Jope. 2001. "The multifaceted roles of glycogen synthase kinase 3beta in cellular signaling." *Prog Neurobiol* 65 (4):391–426. doi: 10.1016/s0301-0082(01)00011-9.

Gu, S., J. Dai, T. Qu, and Z. He. 2020. "Emerging roles of MicroRNAs and long noncoding RNAs in cadmium toxicity." *Biol Trace Elem Res* 195 (2):481–490. doi: 10.1007/s12011-019-01859-4.

Guilarte, T. R., and K. K. Gonzales. 2015. "Manganese-induced parkinsonism is not idiopathic Parkinson's disease: Environmental and genetic evidence." *Toxicol Sci* 146 (2):204–212. doi: 10.1093/toxsci/kfv099.

Guo, Z., Z. Zhang, Q. Wang, J. Zhang, L. Wang, Q. Zhang, H. Li, and S. Wu. 2018. "Manganese chloride induces histone acetylation changes in neuronal cells: Its role in manganese-induced damage." *Neurotoxicology* 65:255–263. doi: 10.1016/j.neuro.2017.11.003.

Hettick, B. E., J. E. Canas-Carrell, A. D. French, and D. M. Klein. 2015. "Arsenic: A review of the element's toxicity, plant interactions, and potential methods of remediation." *Journal of Agricultural and Food Chemistry* 63 (32):7097–7107. doi: 10.1021/acs.jafc.5b02487.

Hintelmann, H. 2010. "Organomercurials: Their formation and pathways in the environment." *Met Ions Life Sci* 7:365–401. doi: 10.1039/BK9781847551771-00365.

Hou, Y., Y. Wang, Q. He, L. Li, H. Xie, Y. Zhao, and J. Zhao. 2018. "NRF2 inhibits NLRP3 inflammasome activation through regulating Trx1/TXNIP complex in cerebral ischemia reperfusion injury." *Behav Brain Res* 336:32–39. doi: 10.1016/j.bbr.2017.06.027.

Huang, Y., W. Li, Z. Y. Su, and A. N. Kong. 2015. "The complexity of the NRF2 pathway: Beyond the antioxidant response." *J Nutr Biochem* 26 (12):1401–1413. doi: 10.1016/j.jnutbio.2015.08.001.

Huisinga, K. L., B. Brower-Toland, and S. C. R. Elgin. 2006. "The contradictory definitions of heterochromatin: Transcription and silencing." *Chromosoma* 115 (2):110–122. doi: 10.1007/s00412-006-0052-x.

Jain, A. K., and A. K. Jaiswal. 2007. "GSK-3beta acts upstream of Fyn kinase in regulation of nuclear export and degradation of NF-E2 related factor 2." *J Biol Chem* 282 (22):16502–16510. doi: 10.1074/jbc.M611336200.

Jiang, T., B. Harder, M. Rojo de la Vega, P. K. Wong, E. Chapman, and D. D. Zhang. 2015. "p62 links autophagy and NRF2 signaling." *Free Radic Biol Med* 88 (Pt B):199–204. doi: 10.1016/j.freeradbiomed.2015.06.014.

Jiang, W.-D., R.-J. Tang, Y. Liu, S.-Y. Kuang, J. Jiang, P. Wu, J. Zhao, Y.-A. Zhang, L. Tang, and W.-N. Tang. 2015. "Manganese deficiency or excess caused the depression of intestinal immunity, induction of inflammation and dysfunction of the intestinal physical barrier, as regulated by NF-κB, TOR and NRF2 signalling, in grass carp (Ctenopharyngodon idella)." *Fish & Shellfish Immunology* 46 (2):406–416. doi: 10.1016/j.fsi.2015.06.007.

Jiang, W.-D., R.-J. Tang, Y. Liu, P. Wu, S.-Y. Kuang, J. Jiang, L. Tang, W.-N. Tang, Y.-A. Zhang, and X.-Q. Zhou. 2017. "Impairment of gill structural integrity by manganese deficiency or excess related to induction of oxidative damage, apoptosis and dysfunction of the physical barrier as regulated by NF-κB, caspase and NRF2 signaling in fish." *Fish & Shellfish Immunology* 70:280–292. doi: 10.1016/j.fsi.2017.09.022.

Kahroba, H., M. Shirmohamadi, M. S. Hejazi, and N. Samadi. 2019. "The role of NRF2 signaling in cancer stem cells: From stemness and self-renewal to tumorigenesis and chemoresistance." *Life Sci*:116986. doi: 10.1016/j.lfs.2019.116986.

Kang, K. A., M. J. Piao, K. C. Kim, H. K. Kang, W. Y. Chang, I. C. Park, Y. S. Keum, Y. J. Surh, and J. W. Hyun. 2014. "Epigenetic modification of NRF2 in 5-fluorouracil-resistant colon cancer cells: Involvement of TET-dependent DNA demethylation." *Cell Death & Disease* 5 (4):e1183. doi: 10.1038/cddis.2014.149.

Ke, T., F. M. Goncalves, C. L. Goncalves, A. A. Dos Santos, J. B. T. Rocha, M. Farina, A. Skalny, A. Tsatsakis, A. B. Bowman, and M. Aschner. 2019. "Post-translational modifications in MeHg-induced neurotoxicity." *Biochim Biophys Acta Mol Basis Dis* 1865 (8):2068–2081. doi: 10.1016/j.bbadis.2018.10.024.

Kerins, M. J., and A. Ooi. 2018. "The roles of NRF2 in modulating cellular iron homeostasis." *Antioxid Redox Signal* 29 (17):1756–1773. doi: 10.1089/ars.2017.7176.

Krajka-Kuzniak, V., J. Paluszczak, and W. Baer-Dubowska. 2017. "The NRF2-ARE signaling pathway: An update on its regulation and possible role in cancer prevention and treatment." *Pharmacol Rep* 69 (3):393–402. doi: 10.1016/j.pharep.2016.12.011.

Lam, H. C., S. M. Cloonan, A. R. Bhashyam, J. A. Haspel, A. Singh, J. F. Sathirapongsasuti, M. Cervo, H. Yao, A. L. Chung, and K. Mizumura. 2013. "Histone deacetylase 6-mediated selective autophagy regulates COPD-associated cilia dysfunction." *The Journal of Clinical Investigation* 123 (12):5212–5230. doi: 10.1172/JCI69636.

Leyssens, L., B. Vinck, C. Van Der Straeten, F. Wuyts, and L. Maes. 2017. "Cobalt toxicity in humans: A review of the potential sources and systemic health effects." *Toxicology* 387:43–56. doi: 10.1016/j.tox.2017.05.015.

Li, H., S. Wu, N. Shi, S. Lian, and W. Lin. 2011. "NRF2/HO-1 pathway activation by manganese is associated with reactive oxygen species and ubiquitin: Proteasome pathway, not MAPKs signaling." *Journal of Applied Toxicology* 31 (7):690–697.

Li, H., S. Wu, N. Shi, W. Lin, J. You, and W. Zhou. 2011. "NF-E2-related factor 2 activation in PC12 cells: Its protective role in manganese-induced damage." *Archives of Toxicology* 85 (8):901–910. doi: 10.1007/s00204-010-0625-6.

Li, L., M. Zhou, B. Chen, Q. Wang, S. Pan, Y. Hou, J. Xia, and X. Zhou. 2021. "ALKBH5 promotes cadmium-induced transformation of human bronchial epithelial cells by regulating PTEN expression in an m6A-dependent manner." *Ecotoxicol Environ Saf* 224:112686. doi: 10.1016/j.ecoenv.2021.112686.

Li, S. G., S. Z. Xu, Q. Niu, Y. S. Ding, L. J. Pang, R. L. Ma, M. X. Jing, K. Wang, X. M. Ma, and G. L. Feng. 2016. "Lutein alleviates arsenic-induced reproductive toxicity in male mice via NRF2 signaling." *Human & Experimental Toxicology* 35 (5):491–500. doi: 10.1177/0960327115595682.

Li, Z., L. Xu, N. Tang, Y. Xu, X. Ye, S. Shen, X. Niu, S. Lu, and Z. Chen. 2014. "The polycomb group protein EZH2 inhibits lung cancer cell growth by repressing the transcription factor NRF2." *FEBS Letters* 588 (17):3000–3007. doi: 10.1016/j.febslet.2014.05.057.

Lin, H.-C., S.-L. Su, C.-Y. Lu, A.-H. Lin, W.-C. Lin, C.-S. Liu, Y.-C. Yang, H.-M. Wang, C.-K. Lii, and H.-W Chen. 2017. "Andrographolide inhibits hypoxia-induced HIF-1α-driven endothelin 1 secretion by activating NRF2/HO-1 and promoting the expression of prolyl hydroxylases 2/3 in human endothelial cells." *Environmental Toxicology* 32 (3):918–930. doi: 10.1002/tox.22293.

Martin, E. M., and R. C. Fry. 2018. "Environmental influences on the epigenome: Exposure-associated DNA methylation in human populations." *Annu Rev Public Health* 39:309–333. doi: 10.1146/annurev-publhealth-040617-014629.

Michaeloudes, C., N. Mercado, C. Clarke, P. K. Bhavsar, I. M. Adcock, P. J. Barnes, and K. F. Chung. 2014. "Bromodomain and extraterminal proteins suppress NF-E2-related factor 2-mediated antioxidant gene expression." *The Journal of Immunology* 192 (10):4913–4920. doi: 10.4049/jimmunol.1301984.

Murphy, M. P., H. Bayir, V. Belousov, C. J. Chang, K. J. A. Davies, M. J. Davies, T. P. Dick, T. Finkel, H. J. Forman, Y. Janssen-Heininger, et al. 2022. "Guidelines for measuring reactive oxygen species and oxidative damage in cells and in vivo." *Nat Metab* 4 (6):651–662. doi: 10.1038/s42255-022-00591-z.

Numazawa, S., M. Ishikawa, A. Yoshida, S. Tanaka, and T. Yoshida. 2003. "Atypical protein kinase C mediates activation of NF-E2-related factor 2 in response to oxidative stress." *Am J Physiol Cell Physiol* 285 (2):C334–C342. doi: 10.1152/ajpcell.00043.2003.

Ong, A. L. C., and T. S. Ramasamy. 2018. "Role of sirtuin1-p53 regulatory axis in aging, cancer and cellular reprogramming." *Ageing Res Rev* 43:64–80. doi: 10.1016/j.arr.2018.02.004.

Qi, Z., C. Mi, F. Wu, X. Yang, Y. Sang, Y. Liu, J. Li, H. Yang, B. Xu, and W. Liu. 2019. "The effect of manganese exposure on GnRH secretion via NRF2/mGluR5/COX-2/PGE2/signaling pathway." *Toxicology and Industrial Health* 35 (3):211–227. doi: 10.1177/0748233719825720.

Rojo de la Vega, M., E. Chapman, and D. D. Zhang. 2018. "NRF2 and the hallmarks of cancer." *Cancer Cell* 34 (1):21–43. doi: 10.1016/j.ccell.2018.03.022.

Ruthenburg, A. J., C. D. Allis, and J. Wysocka. 2007. "Methylation of lysine 4 on histone H3: Intricacy of writing and reading a single epigenetic mark." *Molecular cell* 25 (1):15–30. doi: 10.1016/j.molcel.2006.12.014.

Sanchez-Martin, P., and M. Komatsu. 2018. "p62/SQSTM1: Steering the cell through health and disease." *J Cell Sci* 131 (21). doi: 10.1242/jcs.222836.

Shan, Y., Z. Wei, L. Tao, S. Wang, F. Zhang, C. Shen, H. Wu, Z. Liu, P. Zhu, A. Wang, W. Chen, and Y. Lu. 2016. "Prophylaxis of diallyl disulfide on skin carcinogenic model via p21-dependent NRF2 stabilization." *Sci Rep* 6:35676. doi: 10.1038/srep35676.

Shinkai, Y., D. Sumi, I. Fukami, T. Ishii, and Y. Kumagai. 2006. "Sulforaphane, an activator of NRF2, suppresses cellular accumulation of arsenic and its cytotoxicity in primary mouse hepatocytes." *FEBS Lett* 580 (7):1771–1774. doi: 10.1016/j.febslet.2006.02.031.

Sies, H., V. V. Belousov, N. S. Chandel, M. J. Davies, D. P. Jones, G. E. Mann, M. P. Murphy, M. Yamamoto, and C. Winterbourn. 2022. "Defining roles of specific reactive oxygen species (ROS) in cell biology and physiology." *Nat Rev Mol Cell Biol* 23 (7):499–515. doi: 10.1038/s41580-022-00456-z.

Silva-Islas, C. A., and P. D. Maldonado. 2018. "Canonical and non-canonical mechanisms of NRF2 activation." *Pharmacol Res* 134:92–99. doi: 10.1016/j.phrs.2018.06.013.

Stephenson, A. P., J. A. Schneider, B. C. Nelson, D. H. Atha, A. Jain, K. F. A. Soliman, M. Aschner, E. Mazzio, and R. R. Reams. 2013. "Manganese-induced oxidative DNA damage in neuronal SH-SY5Y cells: Attenuation of thymine base lesions by glutathione and N-acetylcysteine." *Toxicology Letters* 218 (3):299–307. doi: 10.1016/j.toxlet.2012.12.024.

Sun, X., Z. Ou, R. Chen, X. Niu, D. Chen, R. Kang, and D. Tang. 2016. "Activation of the p62-Keap1-NRF2 pathway protects against ferroptosis in hepatocellular carcinoma cells." *Hepatology* 63 (1):173–184. doi: 10.1002/hep.28251.

Taguchi, K., H. Motohashi, and M. Yamamoto. 2011. "Molecular mechanisms of the Keap1-NRF2 pathway in stress response and cancer evolution." *Genes Cells* 16 (2):123–140. doi: 10.1111/j.1365-2443.2010.01473.x.

Tang, J., C. Zheng, F. Zheng, Y. Li, Y. L. Wang, M. Aschner, Z. Guo, G. Yu, S. Wu, and H. Li. 2020. "Global N6-methyladenosine profiling of cobalt-exposed cortex and human neuroblastoma H4 cells presents epitranscriptomics alterations in neurodegenerative disease-associated genes." *Environ Pollut* 266 (Pt 2):115326. doi: 10.1016/j.envpol.2020.115326.

Tarantini, S., M. N. Valcarcel-Ares, A. Yabluchanskiy, Z. Tucsek, P. Hertelendy, T. Kiss, T. Gautam, X. A. Zhang, W. E. Sonntag, R. de Cabo, et al. 2018. "NRF2 deficiency exacerbates obesity-induced oxidative stress, neurovascular dysfunction, blood-brain barrier disruption, neuroinflammation, amyloidogenic gene expression, and cognitive decline in mice, mimicking the aging phenotype." *J Gerontol A Biol Sci Med Sci* 73 (7):853–863. doi: 10.1093/gerona/glx177.

Thevenod, F., and W. K. Lee. 2013. "Cadmium and cellular signaling cascades: Interactions between cell death and survival pathways." *Arch Toxicol* 87 (10):1743–1786. doi: 10.1007/s00204-013-1110-9.

Toyama, T., Y. Shinkai, A. Yasutake, K. Uchida, M. Yamamoto, and Y. Kumagai. 2011. "Isothiocyanates reduce mercury accumulation via an NRF2-dependent mechanism during exposure of mice to methylmercury." *Environ Health Perspect* 119 (8):1117–1122. doi: 10.1289/ehp.1003123.

Tuder, R. M., and I. Petrache. 2012. "Pathogenesis of chronic obstructive pulmonary disease." *The Journal of Clinical Investigation* 122 (8):2749–2755. doi: 10.1172/JCI60324.

Underwood, E. J. 1981. "Trace metals in human and animal health." *Journal of Human Nutrition*.

Unoki, T., M. Akiyama, Y. Kumagai, F. M. Goncalves, M. Farina, J. B. T. da Rocha, and M. Aschner. 2018. "Molecular pathways associated with methylmercury-induced NRF2 modulation." *Front Genet* 9:373. doi: 10.3389/fgene.2018.00373.

Van Jaarsveld, M. T. M., J. Helleman, A. W. M. Boersma, P. F. Van Kuijk, W. F. Van Ijcken, E. Despierre, I. Vergote, R. H. J. Mathijssen, E. M. J. J. Berns, and J. Verweij. 2013. "miR-141 regulates KEAP1 and modulates cisplatin sensitivity in ovarian cancer cells." *Oncogene* 32 (36):4284. doi: 10.1038/onc.2012.433.

Vineetha, R. C., P. Binu, P. Arathi, and R. H. Nair. 2018. "L-ascorbic acid and α-tocopherol attenuate arsenic trioxide-induced toxicity in H9c2 cardiomyocytes by the activation of NRF2 and Bcl2 transcription factors." *Toxicology Mechanisms and Methods* 28 (5):353–360. doi: 10.1080/15376516.2017.1422578.

Vogeley, C., C. Esser, T. Tuting, J. Krutmann, and T. Haarmann-Stemmann. 2019. "Role of the aryl hydrocarbon receptor in environmentally induced skin aging and skin carcinogenesis." *Int J Mol Sci* 20 (23). doi: 10.3390/ijms20236005.
Wang, L., H. Jiang, Z. Yin, M. Aschner, and J. Cai. 2009. "Methylmercury toxicity and NRF2-dependent detoxification in astrocytes." *Toxicol Sci* 107 (1):135–143. doi: 10.1093/toxsci/kfn201.
Wang, L., S. Zhang, H. Cheng, H. Lv, G. Cheng, and X. Ci. 2016. "NRF2-mediated liver protection by esculentoside a against acetaminophen toxicity through the AMPK/Akt/GSK3beta pathway." *Free Radic Biol Med* 101:401–412. doi: 10.1016/j.freeradbiomed.2016.11.009.
White, N. M., C. R. Cabanski, J. M. Silva-Fisher, H. X. Dang, R. Govindan, and C. A. Maher. 2014. "Transcriptome sequencing reveals altered long intergenic non-coding RNAs in lung cancer." *Genome Biology* 15 (8):429. doi: 10.1186/s13059-014-0429-8.
Wild, A. C., H. R. Moinova, and R. T. Mulcahy. 1999. "Regulation of gamma-glutamylcysteine synthetase subunit gene expression by the transcription factor NRF2." *J Biol Chem* 274 (47):33627–33636. doi: 10.1074/jbc.274.47.33627.
Wu, X., B. Yang, Y. Hu, R. Sun, H. Wang, J. Fu, Y. Hou, J. Pi, and Y. Xu. 2017. "NRF2 is a potential modulator of hyperresistance to arsenic toxicity in stem-like keratinocytes." *Oxidative Medicine and Cellular Longevity* 2017. doi: Artn 7417694 10.1155/2017/7417694.
Yan, N., Y. Li, Y. Xing, J. Wu, J. Li, Y. Liang, Y. Tang, Z. Wang, H. Song, H. Wang, S. Xiao, and M. Lu. 2022. "Developmental arsenic exposure impairs cognition, directly targets DNMT3A, and reduces DNA methylation." *EMBO Rep* 23 (6):e54147. doi: 10.15252/embr.202154147.
Yang, M., Y. Yao, G. Eades, Y. Zhang, and Q. Zhou. 2011. "MiR-28 regulates NRF2 expression through a Keap1-independent mechanism." *Breast Cancer Res Treat* 129 (3):983–991. doi: 10.1007/s10549-011-1604-1.
Yu, S., T. O. Khor, K. L. Cheung, W. Li, T. Y. Wu, Y. Huang, B. A. Foster, Y. W. Kan, and A. N. Kong. 2010. "NRF2 expression is regulated by epigenetic mechanisms in prostate cancer of TRAMP mice." *PLoS One* 5 (1):e8579. doi: 10.1371/journal.pone.0008579.
Zeng, F., Y. Lan, N. Wang, X. Huang, Q. Zhou, and Y. Wang. 2022. "Ferroptosis: A new therapeutic target for bladder cancer." *Front Pharmacol* 13:1043283. doi: 10.3389/fphar.2022.1043283.
Zheng, F., F. M. Gonçalves, Y. Abiko, H. Li, Y. Kumagai, and M. Aschner. 2020. "Redox toxicology of environmental chemicals causing oxidative stress." *Redox Biol* 34:101475. doi: 10.1016/j.redox.2020.101475.
Zheng, F., and H. Li. 2019. "Evaluation of NRF2 with exposure to nanoparticles." In *Nanotoxicity*, 229–246. Springer.
Zheng, F., Z. Luo, C. Zheng, J. Li, J. Zeng, H. Yang, J. Chen, Y. Jin, M. Aschner, and S. Wu. 2019. "Comparison of the neurotoxicity associated with cobalt nanoparticles and cobalt chloride in Wistar rats." *Toxicology and Applied Pharmacology* 369:90–99. doi: 10.1016/j.taap.2019.03.003.

10 Insights into the Non-Coding-RNA Regulation of Environmental Stress-Induced Disease

Veronica Miguel and Cristina Espinosa-Diez

10.1 INTRODUCTION

Environmental risk factors, defined as all the external physical, chemical and biological factors that affect human health, such as pollution, radiation, noise, land use patterns, work environment and climate change, are recognized as an important influence on mortality and severity of chronic diseases, including cancer and cardiovascular and respiratory diseases. The World Health Organization estimates that 24% of the global deaths in 2016 were attributed to the environment, amounting to 12.6 million deaths. Air pollution shifted to the fourth leading risk factor of death in 2019 (Dhimal, Neupane, and Lamichhane Dhimal 2021). One of the main mechanisms of environmental stressors to influence disease development is promoting perdurable epigenomic changes in the mammalian genome. Non-coding RNAs (ncRNAs) are epigenetic modifiers susceptible to environmental exposure and have emerged as a new paradigm in understanding gene regulation and cell differentiation. NcRNAs classification is based on their relative size and function: those shorter than 200 nucleotides (nt) are called short non-coding RNAs and include microRNAs (miRNAs), short interfering RNAs (siRNAs), small nuclear RNA (snRNA) and piwi-interacting RNAs (piRNAs), whereas those longer than 200 nt are called long non-coding RNAs (lncRNAs). While the function of piRNAs and siRNAs is related to maintaining genomic stability in the germline and anti-viral responses, multiple microRNAs and lncRNAs have been involved in the development and progression of different diseases. miRNAs post-transcriptionally regulate the expression of target mRNAs. At the same time, lncRNAs can act as positive or negative regulators of gene expression through transcriptional and post-translational regulatory mechanisms targeting chromatin structure, RNA maturation and protein synthesis/transport (Statello et al. 2021; O'Brien et al. 2018). This chapter summarizes the current understanding of the role of miRNAs and lncRNA in pathophysiological changes caused by environmental stressors.

10.2 NON-CODING-RNA BIOGENESIS AND FUNCTION

10.2.1 MICRORNAs

miRNAs comprise a large family of conserved, small ncRNAs of 19–25 nucleotides (nt) long. The human miRNAome comprises 1,917 precursor microRNAs (pre-miRNAs) and 2,693 mature miRNAs, which regulate at least 60% of protein-coding genes. Their mechanism of action is based on the Watson-Crick base-pair complementarity between miRNAs and target sequences mainly located in the 3′ untranslated region (UTR) of mRNAs, repressing their translation and/or inducing degradation. However, functional miRNA binding sites can also be found in the 5′ UTR and open reading

DOI: 10.1201/9781003328100-13

frame regions (ORFs). This feature makes it possible that a single miRNA can potentially target hundreds or thousands of mRNAs and that one mRNA can be regulated by several miRNAs, with cooperative repression achieved by binding closely spaced target sites. They can regulate crucial functions in numerous biological processes, including development, differentiation, stress response and apoptosis. MiRNAs are transcribed by RNA polymerase II (Pol II) as a transcript precursor RNA primary miRNA, and after maturation and being exported to the cytoplasm, they associate with specific mRNAs within the multiprotein complex of Argonaut (AGO) proteins, constituting the core of the RNA-induced silencing complex (RISC) (Gebert and MacRae 2019) (Figure 10.1A).

10.2.2 lncRNAs

lncRNAs are untranslated transcripts longer than 200 nt. While human lncRNAs spans are estimated to be around 127,802 transcripts, some evidence indicates that only 1,000 lncRNAs would be present at greater than one copy per cell (Statello et al. 2021; Derrien et al. 2012). Furthermore, lncRNA expression can be tissue-specific, and they present tight regulation during development. Many lncRNA are evolutionarily conserved. Although the primary sequence is poorly conserved, short functional domains are maintained due to the resemblance of their secondary and tertiary structure (Ross and Ulitsky 2022). LncRNAs are classified by genomic location as i) natural antisense transcripts, ii) bidirectional lncRNAs, iii) long-intergenic RNAs, iv) long-intronic RNAs and v) promoter-associated RNAs or enhancers (Figure 10.1B) (Ma, Bajic, and Zhang 2013). Considering their regulatory function, lncRNAs can be classified into those acting in cis, regulating nearby genes

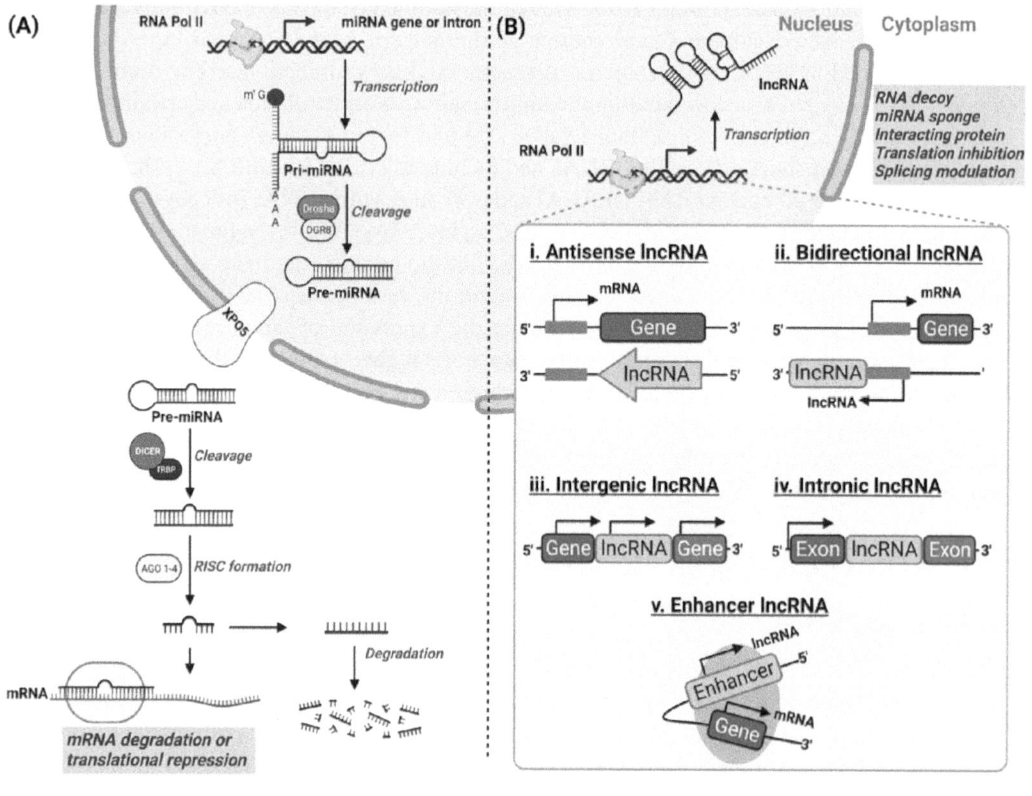

FIGURE 10.1 Overview of non-coding-RNA biogenesis and mechanisms. (A) microRNA biogenesis and functional mechanism. (B) LncRNA biogenesis and classification.

or holding different biological processes in trans, in distant regions from their genomic location (Kopp and Mendell 2018). Likewise, lncRNA can interact with protein or other RNA molecules. They can serve as a decoy for RNA-binding proteins or regulate the abundance of other ncRNA molecules like their counterpart microRNAs; these lncRNAs are known as competing endogenous RNAs (ceRNAs) (Noh et al. 2018).

10.3 ROLES OF NCRNAS IN ENVIRONMENTAL HEALTH

Long-term exposure to environmental hazards triggers molecular and cellular changes, including epigenetic alterations such as the abnormal expression of ncRNAs, which can induce developmental changes or lead to disease progression. We highlight recent advances in the miRNA and lncRNA landscape caused by ultraviolet radiation, persistent environmental chemicals, heavy metals and air pollution and discuss their impact on human health.

10.3.1 Persistent Environmental Chemicals

Long-lasting chemicals constitute a significant public health burden worldwide due to their resistance to degradation, spread capacity and bioaccumulation within the food chain. Although the manufacture and exposure to these compounds have been reduced in recent years, the current levels in the population are still associated with an elevated risk of human disease, altering the function of almost every organ system. Involvement of ncRNAs in xenobiotic-induced pathogenesis has been mostly covered for persistent organic pollutants (POP), such as polychlorinated biphenyls (PCBs), polybrominated diphenyl ethers (PBDEs), perfluoro carboxylic acids (PFCAs) and perfluorooctanesulfonate (PFOS), bisphenol A (BPA) and benzene.

10.3.1.1 Polychlorinated Biphenyls and Polybrominated Diphenyl Ethers

PCBs are a group of 209 congeners with a similar basic structure derived from biphenyl but a different chlorination pattern. Even though PCBs are no longer manufactured, their presence remains high in sediments and soil. They produce a broad spectrum of human adverse effects, including reproductive dysfunction, vascular alterations, neurodevelopmental disorders and obesity (Table 10.1) (Faroon and Ruiz 2016). In primary human endothelial cells, the commercial PCB mixture Aroclor 1260 modifies the expression of 21 miRNAs associated with vascular diseases (Wahlang et al. 2016). This mixture also deregulates miR-155 expression, which alters the immune response (Waugh, Arukwe, and Jaspers 2018). NDL-PCBs can induce neuropsychological dysfunctions by modulating miR-132 (Lesiak et al. 2014). Serum levels of some miRNAs correlate with POP exposure. For example, the levels of four tumor suppressor miRNAs, miR-193a-3p, miR-152, miR-31-5p and miR-34a-5p were downregulated in human blood after exposure to POPs (Krauskopf et al. 2017). In human peripheral blood mononuclear cells, miR-191 expression correlates with higher concentration of the dioxin-like congener PCB169 in the blood (Guida et al. 2013), while serum levels of miR-122-5p, miR-192-5p and miR-99a-5p were associated with hepatotoxicity linked to environmental PCB exposure (Cave et al. 2022).

Polybrominated diphenyl ether (PBDEs) are the bromine analogs from PCBs. They were recently banned because of their association with tumors, neurodevelopmental toxicity and thyroid hormone imbalance. BDE-209 reduces pluripotent gene expression via epigenetic regulation, including the modulation of miR-145 and miR-335, triggering apoptosis through ROS generation (Du et al. 2016). By contrast, PBDE-47 induced the overload of specific immune-related miRNAs (miR-96-5p, miR-30b-5p, miR-223-3p, miR-99-5p, miR-424-5p, miR-302a-3p, miR-130a-3p and let-7a-5p) in derived small extracellular vesicles (sEVs) from M1 macrophages, which exacerbate the LPS-induced pro-inflammatory response (Longo et al. 2021). The liver uses the gut microbiota to biotransform PBDE efficiently. The lack of gut microbiota can increase the number of PBDE-modulated lncRNA in

TABLE 10.1
Relevant Studies in Persistent Environmental Chemicals, Heavy Metals, Air Pollution and microRNAs from 2017 to Present

miRNAs	Target genes	Altered function	Cell type/Tissue/Disease	Reference(s)	
		PCBs and PBDEs			**Persistent environmental chemicals**
miR-155↓	N/A	Inflammation	Chicken fibroblast	(Waugh, Arukwe et al. 2018)	
		PFCAs and PFOS			
miR-200b↑	SIRT-1	Hepatotoxicity	Mouse liver	(Wen 2022)	
miR-16↑ **miR-22↑** **miR-30↑**	BDNF	Neurotoxicity	Neuroblastoma cells	(Guo, He et al. 2017)	
		Benzene and Bisphenol			
miR-222↑	MDM2	leukemogenesis	Lymphocytes	(Wang, Tian et al. 2021)	
miR-451↓ **miR-486↓**	N/A	Erythroid differentiation	Hematopoietic progenitor cells	(Liang, Chen et al. 2018)	
miR-27↓	VEGFB/VEGFC	Endometrial development	Endometrial stromal cell	(Reed, Babayev et al. 2018)	
miR-708↑	NNAT	Obesity	Hypothalamic neurons	(McIlwraith, Lieu et al. 2022)	
		Arsenic			**Heavy metals**
miR-129 ↑	CD200R1	Inflammation	Microglia cell lines	(Singh, Kushwaha et al. 2022)	
miR-122 ↑	PKM2	Autophagy	Primary hepatocytes	(Wang, Zhao et al. 2020)	
miR-372 ↓	LATS2	Cell proliferation and migration	Human prostate cancer cells	(Cao, Feng et al. 2017)	
miR-330 ↑	SKP2	Cell growth, motility, invasion, apoptosis	Pancreatic cancer cells	(Gao, Wang et al. 2019)	
miR-885 ↑	MTPN	Cell survival	Cholangiocarcinoma cells	(Wang, Zhang et al. 2021)	
		Cadmium			
miR-204 ↑	BCL2	Apoptosis	Primary ovarian granulosa cells	(Zhong, Liu et al. 2020)	
		$PM_{2.5}$			**Air pollution**
miR-194 ↓	DAPK1	Apoptosis	Bronchial epithelial injury	(Zhou, Zhong et al. 2018)	
miR-385 ↓	CXCL12	Inflammation	Bronchial epithelial injury	(Zhang, Zhang et al. 2020)	
miR-26a ↓	CD36	Fatty acid uptake	Non-alcoholic fatty liver disease	(Ding, Ye et al. 2019)	
miR-146 ↑	SIRT1	M1 macrophage polarization	Inflammation	(Zhong, Liao et al. 2019)	
		PM_{10}			
miR-19 ↑ **miR-614 ↑**	RORα	M1 macrophage polarization	Inflammation	(Shin, Byun et al. 2020)	

mouse liver. These lncRNA signatures could discriminate between PBDE species such as BDE-47 and BDE-99 (Li and Cui 2018).

10.3.1.2 Perfluoro Carboxylic Acids and Perfluorooctanesulfonate

Perfluorinated compounds had a wide usage as water and stain repellants. Unlikely POPs, these compounds are not lipophilic but are also very persistent by binding to proteins. Thus, they are especially toxic to the nervous and endocrine systems. The perfluoro carboxylic acid PFNA generates hepatomegaly and increases hepatic triglycerides and total cholesterol, while it enhances serum transaminases, at least partially, through the modulation of miR-34a, miR-362-3p and miR-338-3p (Table 10.1) (Wang et al. 2015). PFOS-induced ROS in hepatotoxicity correlates with miR-155-dependent activation of the NRF2 signaling pathway (Wan et al. 2016). Other miRNAs such as miR-200, miR-122-5p and miR-21-5p mediate the PFO-induced hepatotoxicity (Dong et al. 2016; Wen 2022). PFOs also increase the expression of miRNA-16, miRNA-22 and miRNA-30a-5p, which modulate brain-derived neurotrophic factor (BDNF), highlighting the role of PFOS-induced neurotoxic response (Guo et al. 2017). Increased PFOS exposure was also associated with decreased serum levels of miR-101-3p, miR-144-3p, miR-19a-3p, miR-26b-5p and miR-199a-3p, which are closely related to cardiovascular disease, Alzheimer's disease and cancer (Xu et al. 2020).

10.3.1.3 Endocrine Disruptors, Benzene and Bisphenol

Benzene, as well as some of its metabolites, have carcinogenic effects and can cause hematological disorders by leading to genetic, chromosomal or epigenetic abnormalities, and subsequently, altered cellular proliferation and differentiation (Smith 2010). Benzene alters the expression of miRNAs that control pathways involved in vascular endothelial growth factor (VEGF), transforming growth factor-beta (TGF-β) and Wnt signaling (Table 10.1). For example, benzene exposure leads to an increase of miR-222 level in lymphocytes, promoting leukemogenesis by impairing the MDM2-p53 axis (T.-S. Wang et al. 2021). Repression of miRNA-451a and miRNA-486-5p in response to benzene exposure is associated with impaired erythroid cell differentiation (Liang et al. 2018).

Benzene exposure triggers DNA methylation changes by affecting DNMT1 and DNMT3 expression, which has an impact on lncRNA expression. Thus, the lncRNA HOXA Transcript Antisense RNA, Myeloid-Specific 1 (HOTAIRM1) presents decreased expression in long-term benzene-exposed workers due to promoter hypermethylation (Zhang et al. 2020), while other lncRNAs such as Urothelial cancer-associated 1 (UCA1) are upregulated due to hypomethylation (Pan et al. 2021).

Bisphenol A (BPA) is an endocrine disruptor that acts as an agonist or antagonist in estrogen receptor signaling. It has been associated with the development of hormone-dependent cancers such as breast, prostate and ovarian (Shafei et al. 2018). In the estrogen receptor-α (ER)-positive and hormone-sensitive human MCF-7 breast cancer cell line, BPA promotes ER transcriptional activity, which alters the expression profiles of certain miRNAs, including the onco-miR, miR-21 (Tilghman et al. 2012). BPA can also lead to epigenetic modifications in the placenta that may impact the fetus. For example, BPA exposure downregulates miR-27b, affecting vascularization and angiogenesis during endometrial development by targeting *VEGFB* and *VEGFC*, while miR-134 mediates BPA-mediated disturbances in pluripotency in embryonic stem cells and embryoid bodies (Reed et al. 2018; Xiaojiao Chen et al. 2013). MiR-21a-5p promoted BPA-induced obesity *in vivo*, while BPA-induced preadipocyte differentiation is repressed by miR-21a-5p, which targets *map2k3* in the MKK3/p38/mitogen-activated protein kinase (MAPK) pathway (Xie, Song, and Li 2016). BPA controls neuropeptides directly linked to obesity, for example, by inducing the expression of miR-708-5p through an ER stress-mediated mechanism altering neuronatin (Nnat) and neuropeptide Y expression in hypothalamic neuronal models (McIlwraith, Lieu, and Belsham 2022). BPA exposure also alters blood pressure (BP). Thus, serum samples profiling reflects decreased expression of miR-30a-5p, miR-580-3p, miR-627-5p and miR-671-3p but increased miR-636 and miR-1224-3p levels associated with high BPA-induced high blood pressure (Kim, Cho, and Hong 2020).

Among persistent environmental chemicals affecting lncRNAs, BPA exposure has the strongest impact mostly due to changes in DNA methylation in imprinted regions. For example, the cluster H19/IGF2 DNA methylation status is altered in the placenta in response to BPA, inducing severe complications in fetus development (Monnier et al. 2013). It also influences liver and muscle metabolic reprogramming in sheep due to the deregulation of lncRNA signatures (Puttabyatappa et al. 2022). In breast cancer, BPA and diethylstilbestrol (DES) exposure can induce the expression of oncogenic lncRNAs such as HOTAIR *in vitro* and *in vivo*, revealing a new epigenetic mechanism for endocrine disruption in breast cancer (Bhan et al. 2014). BPA also presents adverse effects on the nervous system during development by downregulating the growth arrest-specific 5 lncRNA (GAS5), a regulator of Jun/RAS axis (Pang et al. 2018).

10.3.2 Heavy Metals

Lead, cadmium, mercury, selenium, chromium and arsenic exposure is responsible for heavy metals and metalloids toxicity. However, the observed downstream effects are diverse as they depend on many factors, including the targeted organ, exposure route, time of development, gender and dietary factors. They can disrupt proliferation, differentiation, damage-repairing processes and apoptosis, mainly by ROS generation and epigenetic alterations. While our bodies have developed mechanisms to minimize heavy metal-induced damage, such as excretion through the upregulation of cationic transporters, sulfhydryl-based scavengers (e.g., glutathione and metallothionein), reduction of metabolism and methylation, these pathways can be impaired after continuous exposure (Rüdel et al. 2020).

10.3.2.1 Arsenic

Arsenic poisoning is associated with skin lesions, a broad spectrum of cancers, cardiovascular disease or diabetes through the modulation of epigenetic changes in DNA methylation and histone modifications, also with an impact on miRNAs and lncRNA expression (Bjørklund et al. 2018). For example, miR-21 and miR-221 are negatively associated with arsenic and lead levels and have been proposed as kidney function biomarkers in the context of heavy metal exposure (Kong et al. 2012). There were significant correlations between a number of cord blood miRNAs (let-7a, miR-107, miR-126, miR-16, miR-17, miR-195, miR-20a, miR-20b, miR-454, miR-96 and miR-98) and urinary arsenic (C.-L. Chen et al. 2019). Arsenic exposure exhibits detrimental consequences during pregnancy affecting the mother and the fetus. Higher expression of miR-1290 in the placenta negatively correlated with the gestational age and birthweight under arsenic exposure (Rahman et al. 2018). Arsenic drives DNA demethylation on the miR-129-5p promoter by downregulating DNMT1 expression. This miRNA targets CD200R1, a neuroinflammation regulator (Table 10.1) (Singh et al. 2022).

The role of miRNAs in cellular processes associated with arsenic-induced cancer has gained increasing attention in recent years. For example, the miR-122/PKM2 autophagy axis protects hepatocytes from arsenite stress via the inhibition of PI3K/Akt/mTOR pathway (Yu Wang et al. 2020). Arsenic also inhibits the proliferation and migration of prostate cancer cells through the regulation of large tumor suppressor kinase 2 (LATS2) by repressing miRNA-372 (Cao, Feng, and Chen 2017). In contrast, it promotes the miR-330-mediated S-phase kinase protein 2 (Skp2) inhibition in pancreatic cancer cells, favoring its malignant features (Gao et al. 2019). However, arsenic trioxide can reverse chemotherapy resistance in cholangiocarcinoma, the second most malignant tumor of the hepatobiliary system, through miR-885-5p upregulation and myotrophin (MTPN) targeting (Yuting Wang et al. 2021).

10.3.2.2 Cadmium

Cadmium (Cd) exposure results in deregulated DNA methylation and histone modification. It is associated with liver, bone, kidney and pancreas cancer; it is a known risk factor for cardiovascular disease as well. Cd exposure-induced dysregulated expression of let-7 miRNA family members connects let-7 tumor suppressor roles and cadmium carcinogenesis (Fabbri et al. 2012). In bladder cancer patients, levels of miR-21 correlate with serum Cd concentration (Awadalla et al. 2020). Chronic exposure to Cd induces significant alterations of miRNAs in the kidney. Cd exposure of renal epithelial cells resulted in the upregulation of 38 miRNAs, most of them targeting genes involved in oxidative stress, inflammation and apoptosis (Lemaire et al. 2020). MiRNA-204-5p regulates apoptosis by targeting Bcl2 in ovarian granulosa cells exposed to cadmium (Zhong et al. 2020).

Chronic Cd exposure also triggers changes in lncRNA expression. In this context, the role of MT1DP in Cd-induced liver toxicity through inhibition of NRF2, RhoC pathway and DNA replication stress should be emphasized (Feng et al. 2022; Gao, Li, et al. 2018). Further, the lncRNA ENST00000414355 has been proposed as a biomarker for Cd exposure. Its repression prevents Cd-induced DNA damage by boosting DNA damage/repair pathways. Low-dose Cd exposure represses the lncRNA MEG3 through the activation of DNA methylation pathways and inducing cancer stem-cell-like phenotype, which can be rescued by MEG3 overexpression (Lin et al. 2021).

10.3.2.3 Mercury

Mercury exposure is occupational or by consuming contaminated seafood. In its inorganic form ($Hg2^{+}$) it has deleterious effects in the kidney, whereas the methylated form of mercury (MeHg) targets the central nervous system (CNS). In an *in vitro* model of CNS differentiation, Pallocca et al. showed that MeHg chloride treatment of co-cultured neuronal/glial cells during their differentiation promoted differential regulation of several miRNAs (miR-141, miR-196b, miR-302b, miR-367 and miR-372), whose targets were mapped to pathways involved in axonal guidance, learning and memory (Pallocca et al. 2013). A decrease in the levels of miR-575 and miR-4286 in the human cervix was reported as a response to maternal chemical exposure during pregnancy, while multiple let-7 family members were downregulated with increasing levels of Hg in human placentas. Hg exposure also modifies the profile of vesicle-miRNAs in human milk (Kupsco et al. 2022; Sanders et al. 2015).

MALAT1 responds to $HgCl_2$ and MeHg exposure and could represent a potential biomarker for mercury-induced toxicity (Zhang et al. 2020). Similar to other metals, mercury exposure influences DNA methylation changes in imprinted lncRNA regions (H19, MEG3 and PEG3), which may lead to complications in the fetal growth (Lu et al. 2018).

10.3.3 Air Pollution

Ambient air pollution describes outdoor air pollution, which is mainly a mixture of particulate matter (PM), volatile organic compounds, ozone, oxides of nitrogen and sulfur. Exposure to ambient air pollution has been associated with adverse outcomes in many organ systems, including the lung, the cardiovascular system, the liver and the central nervous system (CNS).

10.3.3.1 Particulate Matter

Particulate matter is separated into three main groups based on their size: coarse particles (PM_{10}), fine particles ($PM_{2.5}$), and ultrafine particles ($PM_{0.1}$). They have a diverse composition of organic or inorganic nature; the smaller the particle, the higher the health risk due to its higher penetrance (K.-H. Kim, Kabir, and Kabir 2015). Exposure to PM has been associated with changes in the miRNA profile, which can mediate the downstream PM-induced deleterious effects (Table 10.1) (Cheng et al. 2020). Most recent literature indicates that PM-induced lung epithelial injury is counteracted by let-7, which modulates oxidative stress through arginase 2. By contrast, it is promoted by miR-194 upregulation, which control the death-associated protein kinase 1 (DAPK1). MiR-382-5p/CXCL12/MMP9 pathway provides a new mechanism, which mediates the inflammatory response to $PM_{2.5}$ exposure (Zhang et al. 2020; Zhou et al. 2018; Song et al. 2016). PM_{10} exposure leads to deregulated expression of miR-8063, miR-4763-3p, miR-4274, miR-181a-5p, miR-6808-5p, miR-30c-5p and miR-3149 in lung epithelial cells, with an impact on apoptosis, autophagy, xenobiotic metabolism and DNA repair (Quezada-Maldonado et al. 2018). MiR-222, miR-21, miR-126-3p, miR-155 and miR-425 have been proposed as biomarkers for identifying human populations exposed to air pollution with a significant risk of developing lung cancer (Sima et al. 2021). The PM also associates with the incidence and progression of hepatocellular carcinoma (HCC). In this disease, miRNA-16 inhibits EMT induced by $PM_{2.5}$ exposure via Twist1. MiR-26a mediates lipid accumulation and inflammation induced by $PM_{2.5}$ liposoluble extracts in hepatocytes by targeting CD36. This process leads to the development of non-alcoholic fatty liver disease (NAFLD), progressively leading to cirrhosis and ultimately to HCC (Ding et al. 2019). Finally, there is evidence that PM is associated with suppressing innate immunity and decreased clearance of viruses. Hou et al. explored the associations of short-term $PM_{2.5}$, elemental carbon (EC) and PM_{10} with miRNA expression, suggesting that latent viral miRNAs are potential mediators of air pollution-associated health effects (Hou et al. 2016). MiR-146a-3p is a novel regulator of $PM_{2.5}$-exerted M1 polarization by targeting SIRT1 (Zhong et al. 2019). MiRNAs also constitute a communication mechanism between nasal epithelial cells and macrophages during PM exposure. Thus, both exosomal miR-19 and miR-614 induce an M1 macrophage phenotype by targeting RORα (Shin et al. 2020).

LncRNA are also susceptible to changes due to exposure to PM. They are also responsible for the development and progression of diseases associated with air pollution. RNAseq profiles have identified circular and lncRNA networks in rat hippocampi in response to $PM_{2.5}$ (Chen et al. 2021). Similarly, some lncRNA signatures could be potentially used as biomarkers in corneal epithelial exposure to $PM_{2.5}$ (Song et al. 2022). Upregulation of some lncRNAs after PM exposure primes the cell to a pro-tumorogenic phenotype. For example, NORAD expression is induced in epithelial cells after exposure to PM_{10}, leading to an aberrant chromosomal segregation (Santibáñez-Andrade et al. 2021). NEAT1 induces EMT in lung epithelial cells exposed to $PM_{2.5}$, driving a tumor-like phenotype (Jiang et al. 2021). Last, the lncRNA MALAT1 affects Zeb1 expression and the EMT process by downregulating miR-204, contributing to lung cancer progression (Luo et al. 2019). Aberrantly expressed long non-coding RNAs due to air pollution can also trigger congenital disabilities. DNA hypermethylation on H19 promoter correlates with increased length and weight at birth after prenatal NO_2 exposure (He et al. 2018).

10.3.4 Pesticides

Besides the need for pesticides to prevent or mitigate pests like insects, rodents or weeds, their exposure can be deleterious for humans (Damalas and Eleftherohorinos 2011). It can cause neurological disorders, reproductive problems and cancer development. Organophosphates, a group of insecticides that inhibit acetylcholinesterase, promote cognitive dysfunction in the learning and memory process. miR-132 and miR-212 have been proposed to mediate the disruption of neurotrophin-mediated cognitive processes after chlorpyrifos exposure (Lee et al. 2016). Urinary and serum miRNAs have been suggested as potential exposure biomarkers (Yuan et al. 2018). This includes some miRNAs with oncogenic function which have been proposed as biomarkers for specific types of pesticide exposure: miR-363 and miR-9 with fungicides; miR-155, miR-181 and miR-21 with insecticides; and let-7, miR-30, miR-126, miR-181 and miR-320 with herbicides (Valencia-Quintana et al. 2022). Several pesticides, including organophosphates, paraquat, rotenone and atrazine, are associated with Parkinson's disease (PD) (Aloizou et al. 2020). Brain-specific signatures result from pesticide exposure, raising the possibility of a likely link between the two. Of note, paraquat-induced miR-153 was dependent on oxidative stress, targeted NRF2 and glutathione-redox homeostasis. Further, the miR-153 level is increased in cerebrospinal fluid samples from PD patients. NRF2-regulated miR-380-3p enhanced PQ-induced neurotoxicity (Cai et al. 2019). Paraquat also produces lung toxicity through superoxide anion formation. MiRNA deregulation in paraquat-treated lung alveolar epithelial cells is related to these mechanism (Zhao et al. 2020). A decrease in miR-146a and miR-141 in the lung has been observed after paraquat exposure, enhancing the inflammatory response to this pesticide (Wu and Li 2018; Zheng et al. 2021). In serum from severe paraquat poisoning patients, decreased miR-27a and miR-219 expression correlates with their respectively inflammation-related target genes IL-10 and TLR4, suggesting their use as biomarkers (Dai et al. 2020).

The role of lncRNAs in response to pesticide exposure is not as well studied as miRNAs. Some lncRNA profiles showed 20 differentially expressed lncRNA in response to N,N-diethyl-m-toluamide (DEET) and fipronil in primary human hepatocytes *in vitro* (Mitchell et al. 2017). The lncRNAs uc.77 and 2700086A05Rik were significantly upregulated in a mouse model of paraquat-induced pulmonary fibrosis and initiated EMT by modulating Zeb2 and HOX3 (Sun et al. 2016). Paraquat also alters lncRNAs in the mouse substantia nigra through the interaction with the transcription factor NRF2 (Wang et al. 2018). However, the functional role of this lncRNA needs further exploration.

10.4 CONCLUDING REMARKS AND FUTURE PERSPECTIVES

A rapidly growing number of ncRNAs has been identified in the last three decades due to developments in genomics and bioinformatics. At the same time, significant efforts have been made to model the complexity of the ncRNAs networks and downstream regulatory mechanisms involved in environmental pollution gene regulation. However, in most cases, definitive or causative roles for these exposure-disease associations have yet to be established. Nevertheless, they have allowed us to explore selected ncRNAs as potential biomarkers of toxic exposure and therapeutic targets.

In the case of miRNAs, strategies directed forward AGO2 immunoprecipitation coupled with deep sequencing can profile the mRNAs from the RISC complex and provide non-biased information on multiple miRNA targets. Moreover, functional annotation and clustering of the targets may unveil the role of miRNA in a particular disease. The establishment of pollutant exposure disease models on cell type-specific conditional miRNA transgenic or knockout animal models would also greatly aid this research. Although a disease-specific blood miRNA signature has been identified in patients and mouse models of these exposure diseases, no definitive data have yet been translated to the clinic. This is partially due to limited knowledge about the cell type of origin and functional role of circulating miRNA and, therefore, if they are tissue/disease-specific or represent more general pathologies like inflammation. Understanding the cell specificity of miRNAs' expression is key to contextualizing whole-tissue miRNA changes and target validation, which would improve the rational selection of biomarkers.

Based on new oligonucleotide therapies under U.S. Food and Drug Administration (FDA) approval and several microRNAs-based therapies clinical trials, we have attempted to summarize the most favorable cases of microRNAs as potential therapeutic agents. Among them, Mesomir is a promising drug for malignant pleural mesothelioma, a lethal disease that lacks effective treatment. It is an intravenously administered Epidermal Growth Factor Receptor-targeted, EnGeneIC Delivery Vehicle (EDV)-packaged, miR-16 mimic in Phase I clinical trial. Results so far have shown an acceptable safety profile. Further, signs of activity support additional studies in combination with chemotherapy or immune checkpoint inhibitors (van Zandwijk et al. 2017). Still, the concern with mesothelioma is the latency state after asbestos exposure, hence the need to identify a good combination of ncRNA biomarkers.

Our understanding of ncRNA biology and its impact on gene expression regulation has grown exponentially. Depending on their topology, interactions and cell types, these molecules may adopt different roles. lncRNA structure transforms RNA molecules into modifiers of gene expression and molecular function. They are versatile molecules that bind either chromatin, nucleic acids or proteins. Moreover, some lncRNAs also encoded functional micropeptides amplifying their regulatory roles. While these molecules have demonstrated essential functions in development and are highly conserved across mammals, their expression is low within the cells and tightly regulated. Fluctuations of lncRNAs due to exposure to environmental toxins in utero or germ line can lead to dramatic developmental and growth defects in the fetus.

A long road lies ahead to clarify the benefits of lncRNA as a potential therapeutic target or biomarker. First, most of the studies and lncRNA screening described here have been done *in vitro*, with some exceptions *in vivo* and few patient studies, lacking more profound functional validation.

Second, even though we described highly conserved lncRNA among mammals due to the different number of mechanisms in several cell lines, it still needs to be determined whether external overexpression or inhibition would be entirely beneficial in pathological progression. However, in-depth knowledge of the downstream mechanisms of these lncRNAs may help to design new therapeutic strategies to prevent and control the deleterious consequences of toxic environmental stressors. Therefore, complementary experimental and population-based epidemiological studies are warranted to clarify lncRNAs' role in the interaction of genetic and environmental factors underpinning the progression of many common diseases.

ACKNOWLEDGMENTS

V.M. is supported by a FEBS Long-Term Fellowship. C.E-D. is supported by a training grant from NHLBI 5T32HL129964-07.

Figures were created with Biorender.com.

REFERENCES

Aloizou, Athina-Maria, Vasileios Siokas, Efstathia-Maria Sapouni, Nikoleta Sita, Ioannis Liampas, Alexandros G. Brotis, Valerii N. Rakitskii, et al. 2020. "Parkinson's Disease and Pesticides: Are MicroRNAs the Missing Link?" *The Science of the Total Environment* 744 (November): 140591. https://doi.org/10.1016/j.scitotenv.2020.140591.

Awadalla, Amira, Wael I. Mortada, Hassan Abol-Enein, and Ahmed A. Shokeir. 2020. "Correlation between Blood Levels of Cadmium and Lead and the Expression of MicroRNA-21 in Egyptian Bladder Cancer Patients." *Heliyon* 6 (12): e05642. https://doi.org/10.1016/j.heliyon.2020.e05642.

Bhan, Arunoday, Imran Hussain, Khairul I. Ansari, Samara A. M. Bobzean, Linda I. Perrotti, and Subhrangsu S. Mandal. 2014. "Bisphenol-A and Diethylstilbestrol Exposure Induces the Expression of Breast Cancer Associated Long Noncoding RNA HOTAIR in Vitro and in Vivo." *The Journal of Steroid Biochemistry and Molecular Biology* 141 (May): 160–170. https://doi.org/10.1016/j.jsbmb.2014.02.002.

Bjørklund, Geir, Jan Aaseth, Salvatore Chirumbolo, Mauricio A. Urbina, and Riaz Uddin. 2018. "Effects of Arsenic Toxicity beyond Epigenetic Modifications." *Environmental Geochemistry and Health* 40 (3): 955–965. https://doi.org/10.1007/s10653-017-9967-9.

Cai, Zhipeng, Fuli Zheng, Yan Ding, Yanting Zhan, Ruijie Gong, Jing Li, Michael Aschner, Qunwei Zhang, Siying Wu, and Huangyuan Li. 2019. "NRF2-Regulated MiR-380-3p Blocks the Translation of Sp3 Protein and Its Mediation of Paraquat-Induced Toxicity in Mouse Neuroblastoma N2a Cells." *Toxicological Sciences* 171 (2): 515–529. https://doi.org/10.1093/toxsci/kfz162.

Cao, Hongwen, Yigeng Feng, and Lei Chen. 2017. "Repression of MicroRNA-372 by Arsenic Sulphide Inhibits Prostate Cancer Cell Proliferation and Migration through Regulation of Large Tumour Suppressor Kinase 2." *Basic & Clinical Pharmacology & Toxicology* 120 (3): 256–263. https://doi.org/10.1111/bcpt.12687.

Cave, Matthew C., Christina M. Pinkston, Shesh N. Rai, Banrida Wahlang, Marian Pavuk, Kimberly Z. Head, Gleta K. Carswell, et al. 2022. "Circulating MicroRNAs, Polychlorinated Biphenyls, and Environmental Liver Disease in the Anniston Community Health Survey." *Environmental Health Perspectives* 130 (1): 17003. https://doi.org/10.1289/EHP9467.

Chen, Chien-Lung, Chen-Huan Lin, An-Lun Li, Chiu-Ching Huang, Biing-Yir Shen, Yun-Ru Chiang, Pei-Luen Fang, et al. 2019. "Plasma MiRNA Profile Is a Biomarker Associated with Urothelial Carcinoma in Chronic Hemodialysis Patients." *American Journal of Physiology. Renal Physiology* 316 (6): F1094–F1102. https://doi.org/10.1152/ajprenal.00014.2019.

Chen, Xiaojiao, Bo Xu, Xiumei Han, Zhilei Mao, Prue Talbot, Minjian Chen, Guizhen Du, et al. 2013. "Effect of Bisphenol A on Pluripotency of Mouse Embryonic Stem Cells and Differentiation Capacity in Mouse Embryoid Bodies." *Toxicology in Vitro : An International Journal Published in Association with BIBRA* 27 (September). https://doi.org/10.1016/j.tiv.2013.09.018.

Chen, Xuewei, Bencheng Lin, Mingzhu Luo, Wenbin Chu, Ping Li, Huanliang Liu, Zhuge Xi, and Rong Fan. 2021. "Identifying CircRNA- and LncRNA-Associated-CeRNA Networks in the Hippocampi of Rats Exposed to PM2.5 Using RNA-Seq Analysis." *Genomics* 113 (1, Part 1): 193–204. https://doi.org/10.1016/j.ygeno.2020.12.025.

Cheng, Man, Bin Wang, Meng Yang, Jixuan Ma, Zi Ye, Li Xie, Min Zhou, and Weihong Chen. 2020. "MicroRNAs Expression in Relation to Particulate Matter Exposure: A Systematic Review." *Environmental Pollution (Barking, Essex: 1987)* 260 (May): 113961. https://doi.org/10.1016/j.envpol.2020.113961.

Dai, Huizhen, Huan Zhang, Xinxia Zu, Hongyan Wang, and Lili Wang. 2020. "Clinical Significance of Serum Levels of MicroRNA-27a and Its Correlation with Interleukin-10 in Patients with Multiple Organ Dysfunction Syndrome Caused by Acute Paraquat Poisoning." *Experimental and Therapeutic Medicine* 19 (3): 1961–1968. https://doi.org/10.3892/etm.2020.8409.

Dai, Yunxiang, Xia Liu, and Yuming Gao. 2020. "Aberrant MiR-219-5p Is Correlated with TLR4 and Serves as a Novel Biomarker in Patients with Multiple Organ Dysfunction Syndrome Caused by Acute Paraquat Poisoning." *International Journal of Immunopathology and Pharmacology* 34 (November): 2058738420974888. https://doi.org/10.1177/2058738420974888.

Damalas, Christos A., and Ilias G. Eleftherohorinos. 2011. "Pesticide Exposure, Safety Issues, and Risk Assessment Indicators." *International Journal of Environmental Research and Public Health* 8 (5): 1402–1419. https://doi.org/10.3390/ijerph8051402.

Derrien, Thomas, Rory Johnson, Giovanni Bussotti, Andrea Tanzer, Sarah Djebali, Hagen Tilgner, Gregory Guernec, et al. 2012. "The GENCODE v7 Catalog of Human Long Noncoding RNAs: Analysis of Their Gene Structure, Evolution, and Expression." *Genome Research* 22 (9): 1775–1789. https://doi.org/10.1101/gr.132159.111.

Dhimal, Meghnath, Tamanna Neupane, and Mandira Lamichhane Dhimal. 2021. "Understanding Linkages between Environmental Risk Factors and Noncommunicable Diseases: A Review." *FASEB BioAdvances* 3 (5): 287–294. https://doi.org/10.1096/fba.2020-00119.

Ding, Dongxiao, Guozhu Ye, Yi Lin, Yanyang Lu, Han Zhang, Xu Zhang, Zhenyu Hong, et al. 2019. "MicroRNA-26a-CD36 Signaling Pathway: Pivotal Role in Lipid Accumulation in Hepatocytes Induced by $PM_{2.5}$ Liposoluble Extracts." *Environmental Pollution (Barking, Essex* 248 (May): 269–278. https://doi.org/10.1016/j.envpol.2019.01.112.

Dong, Hongyan, Ivan Curran, Andrew Williams, Genevieve Bondy, Carole L. Yauk, and Michael G. Wade. 2016. "Hepatic MiRNA Profiles and Thyroid Hormone Homeostasis in Rats Exposed to Dietary Potassium Perfluorooctanesulfonate (PFOS)." *Environmental Toxicology and Pharmacology* 41 (January): 201–210. https://doi.org/10.1016/j.etap.2015.12.009.

Du, Lili, Wen Sun, Huili Zhang, and Dunjin Chen. 2016. "BDE-209 Inhibits Pluripotent Genes Expression and Induces Apoptosis in Human Embryonic Stem Cells." *Journal of Applied Toxicology: JAT* 36 (5): 659–668. https://doi.org/10.1002/jat.3195.

Fabbri, Marco, Chiara Urani, Maria Grazia Sacco, Claudio Procaccianti, and Laura Gribaldo. 2012. "Whole Genome Analysis and MicroRNAs Regulation in HepG2 Cells Exposed to Cadmium." *ALTEX—Alternatives to Animal Experimentation* 29 (2): 173–182. https://doi.org/10.14573/altex.2012.2.173.

Faroon, Obaid, and Patricia Ruiz. 2016. "Polychlorinated Biphenyls: New Evidence from the Last Decade." *Toxicology and Industrial Health* 32 (11): 1825–1847. https://doi.org/10.1177/0748233715587849.

Feng, Wenya, Zijuan Qi, Zheng Dong, Wei Liu, Ming Xu, Ming Gao, and Sijin Liu. 2022. "LncRNA MT1DP Promotes Cadmium-Induced DNA Replication Stress by Inhibiting Chromatin Recruitment of SMARCAL1." *Science of The Total Environment* 807 (February): 151078. https://doi.org/10.1016/j.scitotenv.2021.151078.

Gao, Jiankun, Gu Wang, Jingrong Wu, Yu Zuo, Jing Zhang, and Xintian Jin. 2019. "Skp2 Expression Is Inhibited by Arsenic Trioxide through the Upregulation of MiRNA-330-5p in Pancreatic Cancer Cells." *Molecular Therapy Oncolytics* 12 (February): 214–223. https://doi.org/10.1016/j.omto.2019.01.006.

Gao, Ming, Changying Li, Ming Xu, Yun Liu, Min Cong, and Sijin Liu. 2018. "LncRNA MT1DP Aggravates Cadmium-Induced Oxidative Stress by Repressing the Function of NRF2 and Is Dependent on Interaction with MiR-365." *Advanced Science (Weinheim, Baden-Wurttemberg, Germany)* 5 (7): 1800087. https://doi.org/10.1002/advs.201800087.

Gebert, Luca F. R., and Ian J. MacRae. 2019. "Regulation of MicroRNA Function in Animals." *Nature Reviews. Molecular Cell Biology* 20 (1): 21–37. https://doi.org/10.1038/s41580-018-0045-7.

Guida, Maurizio, Marialuisa Marra, Fulvio Zullo, Marco Guida, Marco Trifuoggi, Elio Biffali, Marco Borra, Giovanna De Mieri, Raffaella D'Alessandro, and Bruna De Felice. 2013. "Association between Exposure to Dioxin-Like Polychlorinated Biphenyls and MiR-191 Expression in Human Peripheral Blood Mononuclear Cells." *Mutation Research/Genetic Toxicology and Environmental Mutagenesis* 753 (1): 36.

Guo, Xin-Xin, Qing-Zhi He, Wu Li, Ding-Xin Long, Xiao-Yuan Pan, Cong Chen, and Huai-Cai Zeng. 2017. "Brain-Derived Neurotrophic Factor Mediated Perfluorooctane Sulfonate Induced-Neurotoxicity via Epigenetics Regulation in SK-N-SH Cells." *International Journal of Molecular Sciences* 18 (4): 893. https://doi.org/10.3390/ijms18040893.

He, Tongkun, Jingyuan Zhu, Jia Wang, Xinmin Ren, Guomei Cheng, Xiaoxue Liu, Qiang Ma, Yanli Zhang, Zhiyuan Li, and Yue Ba. 2018. "Ambient Air Pollution, H19/DMR Methylation in Cord Blood and Newborn Size: A Pilot Study in Zhengzhou City, China." *Chemosphere* 212 (December): 863–871. https://doi.org/10.1016/j.chemosphere.2018.08.140.

Hou, Lifang, Jitendra Barupal, Wei Zhang, Yinan Zheng, Lei Liu, Xiao Zhang, Chang Dou, et al. 2016. "Particulate Air Pollution Exposure and Expression of Viral and Human MicroRNAs in Blood: The Beijing Truck Driver Air Pollution Study." *Environmental Health Perspectives* 124 (3): 344–350. https://doi.org/10.1289/ehp.1408519.

Jiang, Pan, Shengyu Hao, Liang Xie, Guiling Xiang, Weiping Hu, Qinhan Wu, Zilong Liu, and Shanqun Li. 2021. "LncRNA NEAT1 Contributes to the Acquisition of a Tumor Like-Phenotype Induced by PM 2.5 in Lung Bronchial Epithelial Cells via HIF-1α Activation." *Environmental Science and Pollution Research* 28 (32): 43382–43393. https://doi.org/10.1007/s11356-021-13735-7.

Kim, Jin Hee, Yoon Hee Cho, and Yun-Chul Hong. 2020. "MicroRNA Expression in Response to Bisphenol A Is Associated with High Blood Pressure." *Environment International* 141 (August): 105791. https://doi.org/10.1016/j.envint.2020.105791.

Kim, Ki-Hyun, Ehsanul Kabir, and Shamin Kabir. 2015. "A Review on the Human Health Impact of Airborne Particulate Matter." *Environment International* 74 (January): 136–143. https://doi.org/10.1016/j.envint.2014.10.005.

Kong, Alice P. S., Kang Xiao, Kai Chow Choi, Gang Wang, Michael H. M. Chan, Chung Shun Ho, Iris Chan, Chun Kwok Wong, Juliana C. N. Chan, and Cheuk Chun Szeto. 2012. "Associations between MicroRNA (MiR-21, 126, 155 and 221), Albuminuria and Heavy Metals in Hong Kong Chinese Adolescents." *Clinica Chimica Acta; International Journal of Clinical Chemistry* 413 (13–14): 1053–1057. https://doi.org/10.1016/j.cca.2012.02.014.

Kopp, Florian, and Joshua T. Mendell. 2018. "Functional Classification and Experimental Dissection of Long Noncoding RNAs." *Cell* 172 (3): 393–407. https://doi.org/10.1016/j.cell.2018.01.011.

Krauskopf, Julian, Theo M. de Kok, Dennie G. Hebels, Ingvar A. Bergdahl, Anders Johansson, Florentin Spaeth, Hannu Kiviranta, Panu Rantakokko, Soterios A. Kyrtopoulos, and Jos C. Kleinjans. 2017. "MicroRNA Profile for Health Risk Assessment: Environmental Exposure to Persistent Organic Pollutants Strongly Affects the Human Blood MicroRNA Machinery." *Scientific Reports* 7 (1): 9262. https://doi.org/10.1038/s41598-017-10167-7.

Kupsco, Allison, Jenny Jyoung Lee, Diddier Prada, Damaskini Valvi, Lisa Hu, Maria Skaalum Petersen, Brent A. Coull, Pal Weihe, Philippe Grandjean, and Andrea A. Baccarelli. 2022. "Marine Pollutant Exposures and Human Milk Extracellular Vesicle-MicroRNAs in a Mother-Infant Cohort from the Faroe Islands." *Environment International* 158 (January): 106986. https://doi.org/10.1016/j.envint.2021.106986.

Lee, Young S., John A. Lewis, Danielle L. Ippolito, Naissan Hussainzada, Pamela J. Lein, David A. Jackson, and Jonathan D. Stallings. 2016. "Repeated Exposure to Neurotoxic Levels of Chlorpyrifos Alters Hippocampal Expression of Neurotrophins and Neuropeptides." *Toxicology* 340 (January): 53–62. https://doi.org/10.1016/j.tox.2016.01.001.

Lemaire, J., C. Van der Hauwaert, G. Savary, E. Dewaeles, M. Perrais, J. M. Lo Guidice, N. Pottier, F. Glowacki, and C. Cauffiez. 2020. "Cadmium-Induced Renal Cell Toxicity Is Associated with MicroRNA Deregulation." *International Journal of Toxicology* 39 (2): 103–114. https://doi.org/10.1177/1091581819899039.

Lesiak, Adam, Mingyan Zhu, Hao Chen, Suzanne M. Appleyard, Soren Impey, Pamela J. Lein, and Gary A. Wayman. 2014. "The Environmental Neurotoxicant PCB 95 Promotes Synaptogenesis via Ryanodine Receptor-Dependent MiR132 Upregulation." *The Journal of Neuroscience* 34 (3): 717–725. https://doi.org/10.1523/JNEUROSCI.2884-13.2014.

Li, Cindy Yanfei, and Julia Yue Cui. 2018. "Regulation of Protein-Coding Gene and Long Noncoding RNA Pairs in Liver of Conventional and Germ-Free Mice Following Oral PBDE Exposure." *PLoS One* 13 (8): e0201387. https://doi.org/10.1371/journal.pone.0201387.

Liang, Boxuan, Yuncong Chen, Weixin Yuan, Fei Qin, Qian Zhang, Na Deng, Xuexing Liu, et al. 2018. "Down-Regulation of MiRNA-451a and MiRNA-486-5p Involved in Benzene-Induced Inhibition on Erythroid Cell Differentiation in Vitro and in Vivo." *Archives of Toxicology* 92 (1): 259–272. https://doi.org/10.1007/s00204-017-2033-7.

Lin, Hsuan-Pei, Matthew Rea, Zhishan Wang, and Chengfeng Yang. 2021. "Down-Regulation of LncRNA MEG3 Promotes Chronic Low Dose Cadmium Exposure-Induced Cell Transformation and Cancer Stem Cell-Like Property." *Toxicology and Applied Pharmacology* 430 (November): 115724. https://doi.org/10.1016/j.taap.2021.115724.

Longo, Valeria, Alessandra Longo, Giorgia Adamo, Antonino Fiannaca, Sabrina Picciotto, Laura La Paglia, Daniele Romancino, et al. 2021. "2,2'4,4'-Tetrabromodiphenyl Ether (PBDE-47) Modulates the Intracellular MiRNA Profile, SEV Biogenesis and Their MiRNA Cargo Exacerbating the LPS-Induced Pro-Inflammatory Response in THP-1 Macrophages." *Frontiers in Immunology* 12: 664534. https://doi.org/10.3389/fimmu.2021.664534.

Lu, Zhaoxu, Yufeng Ma, Linying Gao, Yingjun Li, Qiang Li, and Mei Qiang. 2018. "Urine Mercury Levels Correlate with DNA Methylation of Imprinting Gene H19 in the Sperm of Reproductive-Aged Men." *PLoS One* 13 (4): e0196314. https://doi.org/10.1371/journal.pone.0196314.

Luo, Fei, Hongying Wei, Huaqi Guo, Yan Li, Yan Feng, Qian Bian, and Yan Wang. 2019. "LncRNA MALAT1, an LncRNA Acting via the MiR-204/ZEB1 Pathway, Mediates the EMT Induced by Organic Extract of PM2.5 in Lung Bronchial Epithelial Cells." *American Journal of Physiology. Lung Cellular and Molecular Physiology* 317 (1): L87–L98. https://doi.org/10.1152/ajplung.00073.2019.

Ma, Lina, Vladimir B. Bajić, and Zhang Zhang. 2013. "On the Classification of Long Non-Coding RNAs." *RNA Biology* 10 (6): 924–933. https://doi.org/10.4161/rna.24604.

McIlwraith, Emma K., Calvin V. Lieu, and Denise D. Belsham. 2022. "Bisphenol A Induces MiR-708-5p through an ER Stress-Mediated Mechanism Altering Neuronatin and Neuropeptide Y Expression in Hypothalamic Neuronal Models." *Molecular and Cellular Endocrinology* 539 (January): 111480. https://doi.org/10.1016/j.mce.2021.111480.

Mitchell, Robert D., Andrew D. Wallace, Ernest Hodgson, and R. Michael Roe. 2017. "Differential Expression Profile of LncRNAs from Primary Human Hepatocytes Following DEET and Fipronil Exposure." *International Journal of Molecular Sciences* 18 (10): 2104. https://doi.org/10.3390/ijms18102104.

Monnier, Paul, Clémence Martinet, Julien Pontis, Irina Stancheva, Slimane Ait-Si-Ali, and Luisa Dandolo. 2013. "H19 LncRNA Controls Gene Expression of the Imprinted Gene Network by Recruiting MBD1." *Proceedings of the National Academy of Sciences* 110 (51): 20693–20698. https://doi.org/10.1073/pnas.1310201110.

Noh, Ji Heon, Kyoung Mi Kim, Waverly G. McClusky, Kotb Abdelmohsen, and Myriam Gorospe. 2018. "Cytoplasmic Functions of Long Noncoding RNAs." *WIREs RNA* 9 (3): e1471. https://doi.org/10.1002/wrna.1471.

O'Brien, Jacob, Heyam Hayder, Yara Zayed, and Chun Peng. 2018. "Overview of MicroRNA Biogenesis, Mechanisms of Actions, and Circulation." *Frontiers in Endocrinology* 9. www.frontiersin.org/articles/10.3389/fendo.2018.00402.

Pallocca, Giorgia, Marco Fabbri, Maria Grazia Sacco, Laura Gribaldo, David Pamies, Incoronata Laurenza, and Anna Bal-Price. 2013. "MiRNA Expression Profiling in a Human Stem Cell-Based Model as a Tool for Developmental Neurotoxicity Testing." *Cell Biology and Toxicology* 29 (4): 239–257. https://doi.org/10.1007/s10565-013-9250-5.

Pan, Zhijie, Bohuan Zhong, Xiaoxuan Ling, Haiqiao Zhang, Qiang Tan, Dongsheng Huang, Jialong Chen, et al. 2021. "The DNMT1-Associated LncRNA UCA1 Was Upregulated in TK6 Cells Transformed by Long-Term Exposure to Hydroquinone and Benzene-Exposed Workers via DNA Hypomethylation." *Journal of Biochemical and Molecular Toxicology* 35 (12): e22920. https://doi.org/10.1002/jbt.22920.

Pang, Wei, Fu-Zhi Lian, Xue Leng, Shu-Min Wang, Yi-Bo Li, Zi-Yu Wang, Kai-Ren Li, Zhi-Xian Gao, and Yu-Gang Jiang. 2018. "Microarray Expression Profiling and Co-Expression Network Analysis of Circulating LncRNAs and MRNAs Associated with Neurotoxicity Induced by BPA." *Environmental Science and Pollution Research International* 25 (15): 15006–15018. https://doi.org/10.1007/s11356-018-1678-y.

Puttabyatappa, Muraly, Nadia Saadat, Venkateswaran Ramamoorthi Elangovan, John Dou, Kelly Bakulski, and Vasantha Padmanabhan. 2022. "Developmental Programming: Impact of Prenatal Bisphenol-A Exposure on Liver and Muscle Transcriptome of Female Sheep." *Toxicology and Applied Pharmacology* 451 (September): 116161. https://doi.org/10.1016/j.taap.2022.116161.

Quezada-Maldonado, Ericka Marel, Yesennia Sánchez-Pérez, Yolanda I. Chirino, Felipe Vaca-Paniagua, and Claudia M. García-Cuellar. 2018. "MiRNAs Deregulation in Lung Cells Exposed to Airborne Particulate Matter (PM10) Is Associated with Pathways Deregulated in Lung Tumors." *Environmental Pollution (Barking, Essex: 1987)* 241 (October): 351–358. https://doi.org/10.1016/j.envpol.2018.05.073.

Rahman, Mohammad L., Liming Liang, Linda Valeri, Li Su, Zhaozhong Zhu, Shangzhi Gao, Golam Mostofa, et al. 2018. "Regulation of Birthweight by Placenta-Derived MiRNAs: Evidence from an Arsenic-Exposed Birth Cohort in Bangladesh." *Epigenetics* 13 (6): 573–590. https://doi.org/10.1080/15592294.2018.1481704.

Reed, Beverly G., Samir N. Babayev, Lucy X. Chen, Bruce R. Carr, R. Ann Word, and Patricia T. Jimenez. 2018. "Estrogen-Regulated MiRNA-27b Is Altered by Bisphenol A in Human Endometrial Stromal Cells." *Reproduction (Cambridge, England)* 156 (6): 559–567. https://doi.org/10.1530/REP-18-0041.

Ross, Caroline Jane, and Igor Ulitsky. 2022. "Discovering Functional Motifs in Long Noncoding RNAs." *WIREs RNA* 13 (4): e1708. https://doi.org/10.1002/wrna.1708.

Rüdel, Heinz, Wolfgang Körner, Thomas Letzel, Michael Neumann, Karsten Nödler, and Thorsten Reemtsma. 2020. "Persistent, Mobile and Toxic Substances in the Environment: A Spotlight on Current Research and Regulatory Activities." *Environmental Sciences Europe* 32 (1): 5. https://doi.org/10.1186/s12302-019-0286-x.

Sanders, Alison P., Heather H. Burris, Allan C. Just, Valeria Motta, Chitra Amarasiriwardena, Katherine Svensson, Emily Oken, et al. 2015. "Altered MiRNA Expression in the Cervix during Pregnancy Associated with Lead and Mercury Exposure." *Epigenomics* 7 (6): 885–896. https://doi.org/10.2217/epi.15.54.

Santibáñez-Andrade, Miguel, Yesennia Sánchez-Pérez, Yolanda I. Chirino, Rocío Morales-Bárcenas, and Claudia M. García-Cuellar. 2021. "Long Non-Coding RNA NORAD Upregulation Induced by Airborne Particulate Matter (PM10) Exposure Leads to Aneuploidy in A549 Lung Cells." *Chemosphere* 266 (March): 128994. https://doi.org/10.1016/j.chemosphere.2020.128994.

Shafei, Ayman, Maggie M. Ramzy, Abdelhares I. Hegazy, Ahmed K. Husseny, Usama G. El-Hadary, Mazen M. Taha, and Ali A. Mosa. 2018. "The Molecular Mechanisms of Action of the Endocrine Disrupting Chemical Bisphenol A in the Development of Cancer." *Gene* 647 (March): 235–243. https://doi.org/10.1016/j.gene.2018.01.016.

Shin, Cheol-Hee, Junhyoung Byun, Kijeong Lee, Byoungjae Kim, Yong Kwan Noh, Na Ly Tran, Kwideok Park, Sang-Heon Kim, Tae Hoon Kim, and Seung Ja Oh. 2020. "Exosomal MiRNA-19a and MiRNA-614 Induced by Air Pollutants Promote Proinflammatory M1 Macrophage Polarization via Regulation of RORα Expression in Human Respiratory Mucosal Microenvironment." *Journal of Immunology (Baltimore, Md.: 1950)* 205 (11): 3179–3190. https://doi.org/10.4049/jimmunol.2000456.

Sima, Michal, Andrea Rossnerova, Zuzana Simova, and Pavel Rossner. 2021. "The Impact of Air Pollution Exposure on the MicroRNA Machinery and Lung Cancer Development." *Journal of Personalized Medicine* 11 (1): 60. https://doi.org/10.3390/jpm11010060.

Singh, Vikas, Shaivya Kushwaha, Jamal Ahmad Ansari, Siddhartha Gangopadhyay, Shubhendra K. Mishra, Rajib K. Dey, Ashok K. Giri, Satyakam Patnaik, and Debabrata Ghosh. 2022. "MicroRNA-129-5p-Regulated Microglial Expression of the Surface Receptor CD200R1 Controls Neuroinflammation." *Journal of Biological Chemistry* 298 (1). https://doi.org/10.1016/j.jbc.2021.101521.

Smith, Martyn T. 2010. "Advances in Understanding Benzene Health Effects and Susceptibility." *Annual Review of Public Health* 31: 133–148 2 p following 148. https://doi.org/10.1146/annurev.publhealth.012809.103646.

Song, Fan, Zhijian Chen, Danni Lyu, Yuzhou Gu, Bing Lu, Shengjie Hao, Yili Xu, Xiuming Jin, Qiuli Fu, and Ke Yao. 2022. "Expression Profiles of Long Noncoding RNAs in Human Corneal Epithelial Cells Exposed to Fine Particulate Matter." *Chemosphere* 287 (January): 131955. https://doi.org/10.1016/j.chemosphere.2021.131955.

Song, Lei, Dan Li, Yue Gu, Xiaoping Li, and Liping Peng. 2016. "Let-7a Modulates Particulate Matter (≤ 2.5 Mm)-Induced Oxidative Stress and Injury in Human Airway Epithelial Cells by Targeting Arginase 2." *Journal of Applied Toxicology: JAT* 36 (10): 1302–1310. https://doi.org/10.1002/jat.3309.

Statello, Luisa, Chun-Jie Guo, Ling-Ling Chen, and Maite Huarte. 2021. "Gene Regulation by Long Non-Coding RNAs and Its Biological Functions." *Nature Reviews Molecular Cell Biology* 22 (2): 96–118. https://doi.org/10.1038/s41580-020-00315-9.

Sun, Hao, Junjie Chen, Wenyi Qian, Jiang Kang, Jun Wang, Lei Jiang, Li Qiao, Wei Chen, and Jinsong Zhang. 2016. "Integrated Long Non-Coding RNA Analyses Identify Novel Regulators of Epithelial-Mesenchymal Transition in the Mouse Model of Pulmonary Fibrosis." *Journal of Cellular and Molecular Medicine* 20 (7): 1234–1246. https://doi.org/10.1111/jcmm.12783.

Tilghman, Syreeta L., Melyssa R. Bratton, H. Chris Segar, Elizabeth C. Martin, Lyndsay V. Rhodes, Meng Li, John A. McLachlan, Thomas E. Wiese, Kenneth P. Nephew, and Matthew E. Burow. 2012. "Endocrine Disruptor Regulation of MicroRNA Expression in Breast Carcinoma Cells." *PLoS One* 7 (3): e32754. https://doi.org/10.1371/journal.pone.0032754.

Valencia-Quintana, Rafael, Iván Uriel Bahena-Ocampo, Gerardo González-Castañeda, Edmundo Bonilla, Mirta Milić, Stefano Bonassi, and Juana Sánchez-Alarcón. 2022. "MiRNAs: A Potentially Valuable Tool in Pesticide Toxicology Assessment-Current Experimental and Epidemiological Data Review." *Chemosphere* 295 (May): 133792. https://doi.org/10.1016/j.chemosphere.2022.133792.

Wahlang, Banrida, Michael C. Petriello, Jordan T. Perkins, Shu Shen, and Bernhard Hennig. 2016. "Polychlorinated Biphenyl Exposure Alters the Expression Profile of MicroRNAs Associated with Vascular Diseases." *Toxicology in Vitro: An International Journal Published in Association with BIBRA* 35 (September): 180–187. https://doi.org/10.1016/j.tiv.2016.06.001.

Wan, Chong, Rui Han, Limin Liu, Fang Zhang, Fang Li, Mingdeng Xiang, and Wenjun Ding. 2016. "Role of MiR-155 in Fluorooctane Sulfonate-Induced Oxidative Hepatic Damage via the NRF2-Dependent Pathway." *Toxicology and Applied Pharmacology* 295 (March): 85–93. https://doi.org/10.1016/j.taap.2016.01.023.

Wang, Jianshe, Shengmin Yan, Wei Zhang, Hongxia Zhang, and Jiayin Dai. 2015. "Integrated Proteomic and MiRNA Transcriptional Analysis Reveals the Hepatotoxicity Mechanism of PFNA Exposure in Mice." *Journal of Proteome Research* 14 (1): 330–341. https://doi.org/10.1021/pr500641b.

Wang, Lijin, Hongyu Yang, Qingqing Wang, Qiaohui Zhang, Zhangjing Wang, Qunwei Zhang, Siying Wu, and Huangyuan Li. 2018. "Paraquat and MPTP Induce Alteration in the Expression Profile of Long Noncoding RNAs in the Substantia Nigra of Mice: Role of the Transcription Factor NRF2." *Toxicology Letters* 291 (July): 11–28. https://doi.org/10.1016/j.toxlet.2018.04.002.

Wang, Tong-Shuai, Wei Tian, Yan Fang, Kong-Rong Guo, An-Qi Li, Yuan Sun, Han-Tian Wu, et al. 2021. "Changes in MiR-222 Expression, DNA Repair Capacity, and MDM2-P53 Axis in Association with Low-Dose Benzene Genotoxicity and Hematotoxicity." *The Science of the Total Environment* 765 (April): 142740. https://doi.org/10.1016/j.scitotenv.2020.142740.

Wang, Yu, Hongjing Zhao, Menghao Guo, Dongxue Fei, Lina Zhang, and Mingwei Xing. 2020. "Targeting the MiR-122/PKM2 Autophagy Axis Relieves Arsenic Stress." *Journal of Hazardous Materials* 383 (February): 121217. https://doi.org/10.1016/j.jhazmat.2019.121217.

Wang, Yuting, Wen Zhang, Lin Chen, Wei Chen, Shufen Xu, Lingyu Tang, Yang Yang, Quanpeng Li, Qi Jiang, and Lin Miao. 2021. "The ATO/MiRNA-885-5p/MTPN Axis Induces Reversal of Drug-Resistance in Cholangiocarcinoma." *Cellular Oncology* 44 (4): 907–916. https://doi.org/10.1007/s13402-021-00610-3.

Waugh, Courtney A., Augustine Arukwe, and Veerle L. B. Jaspers. 2018. "Deregulation of MicroRNA-155 and Its Transcription Factor NF-KB by Polychlorinated Biphenyls during Viral Infections." *APMIS: Acta Pathologica, Microbiologica, et Immunologica Scandinavica* 126 (3): 234–240. https://doi.org/10.1111/apm.12811.

Wen, Ruojian. 2022. "Epigenetic Effect of Perfluorooctane Sulfonate on Lipid Metabolism." In *2022 12th International Conference on Bioscience, Biochemistry and Bioinformatics*, 140–142. ICBBB '22. New York, NY: Association for Computing Machinery. https://doi.org/10.1145/3510427.3511604.

Wu, Wei, and Yong Li. 2018. "Lung Injury Caused by Paraquat Poisoning Results in Increased Interleukin-6 and Decreased MicroRNA-146a Levels." *Experimental and Therapeutic Medicine* 16 (1): 406–412. https://doi.org/10.3892/etm.2018.6153.

Xie, Xiuwen, Jia Song, and Gang Li. 2016. "MiR-21a-5p Suppresses Bisphenol A-Induced Pre-Adipocyte Differentiation by Targeting Map2k3 through MKK3/P38/MAPK." *Biochemical and Biophysical Research Communications* 473 (1): 140–146. https://doi.org/10.1016/j.bbrc.2016.03.066.

Xu, Yiyi, Simona Jurkovic-Mlakar, Ying Li, Karin Wahlberg, Kristin Scott, Daniela Pineda, Christian H. Lindh, Kristina Jakobsson, and Karin Engström. 2020. "Association between Serum Concentrations of Perfluoroalkyl Substances (PFAS) and Expression of Serum MicroRNAs in a Cohort Highly Exposed to PFAS from Drinking Water." *Environment International* 136 (March): 105446. https://doi.org/10.1016/j.envint.2019.105446.

Yuan, Haijun, Mei Yuan, Yonghong Tang, Biao Wang, and Xiangyang Zhan. 2018. "MicroRNA Expression Profiling in Human Acute Organophosphorus Poisoning and Functional Analysis of Dysregulated MiRNAs." *African Health Sciences* 18 (2): 333–342. https://doi.org/10.4314/ahs.v18i2.18.

Zandwijk, Nico van, Nick Pavlakis, Steven C. Kao, Anthony Linton, Michael J. Boyer, Stephen Clarke, Yennie Huynh, et al. 2017. "Safety and Activity of MicroRNA-Loaded Minicells in Patients with Recurrent Malignant Pleural Mesothelioma: A First-in-Man, Phase 1, Open-Label, Dose-Escalation Study." *The Lancet: Oncology* 18 (10): 1386–1396. https://doi.org/10.1016/S1470-2045(17)30621-6.

Zhang, Haiqiao, Qian Yuan, Zhijie Pan, Xiaoxuan Ling, Qiang Tan, Minhua Wu, Dongyan Zheng, Peien Xie, Daxiao Xie, and Linhua Liu. 2020. "Up-Regulation of DNMT3b Contributes to HOTAIRM1 Silencing via DNA Hypermethylation in Cells Transformed by Long-Term Exposure to Hydroquinone and Workers Exposed to Benzene." *Toxicology Letters* 322 (April): 12–19. https://doi.org/10.1016/j.toxlet.2019.12.028.

Zhang, Xinwei, Yanshu Zhang, Qingtao Meng, Hao Sun, Shenshen Wu, Jie Xu, Jun Yun, et al. 2020. "MicroRNA-382-5p Is Involved in Pulmonary Inflammation Induced by Fine Particulate Matter Exposure." *Environmental Pollution (Barking, Essex: 1987)* 262 (July): 114278. https://doi.org/10.1016/j.envpol.2020.114278.

Zhang, Yan, Xing Zhang, Haixia Zhu, Yang Liu, Jian Cao, Dake Li, Bo Ding, Wenjing Yan, Hua Jin, and Shizhi Wang. 2020. "Identification of Potential Prognostic Long Non-Coding RNA Biomarkers for Predicting Recurrence in Patients with Cervical Cancer." *Cancer Management and Research* 12: 719–730. https://doi.org/10.2147/CMAR.S231796.

Zhao, Hua-Wei, Hao Liu, Li-Ying Liu, Zhi Liu, and Xue-Song Dong. 2020. "Analysis of MicroRNA Expression Profiling during Paraquat-Induced Injury of Murine Lung Alveolar Epithelial Cells." *The Journal of Toxicological Sciences* 45 (8): 423–434. https://doi.org/10.2131/jts.45.423.

Zheng, Fenshuang, Junbo Zhu, Wei Zhang, Yangshan Fu, and Zhaoheng Lin. 2021. "Thal Protects against Paraquat-Induced Lung Injury through a MicroRNA-141/HDAC6/IκBα-NF-κB Axis in Rat and Cell Models." *Basic & Clinical Pharmacology & Toxicology* 128 (2): 334–347. https://doi.org/10.1111/bcpt.13505.

Zhong, Ping, Jin Liu, Hong Li, Senbin Lin, Lingfeng Zeng, Lingfeng Luo, Minxia Wu, and Wenchang Zhang. 2020. "MicroRNA-204-5p Regulates Apoptosis by Targeting Bcl2 in Rat Ovarian Granulosa Cells Exposed to Cadmium†." *Biology of Reproduction* 103 (3): 608–619. https://doi.org/10.1093/biolre/ioaa091.

Zhong, Yijue, Jiping Liao, Yan Hu, Yunxia Wang, Chao Sun, Cheng Zhang, and Guangfa Wang. 2019. "PM2.5 Upregulates MicroRNA-146a-3p and Induces M1 Polarization in RAW264.7 Cells by Targeting Sirtuin1." *International Journal of Medical Sciences* 16 (3): 384–393. https://doi.org/10.7150/ijms.30084.

Zhou, Tianyu, Yijue Zhong, Yan Hu, Chao Sun, Yunxia Wang, and Guangfa Wang. 2018. "PM2.5 Downregulates MiR-194-3p and Accelerates Apoptosis in Cigarette-Inflamed Bronchial Epithelium by Targeting Death-Associated Protein Kinase 1." *International Journal of Chronic Obstructive Pulmonary Disease* 13 (August): 2339–2349. https://doi.org/10.2147/COPD.S168629.

11 Oxidative and Inflammatory Potential of Nano/Microplastics in Living Organisms

Konrad Wojnarowski, Moyan Hu, and Dušan Palić

11.1 INTRODUCTION

The awareness of general public and scientific community about the scope and amount of plastic pollution has significantly increased in recent years. Global production of plastics reached 367 million tons in 2021, from which over half was produced in Asia, while Europe contributed 64 million tons (AISBL 2021). Increase in production coupled with chemical stability of plastic materials and shift of consumer preferences toward "single-use" packaging, accelerates generation of plastic waste, leading to contamination of diverse ecosystems and rising concern about environmental and health risks associated with exposure to plastics (Yuan, Nag, and Cummins 2022). Exposure of plastic waste to the elements leads to its degradation through mechanical and chemical processes such as hydrolysis and UV radiation (Corcoran 2022; Duan et al. 2022), to form microplastics (100 nm–5 mm, MPs) and nanoplastics (< 100 nm, NPs) particles. In comparison to lager plastic debris, the smaller size of MPs and NPs presents an opportunity for them to enter different environmental compartments, increases efforts required for their removal or clean up and poses higher risk of exposure and entering food chain.

The number of studies addressing MPs/NPs toxicity is rapidly increasing, with focus on polyethylene (PE), polystyrene (PS) and polyvinyl chloride (PVC), which have the highest production and utilization among many species of plastic materials (AISBL 2021). Rising interest in investigations of MPs/NPs is likely related to their ability to translocate through biological barriers while at the same time maintaining large surface/mass ratio, as well as their potential to accumulate in the higher trophic level organisms via the food chain (Prinz and Korez 2020). It is therefore of high interest of research and regulatory community to evaluate potential ecological as well as human health hazards of MPs/NPs.

A variety of approaches have been used in toxicological studies of MPs/NPs, including *in vivo* (aquatic marine and freshwater and terrestrial organisms including mammals) and *in vitro* (cell culture) models to investigate effects and behavior of MPs/NPs (Lu et al. 2016, 2018; Brun et al. 2018; Pannetier et al. 2019; Nobre et al. 2015). Reports indicate that MPs/NPs are most frequently involved in induction of developmental toxicity, neurotoxicity, cytotoxicity and oxidative stress (Lu et al. 2018, 2016; Barboza et al. 2018; Nobre et al. 2015). Observed toxic effects of MPs/NPs particles suggest that oxidative stress and inflammatory responses are of high importance as critical mechanisms underneath the afore-listed toxicities. However, hazard assessment outcomes from different studies are variable and sometimes contradictory, probably due to the lack of standardized research methodology such as the use of different research models and not the least due to high diversity of the MPs/NPs themselves (different size, shape, surface charging and polymer type). For example, the effects of microplastics in a very high concentration was reported to have no negative effects on crustaceans, which directly contradicted other studies indicating adverse effects on crustaceans after exposure to environmentally relevant MPs concentration (Weber et al. 2018).

DOI: 10.1201/9781003328100-14

Furthermore, observing toxicity endpoints at different biological levels (from molecular to population) can be limited, and sometimes even misleading, based on complexity of the observed model system. The majority of *in vitro* model based studies reported toxicity endpoints only on molecular or cellular level, while *in vivo* studies could detect the adverse effects of MPs/NPs in various organs (Lu et al. 2016; Deng et al. 2017).

In addition to this, addressing possible differences between engineered and environment-derived MPs/NPs has several limitations. Studies have rarely differentiated between the two MPs/NPs sources, and furthermore, the majority of studies have actually used model MPs/NPs (by default engineered, rather than derived from the environment). Therefore, there is very limited information about mechanisms of action specific to environmentally derived MPs/NPs and most of the available information that is related to environment-derived MPs/NPs comes in a form of surveys of wildlife and aquatic organisms, usually only describing current state of situation, or distribution and relying to laboratory studies to explain observed effects. In this chapter, information about the source of nanomaterials is included in the text if it was deemed relevant from the aspects of engineered vs. environment-derived MPs/NPs (specifically in the inflammation section). Unfortunately, not enough studies are available to be used for statistically relevant comparisons between effects of natural vs. engineered nanomaterials. However, as evidenced by recent inflow of data, different origins of nanomaterials should be regarded as different factors regarding their negative impact on living organisms (Figure 11.1 and Figure 11.2) (Gigault et al. 2021). The goal of this chapter is to summarize available data on the possible adverse effects of MPs/NPs related to oxidative metabolism and organism responses on various levels.

11.2 MOLECULAR EFFECTS TRIGGERED BY MICROPLASTICS AND NANOPLASTICS

11.2.1 Oxidative Stress

Molecular events caused by noxious/toxic agent effects in many cases form the foundation for explanations of biological phenomena observed on higher levels of biological complexity. In this section, we summarize molecular effects induced by microplastics and nanoplastics including generation of free radicals, oxidative stress metabolism activation, lipid peroxidation, DNA damage and downstream signaling pathway activation that precede cascade of branching molecular changes and potentially lead to irreparable oxidative damage and exacerbated inflammatory processes.

Oxidative stress occurs when there is an imbalance between the production of ROS and antioxidant-based detoxification/neutralization or when the resulting damage (or adverse redox signaling) cannot be repaired by the cellular repair machinery or compensated by alternative signaling pathways that prevent the detrimental effects of adverse redox signaling. The antioxidant systems are complex and include various antioxidant compounds such as vitamins (e.g. C, E, D3), as well as multiple enzymatic pathways involved in the production of antioxidants or the elimination of oxidants including ROS and RNS. Here, we will focus on oxidative stress pathways most frequently reported in relation to MPs/NPs exposure.

Antioxidant enzymes such as superoxide dismutase (SOD), catalase (CAT), and glutathione peroxidases (GPx) play an important role in protecting tissues against ROS damage (Livingstone 2001). For example, complex ROS scavenging system in innate immune cells such as neutrophils utilizes SOD and CAT enzymatic systems to catalyze the reactions of superoxide anion radical ($O2^{\bullet -}$) and hydrogen peroxide (H_2O_2) into water (H_2O) and oxygen (O_2) as the final metabolites, respectively (Ighodaro and Akinloye 2018). SOD and CAT based antioxidant mechanisms are important cellular tools against adverse effects of ROS, and their activity has frequently been used as biomarker of xenobiotic-mediated oxidative stress (Ighodaro and Akinloye 2018). Exposure to MPs/NPs could increase the concentration of ROS in the organism as we previously stated, and the SOD/CAT antioxidant mechanisms would represent a response to such prooxidative signal (Li et al. 2021).

The activation of antioxidants was found in animal models on different trophic levels following the exposure of MPs including rotifers (*B. koreanus*) (Jeong et al. 2016), water fleas (*D. magna*) (Imhof et al. 2017), fish (*D. rerio*) (Lu et al. 2016) and mammals (*M. musculus*) (Deng et al. 2017; Tang et al. 2022). This phenomenon suggests the existence of common mechanisms to initiate oxidative stress sensing and scavenging system induced by microplastic exposure. These phenomena indicate that both MPs and NPs can induce oxidative stress as fundamental mechanism for further toxic effects induced by MPs/NPs in a majority of organisms.

11.2.2 Reactive Oxygen Species Generation

"Free radicals" are atoms or molecules with one or more unpaired electrons with high potential to engage in different reactions, including oxidative chemical reactions. In biological systems, many radicals are derived from oxygen and are collectively referred to as reactive oxygen species (ROS) (Freeman and Crapo 1982). Reports show that MPs/NPs can induce generation of ROS at different sizes, doses, surface-characteristics and exposure times (Jeong et al. 2017).

Observations of MPs/NPs triggered ROS generation fall in two distinct compartments: extracellular and intracellular. Extracellular ROS generation induced by the MPs relates to the degree of weathering (aging) process during plastic polymers exposure to the elements in the environment (de Souza Machado et al. 2018). Weathering processes of MPs/NPs involves simultaneous or individual action of photo- and thermal oxidation and UV radiation leading to chemical alteration on the surface of plastic polymers (Corcoran 2022). Photo oxidation or UV light radiation can lead to formation of free radicals on MPs/NPs surfaces as primary products via different pathways: subtraction of a hydrogen atom from the macromolecular chain or addition to an unsaturated carbon chain group (crosslinking reaction) (Yousif and Haddad 2013). Once the free radicals are generated along the polymer chain, they can react with atmospheric oxygen and produce polymer peroxy radicals with further generation of secondary polymer alkyl radicals. These weathering induced extracellular free radicals could be one possible explanation as to why a significant increase of ROS was observed after the cellular entrance of the aged MPs (Figure 11.1) (Yousif and Haddad 2013).

However, pristine MPs/NPs can also induce excessive generation of ROS intracellularly. This phenomenon was reported using wide array of model systems, from mammalian cell lines to marine invertebrate and living fish models (Qiao et al. 2019; Schirinzi et al. 2017). MPs/NPs can be engulfed by a (usually phagocytic) cell through endocytosis or pinocytosis (< 150 nm in case of NPs) processes (von Moos, Burkhardt-Holm, and Köhler 2012; Prietl et al. 2014). Once they are inside a phagocyte, these plastic particles appear to trigger the innate immune defense mechanisms to treat them as foreign bodies or substances. During attempts of the cell to neutralize potentially infectious foreign particles, ROS are generated in high quantities as products of NADPH-oxidase or other enzymatic reactions in form of superoxide and hydrogen peroxide. Both $O_2^{\bullet-}$ and H_2O_2 have a role in signal transduction and serve as key mediators driving oxidative stress cascades. Furthermore, when MPs degrade into nano-sized particles, their surface-to-mass ratio increases and gives the NPs ability to penetrate directly through lipid membranes. Changes in surface charge potential and mass/surface ratio in NPs also enable easier absorption of free radicals as well as allows for easier translocation through membranes, presenting a strong correlation between particle size and ROS generation potential: the smaller the particle, the higher the ROS generation potential. Size-dependent differences in toxicity of MPs/NPs are supported by a number of recent studies (Lei et al. 2018; Jeong et al. 2016).

11.2.3 Activation of Oxidative Stress Pathways

It was clearly demonstrated that ROS can induce or mediate the activation of the MAPK (mitogen-activated protein kinase) pathways, but exact mechanisms of MAPK downstream pathway regulations remain unclear (McCubrey, LaHair, and Franklin 2006). Recent studies have reported that activation of MAPKs signaling cascades could induce ARE-mediated gene expression via the NRF2-dependent mechanism, and it appears that MAPK activation by ROS can be downregulated by simultaneous increase in NRF2 activity (Park et al. 2017). In contrast, the prolonged activation of MAPK system with increased ROS can activate various negative consequences (e.g. autophagy), primarily related to the inhibition of ERK and activation of p38MAPK components. It was recently reported that brominated flame retardant induced oxidative stress in two copepod species and activated the MAPK pathway, and it was determined that MAPKs pathways play a synergistic role with NRF2/KEAP1 pathway in the oxidative stress response in the scallops exposed to benzo(a)pyrene (Park et al. 2017).

As a next step in oxidative stress event series, it was observed that microplastics exposure also activated redox sensitive signaling pathways such as MAPKs. MAPK downstream pathways were initiated post MP exposure in marine copepod *P. nana* (Jeong et al. 2017). Jeong et al. (2017) showed that the phosphorylation level increase of ERK and p38 kinase has positive correlation with intracellular ROS generation level post MPs exposure in *P. nana*. Moreover, transcription factor Nrf-2 (nuclear factor erythroid 2-related factor 2) was increased after exposure to MPs suggesting that MPs triggered respiration burst possibly acted via ERK and p38 MAPK pathways by activating Nrf-2 as a master regulator of antioxidant enzymes gene expression (Jeong et al. 2017).

The activation of MAPK pathway appears dependent on the particle size, as initiation of MAPK pathway was more frequently reported after NPs exposure (Jeong et al. 2016). It was suggested that plastic particles with a size of 50 nm could generate more severe oxidative stress when compared to 6 μm plastic particles, causing higher phosphorylation level of p38 MAPKs in both *P. nana* and *B. koreanus* (Jeong et al. 2017). Moreover, the surface charge of nanoplastic particles appears critical in the activation of MAPK cascades, as plain/native polystyrene nanoparticles (PSNPs) induced the activation of p38 and JNK (c-Jun N-terminal kinases) along with ROS induction, while negatively charged PSNPs activated p38 and JNK independently.

11.2.4 Lipid Peroxidation

Lipid peroxidation (LPO) is a self-sustaining chain reaction induced by excessive ROS generation ($O_2^{\bullet -}$ or others) that results in oxidative damage to cell membranes and other lipid-containing structures. Currently there is a debate whether exposure to MPs would universally induce the LPO reaction and cell membrane damage. Barboza et al. (2018) reported significant increase in LPO from brain and muscle tissues of European seabass (*Dicentrarchus labrax*) after the exposure to MPs in concentration of 0.69 mg/L for 24 h. However, another study indicated that the LPO was not significantly induced after 7 days of MPs exposure, along with increased ROS levels, in the hemocytes of the marine mussel *Mytilus spp* (Paul-Pont et al. 2016).

Moreover, in *S. plana*, the LPO levels in gills and digestive gland were either reduced or not significantly increased after 7 days of polystyrene microparticle exposure, and the authors suggested that such result might be explained by sufficient capacity of antioxidant systems to defend the organism from LPO damages. Therefore, based on current literature, it is inconclusive if MPs could induce LPO damage in general (Lin et al. 2019), but it is at least prudent to consider this molecular occurrence as a potential effect triggered by MPs. This approach would be further supported with existing studies indicating that LPO and subsequent damages were induced by NPs exposure, therefore suggesting a possibility that LPO could potentially be a toxicity endpoint induced by plastic particles, especially in the nano-size range.

11.2.5 DNA Damage

Existence of mechanism of MPs/NPs induced DNA damage has been under debate lately, but recently we can observe growing number of studies supporting that such phenomenon occurs in living organisms (Shen et al. 2022; Wang et al. 2022; Ribeiro et al. 2017). Additionally, a lesson learned from wider nano-toxicology research indicated that some nanoparticles (such as titanium dioxide) can induce the DNA damage through two mechanisms: 1) direct interaction between nanoparticles and DNA; or 2) indirect DNA damage caused by nanoparticle generated ROS or other toxic species. Furthermore, it was also indicated that DNA strand breaks occur post nanoparticle exposure. An intracellular dynamic imaging study indicated that cationic functionalized PSNPs could result in a prolonged G0/G1 phase in the cell cycle during mitosis in NIH 3T3 cells, therefore indicating potential for DNA damage and the interference with checkpoint control activation (Liu et al. 2011). Paget et al. indicated that non-functionalized PSNPs did not induce a general genotoxicity except at the highest tested dose of after 8.1 μg/cm^2 exposure for 1 hour (Paget et al. 2015). There is evidence that both MPs and NPs have the ability to induce DNA strands to break in a size and surface charge dependent manner. While the exact mechanism of DNA damage induction by MPs/NPs is currently not fully understood, several studies suggest there is significant role of oxidative stress and physical interaction involved in initiation of DNA strand breaks by NPs.

11.2.6 Acetylcholinesterase (AChE) Inhibition

Acetylcholinesterase (AChE) is an enzyme in the choline esterase family that lyses acetylcholine (ACh) into choline and acetic acid, and it is critical for proper function of the nervous system (Massoulié et al. 1993). As ACh functions mainly as neurotransmitter controlling motor neuron function, the inhibition of AChE activity is causing accumulation of ACh in the synaptic cleft and the severe disturbance of muscle activity (Massoulié et al. 1993). Recently, the inhibition of AChE induced by either MPs or NPs exposure had been identified in multiple studies, and it appears to be shared by a variety of organisms from invertebrates to vertebrates (Deng et al. 2017; Lei et al. 2018). Oliveira et al. (2013) reported that polyethylene MPs in sizes of 1–5 μm induced the inhibition of AChE alone or in combination with pyrene in the brain of goby (*P. microps*). The authors also noted that the AChE inhibition rates induced by MPs and pyrene were not significantly different either as single substances or in combination, suggesting that enzymatic inhibition mechanisms may be shared by MPs and pyrene (Oliveira et al. 2013). The anti-cholinesterases effects of MPs had been also identified in gills of another marine organism (*S. plana*) after exposure to both polystyrene and polyethylene microspheres (Avio et al. 2015). Neurotoxic property of MPs has also been addressed in mammalian models and reported in mice to include MPs (5–20 μm) induced inflammation and AChE inhibition during a sub-chronic exposure period (28 days) (Deng et al. 2017). Recent study indicated a decrease in zebrafish AChE activity in all NPs treatments by 27–40% regardless of their surface modification (Chen et al. 2017). Invertebrate model (*C. elegans* and *M. galloprovincialis*) studies also reported a significant decrease in both level and activity of cholinesterase after the exposure to polystyrene NPs (Lei et al. 2018; Brandts et al. 2018). Taken together, there is a growing body of evidence supporting that MPs have the ability to induce neurotoxic responses and such a toxic characteristic should be considered in environmental risk assessment of MPs/NPs exposure.

The mechanisms involved in the MPs/NPs interference with AChE and possibly other neurotransmitters are largely unknown. Studies using nematode (*C. elegans*) based assay system suggested that due to nematode lack of blood-brain barrier (BBB), the NPs may have more opportunities to interact with *C. elegans* neurons causing more neurotoxic effects (Deng et al. 2017). This assumption is only partially explaining the toxic effects, as NPs have been observed to cross the BBB and eventually reach the brain. However, direct blocking of cholinesterase by NPs has not yet been observed and verified in an *in vitro* system. Presence of the blood-brain barrier is likely excluding micro-sized plastic particles crossing; however, neurotoxicity was nevertheless detected after MPs exposure in organisms with BBB. This strongly suggests that an indirect pathway plays a role in the MPs/NPs neurotoxicity potency (Deng et al. 2017). A possible explanation for this neurotoxicity potency is an involvement of MPs/NPs induced oxidative stress, as anti-cholinesterases induced neurotoxicity can be caused by it. Simultaneous detection of MPs/NPs induced oxidative stress, and AChE inhibition further supports such possibility. Direct relationship between H_2O_2 and modifications of AChE metabolism and activity was established recently when it was determined that H_2O_2 decreased AChE levels through its interference with allosteric AChE effector structure and changes of isoform profile in SH-SY5Y cells, thereby confirming that increase in at least one ROS could promote disturbances in a cholinergic system of neural cells (Garcimartín et al. 2017).

11.3 CELLULAR AND ORGANELLE EFFECTS TRIGGERED BY MICROPLASTICS AND NANOPLASTICS

At the cellular response level, molecular effects sum up and cause observable physiological outcomes within and outside of the cells. Here we will review three adverse cellular effects currently known to be triggered by both microplastic and nanoplastic particles (Figure 11.1 and 11.2).

11.3.1 Mitochondrial Dysfunction

As mitochondria are organelles where most of intracellular production of ROS occurs, an instability in mitochondrial membrane potential can result in excessive generation of ROS through one-electron carriers (e.g. cytochromes and iron-sulphur-proteins) and various oxidases. Reports indicated mitochondria membrane dysfunction after MPs exposure in mice and rotifer *B. koreanus*, which requires further investigations as MPs are unable to accumulate directly in the mitochondria due to their size (Jeong et al. 2016; Lu et al. 2016). The study in rotifer *B. koreanus* reported that 0.5 μm polystyrene microbeads decreased the mitochondrial membrane potential significantly and suggested that MPs may affect the mitochondrial outer membrane indirectly via increased ROS presence in cellular compartments outside of, but in the vicinity of, mitochondria (Jeong et al. 2016). Exposure of mitochondria to excessive oxidative stress in the cytosol can trigger the opening of several mitochondrial Na/K transmembrane channels. Increased membrane channel ionic flux can further lead to a collapse of the mitochondrial membrane potential and releasing reactive oxygen species through "(ROS)-induced ROS-release" (RIRR) mechanism (Zorov, Juhaszova, and Sollott 2006). As it is established that MPs are capable of inducing oxidative stress (Section 11.2.2), the RIRR mechanism could be responsible as mediator between the MPs intracellular presence and observed hampering of the mitochondrial membrane potential.

Mitochondrial membrane damage induced by MPs was correlated with particle size (where smaller MPs induced more effects), leading to the possibility that NPs could impose even more severe damage towards mitochondria in general (Jeong et al. 2016). Supportive to this reasoning, NPs induced mitochondria disruption was reported in *D. magna* and in a variety of human cell lines (Ruenraroengsak and Tetley 2015). However, it is also possible that reported damage could also have been induced by observed direct accumulation of NPs inside mitochondria and not exclusively via indirect pathways as in the case of MPs (Ruenraroengsak and Tetley 2015). Moreover, amine-modified polystyrene nanoparticles induced significantly more severe mitochondrial damage compared to non-modified ones, suggesting that NPs surface charge plays a role in causing adverse effects on mitochondrial function.

11.3.2 Lysosome Disruption

Lysosome is a single membrane cytoplasmic organelle with high sensitivity to xenobiotics including environmental pollutants. Thus, lysosomal membrane stability can be used as a biomarker to estimate potential impact of the environmental pollutants (Moore, Icarus Allen, and McVeigh 2006). The lysosome function disruption was found in blue mussel (*M. galloprovincialis*) after exposure to both MPs and NPs (Avio et al. 2015). The severity of the lysozyme function disturbance after MPs/NPs exposure suggests high biological relevance of these events, especially in context of lysosome function and structure being conserved across species (Lawrence and Zoncu 2019). Furthermore, the link between oxidative metabolism and ROS generation induced stress has been established,

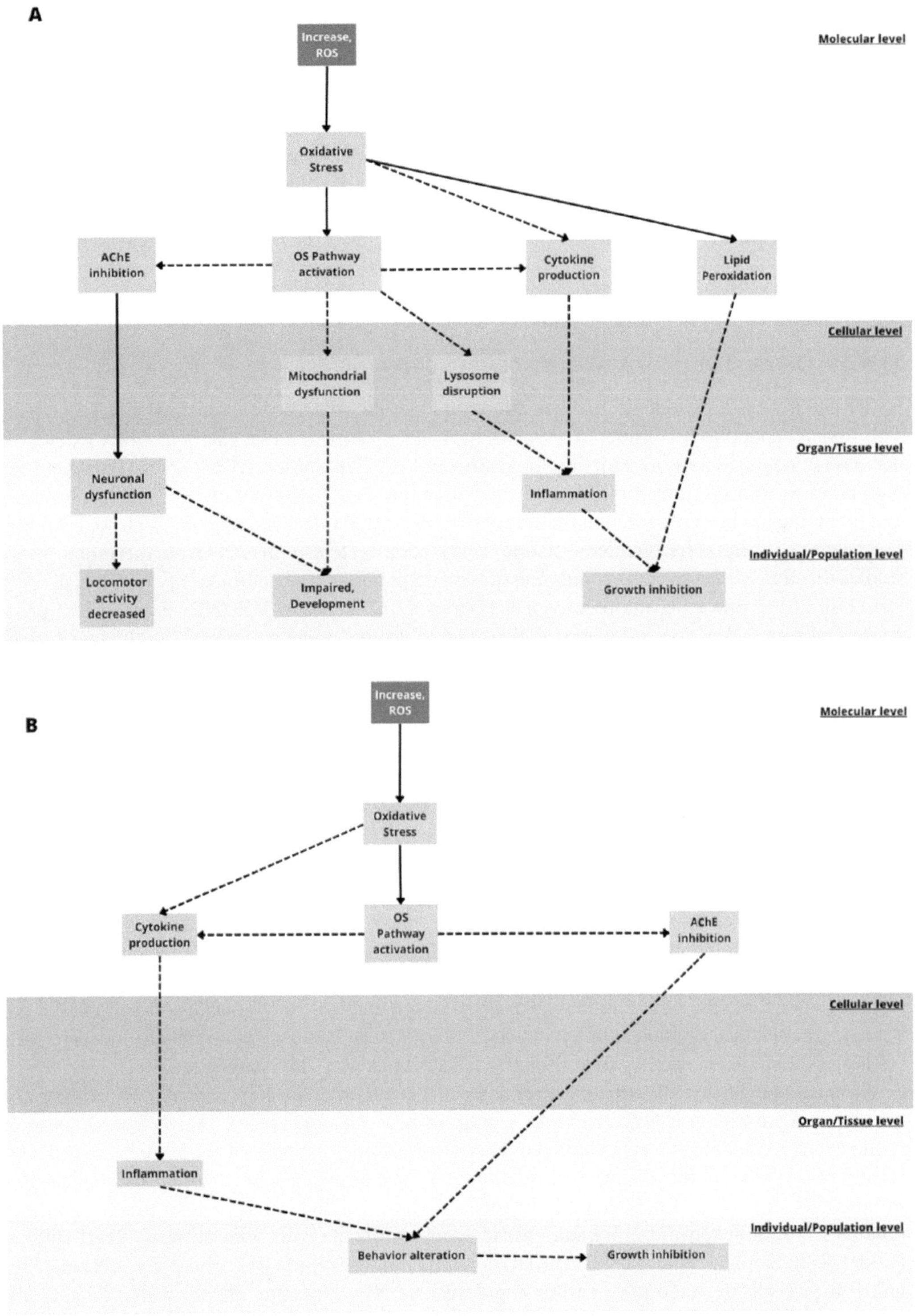

FIGURE 11.1 Schematic diagrams related to possible effects of (A) nanoplastics and (B) combined microplastics and nanoplastics environmental exposure with emphasis on the oxidative stress and inflammation responses. Solid lines: Adjacent or strong evidence relationships; Dashed lines: non-adjacent or weaker evidence supporting the relationship.

including the role of nanomaterials in lysosome function disruption. In summary, two probable MPs/NPs induced mechanisms can lead to lysosomal damage. First is a direct damage after ingestion of plastic particles in the cell via endocytosis or permeation (< 50 nm) and attempts to digest the foreign body which may result in the lysosome disruption (Weber et al. 2018). Another possible mechanism would be an excessive production of ROS induced by MPs and NPs as lysosomal membranes are highly susceptible to oxidative effects of ROS (Regoli and Giuliani 2014). The exact mechanisms of lysosome disruption by MPs/NPs are not fully understood, and this knowledge gap calls for further investigation.

11.4 MICROPLASTICS AND NANOPLASTICS EFFECTS ON TISSUES AND ORGANS OF LIVING ORGANISMS

11.4.1 Decrease of Neuronal Network Function

Inhibition of AChE (Section 11.2.6) was found in a variety of organisms post MPs/NPs exposure. This toxicity endpoint is considered a type of neuronal dysfunction that can potentially extend into decrease of neuronal network function. In addition to AChE inhibition, MPs/NPs may also interfere with other neurotransmitter functions. For example, increased concentration of taurine, aspartate and threonine has been observed in the serum of MPs treated mice, and together with AChE inhibition suggests potential for broader systemic neurotoxicity (Deng et al. 2017). Furthermore, several studies illustrate simultaneous suppression of locomotor ability and neurotoxicity (Brun et al. 2018; Lei et al. 2018), providing indirect evidence about a link between MPs/NPs induced changes in neurotransmitters function and behavior alteration of testing model organisms. As the mechanisms of decreased neuronal network function are far from understood, detailed studies are needed to increase our understanding of MPs/NPs involvement in this key event.

11.4.2 Inflammation

Inflammation is a potent mechanism responsible for defenses against pathogenic and other noxic agents and can be elicited by exposure to, or presence of, xenobiotics or toxicants. In recent years, engineered nanomaterials have entered the environment and can often be found in living organisms and could be considered as a special case of "particulate xenobiotics" that can interact with the immune system (Ng et al. 2018). Inflammatory processes can be very destructive and must be carefully balanced and controlled at multiple levels of complexity. We will focus on three levels of biological organization (molecular, cellular and tissue) of responses involved in inflammatory responses to MPs/NPs. As immune system generally recognizes microplastic and nanoplastic particles as xenobiotics, multiple studies include inflammatory response endpoints to measure MPs/NPs effects both *in vivo* and *in vitro* (Figure 11.2) (Brun et al. 2018; Deng et al. 2017).

At molecular level, inflammatory responses triggered by MPs/NPs exposure mostly involve activation of pro-inflammatory cytokines, signaling molecules secreted primarily by immune cells (leukocytes). Available reports clearly demonstrate the ability of nanoplastic particles to influence release of cytokines or the alteration of inflammatory response genes *in vitro* cell line models. Pro-inflammatory cytokine IL6, IL8 and IL1β genes are up-regulated in response to 10 μg/ml polystyrene NPs (44 nm) exposure in human gastric adenocarcinoma cells with decreased cell viability (Inkielewicz-Stepniak et al. 2018). Similarly, Prietl et al. (2014) report increased secretion of IL6 and IL8 that was detected after carboxyl polystyrene NPs exposure. However, the authors also noted that IL6 production increased with particle size while IL8 secretion showed the opposite trend in THP-1 monocytic cell line, likely due to regulation of immune responses against pathogens of different sizes (Prietl et al. 2014). Details of mechanisms involved in the NPs exposure and cytokines production are not yet clear, however, observed pro-inflammatory responses could relate to the oxidative stress and its corresponding lysosome membrane disintegration (Avio et al.

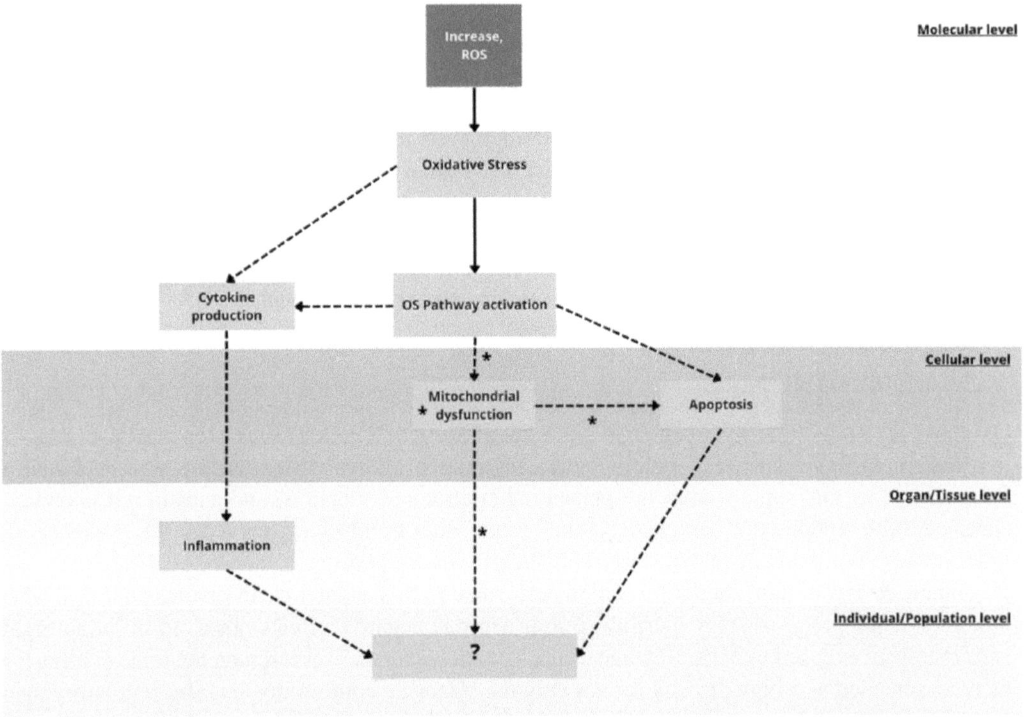

FIGURE 11.2 Schematic diagrams related to possible effects of nanoplastics and combined microplastics and nanoplastics from human/public health perspective with emphasis on the oxidative stress and inflammation responses. Solid lines: Adjacent or strong evidence relationships; Dashed lines: non-adjacent or weaker evidence supporting the relationship. Asterisk (*) indicates mitochondrial dysfunction and related mechanistic relationships that apply to nanoplastics only.

2015). As for the possible effects of MPs, it was reported that 0.5 μm polystyrene MPs increased both mRNA levels of IL1α, IL1β and IFN and their respective protein levels in zebrafish gut (Jin et al. 2018).

The second and third level of inflammatory responses induced by MPs/NPs would be cellular responses and tissue damage, respectively. Innate immune cellular effectors such as phagocytic cells appear to interact frequently with MPs/NPs, as expected, considering they represent the first line of organismal defenses against particulate matter agents. Increase of degranulation of neutrophil primary granules and neutrophil extracellular trap (NET) release was detected after the exposure of both polystyrene and polycarbonate NPs in the fish model (Greven et al. 2016). Release of granular contents as well as NETs release suggested that NPs are capable of phagocyte activation and could function as a potential stressor of innate immunity. Significant influx of neutrophils into the rat lung was observed after inhalation exposure to polystyrene NPs and MPs and the influx was significantly greater for smaller particles (64 nm) compared to larger ones (535 nm), while larger particles caused higher levels of interleukin secretion (Brown et al. 2001). These findings demonstrate the potency of MPs/NPs to induce inflammatory cellular responses.

Inflammatory tissue damage was detected in different *in vivo* models after exposure to either MPs or NPs (Lu et al. 2016; Deng et al. 2017). Polystyrene micro particles (5 μm) and nanoparticles (70 nm) were both accumulating in the gill, liver and gut of zebrafish after 7 days of exposure, causing typical inflammatory damage in liver (vacuolation, leukocyte/neutrophil infiltration, necrosis and lipidosis) (Lu et al. 2016). Liver histopathology findings were supported by increased

SOD and CAT activity, indicating the possibility of underlying oxidative stress as triggering inflammatory response. Studies by Capó et al. (2021) where long-term *Sparus aurata* exposure to both degraded and non-degraded microplastic have resulted in similar increase in SOD and CAT activity, related to the liver inflammation, thus additionally strengthening link between MPs exposure and liver inflammation. Similar hepatotoxicity was histologically detected in mice model when both 5 μm and 20 μm polystyrene MPs induced liver lipid accumulation and inflammation (Deng et al. 2017).

Published findings support that both MPs and NPs exposure can promote inflammation at all three levels (molecular, cellular and tissue), possibly through oxidative stress and lysosome dysfunction but more evidence is required for better mechanistic understanding.

11.5 CONCLUSIONS

Oxidative and inflammatory potential of nano/microplastics (NPs/MPs) has been intensively studied in recent years, as evidenced by rapidly increasing number of available peer-reviewed publications. Despite the fact that some studies are presenting contradictory results, and gaps in our knowledge continue to be obvious, we can clearly notice a prevalent trend to acknowledge the existence of a link between living organism exposures to MPs/NPs and inflammatory and/or oxidative processes. Systematic review of published information performed in this chapter clearly paints an image where aforementioned processes play a critical role in various negative effects observed in living organisms that were exposed to MPs/NPs materials. Therefore, further research on the topic continues to be recommended as a high priority for not only the scientific community, but also regulatory agencies with a role in environmental and health protection.

REFERENCES

AISBL. 2021. "Plastics—the Facts 2021: An analysis of European plastics production, demand and waste data." accessed 05.12.2022. https://plasticseurope.org/knowledge-hub/plastics-the-facts-2021/.

Avio, Carlo Giacomo, Stefania Gorbi, Massimo Milan, Maura Benedetti, Daniele Fattorini, Giuseppe d'Errico, Marianna Pauletto, Luca Bargelloni, and Francesco Regoli. 2015. "Pollutants bioavailability and toxicological risk from microplastics to marine mussels." *Environmental Pollution* 198:211–222. https://doi.org/10.1016/j.envpol.2014.12.021.

Barboza, Luís Gabriel Antão, Luís Russo Vieira, Vasco Branco, Neusa Figueiredo, Felix Carvalho, Cristina Carvalho, and Lúcia Guilhermino. 2018. "Microplastics cause neurotoxicity, oxidative damage and energy-related changes and interact with the bioaccumulation of mercury in the European seabass, Dicentrarchus labrax (Linnaeus, 1758)." *Aquatic Toxicology* 195:49–57. https://doi.org/10.1016/j.aquatox.2017.12.008.

Brandts, I., M. Teles, A. P. Gonçalves, A. Barreto, L. Franco-Martinez, A. Tvarijonaviciute, M. A. Martins, A. M. V. M. Soares, L. Tort, and M. Oliveira. 2018. "Effects of nanoplastics on Mytilus galloprovincialis after individual and combined exposure with carbamazepine." *Science of The Total Environment* 643:775–784. https://doi.org/10.1016/j.scitotenv.2018.06.257.

Brown, D. M., M. R. Wilson, W. MacNee, V. Stone, and K. Donaldson. 2001. "Size-dependent proinflammatory effects of ultrafine polystyrene particles: A role for surface area and oxidative stress in the enhanced activity of ultrafines." *Toxicology and Applied Pharmacology* 175 (3):191–199. https://doi.org/10.1006/taap.2001.9240.

Brun, Nadja R., Bjørn E. V. Koch, Mónica Varela, Willie J. G. M. Peijnenburg, Herman P. Spaink, and Martina G. Vijver. 2018. "Nanoparticles induce dermal and intestinal innate immune system responses in zebrafish embryos." *Environmental Science: Nano* 5 (4):904–916. doi: 10.1039/C8EN00002F.

Capó, X., J. J. Company, C. Alomar, M. Compa, A. Sureda, A. Grau, B. Hansjosten, J. López-Vázquez, J. B. Quintana, R. Rodil, and S. Deudero. 2021. "Long-term exposure to virgin and seawater exposed microplastic enriched-diet causes liver oxidative stress and inflammation in gilthead seabream Sparus aurata, Linnaeus 1758." *Science of The Total Environment* 767:144976. https://doi.org/10.1016/j.scitotenv.2021.144976.

Chen, Qiqing, Michael Gundlach, Shouye Yang, Jing Jiang, Mirna Velki, Daqiang Yin, and Henner Hollert. 2017. "Quantitative investigation of the mechanisms of microplastics and nanoplastics toward zebrafish larvae locomotor activity." *Science of The Total Environment* 584–585:1022–1031. https://doi.org/10.1016/j.scitotenv.2017.01.156.

Corcoran, Patricia L. 2022. "Degradation of microplastics in the environment." In *Handbook of Microplastics in the Environment*, edited by Teresa Rocha-Santos, Monica F. Costa and Catherine Mouneyrac, 531–542. Cham: Springer International Publishing.

de Souza Machado, Anderson Abel, Werner Kloas, Christiane Zarfl, Stefan Hempel, and Matthias C. Rillig. 2018. "Microplastics as an emerging threat to terrestrial ecosystems." 24 (4):1405–1416. https://doi.org/10.1111/gcb.14020.

Deng, Yongfeng, Yan Zhang, Bernardo Lemos, and Hongqiang Ren. 2017. "Tissue accumulation of microplastics in mice and biomarker responses suggest widespread health risks of exposure." *Scientific Reports* 7 (1):46687. doi: 10.1038/srep46687.

Duan, Jiajun, Yang Li, Jianan Gao, Runzi Cao, Enxiang Shang, and Wen Zhang. 2022. "ROS-mediated photoaging pathways of nano- and micro-plastic particles under UV irradiation." *Water Research* 216:118320. https://doi.org/10.1016/j.watres.2022.118320.

Freeman, B. A., and J. D. Crapo. 1982. "Biology of disease: Free radicals and tissue injury." *Laboratory Investigation: A Journal of Technical Methods and Pathology* 47 (5):412–426.

Garcimartín, Alba, M. Elvira López-Oliva, M. Pilar González, Francisco J. Sánchez-Muniz, and Juana Benedí. 2017. "Hydrogen peroxide modifies both activity and isoforms of acetylcholinesterase in human neuroblastoma SH-SY5Y cells." *Redox Biology* 12:719–726. https://doi.org/10.1016/j.redox.2017.04.004.

Gigault, Julien, Hind El Hadri, Brian Nguyen, Bruno Grassl, Laura Rowenczyk, Nathalie Tufenkji, Siyuan Feng, and Mark Wiesner. 2021. "Nanoplastics are neither microplastics nor engineered nanoparticles." *Nature Nanotechnology* 16 (5):501–507. doi: 10.1038/s41565-021-00886-4.

Greven, Anne-Catherine, Teresa Merk, Filiz Karagöz, Kristin Mohr, Markus Klapper, Boris Jovanović, and Dušan Palić. 2016. "Polycarbonate and polystyrene nanoplastic particles act as stressors to the innate immune system of fathead minnow (Pimephales promelas)." 35 (12):3093–3100. https://doi.org/10.1002/etc.3501.

Ighodaro, O. M., and O. A. Akinloye. 2018. "First line defence antioxidants-superoxide dismutase (SOD), catalase (CAT) and glutathione peroxidase (GPX): Their fundamental role in the entire antioxidant defence grid." *Alexandria Journal of Medicine* 54 (4):287–293. doi: 10.1016/j.ajme.2017.09.001.

Imhof, Hannes K., Jakub Rusek, Michaela Thiel, Justyna Wolinska, and Christian Laforsch. 2017. "Do microplastic particles affect Daphnia magna at the morphological, life history and molecular level?" *PLoS One* 12 (11):e0187590. doi: 10.1371/journal.pone.0187590.

Inkielewicz-Stepniak, Iwona, Lidia Tajber, Gavin Behan, Hongzhou Zhang, Marek W. Radomski, Carlos Medina, and Maria J. Santos-Martinez. 2018. "The role of mucin in the toxicological impact of polystyrene nanoparticles." *Materials (Basel)* 11 (5):724. doi: 10.3390/ma11050724.

Jeong, Chang-Bum, Hye-Min Kang, Min-Chul Lee, Duck-Hyun Kim, Jeonghoon Han, Dae-Sik Hwang, Sami Souissi, Su-Jae Lee, Kyung-Hoon Shin, Heum Gi Park, and Jae-Seong Lee. 2017. "Adverse effects of microplastics and oxidative stress-induced MAPK/NRF2 pathway-mediated defense mechanisms in the marine copepod Paracyclopina nana." *Scientific Reports* 7 (1):41323. doi: 10.1038/srep41323.

Jeong, Chang-Bum, Eun-Ji Won, Hye-Min Kang, Min-Chul Lee, Dae-Sik Hwang, Un-Ki Hwang, Bingsheng Zhou, Sami Souissi, Su-Jae Lee, and Jae-Seong Lee. 2016. "Microplastic size-dependent toxicity, oxidative stress induction, and p-JNK and p-p38 activation in the Monogonont Rotifer (Brachionus koreanus)." *Environmental Science & Technology* 50 (16):8849–8857. doi: 10.1021/acs.est.6b01441.

Jin, Yuanxiang, Jizhou Xia, Zihong Pan, Jiajing Yang, Wenchao Wang, and Zhengwei Fu. 2018. "Polystyrene microplastics induce microbiota dysbiosis and inflammation in the gut of adult zebrafish." *Environmental Pollution* 235:322–329. https://doi.org/10.1016/j.envpol.2017.12.088.

Lawrence, Rosalie E., and Roberto Zoncu. 2019. "The lysosome as a cellular centre for signalling, metabolism and quality control." *Nature Cell Biology* 21 (2):133–142. doi: 10.1038/s41556-018-0244-7.

Lei, Lili, Mengting Liu, Yang Song, Shibo Lu, Jiani Hu, Chengjin Cao, Bing Xie, Huahong Shi, and Defu He. 2018. "Polystyrene (nano)microplastics cause size-dependent neurotoxicity, oxidative damage and other adverse effects in Caenorhabditis elegans." *Environmental Science: Nano* 5 (8):2009–2020. doi: 10.1039/C8EN00412A.

Li, Siwen, Yu Ma, Shuzi Ye, Sixuan Tang, Ningjuan Liang, Yuehui Liang, and Fang Xiao. 2021. "Polystyrene microplastics trigger hepatocyte apoptosis and abnormal glycolytic flux via ROS-driven calcium overload." *Journal of Hazardous Materials* 417:126025. https://doi.org/10.1016/j.jhazmat.2021.126025.

Lin, Wei, Ruifen Jiang, Sizi Hu, Xiaoying Xiao, Jiayi Wu, Songbo Wei, Yaxin Xiong, and Gangfeng Ouyang. 2019. "Investigating the toxicities of different functionalized polystyrene nanoplastics on Daphnia magna." *Ecotoxicology and Environmental Safety* 180:509–516. https://doi.org/10.1016/j.ecoenv.2019.05.036.

Liu, Yuexian, Wei Li, Fang Lao, Ying Liu, Liming Wang, Ru Bai, Yuliang Zhao, and Chunying Chen. 2011. "Intracellular dynamics of cationic and anionic polystyrene nanoparticles without direct interaction with mitotic spindle and chromosomes." *Biomaterials* 32 (32):8291–8303. https://doi.org/10.1016/j.biomaterials.2011.07.037.

Livingstone, D. R. 2001. "Contaminant-stimulated reactive oxygen species production and oxidative damage in aquatic organisms." *Marine Pollution Bulletin* 42 (8):656–666. doi: 10.1016/s0025-326x(01)00060-1.

Lu, Liang, Zhiqin Wan, Ting Luo, Zhengwei Fu, and Yuanxiang Jin. 2018. "Polystyrene microplastics induce gut microbiota dysbiosis and hepatic lipid metabolism disorder in mice." *Science of The Total Environment* 631:449–458.

Lu, Yifeng, Yan Zhang, Yongfeng Deng, Wei Jiang, Yanping Zhao, Jinju Geng, Lili Ding, Hongqiang Ren. 2016. "Uptake and accumulation of polystyrene microplastics in zebrafish (Danio rerio) and toxic effects in liver." *Journal of Environmental Science and Technology* 50 (7):4054–4060.

Massoulié, Jean, Leo Pezzementi, Suzanne Bon, Eric Krejci, and François-Marie Vallette. 1993. "Molecular and cellular biology of cholinesterases." *Progress in Neurobiology* 41 (1):31–91. https://doi.org/10.1016/0301-0082(93)90040-Y.

McCubrey, James A., Michelle M. LaHair, and Richard A. Franklin. 2006. "Reactive oxygen species-induced activation of the MAP kinase signaling pathways." *Antioxidants & Redox Signaling* 8 (9–10):1775–1789.

Moore, Michael N., J. Icarus Allen, and Allan McVeigh. 2006. "Environmental prognostics: An integrated model supporting lysosomal stress responses as predictive biomarkers of animal health status." *Marine Environmental Research* 61 (3):278–304. https://doi.org/10.1016/j.marenvres.2005.10.005.

Ng, Ee-Ling, Esperanza Huerta Lwanga, Simon M. Eldridge, Priscilla Johnston, Hang-Wei Hu, Violette Geissen, and Deli Chen. 2018. "An overview of microplastic and nanoplastic pollution in agroecosystems." *Science of The Total Environment* 627:1377–1388. https://doi.org/10.1016/j.scitotenv.2018.01.341.

Nobre, C. R., M. F. M. Santana, A. Maluf, F. S. Cortez, A. Cesar, C. D. S. Pereira, and A. Turra. 2015. "Assessment of microplastic toxicity to embryonic development of the sea urchin Lytechinus variegatus (Echinodermata: Echinoidea)." *Marine Pollution Bulletin* 92 (1):99–104. https://doi.org/10.1016/j.marpolbul.2014.12.050.

Oliveira, M., A. Ribeiro, K. Hylland, and L. Guilhermino. 2013. "Single and combined effects of microplastics and pyrene on juveniles (0+ group) of the common goby Pomatoschistus microps (Teleostei, Gobiidae)." *Ecological Indicators* 34:641–647. https://doi.org/10.1016/j.ecolind.2013.06.019.

Paget, Vincent, Samir Dekali, Thierry Kortulewski, Romain Grall, Christelle Gamez, Kelly Blazy, Olivier Aguerre-Chariol, Sylvie Chevillard, Anne Braun, et al. 2015. "Specific uptake and genotoxicity induced by polystyrene nanobeads with distinct surface chemistry on human lung epithelial cells and macrophages." *PLoS One* 10 (4):e0123297. doi: 10.1371/journal.pone.0123297.

Pannetier, Pauline, Jérôme Cachot, Christelle Clérandeau, Florian Faure, Kim Van Arkel, Luiz F. de Alencastro, Clément Levasseur, Frédéric Sciacca, Jean-Pascal Bourgeois, and Bénédicte Morin. 2019. "Toxicity assessment of pollutants sorbed on environmental sample microplastics collected on beaches: Part I-adverse effects on fish cell line." *Environmental Pollution* 248:1088–1097. https://doi.org/10.1016/j.envpol.2018.12.091.

Park, Jun Chul, Jeonghoon Han, Min-Chul Lee, Hye-Min Kang, Chang-Bum Jeong, Dae-Sik Hwang, Minghua Wang, and Jae-Seong Lee. 2017. "Adverse effects of BDE-47 on life cycle parameters, antioxidant system, and activation of MAPK signaling pathway in the rotifer Brachionus koreanus." *Aquatic Toxicology* 186:105–112. https://doi.org/10.1016/j.aquatox.2017.02.025.

Paul-Pont, Ika, Camille Lacroix, Carmen González Fernández, Hélène Hégaret, Christophe Lambert, Nelly Le Goïc, Laura Frère, Anne-Laure Cassone, Rossana Sussarellu, Caroline Fabioux, et al. 2016. "Exposure of marine mussels Mytilus spp. to polystyrene microplastics: Toxicity and influence on fluoranthene bioaccumulation." *Environmental Pollution* 216:724–737. https://doi.org/10.1016/j.envpol.2016.06.039.

Prietl, B., C. Meindl, E. Roblegg, T. R. Pieber, G. Lanzer, and E. Fröhlich. 2014. "Nano-sized and micro-sized polystyrene particles affect phagocyte function." *Cell Biology and Toxicology* 30 (1):1–16. doi: 10.1007/s10565-013-9265-y.

Prinz, Natalie, and Špela Korez. 2020. "Understanding how microplastics affect marine biota on the cellular level is important for assessing ecosystem function: a review." YOUMARES 9-The Oceans: Our Research, Our Future: Proceedings of the 2018 Conference for YOUng MArine RESearcher in Oldenburg, Germany.

Qiao, Ruxia, Cheng Sheng, Yifeng Lu, Yan Zhang, Hongqiang Ren, and Bernardo Lemos. 2019. "Microplastics induce intestinal inflammation, oxidative stress, and disorders of metabolome and microbiome in zebrafish." *Science of The Total Environment* 662:246–253. https://doi.org/10.1016/j.scitotenv.2019.01.245.

Regoli, Francesco, and Maria Elisa Giuliani. 2014. "Oxidative pathways of chemical toxicity and oxidative stress biomarkers in marine organisms." *Marine Environmental Research* 93:106–117. https://doi.org/10.1016/j.marenvres.2013.07.006.

Ribeiro, Francisca, Ana R. Garcia, Beatriz P. Pereira, Maria Fonseca, Nélia C. Mestre, Tainá G. Fonseca, Laura M. Ilharco, and Maria João Bebianno. 2017. "Microplastics effects in Scrobicularia plana." *Marine Pollution Bulletin* 122 (1):379–391. https://doi.org/10.1016/j.marpolbul.2017.06.078.

Ruenraroengsak, Pakatip, and Teresa D. Tetley. 2015. "Differential bioreactivity of neutral, cationic and anionic polystyrene nanoparticles with cells from the human alveolar compartment: Robust response of alveolar type 1 epithelial cells." *Particle and Fibre Toxicology* 12 (1):19. doi: 10.1186/s12989-015-0091-7.

Schirinzi, Gabriella F., Ignacio Pérez-Pomeda, Josep Sanchís, Cesare Rossini, Marinella Farré, and Damià Barceló. 2017. "Cytotoxic effects of commonly used nanomaterials and microplastics on cerebral and epithelial human cells." *Environmental Research* 159:579–587. https://doi.org/10.1016/j.envres.2017.08.043.

Shen, Fanglin, Dan Li, Jianhua Guo, and Jianmin Chen. 2022. "Mechanistic toxicity assessment of differently sized and charged polystyrene nanoparticles based on human placental cells." *Water Research* 223:118960. https://doi.org/10.1016/j.watres.2022.118960.

Tang, Xinyu, Xue Fan, Tong Xu, Yujiao He, Qianru Chi, Zhe Li, and Shu Li. 2022. "Polystyrene nanoplastics exacerbated lipopolysaccharide-induced necroptosis and inflammation via the ROS/MAPK pathway in mice spleen." 37 (10):2552–2565. https://doi.org/10.1002/tox.23618.

von Moos, Nadia, Patricia Burkhardt-Holm, and Angela Köhler. 2012. "Uptake and effects of microplastics on cells and tissue of the Blue Mussel Mytilus edulis L. after an experimental exposure." *Environmental Science & Technology* 46 (20):11327–11335. doi: 10.1021/es302332w.

Wang, Quanlong, Catharine A. Adams, Fayuan Wang, Yuhuan Sun, and Shuwu Zhang. 2022. "Interactions between microplastics and soil fauna: A critical review." *Critical Reviews in Environmental Science and Technology* 52 (18):3211–3243. doi: 10.1080/10643389.2021.1915035.

Weber, Annkatrin, Christian Scherer, Nicole Brennholt, Georg Reifferscheid, and Martin Wagner. 2018. "PET microplastics do not negatively affect the survival, development, metabolism and feeding activity of the freshwater invertebrate Gammarus pulex." *Environmental Pollution* 234:181–189. https://doi.org/10.1016/j.envpol.2017.11.014.

Yousif, Emad, and Raghad Haddad. 2013. "Photodegradation and photostabilization of polymers, especially polystyrene: Review." *SpringerPlus* 2 (1):398. doi: 10.1186/2193-1801-2-398.

Yuan, Zhihao, Rajat Nag, and Enda Cummins. 2022. "Human health concerns regarding microplastics in the aquatic environment: From marine to food systems." *Science of The Total Environment* 823:153730. https://doi.org/10.1016/j.scitotenv.2022.153730.

Zorov, Dmitry B., Magdalena Juhaszova, and Steven J. Sollott. 2006. "Mitochondrial ROS-induced ROS release: An update and review." *Biochimica et Biophysica Acta (BBA)—Bioenergetics* 1757 (5):509–517. https://doi.org/10.1016/j.bbabio.2006.04.029.

12 Microplastics and Nanoplastics Contamination

An Emerging Environmental Issue for Skin Health

Alessandra Pecorelli and Giuseppe Valacchi

12.1 INTRODUCTION

Plastic pollution is an emerging concern worldwide (Lamichhane et al. 2022). Plastics are everywhere in our life, and, therefore, plastic pollution is now considered ubiquitous in every corner of the globe, as its contamination involves not only marine ecosystems but also freshwater systems, soils, sediments, atmosphere and, clearly, plants and animals (Wu et al. 2019; Jiang et al. 2020).

Since the 1950s, production and consumption of plastic have grown exponentially from two million tons to around 370 million tons in 2019 (Lamichhane et al. 2022). Furthermore, based on a recent report by the Organization for Economic Co-operation and Development, the amount of plastic waste produced globally is projected to almost triple in the coming decades, reaching 1,231 million metric tons by 2050 (Lamichhane et al. 2022).

Currently, a significant contributor to the increase in plastic consumption and disposal can be directly attributed to the COVID-19 pandemic, which has sparked a rush for personal protective equipment including face masks, gloves, face shields, and gowns, but also plastic-packaged food and groceries and disposable utensils (Lamichhane et al. 2022). According to Research and Markets report, as of July 2022, higher investments in the healthcare sector coupled with the growing consumer concerns about their own health, especially in the Asia-Pacific region, are expected to drive the market of the global medical plastics from USD 45.89 billion in 2021 to a value of USD 76.13 billion by 2028. Inadequate disposal of these materials, which are typically non-recyclable and individually treated to prevent infection spread, can become a severe threat to natural ecosystems and human health (Lamichhane et al. 2022).

In a dynamic cycle that includes all of Earth's ecosystems—atmosphere, terrestrial biosphere and aquatic systems—plastics are transported by winds and ocean currents, while being eroded and weathered, breaking down over time into progressively smaller pieces known as microplastics (MPs) and nanoplastics (NPs). Under physical, mechanical and chemical insults and microbial colonization, plastic fragments pollute entire ecosystems, contaminating soil, water and atmosphere and enter the food chain threatening the environment and human health (Lamichhane et al. 2022; Wang et al. 2021).

Recently, research on MPs and NPs has grown exponentially revealing potential harmful implications of long-term, chronic exposure to MPs/NPs on humans. However, knowledge gaps in this thematic area are still relatively broad, especially regarding real levels of human exposure and toxicological effect mechanisms. To better understand the ability of MPs/NPs to cross epithelial barriers, accumulate in tissues/organs, and affect human physiology/homeostasis, further investigations in more reliable models and additional detailed clinical-epidemiological studies are recommended (O'Neill and Lawler 2021; Rahman et al. 2021). Therefore, this chapter summarizes the current

DOI: 10.1201/9781003328100-15

knowledge on the characteristic physical–chemical properties of MPs/NPs and their interactions with other environmental contaminants such as chemicals and pathogens. Existing evidence on routes of human exposure, possible human health risks and toxicological mechanisms of MPs/NPs, especially on skin tissue, are also reviewed.

12.2 MICRO AND NANO-SIZED PLASTICS FEATURES

Plastic debris enters the different ecosystems in a wide variety of sizes, shapes, chemical composition, specific density and color. According to their size, plastic debris can be categorized as mega- (> 50 cm), macro- (between 5 cm and < 50 cm), meso- (between 5 mm and < 5 cm), micro- (between 100 nm and < 5 mm) and nano- (< 100 nm) plastics (Figure 12.1) (Hirt and Body-Malapel 2020).

In particular, primary MPs and NPs comprise plastic fragments, small pellets and beads of microscopic size intentionally produced in their size and shape for specific applications. Some examples include microbeads used as additive in cosmetics, personal care and medicine products (i.e. makeup, toothpastes, exfoliants, scrubs, soaps, lotions, deodorants, drug delivery particles in medicines, etc.) or pellets utilized as raw material in the plastics manufacturing (Figure 12.1) (Lamichhane et al. 2022; Wu et al. 2019; Yee et al. 2021). Secondary MPs and NPs, instead, arise from the breakdown of larger plastic products by the action of chemical, thermal, biological, photolytic, hydrolytic and

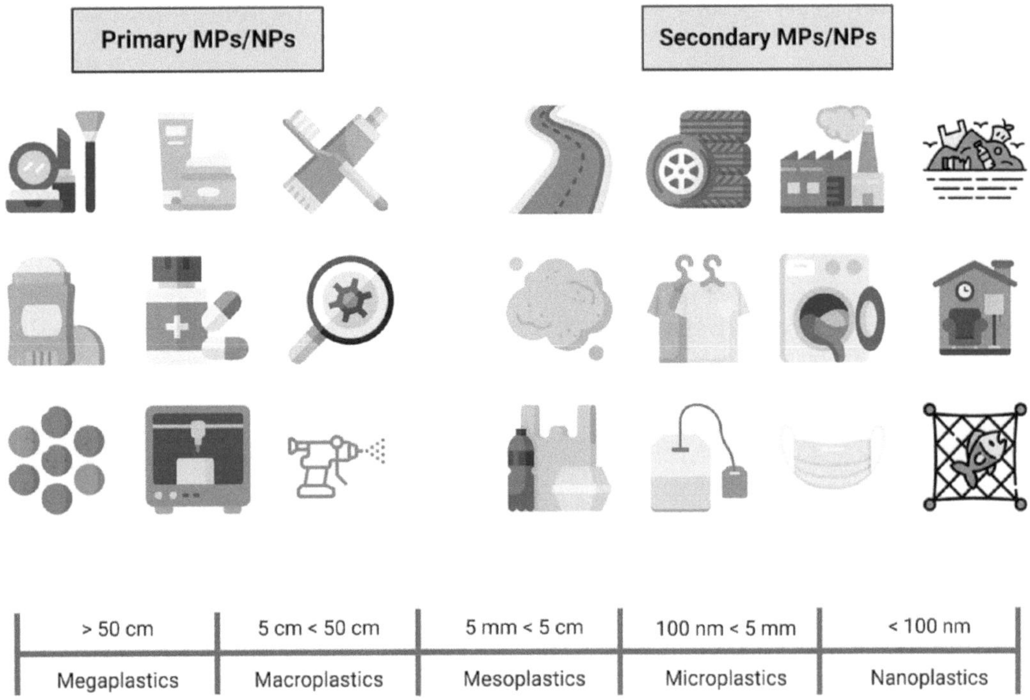

FIGURE 12.1 Sources and sizes of plastics and MPs/NPs into the Earth's ecosystems. Primary MPs are originally manufactured in micrometer-size (0.1 μm–5 mm). Primary MPs/NPs include microbeads used in pharmaceuticals and personal care products, nanoparticles in the field of nanoscience and nanotechnology, industrial plastic powders and pellets and particles in paints and industrial blasting abrasives. Secondary MPs/NPs originate from the breakdown of larger plastics by the action of chemical, thermal, biological, photolytic, hydrolytic and mechanical agents. Main sources include car tires, road markings, fibers in synthetic textiles, marine coatings, city dust, municipal solid waste landfills, face masks, food containers, teabags, cigarette butts and disposable cups. Icons downloaded from Flaticon.com.

mechanical agents such as UV-visible light degradation, photodegradation, mechanical abrasion, wave action, temperature fluctuations and microbial degradation (Figure 12.1) (Wu et al. 2019; Yee et al. 2021). Importantly, multiple processes can act simultaneously with different alternative fates for plastic degradation in different environmental compartments (Chamas et al. 2020). Moreover, secondary MPs and NPs show different features compared to their corresponding starting materials, which further exacerbates concerns about these emerging contaminants.

Plastics degrade at a very slow rate making them pervasive and permanent pollutants capable of creating a serious threat to the different ecosystems (Yee et al. 2021; Chamas et al. 2020). The thermo-oxidative degradation mediated by UV light occurs more rapidly in the coastal areas than in the ocean. However, due to lower temperatures and oxygen concentrations in the marine environment, other studies have reported a slightly higher degradation rate of plastics on land than in aquatic ecosystems (Chamas et al. 2020).

A basic property of MPs and NPs is their shape that varies from microbeads, to microfibers, granules, films, foam and fragments (Wu et al. 2019). Another typical form of primary MPs is the lentil-sized shape of the plastic pellets used in the manufacturing of plastic products. In particular, primary MPs occur in the environment as pellets, microbeads and fragments. Secondary MPs—generated by the degradation and fragmentation of macroplastics through environmental weathering and aging—are mainly shaped as fragments and films.

Fibers and fragments are generally the most abundant morphotypes of MPs found to pollute the marine environment where their abundance is significantly influenced by anthropogenic activities such as commercial fishing, aquaculture, shipping and tourism (Wu et al. 2019). Depending on their density, these contaminants can pollute both the water columns and the sediments, with low-density MPs floating predominantly in seawater, while high-density MPs accumulating in seafloor sediments. Furthermore, a bidirectional exchange between sediments and seawater as well as with the atmosphere contributes to a recirculation of low-density MPs to a greater extent than higher-density debris (Wu et al. 2019).

While MPs/NPs pollution in aquatic ecosystems has been extensively studied, interest in their presence in the atmosphere has only recently gained attention. MPs and NPs have been found in the air of urban, suburban and rural areas; in addition, tropospheric transport of MPs and NPs is considered as major pathway to pollute not only continents but also oceans and even pristine polar regions (Zhang et al. 2020; Evangeliou et al. 2020; Allen et al. 2022). Sizes, shapes and densities affect MPs' aerodynamics and, therefore, their ability to be transported over longer distances from their sources. Different types of shapes have been detected in atmospheric MPs such as fibers, foam, fragments and films; among them, fibers and fragments are originated, respectively, from fabric such as clothes and textiles and from disposable plastics via fragmentation (Zhang et al. 2020). Therefore, fibers are more abundant in the urban areas while fragments are the prevalent type in industrial areas (Zhang et al. 2020; Wang et al. 2021).

MPs and NPs are also considerably heterogeneous in terms of chemical composition and the abundance of the various polymers is highly variable over different geographical areas. However, this divergence may be partly related to the difference in sampling techniques and identification approaches (Cowger et al. 2020). Currently, six main plastics groups dominate the production and consumption patterns of plastics around the world: polypropylene (PP), high and low density polyethylene (PE), polyvinyl chloride (PVC), polystyrene (PS), polyurethane (PUR) and polyethylene terephthalate (PET) (Wu et al. 2019; Zhang et al. 2020). Accordingly, polyethylene, polypropylene and polystyrene are the prevalent polymers in MPs/NPs found in marine ecosystems (Erni-Cassola et al. 2019). Likewise, chemical composition of airborne MPs is mainly characterized by polymers like polyethylene, polyethylene terephthalate, polypropylene and polystyrene, although at least 20 different types of polymers have been detected in atmospheric samples (Munyaneza et al. 2022; Wang et al. 2021).

Different polymer densities affect the distribution of MPs in water as well as in air. For example, polyethylene, polypropylene and some forms of polystyrene, which are less dense polymers, are found in the surface water and atmospheric fallout. On the other hand, high-density polymers such as polystyrene, polyvinyl chloride and polyesters (PES) tend to sink in the seafloor sediments and are observed in atmospheric deposition samples (Munyaneza et al. 2022; Zhang et al. 2020). Biofouling and aggregation can also increase the density of less dense polymers causing an alteration of their buoyancy and consequent sinking below the sea surface (Wu et al. 2019; Erni-Cassola et al. 2019).

Besides heterogeneity in size, shape and chemical composition, MPs/NPs recovered from water and air samples show different colors ranging from white and transparent/translucent to black, red, pink, blue/green, brown, grey, purple and yellow. The most frequent colors in air samples are transparent, white, black, blue and yellow; while red, green, brown and gray MPs show lower prevalence (Joana C. Prata et al. 2020; Wang et al. 2021). In an analysis on nearly 9,000 plastic debris collected from the surface water in ocean basins and semi-enclosed seas, Martí and colleagues identified white and transparent/translucent (47%), yellow and brown (26%) and blue-like (9%) as the most abundant colors of MPs in aquatic ecosystems (Martí et al. 2020).

The color of microplastics can be a possible tool for the identification of plastic sources of origin. However, photo-oxidation and degradation degree due to weathering produce changes in plastic color over time, causing the appearance of a yellowish-brown color and making the identification of polymer sources quite misleading (Martí et al. 2020). Nevertheless, color as well as size of plastic debris reveal the exposure time of the sample in the environment and, therefore may serve as useful indicator of the age of plastics (Zhao et al. 2022; Martí et al. 2020).

12.3 MICROPLASTICS AND NANOPLASTICS AS CARRIERS FOR CHEMICALS AND PATHOGENS

A further aggravating feature of MPs/NPs is connected to their large and hydrophobic surface area associated with considerable sorption capacity, making them potential carriers of various other contaminants and generating a secondary pollution (Kinigopoulou et al. 2022). The combination with other stressors, which can be both chemicals and pathogens, changes the behaviors of MPs/NPs as well as adsorbed contaminants (Jiang et al. 2020). In addition, the components adsorbed to the MPs/NPs are exposed to UV radiation and oxidation, which can increase their toxicity.

Several factors can contribute to the sorption ability of MPs/NPs such as the morphological features, chemical composition and aging degree as well as environmental surrounding factors including sunlight, pH, temperature, ionic strength, microorganisms and long exposure times (Joo et al. 2021; Menéndez-Pedriza and Jaumot 2020). In this context, for example, NPs, which due their small size present higher specific surface area for sorption of other contaminants, pose different risks than MPs, especially for human health, also considering their ability to cross biological barriers and translocate in the body tissues after inhalation and/or ingestion (Lai, Liu, and Qu 2022).

Chemical contaminants carried by MPs and NPs include both inorganic and organic compounds. Hazardous chemical ingredients are typically introduced in plastics during the manufacturing as additives or can be adsorbed from the surrounding environment such as, for example, persistent organic pollutants (POPs), pharmaceuticals, pesticides or insecticides, herbicides or heavy metals.

Additives are added to provide the materials certain properties such as color, transparency and to enhance qualities including flexibility, thermoresistance, mechanical, electrical and UV resistance for prolonging their life; however they are in most cases toxic, oncogenic or endocrine active substances (Campanale et al. 2020). Most common additives found in plastics comprise UV filters, plasticizers, flame retardants, antioxidants, acid scavengers, light stabilizers, lubricants, colorants, antistatic compounds, slip agents and thermal stabilizers (Campanale et al. 2020; Hahladakis et al. 2018). The cocktail generated by the combination of MPs of different sizes, shapes and chemical compositions together with a multitude of additives contributes to contaminate water, air, soil and food, leading humans to be exposed to multiple chemicals at the same time.

Several studies have shown higher sorption affinity of MPs for heavy metals such as, among other, chromium, lead, zinc, arsenic, cadmium, copper and nickel (Liu et al. 2022). Metals can be added to plastic products as catalysts, flame retardants, fillers, pigments and stabilizers (Campanale et al. 2020; Godoy et al. 2019). Furthermore, MPs can easily adsorb metals on their surface and processes related to aging and weathering increase their sorption capacity. In addition, biofilms formed on MPs surface can enhance the accumulation of various metals (Liu et al. 2022; Menéndez-Pedriza and Jaumot 2020). Although the interaction between MPs and heavy metals is mainly studied in aquatic systems, it should be noted that heavy metals are common air pollutants, which, therefore, can also be adsorbed by atmospheric MPs/NPs (Chen, Feng, and Wang 2020). However, the contamination of airborne MPs with metals is only marginally addressed; in fact, few papers are available on this topic which needs to be better explored in future research (Facciolà et al. 2021).

Due their hydrophobic nature, MPs easily adsorbed POPs from the environment such as, for example, polychlorinated biphenyl (PCBs), polybrominated diphenyl ethers (PBDEs), dichlorobiphenyl trichloroethane (DDTs), hexachlorocyclohexanes (HCHs) and polycyclic aromatic hydrocarbon (PAHs). Their interactions (i.e. hydrophobic interactions, partition, electrostatic interactions and other non-covalent forces) are dependent on many factors connected to the physicochemical properties of MPs (e.g. size, shape, polymer, functional groups, etc.) as well as POPs (e.g. functional groups and hydrophobicity) with an important contribution of the surrounding environmental conditions (e.g. temperature, pH, ionic strength, colonization by biofilms, etc.). Any factor that affects the interactions between MPs and POPs, including plastic aging, can promote the release of these toxic compounds into the environment. As for metals, to date most of the studies have focused on the role of MPs as POPs carriers in aquatic and coastal ecosystems, however MPs can contribute to the distribution and desorption of these co-contaminants even in the terrestrial environment. Nevertheless, due to the low concentrations of contaminants implicated, in particular the levels of MPs compared to those of POPs, the extent of this phenomenon is generally considered negligible in terms of negative impacts on natural ecosystems. On the other hand, interaction of MPs with POPs can have a more significant impact in specific areas where these pollutants are present in higher levels, contributing to their transfer over long distances (Fu et al. 2021; Gateuille and Naffrechoux 2022).

The mechanisms and dynamics of sorption involved in the interactions between MPs and hydrophilic compounds such as pharmaceuticals are similar to those previously described for other chemicals and dependent on the characteristics of pharmaceutical compounds and polymeric features of MPs as well as effects of the surrounding matrix. The high detection of specific pharmaceuticals into water bodies, including anti-inflammatory drugs, antibiotics, analgesics, antiepileptics, antidepressants, lipid lowering agents, etc., is a growing global concern. The formation of complexes of MPs with pharmaceuticals appears to contribute to the transfer of these contaminants to humans by direct ingestion or through the consumption of contaminated seafood and terrestrial food. Once adsorbed, their biological effects still remain largely unknown and further studies are required to better understand the effective human health risks (Atugoda et al. 2021; Klavins et al. 2022).

Besides accumulating chemical contaminants, MPs can represent a carrier for pathogens that are able to colonize the plastic surfaces. The formation of biofilms, in turn, induces changes in MPs' behaviors and their associated ecological and biological risks. For example, the presence of biofilms on the MPs can promote the sorption and transport of other pollutants such as metals and organic pollutants. Biofilm formation has mainly been studied in aquatic matrix and, as with other MPs interactions, appears to be influenced by MPs properties and surrounding environmental conditions. Colonized MPs serve as a vector for the spread of microbial pathogens such as *Vibrio* spp., which can pose a threat to human health. In this context, a major concern for MPs biofouling is the spread of antibiotic resistance due the increase in horizontal gene transfer of antibiotic-resistant genes (ARGs) (Marathe and Bank 2022). In addition, airborne MPs have been recognized as potential physical vectors of SARS-CoV-2, contributing to spread the COVID-19 pandemic (Alava et al. 2022; Amato-Lourenço et al. 2022). Moreover, even if not pathogenic, microorganisms absorbed on

MPs can affect human microbiome with health implications (Celebi Sozener et al. 2022; Lear et al. 2021; Santos et al. 2022). However, more research is needed in this field.

Although research on MPs has gained significant interest in recent years, there are still aspects in need of further attention about their role as vectors for other pollutants and microorganisms. As an example, beyond the aquatic compartments where the phenomenon is already widely investigated, there is limited knowledge regarding the interactions of MPs/NPs with other common contaminants in terrestrial and atmospheric environmental matrices (Joana C. Prata et al. 2020). Furthermore, the role of MPs in the transfer of other contaminants to human body, the involved mechanisms and the resulting human health risks are areas needing further evaluation (Chen et al. 2022; Munyaneza et al. 2022).

12.4 ROUTES FOR MPS AND NPS ENTRY IN HUMAN BODY

As persistent pollutants that are widespread in a wide range of ecosystems around the world, MPs/NPs inevitably affect living organisms. Specifically, humans can be exposed to MPs/NPs through at least three routes of exposure, which include ingestion, inhalation and skin contact (Figure 12.2) (Amelia et al. 2021; Bhat, Gedik, and Gaga 2022; Facciolà et al. 2021; Joana Correia Prata et al. 2020; Yee et al. 2021). In a recent comprehensive review article, Ramsperger and colleagues discussed the existing literature on MPs/NPs in relation to routes of exposure, mechanisms of translocation across biological barriers and fate in human tissues and fluids. The discrepancies between

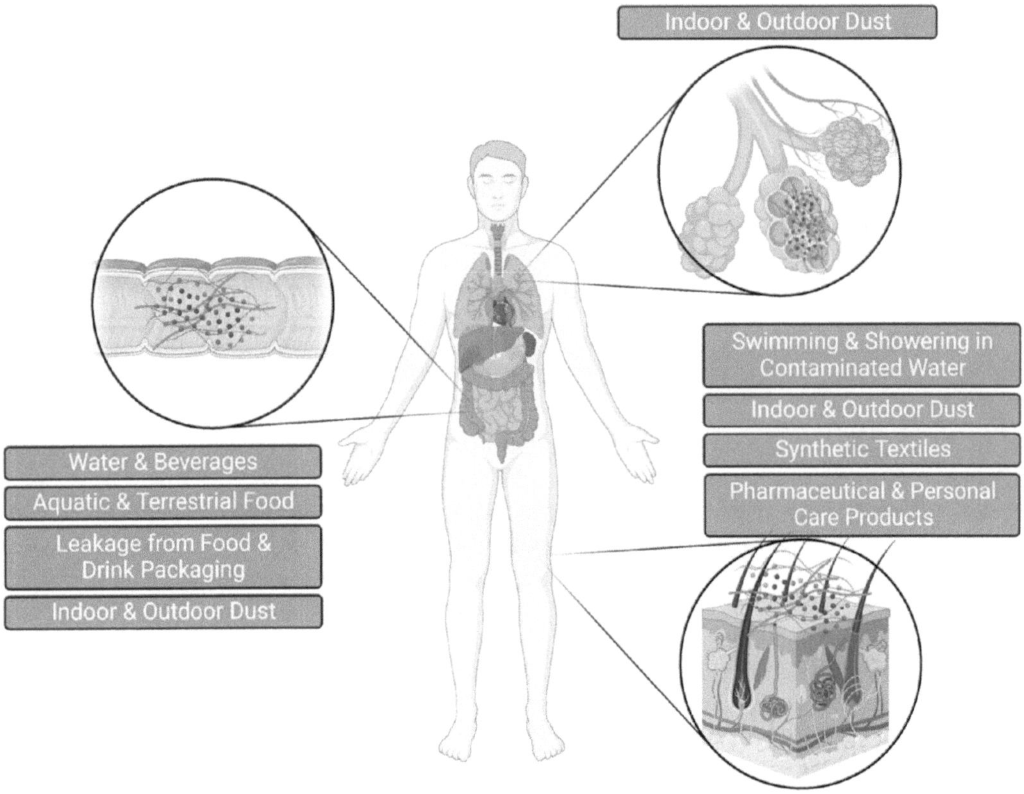

FIGURE 12.2 Summary of the exposure routes for MPs and NPs in human body. A combination of inhalation, ingestion and skin contact allows MPs and NPs enter human body via polluted/contaminated atmosphere, food/drinking water, dust, personal care products or fabrics. Created with BioRender.com.

sizes and properties of particles and characteristics of biological barriers, as well as analytical and methodological limitations in the identification of MPs/NPs in environmental and biological matrices underline the importance of a more critical analysis of current knowledge in this research field that is still in its infancy and the urgent need for further development of reliable methods for studying MP/NP (Ramsperger et al. 2023).

12.4.1 Ingestion

Ingestion of MPs and NPs should probably be regarded as the major route of entry into the human body. MPs pollute drinking water and enter the food chain, as organisms can mistake MPs for food. In addition, it has been demonstrated that not only aquatic organisms but also plants can accumulate MPs and especially NPs that, due to their small size, can enter directly into the plant cell walls (Azeem et al. 2021). Consequently, human exposure can easily occurs by drinking contaminated water and via ingestion of aquatic and terrestrial food products (Jin et al. 2021). A further contribution to ingested MPs can come also from the release and leakage of MPs from food-packaging materials such as, for example, take-out containers, bottles and plastic teabags (Udovicki et al. 2022). Finally, ingestion of dust from indoor and ambient air, especially for children, represents another important source of MPs and NPs entry in human body (Salthammer 2022). Accordingly, MPs and NPs have been detected in human stool samples and, in addition, infant stool seems to have significantly higher concentrations of MPs compared with adult stool, suggesting the infants and children are more likely to be affected by exposures to these contaminants (Yee et al. 2021; Yong, Valiyaveetill, and Tang 2020).

Several studies report estimates for human MPs/NPs ingestion. Based on the American diet, the annual consumption of MPs from food and water ranges from 39,000 to 52,000 particles. If the inhalation contribution is also considered, the estimate rises to over 74,000 and 121,000 MPs each year (Udovicki et al. 2022). Research focused on children estimate that a single infant could intake a range of 14,600–4,550,000 particles/day of MPs from feeding bottles (Amran et al. 2022). Considering various exposure pathways, it has been proposed that people may ingest 0.1–5 g of MPs weekly, i.e. one credit card a week. However, another paper revisited these data and found errors in the calculation leading to a vast overestimation of the total ingested MPs mass by several orders of magnitude. Indeed, based on a more recent analysis, a probable amount of 4.1 μg/week of ingested MPs was calculated for adults (Pletz 2022). In general, what emerges is the lack of accurate information on human exposure concentrations to MPs, mainly due to the continuous evolution of techniques for the quantitative evaluation of MPs.

12.4.2 Inhalation

Because MPs and NPs are ubiquitous in the global ecosystem, including the atmosphere, people are at risk of exposure and potential toxicity including via inhalation or skin contact. Recent evidence suggests that indoor MP inhalation may even be the predominant route of human exposure (Facciolà et al. 2021). However, while MPs contamination of aquatic compartments has been well documented and studied, there is still limited attention to continental contamination; our knowledge of the quantities, characteristics and pathways of atmospheric MPs as well as the health risks for humans needs to be addressed in more detail (Munyaneza et al. 2022).

Atmospheric MPs originate mainly from car tires, brake and road wear, synthetic textiles used in clothes and home furniture and city dust resuspension. Other possible sources are related to waste from incinerators and landfills, industrial emissions and synthetic particles used in horticultural fields (Munyaneza et al. 2022). Interestingly, studies report that MPs fibers shedding from synthetic clothes and textiles during use and washing represent a prevalent source of airborne MPs in both indoor and outdoor air. Furthermore, it is interesting to note that MPs levels are significantly higher in indoor (0.4 to 59.4 fibers/m^3) than in outdoor environments (0 to 1.5 particles/m^3); in addition, indoor MPs show a different composition when compared to outdoor MPs (Munyaneza et al. 2022; Joana C. Prata et al. 2020). A study on three indoor environments found the higher MPs concentrations in buses (17.3 ± 2.4 MPs/m^3) followed by 5.8 ± 1.9 MPs/m^3 in subways, 4.8 ± 1.6 MPs/m^3 in houses, while the lower levels are present in the workplaces (4.2 ± 1.6 MPs/m^3) (Torres-Agullo et al. 2022). Poor ventilation, airflow and room spacing as well as the presence of furniture, textiles, and building materials associated with human activities that continuously produce MPs are considered factors increasing their concentrations in indoor air. On the other hand, dilution and resuspension of outdoor atmosphere contribute to lowering MPs levels in external environments (Munyaneza et al. 2022; Joana C. Prata et al. 2020). Therefore, based on statistics, it is estimated that a person may inhale between 26 and 130 airborne MPs per day, while the total number of MPs inhaled can reach up to 272 MPs per day in an adult male performing light activity (Joana C. Prata et al. 2020).

Considering that people in developed countries spend around 90% of their time indoors, inhaling high concentrations of indoor MPs creates a greater concern in terms of possible risks to human health. Furthermore, alarming evidence highlighted how the large use of masks for reducing the spread of SARS-CoV-2 virus represents not only a critical source of MPs/NPs in the environment but also increases the MP inhalation risk by the wearer. Indeed, while wearing surgical and N95 facemasks could reduce the inhalation and ingestion of airborne MPs by acting as a barrier, on the other hand, it has been shown that, after disinfection treatments, reused masks can release fiber-like MPs that can be inhaled or ingested with possible harmful effects to the wearer's health (Li et al. 2021).

12.4.3 Skin Contact

Due its location and extent, skin is directly exposed to various environmental stressors (i.e. cigarette smoke, ozone, diesel exhaust, particulate matter and UV), which are able to exacerbate cutaneous conditions and aging, also affecting cutaneous microbiome (McDaniel, Farris, and Valacchi 2018; Valacchi et al. 2012). Therefore, it is easy to believe that the skin health can also be affected by MPs/NPs. Cutaneous exposure can occur while swimming or showering in water that contains MPs/NPs or when humans use personal care products containing MPs/NPs (Morgan and DeLouise 2020). The presence of these contaminants in both indoor and outdoor air implies that people may be at risk of potential exposure to MPs/NPs even through indoor dust and atmospheric fallout (Khalid Ageel, Harrad, and Abou-Elwafa Abdallah 2022). However, because these pollutants have emerged as a more recent problem compared to other environmental contaminants, current data on this thematic are still limited.

In general, among the possible pathways of human exposure to MPs/NPs, so far skin contact is considered the least likely route. In fact, the epidermis, i.e. the outermost protective layer of the skin, is a specialized epithelial tissue that functions as an efficient barrier protecting the body against physical, chemical and microbial aggressions. Therefore, large particles cannot cross the stratum corneum, which is composed of cornified dead cells embedded in a matrix of lipid lamellae and plays a crucial role in the skin barrier function. In addition, the hydrophobic properties of plastic particles prevent their entry into the skin via water. The main limitation for MPs/NPs to enter the skin rely on their size; based on that, particles smaller than 100 nm, i.e. NPs, have the potential to permeate the stratum corneum and infiltrate the top layers of the skin at a depth of 2–3 μm (Campanale et al. 2020; Yee et al. 2021). Furthermore, ingredients of health and beauty products (e.g. oleic acid and ethanol) or contaminants such as highly lipophilic organics (e.g. PAHs) can facilitate MPs/NPs penetration across the cutaneous barrier (Dong et al. 2023; Yee et al. 2021).

In a healthy skin, possible routes for the penetration include intercellular, transcellular and through skin pores such as sweat glands and hair follicles as well as skin wounds. Infants and young children are at greater risk for MPs/NPs exposure as the skin barrier may not be fully formed in premature neonates and up to 4 years of age (Amran et al. 2022; Sripada et al. n.d.). Additionally, cutaneous conditions such as diseases (e.g. eczema) or exposure to other external stressors such as

UV light could promote MPs/NPs permeation into a damaged skin barrier with a perturbed expression of tight-junction-related proteins (Yee et al. 2021).

12.5 IMPLICATIONS FOR HUMAN HEALTH: FOCUS ON SKIN

With regard to human health, so far, most of the research conducted on the exposure to MPs/NPs focused on their harmful effects induced on gut and lungs (Figure 12.3). Indeed, after inhalation and ingestion, MPs/NPs are able to rapidly deposit in the respiratory tract and be swallowed, affecting lung and gut epithelia with physical injury by abrasion as well as chemical damage (Joana Correia Prata et al. 2020; Vethaak and Legler 2021; Yee et al. 2021). Besides the disruption of epithelial permeability, MPs/NPs can cause microbiota dysbiosis, leading to lung, gastrointestinal, immune and other system diseases (Hirt and Body-Malapel 2020; Lear et al. 2021; Santos et al. 2022). Once in the intestinal and airway lumens, the smaller-sized fraction of inhaled and ingested MPs (0.1–10 μm) and NPs (< 100 nm) can be internalized and translocate, via phagocytosis or endocytosis, across living cells in the airway and gastrointestinal epithelia to the lymphatic and/or circulatory system and become trapped in the human body throughout the lifetime. In addition, MPs larger than 130 μm can cross the epithelial cell layers via paracellular transport that implies

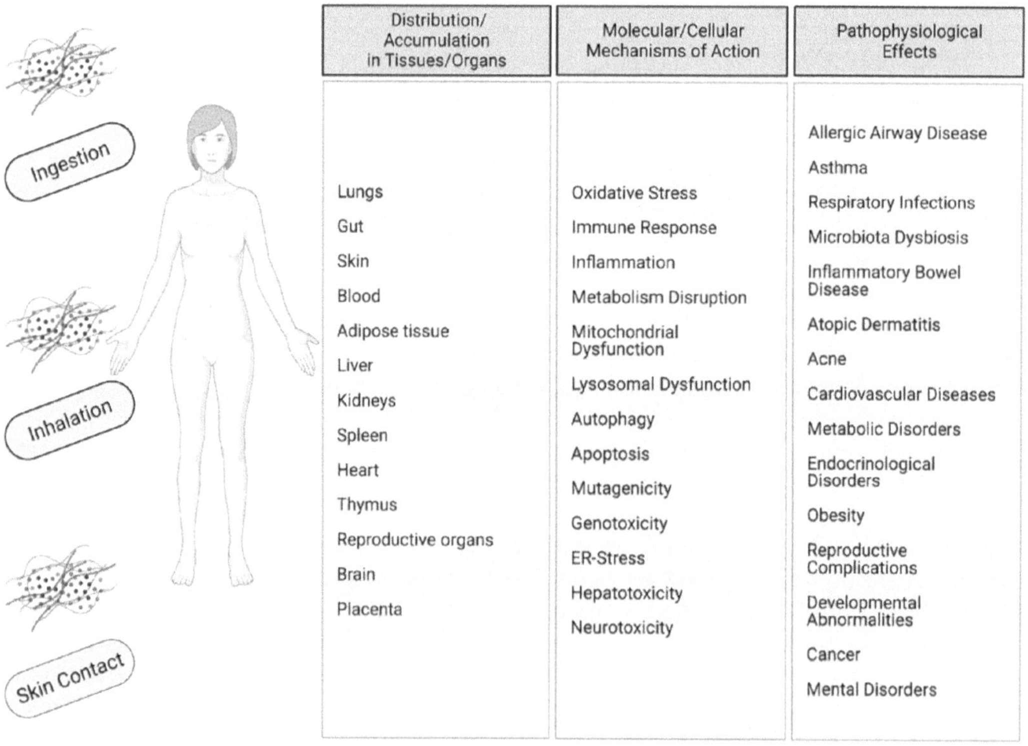

FIGURE 12.3 Impacts of MPs/NPs and associated contaminants on human health. Following inhalation, ingestion and skin contact, small sized MPs and NPs can translocate across living cells and enter the circulatory system to reach distant target tissues and organs into the human body. Harmful effects of MPs and NPs may occur through several mechanisms of action including oxidative stress, inflammation and metabolism disruption, among others. MPs and NPs could contribute to the initiation and exacerbations of human diseases, such as asthma, allergic rhinitis, inflammatory bowel disease, microbiota dysbiosis, atopic dermatitis and some types of cancer. Created with BioRender.com.

mechanical kneading of particles through gaps created in the single-layer epithelium at the villus tips (Campanale et al. 2020).

A recent study found MPs circulating in human blood with an average concentration of 1.6 μg/ml (Leslie et al. 2022). This is the first evidence of their bioavailability for uptake into the human bloodstream that also implies the ability of MPs/NPs to reach any other tissue/organ of our body including, not only in lungs and gut, but also liver, kidneys, spleen, brain and placenta (Joana Correia Prata et al. 2020; Vethaak and Legler 2021; Yee et al. 2021). Due their persistent nature, MPs/NPs induce chronic and intense localized biological responses such as inflammation and oxidative stress, mitochondrial and lysosomal dysfunction, autophagy as well as mutagenicity and genotoxicity, also due to the carried contaminants (Munyaneza et al. 2022; Joana Correia Prata et al. 2020; Vethaak and Legler 2021). Based on individual susceptibility and levels of exposure, MPs/NPs have been associated with respiratory infections, asthma-like reactions, inflammatory bowel disease, cardiovascular diseases, impaired immune function, metabolic and endocrinological disorders, obesity, developmental abnormalities and reproductive complications, neurotoxicity and cancer (Figure 12.3) (Campanale et al. 2020; Lamichhane et al. 2022; Vethaak and Legler 2021; Munyaneza et al. 2022).

Since it is in constant contact with the external environment, it is clear that skin can be affected by MPs/NPs, especially when its barrier properties are compromised by other stressors (Figure 12.4A and 12.4B) (Yee et al. 2021). However, the health impacts of MPs/NPs exposure on skin remain poorly understood. In general, it is believed that, after the contact with the stratum corneum, the potentially harmful effects of MPs/NPs are limited only to the outer layers of the epidermis (Campanale et al. 2020; Yee et al. 2021). Indeed, in a study on porcine skin, confocal images revealed a size-dependent accumulation of polystyrene NPs (20 and 200 nm) preferentially in the follicular openings without penetration in the stratum corneum (Yee et al. 2021). Similarly, after topical application onto hairless rat or human skin, microspheres (5 μm) were specifically visualized within pilosebaceous ducts, while penetration across the stratum corneum was not detected (Rolland et al. 1993). However, using *ex vivo* full-thickness human skin samples, Döge and colleagues reported that 20 and 200 nm polystyrene NPs are able to penetrate deeper into the stratum corneum, in particular via skin furrows, lipid channels and vellus hair follicles, and accumulate within the epidermis cells (Döge et al. 2018). In another study, polystyrene NPs, topically applied on tape-stripped mouse skin, penetrated along the follicular duct and translocated to the surrounding tissue where they were internalized by epidermal and dermal antigen-presenting cells and transported to the proximal lymph nodes (Mahe et al. 2009). Moreover, as demonstrated in a model of dithranol-induced irritative dermatitis in mice, cutaneous inflammation enhances the deeper penetration of NPs with preferential accumulation in hair follicles and sebaceous glands (Abdel-Mottaleb et al. 2012a).

Thus, despite some conflicting results, it is highly probable that the smaller NPs penetrate the stratum corneum and accumulate in viable epidermal cells. Once internalized, MPs/NPs can cause alterations of several signaling pathways and cellular processes (Figure 12.4C). In the study by Gopinath et al., after protein-corona formation, NPs (30–300 nm) isolated from commercial face-scrubs were rapidly recognized by keratinocytes and internalized via micropinocytosis (Gopinath et al. 2021). The internalization triggered lysosomal activity resulting in particle degradation and release of additives within cells. In addition, in a dose-dependent manner, internalized NPs caused reactive oxygen species (ROS) production, proliferation inhibition, senescence and autophagy activity (Gopinath et al. 2021). Accordingly, polyethylene MPs (circa 30 and 6 μm) induced high levels of nitrite associated with a proinflammatory response in HaCaT keratinocyte cell line (i.e. increased interleukin 6 release) (Gautam et al. 2022). Polystyrene MPs were not toxic to human dermal fibroblasts at an experimental concentration of circa 500 μg/mL, while they were mostly located within the cytoplasm (Hwang et al. 2020). Nevertheless, when dermal fibroblasts were treated with secondary MPs (i.e. randomly-shaped polystyrene microfragments) ROS production, apoptotic structures and autophagosomes and cytotoxicity by physical disruption of the plasma membrane were observed (Choi et al. 2020). Furthermore, polystyrene NPs caused increase in ROS production and genotoxic

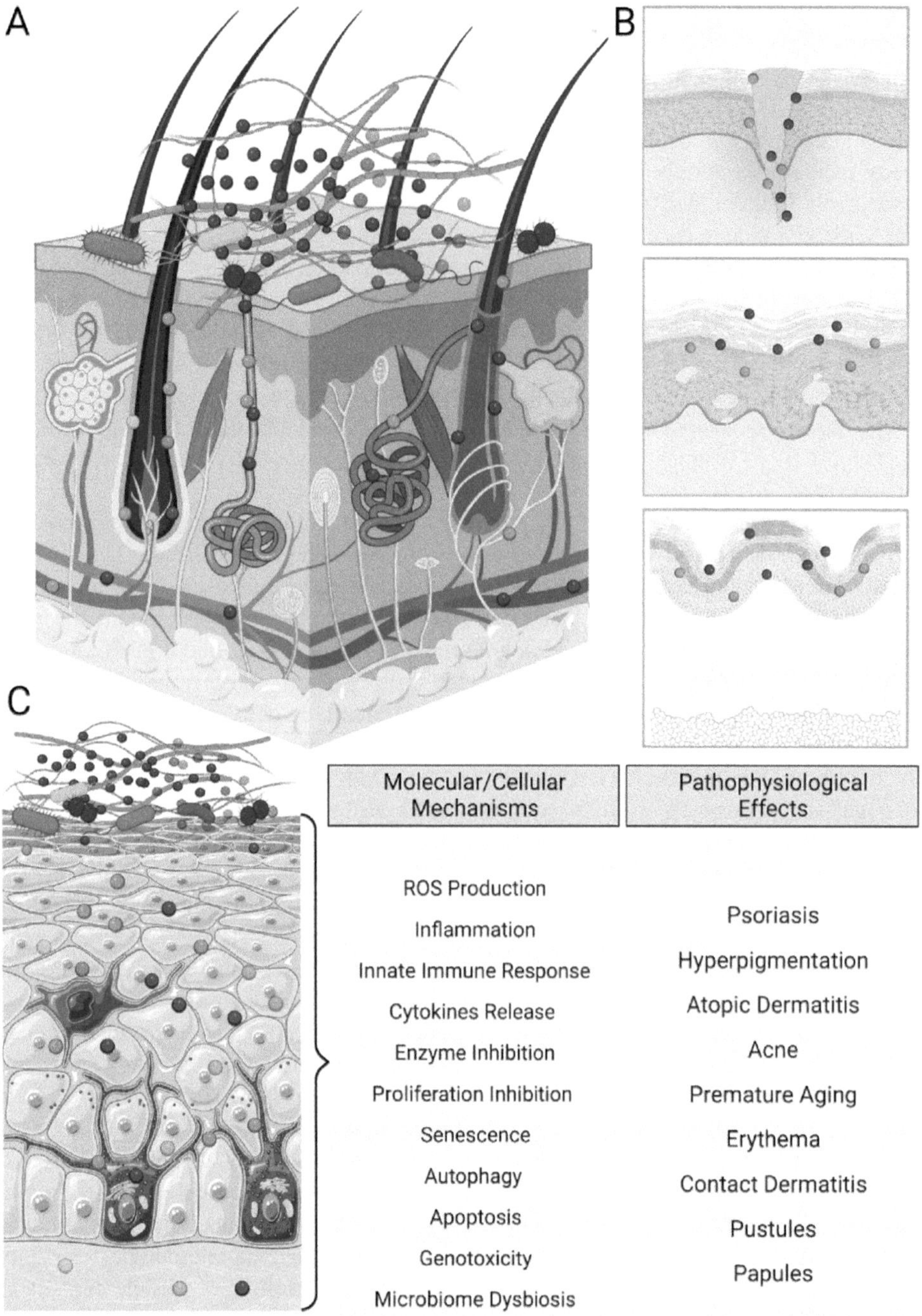

FIGURE 12.4 Mechanisms of exposure to MPs/NPs through the skin tissue and potential health effects. (A) Hair follicles and sweat and sebaceous glands provide MPs/NPs with potential pathways for penetration of the stratum corneum into intact, healthy skin. In the dermis, MPs/NPs could gain access to the lymphatic and/or circulatory system and could be translocated to other tissues and organs. MPs/NPs unable to penetrate the epidermal barrier can affect skin microbiome. (B) Wounds and damage to skin (e.g. eczema and photoaging) accelerates penetration of MPs/NPs into deeper skin. (C) Once internalized, MPs/NPs could cause alterations of several signaling pathways and cellular processes, contributing to skin conditions including psoriasis, acne, premature aging, etc. Created with BioRender.com.

effects in the human fibroblast foreskin Hs27 cell line (Poma et al. 2019). Furthermore, other effects of MPs/NPs on human dermal fibroblasts are summarized in the review article by Danopoulos et al. confirming their ability to affect skin cell homeostasis, at least in in vitro approaches (Danopoulos et al. 2022).

Because MPs/NPs can cross the epidermal barrier and then reach the dermis, they could activate skin-resident immune cells, for example, dermal dendritic cells, Langerhans cells and resident T cells and B cells, eliciting innate immune responses (Yee et al. 2021). Internalization of MPs/NPs by immune cells triggers ROS production, disturbs cellular signaling pathways, modulates the transcriptional regulation and enzymatic activities and induces release of cytokines (Yang et al. 2022). Collectively, the disruption of these cellular processes could initiate a series of inflammatory responses and even lead to allergic reactions and other pathological conditions. Such phenomena have already been demonstrated in the gastrointestinal and respiratory systems and associated with pathological disorders including inflammatory bowel disease and interstitial lung disease (Celebi Sozener et al. 2022; Hirt and Body-Malapel 2020; Lamichhane et al. 2022; Wang et al. 2021). Nonetheless, a role for MPs/NPs in the development of inflammatory skin conditions still needs to be explored.

Nonetheless, even MPs/NPs that are unable to penetrate the epidermal barrier and remain attached/packed on the skin surface (i.e. entanglement) can cause harmful effects on the tissue, for example by affecting skin microbiome (Celebi Sozener et al. 2022; Santos et al. 2022). In this context, it is possible that the higher incidence of skin manifestations such as acne, erythema, contact dermatitis, pustules, papules and pigmentation related to the prolonged use of the mask during the COVID-19 pandemic may in part to be linked to MPs/NPs-induced skin dysbiosis (Ganesapillai et al. 2022).

Finally, another concern about the impact of MPs/NPs exposure on human health is that the cutaneous tissue could have a significant role in the systemic effects of these pollutants. In fact, once penetrated into deeper skin layers, MPs/NPs internalized by skin macrophages and Langerhans cells can migrate to lymph nodes and be transported to other tissues and organs through the circulatory system (Campanale et al. 2020). This phenomenon could be particularly relevant, taking into account that the skin is the largest organ of our body. In fact, considering the surface area of the interfollicular epithelium and other appendages such as sweat and sebaceous glands, the total area of the skin in adults is at least 30 m^2 (Gallo 2017), a surface that provides a vast interface for the interaction and penetration of MPs/NPs into the body. On the other hand, given the potential ability of ingested/inhaled MPs/NPs to produce systemic consequences after the entry in circulation and accumulation in different target tissues, it is also likely that MPs/NPs and their contaminants could disturb skin physiology also acting from the inside our body, for example by inducing skin irritations (Atugoda et al. 2021). In particular, circulating MPs/NPs could perturb redox and immune homeostasis in the vascularized dermal layer of the skin with consequent repercussions to the epidermal keratinocytes and their ability to act as a shield against external stressors including pathogens.

However, to date, there are not enough speculative data on the hypothesis of internal and external cutaneous exposure to MPs/NPs, therefore, it is urgent to pay more attention to the toxic effects of MPs/NPs to the skin and the role played by this tissue as entry route of MPs/NPs into the human body. Overall, based on the current evidence, it is possible assume that the long-term exposures, accumulation and persistent nature of MPs/NPs can promote subclinical and chronic detrimental processes, leading to the development or exacerbation of preexistent skin disorders including psoriasis, hyperpigmentation, atopic dermatitis, acne and even premature aging (Celebi Sozener et al. 2022; Enyoh et al. 2020).

12.6 CONCLUSIONS

Considering the expected surge in plastic consumption and global production, especially due to the COVID-19 pandemic, the environmental levels of MPs/NPs are projected to increase significantly in the near future. Therefore, as an integral part of the exposome/pollutome to which we are exposed,

MPs/NPs pollution can no longer be ignored, and more research is needed to evaluate its pathological mechanisms and the long-term effects on tissues/organs including skin (Munyaneza et al. 2022).

Meanwhile, despite the lack of data directly relevant to humans, investigations on *in vitro* cell systems and in animals have clearly highlighted the harmful effects of MPs/NPs on skin cells and the ability of these pollutants to deeply penetrate the epidermal barrier (Yee et al. 2021). Since the skin is a complex organ with multiple functions and a highly dynamic tissue, which is constantly renewed throughout life, future studies using models such as 3D skin tissue equivalents or human skin explants could provide more reliable evidence than a single type of cell such as keratinocytes or dermal fibroblasts (Morgan and DeLouise 2020). Epidemiological and clinical studies using techniques such as tape strips and skin biopsies could be useful to provide further direct evidence and pathological mechanisms of MPs/NPs exposure in human skin, especially in vulnerable individuals, including children, pregnant women and workers in the plastic production and waste management sectors. Furthermore, MPs/NPs are also contaminated by other chemicals and pathogens, whose contribution to the impact of MPs/NPs on skin health has yet to be explored. Finally, human health concerns arising from the presence of MPs/NPs in beauty and personal care products also need to be clarified.

In other tissues, MPs/NPs can lead to a suite of pathological responses including inflammation, genotoxicity, oxidative stress, apoptosis, pyroptosis and necrosis with a broad range of outcomes including tissue damage, fibrosis and carcinogenesis (Yee et al. 2021). Some of these effects could be also related to the ability of MPs/NPs to influence the tissue-specific microbiome, as already observed in the gut (Hirt and Body-Malapel 2020; Lear et al. 2021). Moreover, in a vicious circle, MPs-induced dysbiosis can fuel inflammatory/oxidative responses, triggering the onset of chronic diseases. However, to date, there is no real knowledge whether MPs/NPs can also have an impact on cutaneous microbiome, a topic that is worth further investigation. Furthermore, more research is needed to better understand the possibility of synergistic interactions between MPs/NPs and other environmental stressors and their role in the development and/or exacerbation of skin conditions including skin aging.

REFERENCES

Abdel-Mottaleb, Mona M. A., Brice Moulari, Arnaud Beduneau, Yann Pellequer, and Alf Lamprecht. 2012a. "Nanoparticles Enhance Therapeutic Outcome in Inflamed Skin Therapy." *European Journal of Pharmaceutics and Biopharmaceutics* 82 (1): 151–157. https://doi.org/10.1016/j.ejpb.2012.06.006.

Alava, Juan José, Ana Tirapé, Karly McMullen, Miguel Uyaguari, and Gustavo A. Domínguez. 2022. "Microplastics and Macroplastic Debris as Potential Physical Vectors of SARS-CoV-2: A Hypothetical Overview with Implications for Public Health." *Microplastics* 1 (1): 156–166. https://doi.org/10.3390/microplastics1010010.

Allen, Deonie, Steve Allen, Sajjad Abbasi, Alex Baker, Melanie Bergmann, Janice Brahney, Tim Butler, et al. 2022. "Microplastics and Nanoplastics in the Marine-Atmosphere Environment." *Nature Reviews Earth & Environment* 3 (6): 393–405. https://doi.org/10.1038/s43017-022-00292-x.

Amato-Lourenço, Luís Fernando, Natália de Souza Xavier Costa, Kátia Cristina Dantas, Luciana Dos Santos Galvão, Fernando Negri Moralles, Suzette Cleuza Ferreira Spina Lombardi, Alfredo Mendroni Júnior, et al. 2022. "Airborne Microplastics and SARS-CoV-2 in Total Suspended Particles in the Area Surrounding the Largest Medical Centre in Latin America." *Environmental Pollution (Barking, Essex: 1987)* 292 (Pt A): 118299. https://doi.org/10.1016/j.envpol.2021.118299.

Amelia, Tan Suet May, Wan Mohd Afiq Wan Mohd Khalik, Meng Chuan Ong, Yi Ta Shao, Hui-Juan Pan, and Kesaven Bhubalan. 2021. "Marine Microplastics as Vectors of Major Ocean Pollutants and Its Hazards to the Marine Ecosystem and Humans." *Progress in Earth and Planetary Science* 8 (1): 12. https://doi.org/10.1186/s40645-020-00405-4.

Amran, Nur Hanisah, Siti Sarah Mohamad Zaid, Mohd Helmy Mokhtar, Latifah Abd Manaf, and Shatrah Othman. 2022. "Exposure to Microplastics during Early Developmental Stage: Review of Current Evidence." *Toxics* 10 (10): 597. https://doi.org/10.3390/toxics10100597.

Atugoda, Thilakshani, Meththika Vithanage, Hasintha Wijesekara, Nanthi Bolan, Ajit K. Sarmah, Michael S. Bank, Siming You, and Yong Sik Ok. 2021. "Interactions between Microplastics, Pharmaceuticals and Personal Care Products: Implications for Vector Transport." *Environment International* 149 (April): 106367. https://doi.org/10.1016/j.envint.2020.106367.

Azeem, Imran, Muhammad Adeel, Muhammad Arslan Ahmad, Noman Shakoor, Gama Dingba Jiangcuo, Kamran Azeem, Muhammad Ishfaq, et al. 2021. "Uptake and Accumulation of Nano/Microplastics in Plants: A Critical Review." *Nanomaterials* 11 (11): 2935. https://doi.org/10.3390/nano11112935.

Bhat, Mansoor Ahmad, Kadir Gedik, and Eftade O. Gaga. 2022. "Atmospheric Micro (Nano) Plastics: Future Growing Concerns for Human Health." *Air Quality, Atmosphere & Health*, October. https://doi.org/10.1007/s11869-022-01272-2.

Campanale, Claudia, Carmine Massarelli, Ilaria Savino, Vito Locaputo, and Vito Felice Uricchio. 2020. "A Detailed Review Study on Potential Effects of Microplastics and Additives of Concern on Human Health." *International Journal of Environmental Research and Public Health* 17 (4). https://doi.org/10.3390/ijerph17041212.

Celebi Sozener, Zeynep, Betul Ozdel Ozturk, Pamir Cerci, Murat Turk, Begum Gorgulu Akin, Mubeccel Akdis, Seda Altiner, et al. 2022. "Epithelial Barrier Hypothesis: Effect of the External Exposome on the Microbiome and Epithelial Barriers in Allergic Disease." *Allergy* 77 (5): 1418–1449. https://doi.org/10.1111/all.15240.

Chamas, Ali, Hyunjin Moon, Jiajia Zheng, Yang Qiu, Tarnuma Tabassum, Jun Hee Jang, Mahdi Abu-Omar, Susannah L. Scott, and Sangwon Suh. 2020. "Degradation Rates of Plastics in the Environment." *ACS Sustainable Chemistry & Engineering* 8 (9): 3494–3511. https://doi.org/10.1021/acssuschemeng.9b06635.

Chen, Guanglong, Qingyuan Feng, and Jun Wang. 2020. "Mini-Review of Microplastics in the Atmosphere and Their Risks to Humans." *Science of The Total Environment* 703 (February): 135504. https://doi.org/10.1016/j.scitotenv.2019.135504.

Chen, Xue-juan, Jin-jin Ma, Rui-lian Yu, Gong-ren Hu, and Yu Yan. 2022. "Bioaccessibility of Microplastic-Associated Heavy Metals Using an in Vitro Digestion Model and Its Implications for Human Health Risk Assessment." *Environmental Science and Pollution Research* 29 (51): 76983–76991. https://doi.org/10.1007/s11356-022-20983-8.

Choi, Daheui, Junah Bang, Taeho Kim, Yoogyeong Oh, Youngdeok Hwang, and Jinkee Hong. 2020. "In Vitro Chemical and Physical Toxicities of Polystyrene Microfragments in Human-Derived Cells." *Journal of Hazardous Materials* 400 (December): 123308. https://doi.org/10.1016/j.jhazmat.2020.123308.

Cowger, Win, Andy M. Booth, Bonnie M. Hamilton, Clara Thaysen, Sebastian Primpke, Keenan Munno, Amy L. Lusher, et al. 2020. "Reporting Guidelines to Increase the Reproducibility and Comparability of Research on Microplastics." *Applied Spectroscopy* 74 (9): 1066–1077. https://doi.org/10.1177/0003702820930292.

Danopoulos, Evangelos, Maureen Twiddy, Robert West, and Jeanette M. Rotchell. 2022. "A Rapid Review and Meta-Regression Analyses of the Toxicological Impacts of Microplastic Exposure in Human Cells." *Journal of Hazardous Materials* 427 (April): 127861. https://doi.org/10.1016/j.jhazmat.2021.127861.

Döge, Nadine, Sabrina Hadam, Pierre Volz, Alexander Wolf, Karl-Heinz Schönborn, Ulrike Blume-Peytavi, Ulrike Alexiev, and Annika Vogt. 2018. "Identification of Polystyrene Nanoparticle Penetration across Intact Skin Barrier as Rare Event at Sites of Focal Particle Aggregations." *Journal of Biophotonics* 11 (4): e201700169. https://doi.org/10.1002/jbio.201700169.

Dong, Xusheng, Xinbei Liu, Qiuling Hou, and Zhonghua Wang. 2023. "From Natural Environment to Animal Tissues: A Review of Microplastics (Nanoplastics) Translocation and Hazards Studies." *Science of The Total Environment* 855 (January): 158686. https://doi.org/10.1016/j.scitotenv.2022.158686.

Enyoh, Christian Ebere, Leila Shafea, Andrew Wirnkor Verla, Evelyn Ngozi Verla, Wang Qingyue, Tanzin Chowdhury, and Marcel Paredes. 2020. "Microplastics Exposure Routes and Toxicity Studies to Ecosystems: An Overview." *Environmental Analysis, Health and Toxicology* 35 (1): e2020004. https://doi.org/10.5620/eaht.e2020004.

Erni-Cassola, Gabriel, Vinko Zadjelovic, Matthew I. Gibson, and Joseph A. Christie-Oleza. 2019. "Distribution of Plastic Polymer Types in the Marine Environment; A Meta-Analysis." *Journal of Hazardous Materials* 369 (May): 691–698. https://doi.org/10.1016/j.jhazmat.2019.02.067.

Evangeliou, N., H. Grythe, Z. Klimont, C. Heyes, S. Eckhardt, S. Lopez-Aparicio, and A. Stohl. 2020. "Atmospheric Transport Is a Major Pathway of Microplastics to Remote Regions." *Nature Communications* 11 (1): 3381. https://doi.org/10.1038/s41467-020-17201-9.

Facciolà, Alessio, Giuseppa Visalli, Marianna Pruiti Ciarello, and Angela Di Pietro. 2021. "Newly Emerging Airborne Pollutants: Current Knowledge of Health Impact of Micro and Nanoplastics." *International Journal of Environmental Research and Public Health* 18 (6): 2997. https://doi.org/10.3390/ijerph18062997.

Fu, Lina, Jing Li, Guoyu Wang, Yaning Luan, and Wei Dai. 2021. "Adsorption Behavior of Organic Pollutants on Microplastics." *Ecotoxicology and Environmental Safety* 217 (July): 112207. https://doi.org/10.1016/j.ecoenv.2021.112207.

Gallo, Richard L. 2017. "Human Skin Is the Largest Epithelial Surface for Interaction with Microbes." *The Journal of Investigative Dermatology* 137 (6): 1213–1214. https://doi.org/10.1016/j.jid.2016.11.045.

Ganesapillai, Mahesh, Bidisha Mondal, Ishita Sarkar, Aritro Sinha, Saikat Sinha Ray, Young-Nam Kwon, Kazuho Nakamura, and K. Govardhan. 2022. "The Face behind the Covid-19 Mask: A Comprehensive Review." *Environmental Technology & Innovation* 28 (November): 102837. https://doi.org/10.1016/j.eti.2022.102837.

Gateuille, David, and Emmanuel Naffrechoux. 2022. "Transport of Persistent Organic Pollutants: Another Effect of Microplastic Pollution?" *WIREs Water* 9 (5): e1600. https://doi.org/10.1002/wat2.1600.

Gautam, Ravi, JiHun Jo, Manju Acharya, Anju Maharjan, DaEun Lee, Pramod Bahadur K.c., ChangYul Kim, KilSoo Kim, HyoungAh Kim, and Yong Heo. 2022. "Evaluation of Potential Toxicity of Polyethylene Microplastics on Human Derived Cell Lines." *Science of The Total Environment* 838 (September): 156089. https://doi.org/10.1016/j.scitotenv.2022.156089.

Godoy, V., G. Blázquez, M. Calero, L. Quesada, and M. A. Martín-Lara. 2019. "The Potential of Microplastics as Carriers of Metals." *Environmental Pollution* 255 (December): 113363. https://doi.org/10.1016/j.envpol.2019.113363.

Gopinath, Ponnusamy Manogaran, Krishna Sundar Twayana, Palaniyandi Ravanan, John Thomas, Amitava Mukherjee, David F. Jenkins, and Natarajan Chandrasekaran. 2021. "Prospects on the Nano-Plastic Particles Internalization and Induction of Cellular Response in Human Keratinocytes." *Particle and Fibre Toxicology* 18 (1): 35. https://doi.org/10.1186/s12989-021-00428-9.

Hahladakis, John N., Costas A. Velis, Roland Weber, Eleni Iacovidou, and Phil Purnell. 2018. "An Overview of Chemical Additives Present in Plastics: Migration, Release, Fate and Environmental Impact during Their Use, Disposal and Recycling." *Journal of Hazardous Materials* 344 (February): 179–199. https://doi.org/10.1016/j.jhazmat.2017.10.014.

Hirt, Nell, and Mathilde Body-Malapel. 2020. "Immunotoxicity and Intestinal Effects of Nano- and Microplastics: A Review of the Literature." *Particle and Fibre Toxicology* 17 (1): 57. https://doi.org/10.1186/s12989-020-00387-7.

Hwang, Jangsun, Daheui Choi, Seora Han, Se Yong Jung, Jonghoon Choi, and Jinkee Hong. 2020. "Potential Toxicity of Polystyrene Microplastic Particles." *Scientific Reports* 10 (1): 7391. https://doi.org/10.1038/s41598-020-64464-9.

Jiang, Baorong, Alexandra E. Kauffman, Lei Li, Wayne McFee, Bo Cai, John Weinstein, Jamie R. Lead, Saurabh Chatterjee, Geoffrey I. Scott, and Shuo Xiao. 2020. "Health Impacts of Environmental Contamination of Micro- and Nanoplastics: A Review." *Environmental Health and Preventive Medicine* 25 (1): 29. https://doi.org/10.1186/s12199-020-00870-9.

Jin, Mengke, Xue Wang, Tao Ren, Jian Wang, and Jiajia Shan. 2021. "Microplastics Contamination in Food and Beverages: Direct Exposure to Humans." *Journal of Food Science* 86 (7): 2816–2837. https://doi.org/10.1111/1750-3841.15802.

Joo, Sung Hee, Yejin Liang, Minbeom Kim, Jaehyun Byun, and Heechul Choi. 2021. "Microplastics with Adsorbed Contaminants: Mechanisms and Treatment." *Environmental Challenges* 3 (April): 100042. https://doi.org/10.1016/j.envc.2021.100042.

Khalid Ageel, Hassan, Stuart Harrad, and Mohamed Abou-Elwafa Abdallah. 2022. "Occurrence, Human Exposure, and Risk of Microplastics in the Indoor Environment." *Environmental Science: Processes & Impacts* 24 (1): 17–31. https://doi.org/10.1039/D1EM00301A.

Kinigopoulou, Vasiliki, Ioannis Pashalidis, Dimitrios Kalderis, and Ioannis Anastopoulos. 2022. "Microplastics as Carriers of Inorganic and Organic Contaminants in the Environment: A Review of Recent Progress." *Journal of Molecular Liquids* 350 (March): 118580. https://doi.org/10.1016/j.molliq.2022.118580.

Klavins, Maris, Linards Klavins, Olena Stabnikova, Viktor Stabnikov, Andrii Marynin, Linda Ansone-Bertina, Marcis Mezulis, and Ashok Vaseashta. 2022. "Interaction between Microplastics and Pharmaceuticals Depending on the Composition of Aquatic Environment." *Microplastics* 1 (3): 520–535. https://doi.org/10.3390/microplastics1030037.

Lai, Hanpeng, Xing Liu, and Man Qu. 2022. "Nanoplastics and Human Health: Hazard Identification and Biointerface." *Nanomaterials* 12 (8): 1298. https://doi.org/10.3390/nano12081298.

Lamichhane, G., A. Acharya, R. Marahatha, B. Modi, R. Paudel, A. Adhikari, B. K. Raut, S. Aryal, and N. Parajuli. 2022. "Microplastics in Environment: Global Concern, Challenges, and Controlling Measures." *International Journal of Environmental Science and Technology*, May. https://doi.org/10.1007/s13762-022-04261-1.

Lear, G., J. M. Kingsbury, S. Franchini, V. Gambarini, S. D. M. Maday, J. A. Wallbank, L. Weaver, and O. Pantos. 2021. "Plastics and the Microbiome: Impacts and Solutions." *Environmental Microbiome* 16 (1): 2. https://doi.org/10.1186/s40793-020-00371-w.

Leslie, Heather A., Martin J. M. van Velzen, Sicco H. Brandsma, A. Dick Vethaak, Juan J. Garcia-Vallejo, and Marja H. Lamoree. 2022. "Discovery and Quantification of Plastic Particle Pollution in Human Blood." *Environment International* 163 (May): 107199. https://doi.org/10.1016/j.envint.2022.107199.

Li, Lu, Xiaoli Zhao, Zhouyang Li, and Kang Song. 2021. "COVID-19: Performance Study of Microplastic Inhalation Risk Posed by Wearing Masks." *Journal of Hazardous Materials* 411 (June): 124955. https://doi.org/10.1016/j.jhazmat.2020.124955.

Liu, Si, JinHui Huang, Wei Zhang, LiXiu Shi, KaiXin Yi, HanBo Yu, ChenYu Zhang, SuZhou Li, and JiaoNi Li. 2022. "Microplastics as a Vehicle of Heavy Metals in Aquatic Environments: A Review of Adsorption Factors, Mechanisms, and Biological Effects." *Journal of Environmental Management* 302 (January): 113995. https://doi.org/10.1016/j.jenvman.2021.113995.

Mahe, Brice, Annika Vogt, Christelle Liard, Darragh Duffy, Valérie Abadie, Olivia Bonduelle, Alexandre Boissonnas, et al. 2009. "Nanoparticle-Based Targeting of Vaccine Compounds to Skin Antigen-Presenting Cells by Hair Follicles and Their Transport in Mice." *Journal of Investigative Dermatology* 129 (5): 1156–1164. https://doi.org/10.1038/jid.2008.356.

Marathe, Nachiket P., and Michael S. Bank. 2022. "The Microplastic-Antibiotic Resistance Connection." In *Microplastic in the Environment: Pattern and Process*, edited by Michael S. Bank, 311–322. Environmental Contamination Remediation and Management. Cham: Springer International Publishing. https://doi.org/10.1007/978-3-030-78627-4_9.

Martí, Elisa, Cecilia Martin, Matteo Galli, Fidel Echevarría, Carlos M. Duarte, and Andrés Cózar. 2020. "The Colors of the Ocean Plastics." *Environmental Science & Technology* 54 (11): 6594–6601. https://doi.org/10.1021/acs.est.9b06400.

McDaniel, David, Patricia Farris, and Giuseppe Valacchi. 2018. "Atmospheric Skin Aging: Contributors and Inhibitors." *Journal of Cosmetic Dermatology* 17 (2): 124–137. https://doi.org/10.1111/jocd.12518.

Menéndez-Pedriza, Albert, and Joaquim Jaumot. 2020. "Interaction of Environmental Pollutants with Microplastics: A Critical Review of Sorption Factors, Bioaccumulation and Ecotoxicological Effects." *Toxics* 8 (2): 40. https://doi.org/10.3390/toxics8020040.

Morgan, Sarah E., and Lisa A. DeLouise. 2020. "Further Studies in Translatable Model Systems Are Needed to Predict the Impacts of Human Microplastic Exposure." *Open Access Journal of Toxicology* 4 (3): 79–82.

Munyaneza, Janvier, Qilong Jia, Fahim A. Qaraah, Md Faysal Hossain, Chengzi Wu, Huajun Zhen, and Guangli Xiu. 2022. "A Review of Atmospheric Microplastics Pollution: In-Depth Sighting of Sources, Analytical Methods, Physiognomies, Transport and Risks." *Science of The Total Environment* 822 (May): 153339. https://doi.org/10.1016/j.scitotenv.2022.153339.

O'Neill, Sandra M., and Jenny Lawler. 2021. "Knowledge Gaps on Micro and Nanoplastics and Human Health: A Critical Review." *Case Studies in Chemical and Environmental Engineering* 3 (June): 100091. https://doi.org/10.1016/j.cscee.2021.100091.

Pletz, Martin. 2022. "Ingested Microplastics: Do Humans Eat One Credit Card per Week?" *Journal of Hazardous Materials Letters* 3 (November): 100071. https://doi.org/10.1016/j.hazl.2022.100071.

Poma, Anna, Giulia Vecchiotti, Sabrina Colafarina, Osvaldo Zarivi, Massimo Aloisi, Lorenzo Arrizza, Giuseppe Chichiriccò, and Piero Di Carlo. 2019. "In Vitro Genotoxicity of Polystyrene Nanoparticles on the Human Fibroblast Hs27 Cell Line." *Nanomaterials* 9 (9): 1299. https://doi.org/10.3390/nano9091299.

Prata, Joana C., Joana L. Castro, João P. da Costa, Mário Cerqueira, Armando C. Duarte, and Teresa Rocha-Santos. 2020. "Airborne Microplastics." In *Handbook of Microplastics in the Environment*, edited by Teresa Rocha-Santos, Mónica Costa, and Catherine Mouneyrac, 1–25. Cham: Springer International Publishing. https://doi.org/10.1007/978-3-030-10618-8_37-1.

Prata, Joana Correia, João P. da Costa, Isabel Lopes, Armando C. Duarte, and Teresa Rocha-Santos. 2020. "Environmental Exposure to Microplastics: An Overview on Possible Human Health Effects." *Science of The Total Environment* 702 (February): 134455. https://doi.org/10.1016/j.scitotenv.2019.134455.

Rahman, Arifur, Atanu Sarkar, Om Prakash Yadav, Gopal Achari, and Jaroslav Slobodnik. 2021. "Potential Human Health Risks Due to Environmental Exposure to Nano- and Microplastics and Knowledge Gaps: A Scoping Review." *Science of The Total Environment* 757 (February): 143872. https://doi.org/10.1016/j.scitotenv.2020.143872.

Ramsperger, Anja F. R. M., Enrico Bergamaschi, Marco Panizzolo, Ivana Fenoglio, Francesco Barbero, Ruud Peters, Anna Undas, et al. 2023. "Nano- and Microplastics: A Comprehensive Review on Their Exposure Routes, Translocation, and Fate in Humans." *NanoImpact* 29 (January): 100441. https://doi.org/10.1016/j.impact.2022.100441.

Rolland, Alain, Nathalie Wagner, Alain Chatelus, Braham Shroot, and Hans Schaefer. 1993. "Site-Specific Drug Delivery to Pilosebaceous Structures Using Polymeric Microspheres." *Pharmaceutical Research* 10 (12): 1738–1744. https://doi.org/10.1023/A:1018922114398.

Salthammer, Tunga. 2022. "Microplastics and Their Additives in the Indoor Environment." *Angewandte Chemie International Edition* 61 (32): e202205713. https://doi.org/10.1002/anie.202205713.

Santos, Andressa Liberal, Cândido Carvalho Rodrigues, Miguel Oliveira, and Thiago Lopes Rocha. 2022. "Microbiome: A Forgotten Target of Environmental Micro(Nano)Plastics?" *Science of The Total Environment* 822 (May): 153628. https://doi.org/10.1016/j.scitotenv.2022.153628.

Sripada, Kam, Aneta Wierzbicka, Khaled Abass, Joan O. Grimalt, Andreas Erbe, Halina B. Röllin, Pál Weihe, et al. n.d. "A Children's Health Perspective on Nano- and Microplastics." *Environmental Health Perspectives* 130 (1): 015001. https://doi.org/10.1289/EHP9086.

Torres-Agullo, A., A. Karanasiou, T. Moreno, and S. Lacorte. 2022. "Airborne Microplastic Particle Concentrations and Characterization in Indoor Urban Microenvironments." *Environmental Pollution* 308 (September): 119707. https://doi.org/10.1016/j.envpol.2022.119707.

Udovicki, Bozidar, Mirjana Andjelkovic, Tanja Cirkovic-Velickovic, and Andreja Rajkovic. 2022. "Microplastics in Food: Scoping Review on Health Effects, Occurrence, and Human Exposure." *International Journal of Food Contamination* 9 (1): 7. https://doi.org/10.1186/s40550-022-00093-6.

Valacchi, Giuseppe, Claudia Sticozzi, Alessandra Pecorelli, Franco Cervellati, Carlo Cervellati, and Emanuela Maioli. 2012. "Cutaneous Responses to Environmental Stressors." *Annals of the New York Academy of Sciences* 1271 (October): 75–81. https://doi.org/10.1111/j.1749-6632.2012.06724.x.

Vethaak, A. Dick, and Juliette Legler. 2021. "Microplastics and Human Health." *Science* 371 (6530): 672–674. https://doi.org/10.1126/science.abe5041.

Wang, Yi, Jian Huang, Fengxiao Zhu, and Shaoda Zhou. 2021. "Airborne Microplastics: A Review on the Occurrence, Migration and Risks to Humans." *Bulletin of Environmental Contamination and Toxicology* 107 (4): 657–664. https://doi.org/10.1007/s00128-021-03180-0.

Wu, Panfeng, Jinsheng Huang, Yuling Zheng, Yicheng Yang, Yue Zhang, Feng He, Hao Chen, et al. 2019. "Environmental Occurrences, Fate, and Impacts of Microplastics." *Ecotoxicology and Environmental Safety* 184 (November): 109612. https://doi.org/10.1016/j.ecoenv.2019.109612.

Yang, Wenjie, Nahar Jannatun, Yanqiao Zeng, Tinghao Liu, Guofang Zhang, Chunying Chen, and Yang Li. 2022. "Impacts of Microplastics on Immunity." *Frontiers in Toxicology* 4 (September): 956885. https://doi.org/10.3389/ftox.2022.956885.

Yee, Maxine Swee-Li, Ling-Wei Hii, Chin King Looi, Wei-Meng Lim, Shew-Fung Wong, Yih-Yih Kok, Boon-Keat Tan, Chiew-Yen Wong, and Chee-Onn Leong. 2021. "Impact of Microplastics and Nanoplastics on Human Health." *Nanomaterials* 11 (2). https://doi.org/10.3390/nano11020496.

Yong, Cheryl Qian Ying, Suresh Valiyaveetill, and Bor Luen Tang. 2020. "Toxicity of Microplastics and Nanoplastics in Mammalian Systems." *International Journal of Environmental Research and Public Health* 17 (5): 1509. https://doi.org/10.3390/ijerph17051509.

Zhang, Yulan, Shichang Kang, Steve Allen, Deonie Allen, Tanguang Gao, and Mika Sillanpää. 2020. "Atmospheric Microplastics: A Review on Current Status and Perspectives." *Earth-Science Reviews* 203 (April): 103118. https://doi.org/10.1016/j.earscirev.2020.103118.

Zhao, Xiaoli, Junyu Wang, Kenneth Mei Yee Leung, and Fengchang Wu. 2022. "Color: An Important But Overlooked Factor for Plastic Photoaging and Microplastic Formation." *Environmental Science & Technology* 56 (13): 9161–9163. https://doi.org/10.1021/acs.est.2c02402.

13 Water Pipe Smoking and E-Cigarettes

A Safer Alternative to Combustible Cigarettes?

Andre Faria, Arthur Faria, and Ismail Laher

13.1 INTRODUCTION

Tobacco use remains the world's leading preventable cause of death and a risk factor for cardiovascular diseases (CVDs) ("Tobacco" n.d.). It is estimated that ~1.1 billion people use tobacco products worldwide, with cigarettes being the preferred product (82% of users) (Ezzati and Lopez 2003). This global pandemic is one of the most serious public health threats the world faces, killing more than 8 million people a year, including ~1.2 million deaths from exposure to second-hand smoke ("Tobacco" n.d.).

Several studies report that cigarette smoking is related to an array of illnesses such as CVDs, type 2 diabetes, pulmonary and renal disorders and cancer (Caliri, Tommasi, and Besaratinia 2021; Ambrose and Barua 2004; Badran and Laher 2020; Samet 2013). The large negative impact of tobacco use on general health has prompted efforts by governments and healthcare authorities to enact global measures to reduce the use and availability of cigarettes. Thus, many countries currently implement comprehensive smoke-free laws as well as other population-based interventions that have proven to be effective based on morbidity and mortality related to cigarette smoking. As an example, New Zealand recently passed a law banning the sale of tobacco products to anybody born after 1 January 2009, leading to an estimated reduction of $5 billion to the healthcare system. It is hoped that this will lead to a smoke free country by 2025 (Dyer 2021). However, although tobacco product use has significantly decreased in some nations, the number of smokers worldwide has increased from 721 million in 1980 to 1.1 billion in 2020 due to population growth and the unrestricted availability of tobacco products in some countries (Ng et al. 2014; "Tobacco" n.d.). For example, smoking is more prevalent in middle-income countries such as China, India, and Russia, where restriction on the use and availability of tobacco products is less well established (Giovino et al. 2012).

The awareness of the harms of tobacco use has led to an increasing use of e-cigarettes (also known as vapers) and water pipe tobacco (also known as hookah and shisha), particularly by youth. Many cultural and social aspects underlie the use e-cigarettes and water pipes, but one of the strongest reasons is the misbelief that these alternative options are less harmful than regular cigarettes (Badran and Laher 2020). A growing body of evidence indicates deleterious effects of vaping and smoking shisha, but epidemiological data is needed to analyze this increasing use of these alternatives.

13.2 WATER PIPE SMOKING

Although its origins are unclear, India and China used to trade water pipes (hookah, shisha) across Asia and Africa in the sixteenth century (Bello 2012). Since the smoke first passes through a container of water when using a water pipe (Figure 13.1), it was initially thought to lessen the negative

DOI: 10.1201/9781003328100-16

effects of traditional tobacco use—a belief that is still widely held by many water pipe users even today. This misperception of the addictive potential and potential adverse health effects of water pipe smoking has led to increases in its use.

The water pipe has been used in many different ways over the years, but in its current form, it consists of a body, a water base, a head or bowl (where tobacco is placed), and a hose that terminates in a mouthpiece (Figure 13.1). Charcoal briquettes are placed on top of the tobacco-filled bowl, which is typically made of clay, marble, or glass. The charcoal pieces are frequently separated from the tobacco by a perforated aluminum foil to allow for heated air to pass through the tobacco and for the smoke to pass through the stem. Smoke is cooled and made more humid by bubbling through the exit stem that is submerged in water. The water can occasionally have crushed ice, dried fruit, or mint leaves added. Smoke that emerges from the water is drawn by the user through a hose that is typically made of leather, vinyl, or plastic. While some water pipes have rigid mouthpiece reeds, others could have multiple hose ports to allow for multiple users.

Data from the Global Youth Tobacco Survey, which included 13 to 15-year-olds from seven Middle Eastern nations, indicates that the rates of water pipe smoking ranges from 9% to 15%, which is even higher than the rates of cigarette smoking in nearly every nation studied (Moh'D Al-Mulla et al. 2008). In several countries, smoking tobacco with a water pipe has become even more popular than smoking cigarettes. For instance, a 2012 study of students in the United Kingdom revealed that current use of a water pipe was more prevalent than cigarette smoking (7.6% versus

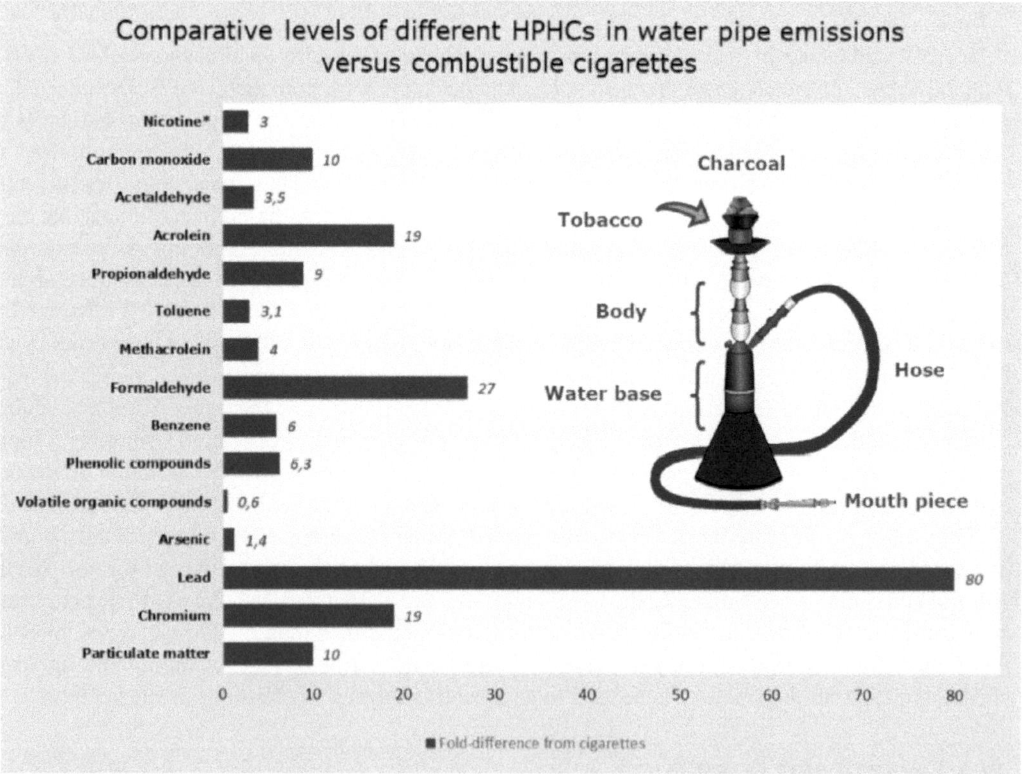

FIGURE 13.1 Illustrative scheme of water pipe basic components and a comparative chart showing the relative abundance of harmful or potentially harmful constituents (HPHCs) in water pipe vs. cigarette use. The data are shown as the fold difference between smoking a single cigarette and a typical session of water pipe smoking. Adapted from Shihadeh (2003); Monn et al. (2007); Schubert et al. (2015); Bhatnagar et al. (2019).

3.4%) (Jawad et al. 2013). In a more recent study, regular and occasional use by adults reached 7.25% in the Eastern Mediterranean and 3.8% in the Americas, while latest use (within the last 30 days) by youths is estimated at 10.6% in Europe, 10.3% in the Eastern Mediterranean, and 6.8% in the Americas (Jawad et al. 2018).

The social status of girls and women has changed in the Middle East, largely due to social transformation, employment opportunities, and educational advancements, causing behavioral changes in many women (Bashirian et al. 2021). This transformation and increased gender parity has led to some shifts in drug use, from mono- to bi-gender, including the increased popularity of water pipe use among women. Shisha smoking is more common among women than cigarette use, which is related to the fact that water pipe use is seen as a socially acceptable activity and has fewer negative connotations for females. A recent study reports that women believe that waterpipe smoking causes them to feel more attractive and liberated (Labib et al. 2007).

There are several reasons for the growing epidemic of water pipe smoking: i) deceptive marketing tactics that promote flavored water pipe tobacco smoking as a tasty and safer alternative to cigarettes; ii) claims that it contains less nicotine; iii) social acceptance in homes, coffee shops, and restaurants; iv) the influence of social media; v) the absence of water pipe-specific policy and regulations; and vi) the social connotation of freedom and independence in female users. However, current clinical and experimental research suggests that water pipe smoking is just as harmful as smoking cigarettes, if not more so. However, theses analysis usually don't account for differences in patterns of consumption and subsequent impacts in human health. The water pipe differs from cigarettes in the method of consumption but has similar harmful constituents found in cigarettes but is generated in greater amounts (Bhatnagar et al. 2019; Shihadeh 2003; Monn et al. 2007; Schubert et al. 2015). However, cigarette smokers often consume ten cigarettes a day, while shisha sessions occur once or twice a day. Furthermore, water pipes are frequently shared among multiple users, reducing the final exposure to toxins and chemicals per individual. Theses individualities of use should also be factored in for hazard comparisons. A correlation of some harmful or potentially harmful constituents (HPHCs) between smoking one cigarette (1 g of tobacco) and one water pipe session (8–12 g of tobacco) is shown in Figure 13.1.

The myriad of HPHCs present in water pipe tobacco smoke is fairly similar to that of the chemicals found in combustible cigarette smoke. Differences arise from variations in the combustion mode, the temperature at which the tobacco burns, and the smoke volume delivered. Heavy metals, particulate matter (PM), nicotine, carbon monoxide, and nicotine-derived volatile organic chemicals (VOCs) are important HPHCs. One session of water pipe use is thought to produce plasma nicotine levels comparable to smoking two to three cigarettes (Jacob et al. 2011). Water pipe smoking also produces more carbon monoxide (CO), primarily due to the use of charcoal. When charcoal is replaced with an electric heater, the CO released by water pipes is reduced by 90% (Daher et al. 2010). It is also estimated that the amount of CO exhaled after a water pipe session is greater than the amount exhaled after smoking an entire pack of cigarettes.

In most cases for water pipe smoking a highly flavored mix called "moassel" is used—sweetened, flavored tobacco composed of glycerol, shredded tobacco, molasses, other additives, and flavors such as candy and fruit. This contributes to the rising popularity of hookah, especially among the younger public, since these flavors mask the traditional smell of tobacco and are associated with the notion of a healthier choice than cigarettes (Owens, Ha, and Soulakova 2019). Nonetheless, even with the more palatable taste and improved smell, many health risks are still involved in these practices. Flavored tobacco has several volatile chemicals added to them, with esters amounting to the major component up to 40% (Farag et al, 2018). Superheating flavored hookah tobacco at 190 °C results in the release of a mixture of phenol derivatives and polycyclic aromatic compounds that are related to an increase in inflammation and oxidative stress, with special harmful effects on the brain, lungs, and colon (Farag et al, 2018; Owens et al, 2019). Studies analyzing the specific influence of added flavors on human health as well as if they may present as an independent risk factor are still lacking.

Regarding particulate matter (PM), although the size of PM delivered when using a water pipe is smaller than that of PM delivered when smoking cigarettes (0.04–0.15 vs. 0.15–0.5 m), the amount of PM delivered in one water pipe inhalation (1L) can be up to 70 particles in comparison to 9.2 particles in one cigarette breath (45 ml) (Monn et al. 2007). Therefore, 100 puffs from a water pipe during an average 1-hour session can produce up to ten times as much PM as one cigarette (10 puffs) (Daher et al. 2010). The "Pathophysiology and Health Effects" section of this chapter provides additional information on the various mechanisms by which the components of water pipe tobacco cause oxidative stress and increases the risk for several health conditions.

13.3 E-CIGARETTES

The first registered patent for a "smokeless nontobacco cigarette" was made by Herbert A. Gilbert in 1963, which is the earliest known proposal for an e-cigarette. The goal of this invention was to make smoking safer by using heated, flavor-infused air in place of burning tobacco (Fadus et al, 2019). The e-cigarette was produced in its current form 40 years later (in 2003) by Hon Lik, a Chinese pharmacist, whose patents are held by the Ruyan company in China (Adkison et al. 2013). Since then, e-cigarettes have spread rapidly throughout the world and are increasingly well-known and used (Schraufnagel et al. 2014; Adkison et al. 2013). Numerous studies have examined the usage patterns, beneficial effects, safety, and risks associated with e-cigarettes.

The first generation of e-cigarettes were made to resemble tobacco products (e.g., cigarettes, cigars, pipes, etc.). However, the new generation are manufactured as pens, USB sticks, or other commonplace items. There are usually three main parts of an e-cigarette: the cartridge, atomizer, and battery that are all housed in a stainless-steel body (Figure 13.2). The cartridge holds a liquid mixture that frequently consists of propylene glycol and/or vegetable glycerin, water, and flavors (e.g., tobacco, mint, fruit, coffee, chocolate, etc.). The liquid mixture contains either nicotine in varying amounts (0–36 mg/mL) or can be nicotine free. When a user inhales through the mouthpiece, a pressure-sensitive circuit in an e-cigarette activates the atomizer consisting of a small heating element. The liquid in the cartridge is then heated by the atomizer, which traditionally is made of a wick to transport the e-liquid from the reservoir to an electrically heated metal coil, creating a vapor that resembles smoke. The amount of e-liquid in the wick is critical in dictating the temperatures reached within the atomizer when the coil is heated. The user of an e-cigarette draws vapor through the mouthpiece and inhales it. The power source to this system, which is typically a rechargeable lithium battery, occupies the biggest part inside the device.

Although the use of combustible cigarettes has decreased significantly in young people recently, the introduction of e-cigarettes and pod mod devices has led to increases in the overall use of nicotine products by young people. The percentage of high school students in the United States who smoked vape cigarettes in the past 30 days has increased from 1.5% in 2011 to 11.7% in 2017 and 20.8% in 2018, while the 30-day prevalence rates of combustible cigarette decreases over the same period was 18.7% in 2011, 9.7% in 2017, and 7.6% in 2018 (Cullen et al. 2018; Mirbolouk et al. 2018). Global use followed the same trend. A 2017 study in France reported that 3.6% of university students used e-cigarettes daily, whereas in Switzerland, researchers found a smoking rate of 24% among 8th-grade and high school students (Filippidis et al. 2017; Kinouani, Pereira, and Tzourio 2017).

Recent research suggests that e-cigarettes can potentially assist smokers to quit or reduce smoking levels. However, while studies have reported that e-cigarette vapor has a significantly lower level of toxic substances than combustible cigarettes (between 9 and 450 times lower), there is still concern about the presence of carcinogens in e-cigarette vapors (Goniewicz et al. 2014; Rubinstein et al. 2018). There is currently scant data on the presence or absence of toxic chemicals in the vapor produced by e-cigarettes and inhaled by users. A chart comparing some fold-differences of HPHCs after smoking one cigarette (1 g of tobacco) and an e-cigarette (15 puffs of 2 sec, with intervals between puffs of 30 sec, and average puff volume of 55 mL) is shown in Figure 13.2.

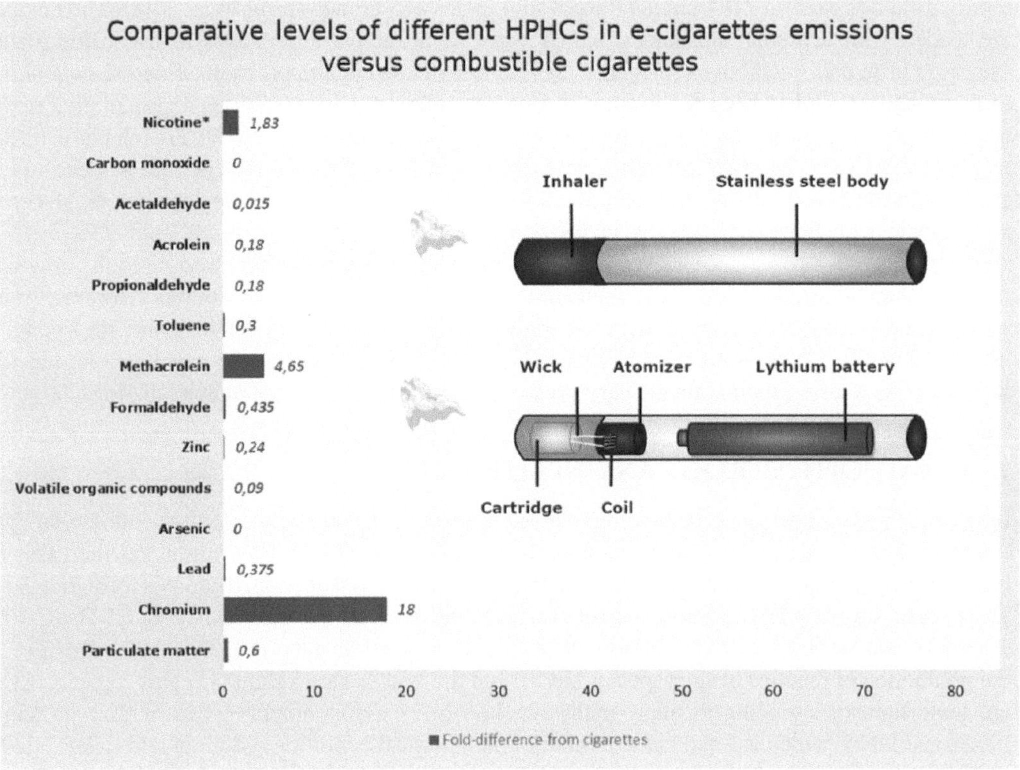

FIGURE 13.2 Illustrative scheme of an e-cigarette basic components and a comparative chart showing the relative abundance of harmful or potentially harmful substances in an e-cigarette vs. a cigarette. The data are shown as the fold difference between one single cigarette and a typical use of vaping. Adapted from Goniewicz et al. (2014); Cunningham et al. (2020).

Source: *Nicotine levels depend on the initial concentration of the liquid inside the cartridge. The displayed value is from the most concentrated solution studied (30mg/ml).

Few studies compared the components of tobacco smoke and vapor from e-cigarettes; however a comprehensive comparison of the toxic constituents and adverse health effects was previously presented as an overview (Munzel et al, 2020a). Data from a 2020 study shows that, in almost every case, toxicants examined for both types of products were substantially higher in combustible cigarettes compared to the electronic alternative (Cunningham et al. 2020). Nicotine levels varied widely and depended on the e-cigarette brand and the liquid concentration used in the cartridge. The CO emissions from all e-cigarettes were below detection limits and, therefore, were >99% lower than from regular cigarettes. None of the volatile organic toxicants examined were detected in relevant amounts in e-cigarette vapor. Most carbonyls (except for methacrolein) and dicarbonyls concentrations (acetaldehyde, acrolein, propionaldehyde, toluene, and formaldehyde) were lower in e-cigarette aerosols than in cigarette smoke. The reasons for this are unclear but could be related to decomposition of the cotton wick during the heating process. Higher chromium concentrations occur in e-cigarette vapor, while its levels in cigarette smoke were negligible, likely due to the metal composition of the coil and subsequent release during e-cigarette use (Cunningham et al. 2020; Goniewicz et al. 2014).

Furthermore, the use of e-cigarettes presents another potential health hazard. One of the unique selling points of e-cigarettes is the variety of flavors and scents available. According to a 2014 report, e-liquids exist in 7764 unique flavors sold under 466 brands (Kaur et al. 2018). The devices are loaded with artificial substances, which serve as attractive alternatives to users due to the improved taste and smell but can nonetheless lead to extended health problems. Excessive smoking of these substances has been reported to induce pro-inflammatory responses and oxidative stress in the brain, lung, heart, and colon (Canistro et al. 2017; Been et al. 2022). The changes in the expression of genes encoding inflammatory molecules are likely due to the chemicals added to the e-liquids used to create the flavors such as plant extracts, nicotinic salts, propylene glycol, glycerol, and benzoic acid (Been et al. 2022; Canistro et al. 2017).

E-cigarettes do not produce smoke similar to that from burning tobacco, leading to the assumption that using an e-cigarette is safer than smoking tobacco (Munzel et al, 2020b). However, some safety related issues remain from using e-cigarettes, and the chemicals and toxicants they contain have not been fully disclosed. E-cigarettes are produced by a number of companies that are not regulated. As a result, their contents vary greatly and occasionally do not match labeling.

13.4 PATHOPHYSIOLOGY AND HEALTH EFFECTS

Both active and passive tobacco smoking increases the incidence of atherosclerosis, endothelial dysfunction, and various CVDs. The generation of oxidative stress by HPHCs stimulates increases in heart rate, myocardial contractility, inflammation, endothelial dysfunction, thrombus formation, and lower serum levels of HDL cholesterol, all of which contribute to the development of CVDs (Caliri, Tommasi, and Besaratinia 2021; Samet 2013). Oxidative stress results from an imbalance between the production of reactive oxygen species (ROS) and the body's antioxidant or repair capacity. ROS are important for signaling in many pathways, but, in excessive amounts, can oxidize proteins, lipids, and DNA, which is linked to a number of pathological conditions (Badran and Laher 2020).

The increasing popularity in the use of alternatives to cigarette smoking by youth raises serious concerns on the potential for long-term effects on brain development and behavior (Squeglia and Gray 2016; Yuan et al. 2015). Critical aspects of brain development are regulated by nicotinic receptors, where even low doses of nicotine exposure can negatively impact brain development, resulting in cognitive impairment and chemical dependence. Nicotine's pathological effects primarily result from receptor activation and catecholamine release, and several studies suggest that nicotine-dependent oxidative stress damages cells mediated either by increasing ROS production and/or decreasing antioxidant capacity in various experimental models (see Figure 13.3) (Steven et al. 2019; Arany et al. 2016; Wu et al. 2018). It is important to note that nicotine exposure, especially in e-cigarettes, can also have an acute poisoning potential (Bhatt et al, 2020). Nicotine concentrations in e-cigarette cartridges can range from 0 to 36 mg/mL and can be lethal in large doses (0.5–1 mg/kg body weight in adults and 0.1–0.2 mg/kg body weight in children) (Schraufnagel et al. 2014; Rom et al. 2015). High levels of nicotine, such as those found in e-cigarette cartridges, can cause severe intoxication after ingestion, inhalation, or direct skin contact while the cartridges are being replaced or handled. Children, pregnant women, and nursing mothers are population groups particularly at risk.

Smoking a water pipe results in a significant amount of carbon monoxide (CO) being inhaled in a single session. CO is readily absorbed in the lungs and can form a tight but reversible bond with hemoglobin (250 times stronger than with oxygen), leading to tissue hypoxia. After the elimination of CO, reperfusion takes place and results in a burst of formation of ROS from various sources (Akyol et al. 2014).

Heavy metals, particulate matter, and aerosol collected matter increases inflammation and oxidative stress (Margham et al. 2021; Rao et al. 2018; Cunningham et al. 2020; Badran and Laher 2020). Exposure to these components stimulates the expression of inducible nitric oxide synthase (iNOS), an important source of RNS, and culminates in endoplasmic reticulum stress and subsequent recruitment and activation of leukocytes. After being inhaled, the molecular events by which particulate matter and pulmonary oxidative stress triggers systemic inflammation, along with

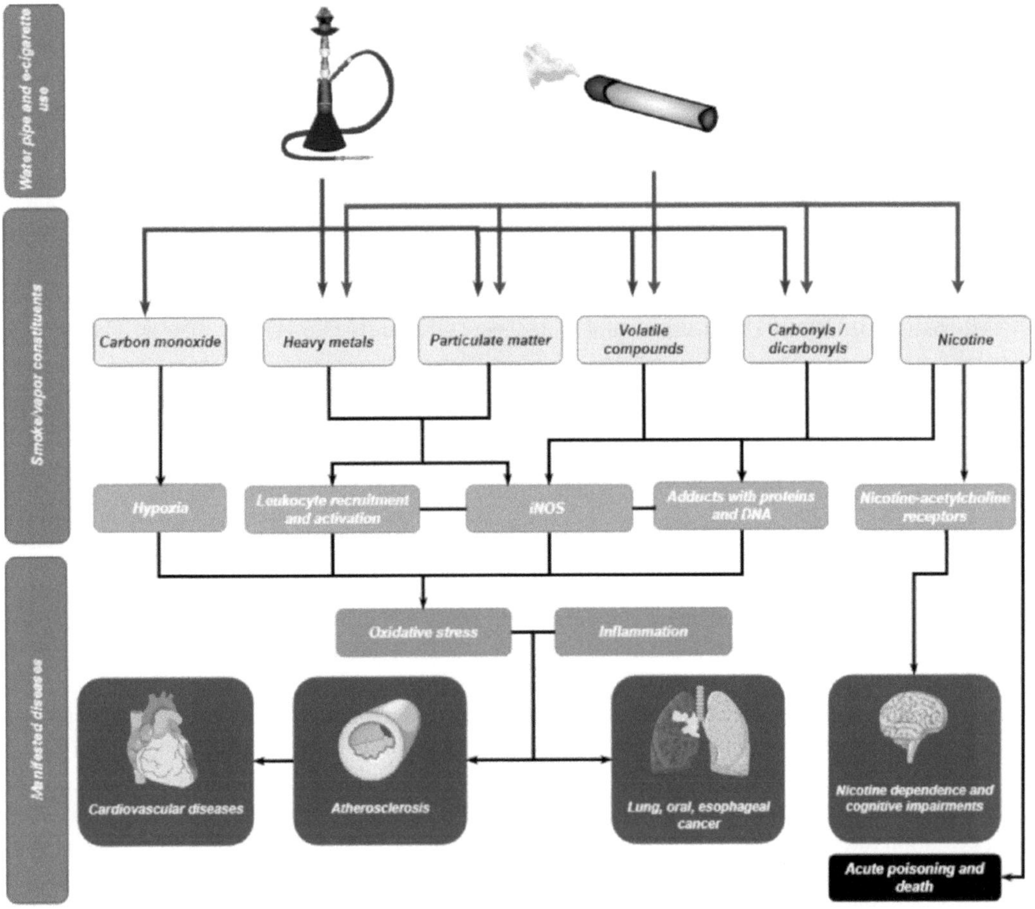

FIGURE 13.3 Pathophysiological mechanisms underlying harmful or potentially harmful constituents' exposure, leading to various diseases for both water pipe and e-cigarette smoking.

Source: The figure was partly generated using Servier Medical Art, provided by Servier, licensed under a Creative Commons Attribution 3.0 unported license.

the contributions and interactions between nonimmune and immune cells, is highly complex and can vary widely (Rao et al. 2018). The situation in humans is decidedly more complex, with the ultimate effect depending on particle fate (e.g., lung clearance vs. retention), sequestration, intracellular distribution, pathways of potential systemic transmission, and ultimately on multiple host factors, including susceptibility. However, despite the difficulty in studying isolated molecules, there is growing evidence showing that particles arising from both e-cigarettes and water pipe smoking enhances oxidative stress, atherosclerosis, and subsequent CVDs (Steven et al. 2019; Bhatt, Ramphul, and Bush 2020; Rom et al. 2015; Daher et al. 2010; Monn et al. 2007). This is confirmed by recent epidemiological data showing that e-cigarettes increase the risk of cardiovascular events by 1.27-fold compared to nonsmokers, and in water pipe users, by 3-fold (Yatsuya and Folsom 2010; Sibai et al. 2014).

Volatile organic compounds, carbonyls, and dicarbonyls are present in significantly higher amounts in shisha smoke when compared to combustible cigarettes and have been shown to have carcinogenic properties in many studies (Badran and Laher 2020; Jacob et al. 2011; Schubert et al. 2015; Daher et al. 2010). These compounds can cause protein and DNA damage, induce oxidative stress, and trigger or suppress immune response to increase the risk of neoplasias. This latter was reported in a systematic review indicating a positive association between water pipe smoke and lung cancer [OR = 4.58; 95% CI: 2.61–8.03], oral cancer [OR = 4.17; 95% CI: 2.53–6.89], and esophageal cancer [OR = 3.63; 95% CI: 1.39–9.44] (Waziry et al. 2017). The levels of toxic compounds in e-cigarette vapors are 9- to 450-fold lower compared with smoke generated from conventional cigarettes (Goniewicz et al. 2014). This small exposure to carbonyl compounds can cause mouth and throat irritation which are the most frequently adverse events reported by e-cigarette users (Goniewicz, Lingas, and Hajek 2013). Studies evaluating possible adverse health outcomes of short- and long-term exposure to these compounds are still needed.

Women are more likely than men to develop respiratory conditions when smoking waterpipes (Bashirian et al. 2021). Using water pipes increases adverse health effects such as altered menstrual function, decreased bone density, disorders caused by estrogen deficiency, and pregnancy issues (Salameh, Khayat, and Waked 2012). Concerns need to be addressed regarding the dangers of e-cigarette use by women, particularly during pregnancy. Unfortunately, there is a lack of studies examining the relationship between maternal e-cigarette exposure and offspring health in humans. Although inconsistent, some evidence from laboratory and animal studies suggested that maternal exposure to e-cigarette vapor may limit the growth of offspring. The metabolic health of offspring may also be impacted by exposure to e-cigarettes, which can manifest as compromised placental function, resulting in fetal structural abnormalities, as well as altered glucose homeostasis. Additionally, exposure during pregnancy may result in defects in the development of the nervous, vascular, and respiratory systems (Zhang et al. 2022; Mescolo, Ferrante, and la Grutta 2021).

There is a growing body of evidence supporting the use of e-cigarettes for tobacco cessation. According to a recent metanalysis, daily e-cigarette use led to an increase in tobacco quitting rates [OR = 1.529, 95% CI: 1.158–2.019] (Wang et al, 2021). However, compared to studies of widely used smoking cessation therapies, the abstinence rates reported in the aforementioned study are significantly lower. There is recent persuasive evidence that bupropion increases long-term smoking cessation rates [OR 1.64, 95% CI: 1.52–1.77]. Varenicline, a partial agonist at some heteromeric neuronal nicotinic receptors widely used as a smoking cessation aid, resulted in even greater smoking cessation rates [OR 2.31, 95% CI: 2.24–2.37] (Howes et al. 2020).

Claims that e-cigarette use can lower overall health risks in smokers should be interpreted with caution. Use stimulation and large consumer offerings could have detrimental effects at a populational level. E-cigarettes increase smoking initiation and decrease smoking cessation in individual unwilling to quit. Furthermore, e-cigarettes may act as a gateway to combustible tobacco products and increase nicotine use and dependence, especially in teenagers (Wang et al, 2021; Goniewicz et al, 2013; Kinouani et al, 2017).

13.5 CONCLUSION

There is concern that a significant number of people who would not have otherwise used tobacco are instead exposed to nicotine through electronic cigarettes and water pipe smoking (Münzel, Hahad, et al. 2020). This is of concern even though e-cigarette use is generally regarded as safer than smoking conventional cigarettes. The damaging effects of nicotine can affect the developing adolescent brain, and past nicotine use indicates unsettling patterns of future nicotine use. Water pipe users are at higher risk of harm as compared to non-smokers. The misconception that shisha has lower negative effects compared to traditional tobacco use needs to be considered in view of the presence of components that enhance the risk to cardiovascular diseases, cancer, and death.

E-cigarettes could assist in smoking reduction or cessation, but the evidence of support of this is not robust. Studies have yet to demonstrate that e-cigarettes are superior to other popular smoking cessation methods. There is also a need to evaluate the potential benefits of e-cigarettes in assisting smokers to quit. The population's potential harm from increased social acceptance of smoking and nicotine use, which afflicts a high potential for addiction, should also be considered. Electronic nicotine delivery devices should be restricted or banned until more data about their acute and long-term health risks are available, e.g., in young children pregnant females and their offspring, etc.

Molecular and histological studies on the differences in pathology (onset, regional variations, extent) caused by smoking cigarettes, water pipes, and e-cigarettes in target organs (lung, brain, heart, kidney, etc.) are warranted. Another unknown factor is the impact on pathology after cessation of smoking these various products—do the changes in structure and function of target organs occur at the same rates and to a similar extent?

REFERENCES

Adkison, Sarah E., Richard J. O'Connor, Maansi Bansal-Travers, Andrew Hyland, Ron Borland, Hua Hie Yong, K. Michael Cummings, et al. 2013. "Electronic Nicotine Delivery Systems: International Tobacco Control Four-Country Survey." *American Journal of Preventive Medicine* 44 (3): 207–215. https://doi.org/10.1016/J.AMEPRE.2012.10.018.

Akyol, Sumeyya, Serpil Erdogan, Nuri Idiz, Safa Celik, Mehmet Kaya, Fatma Ucar, Senol Dane, and Omer Akyol. 2014. "The Role of Reactive Oxygen Species and Oxidative Stress in Carbon Monoxide Toxicity: An in-Depth Analysis." *Redox Report : Communications in Free Radical Research* 19 (5): 180–189. https://doi.org/10.1179/1351000214Y.0000000094.

Ambrose, John A., and Rajat S. Barua. 2004. "The Pathophysiology of Cigarette Smoking and Cardiovascular Disease: An Update." *Journal of the American College of Cardiology* 43 (10): 1731–1737. https://doi.org/10.1016/j.jacc.2003.12.047.

Arany, Istvan, Samuel Hall, Dustin K. Reed, Caitlyn T. Reed, and Mehul Dixit. 2016. "Nicotine Enhances High-Fat Diet-Induced Oxidative Stress in the Kidney." *Nicotine & Tobacco Research : Official Journal of the Society for Research on Nicotine and Tobacco* 18 (7): 1628–1634. https://doi.org/10.1093/NTR/NTW029.

Badran, Mohammad, and Ismail Laher. 2020. "Waterpipe (Shisha, Hookah) Smoking, Oxidative Stress and Hidden Disease Potential." *Redox Biology* 34 (July): 101455. https://doi.org/10.1016/J.REDOX.2020.101455.

Bashirian, Saeed, Majid Barati, Manoochehr Karami, Behrooz Hamzeh, Maryam Afshari, and Elahe Ezati. 2021. "Determinants of Waterpipe Smoking among Women: A Systematic Review." *International Journal of Preventive Medicine* 12 (1): 25. https://doi.org/10.4103/IJPVM.IJPVM_116_20.

Been, Terek, Hussein Traboulsi, Sofia Paoli, Bayan Alakhtar, Koren K. Mann, David H. Eidelman, and Carolyn J. Baglole. 2022. "Differential Impact of JUUL Flavors on Pulmonary Immune Modulation and Oxidative Stress Responses in Male and Female Mice." *Archives of Toxicology* 96 (6): 1783–1798. https://doi.org/10.1007/S00204-022-03269-3.

Bello, David A. 2012. "Golden-Silk Smoke: A History of Tobacco in China, 1550–2010. By Carol Benedict (Berkeley, University of California Press, 2011) 334 Pp. 49.95." *The Journal of Interdisciplinary History* 43 (1): 158–159. https://doi.org/10.1162/JINH_R_00373.

Bhatnagar, Aruni, Wasim Maziak, Thomas Eissenberg, Kenneth D. Ward, George Thurston, Brian A. King, Erin L. Sutfin, et al. 2019. "Water Pipe (Hookah) Smoking and Cardiovascular Disease: A Scientific Statement from the American Heart Association." *Circulation* 139 (19): e917. https://doi.org/10.1161/CIR.0000000000000671.

Bhatt, Jayesh Mahendra, Manisha Ramphul, and Andrew Bush. 2020. "An Update on Controversies in E-Cigarettes." *Paediatric Respiratory Reviews* 36 (November): 75. https://doi.org/10.1016/J.PRRV.2020.09.003.

Caliri, Andrew W., Stella Tommasi, and Ahmad Besaratinia. 2021. "Relationships among Smoking, Oxidative Stress, Inflammation, Macromolecular Damage, and Cancer." *Mutation Research* 787 (January): 108365. https://doi.org/10.1016/J.MRREV.2021.108365.

Canistro, Donatella, Fabio Vivarelli, Silvia Cirillo, Clara Babot Marquillas, Annamaria Buschini, Mirca Lazzaretti, Laura Marchi, et al. 2017. "E-Cigarettes Induce Toxicological Effects That Can Raise the Cancer Risk." *Scientific Reports 2017 7:1* 7 (1): 1–9. https://doi.org/10.1038/s41598-017-02317-8.

Cullen, Karen A., Bridget K. Ambrose, Andrea S. Gentzke, Benjamin J. Apelberg, Ahmed Jamal, and Brian A. King. 2018. "Notes from the Field: Use of Electronic Cigarettes and Any Tobacco Product among Middle and High School Students—United States, 2011–2018." *Morbidity and Mortality Weekly Report* 67 (45): 1276. https://doi.org/10.15585/MMWR.MM6745A5.

Cunningham, Anthony, Kevin McAdam, Jesse Thissen, and Helena Digard. 2020. "The Evolving E-Cigarette: Comparative Chemical Analyses of E-Cigarette Vapor and Cigarette Smoke." *Frontiers in Toxicology* 2 (December): 7. https://doi.org/10.3389/FTOX.2020.586674.

Daher, Nancy, Rawad Saleh, Ezzat Jaroudi, Hiba Sheheitli, Thérèse Badr, Elizabeth Sepetdjian, Mariam al Rashidi, Najat Saliba, and Alan Shihadeh. 2010. "Comparison of Carcinogen, Carbon Monoxide, and Ultrafine Particle Emissions from Narghile Waterpipe and Cigarette Smoking: Sidestream Smoke Measurements and Assessment of Second-Hand Smoke Emission Factors." *Atmospheric Environment (Oxford, England : 1994)* 44 (1): 8–14. https://doi.org/10.1016/J.ATMOSENV.2009.10.004.

Dyer, Owen. 2021. "New Zealand Plans to Outlaw Tobacco Sales to Citizens Born after 2008." *BMJ* 375 (December): n3057. https://doi.org/10.1136/BMJ.N3057.

Ezzati, Majid, and Alan D. Lopez. 2003. "Estimates of Global Mortality Attributable to Smoking in 2000." *Lancet (London, England)* 362 (9387): 847–852. https://doi.org/10.1016/S0140-6736(03)14338-3.

Fadus, Matthew C., Tracy T. Smith, and Lindsay M. Squeglia. 2019. "The Rise of E-Cigarettes, Pod Mod Devices, and JUUL among Youth: Factors Influencing Use, Health Implications, and Downstream Effects." *Drug and Alcohol Dependence* 201 (August): 85. https://doi.org/10.1016/J.DRUGALCDEP.2019.04.011.

Farag, Mohamed A., Moamen M. Elmassry, and Sherweit H. El-Ahmady. 2018. "The Characterization of Flavored Hookahs Aroma Profile and in Response to Heating as Analyzed via Headspace Solid-Phase Microextraction (SPME) and Chemometrics." *Scientific Reports* 8 (1). https://doi.org/10.1038/S41598-018-35368-6.

Filippidis, Filippos T., Anthony A. Laverty, Vasiliki Gerovasili, and Constantine I. Vardavas. 2017. "Two-Year Trends and Predictors of e-Cigarette Use in 27 European Union Member States." *Tobacco Control* 26 (1): 98–104. https://doi.org/10.1136/TOBACCOCONTROL-2015-052771.

Giovino, Gary A., Sara A. Mirza, Jonathan M. Samet, Prakash C. Gupta, Martin J. Jarvis, Neeraj Bhala, Richard Peto, et al. 2012. "Tobacco Use in 3 Billion Individuals from 16 Countries: An Analysis of Nationally Representative Cross-Sectional Household Surveys." *Lancet (London, England)* 380 (9842): 668–679. https://doi.org/10.1016/S0140-6736(12)61085-X.

Goniewicz, Maciej L., Elena O. Lingas, and Peter Hajek. 2013. "Patterns of Electronic Cigarette Use and User Beliefs about Their Safety and Benefits: An Internet Survey." *Drug and Alcohol Review* 32 (2): 133–140. https://doi.org/10.1111/J.1465-3362.2012.00512.X.

Goniewicz, Maciej Lukasz, Jakub Knysak, Michal Gawron, Leon Kosmider, Andrzej Sobczak, Jolanta Kurek, Adam Prokopowicz, et al. 2014. "Levels of Selected Carcinogens and Toxicants in Vapor from Electronic Cigarettes." *Tobacco Control* 23 (2): 133. https://doi.org/10.1136/TOBACCOCONTROL-2012-050859.

Howes, Seth, Jamie Hartmann-Boyce, Jonathan Livingstone-Banks, Bosun Hong, and Nicola Lindson. 2020. "Antidepressants for Smoking Cessation." *The Cochrane Database of Systematic Reviews* 4 (4). https://doi.org/10.1002/14651858.CD000031.PUB5.

Jacob, Peyton, Ahmad H. Abu Raddaha, Delia Dempsey, Christopher Havel, Margaret Peng, Lisa Yu, and Neal L. Benowitz. 2011. "Nicotine, Carbon Monoxide, and Carcinogen Exposure after a Single Use of a Water Pipe." *Cancer Epidemiology, Biomarkers & Prevention : A Publication of the American Association for Cancer Research, Cosponsored by the American Society of Preventive Oncology* 20 (11): 2345–2353. https://doi.org/10.1158/1055-9965.EPI-11-0545.

Jawad, Mohammed, Rana Charide, Reem Waziry, Andrea Darzi, Rami A. Ballout, and Elie A. Akl. 2018. "The Prevalence and Trends of Waterpipe Tobacco Smoking: A Systematic Review." *PloS One* 13 (2). https://doi.org/10.1371/JOURNAL.PONE.0192191.

Jawad, Mohammed, Amanda Wilson, John Tayu Lee, Sena Jawad, Fiona L. Hamilton, and Christopher Millett. 2013. "Prevalence and Predictors of Water Pipe and Cigarette Smoking among Secondary School Students in London." *Nicotine & Tobacco Research : Official Journal of the Society for Research on Nicotine and Tobacco* 15 (12): 2069–2075. https://doi.org/10.1093/NTR/NTT103.

Kaur, Gagandeep, Rakeysha Pinkston, Benathel McLemore, Waneene C. Dorsey, and Sanjay Batra. 2018. "Immunological and Toxicological Risk Assessment of E-Cigarettes." *European Respiratory Review : An Official Journal of the European Respiratory Society* 27 (147). https://doi.org/10.1183/16000617.0119-2017.

Kinouani, Shérazade, Edwige Pereira, and Christophe Tzourio. 2017. "Electronic Cigarette Use in Students and Its Relation with Tobacco-Smoking: A Cross-Sectional Analysis of the i-Share Study." *International Journal of Environmental Research and Public Health* 14 (11). https://doi.org/10.3390/IJERPH14111345.

Labib, Nargis, Ghada Radwan, Nabiel Mikhail, Mostafa K. Mohamed, Maged el Setouhy, Christopher Loffredo, and Ebenezer Israel. 2007. "Comparison of Cigarette and Water Pipe Smoking among Female University Students in Egypt." *Nicotine & Tobacco Research : Official Journal of the Society for Research on Nicotine and Tobacco* 9 (5): 591–96. https://doi.org/10.1080/14622200701239696.

Margham, J., K. McAdam, A. Cunningham, A. Porter, S. Fiebelkorn, D. Mariner, H. Digard, and C. Proctor. 2021. "The Chemical Complexity of E-Cigarette Aerosols Compared with the Smoke from a Tobacco Burning Cigarette." *Frontiers in Chemistry* 9 (September): 796. https://doi.org/10.3389/FCHEM.2021.743060/BIBTEX.

Mescolo, Federica, Giuliana Ferrante, and Stefania la Grutta. 2021. "Effects of E-Cigarette Exposure on Prenatal Life and Childhood Respiratory Health: A Review of Current Evidence." *Frontiers in Pediatrics* 9 (August): 878. https://doi.org/10.3389/FPED.2021.711573/BIBTEX.

Mirbolouk, Mohammadhassan, Paniz Charkhchi, Sina Kianoush, S. M. Iftekhar Uddin, Olusola A. Orimoloye, Rana Jaber, Aruni Bhatnagar, et al. 2018. "Prevalence and Distribution of E-Cigarette Use among U.S. Adults: Behavioral Risk Factor Surveillance System, 2016." *Annals of Internal Medicine* 169 (7): 429–438. https://doi.org/10.7326/M17-3440.

Moh'D Al-Mulla, Ahmad, Sahar Abdou Helmy, Jawad Al-Lawati, Sami al Nasser, Salah Ali Abdel Rahman, Ayesha Almutawa, Bassam Abi Saab, et al. 2008. "Prevalence of Tobacco Use among Students Aged 13–15 Years in Health Ministers' Council/Gulf Cooperation Council Member States, 2001–2004." *The Journal of School Health* 78 (6): 337–343. https://doi.org/10.1111/J.1746-1561.2008.00311.X.

Monn, Christian, Ph Kindler, A. Meile, and O. Brändli. 2007. "Ultrafine Particle Emissions from Waterpipes." *Tobacco Control* 16 (6): 390–393. https://doi.org/10.1136/TC.2007.021097.

Münzel, Thomas, Omar Hahad, Marin Kuntic, John F. Keaney, John E. Deanfield, and Andreas Daiber. 2020a. "Effects of Tobacco Cigarettes, e-Cigarettes, and Waterpipe Smoking on Endothelial Function and Clinical Outcomes." *European Heart Journal* 41 (41): 4057–4070. https://doi.org/10.1093/EURHEARTJ/EHAA460.

Münzel, Thomas, Marin Kuntic, Sebastian Steven, Omar Hahad, and Andreas Daibe. 2020b. "Is Vaping Better Than Smoking Cigarettes?" *European Heart Journal* 41 (28): 2612–2614. https://doi.org/10.1093/EURHEARTJ/EHAA267.

Ng, Marie, Michael K. Freeman, Thomas D. Fleming, Margaret Robinson, Laura Dwyer-Lindgren, Blake Thomson, Alexandra Wollum, et al. 2014. "Smoking Prevalence and Cigarette Consumption in 187 Countries, 1980–2012." *JAMA* 311 (2): 183–192. https://doi.org/10.1001/JAMA.2013.284692.

Owens, Victoria L., Trung Ha, and Julia N. Soulakova. 2019. "Widespread Use of Flavored E-Cigarettes and Hookah Tobacco in the United States." *Preventive Medicine Reports* 14 (June): 100854. https://doi.org/10.1016/J.PMEDR.2019.100854.

Rao, Xiaoquan, Jixin Zhong, Robert D. Brook, and Sanjay Rajagopalan. 2018. "Effect of Particulate Matter Air Pollution on Cardiovascular Oxidative Stress Pathways." *Antioxidants & Redox Signaling* 28 (9): 797. https://doi.org/10.1089/ARS.2017.7394.

Rom, Oren, Alessandra Pecorelli, Giuseppe Valacchi, and Abraham Z. Reznick. 2015. "Are E-Cigarettes a Safe and Good Alternative to Cigarette Smoking?" *Annals of the New York Academy of Sciences* 1340 (1): 65–74. https://doi.org/10.1111/NYAS.12609.

Rubinstein, Mark L., Kevin Delucchi, Neal L. Benowitz, and Danielle E. Ramo. 2018. "Adolescent Exposure to Toxic Volatile Organic Chemicals From E-Cigarettes." *Pediatrics* 141 (4). https://doi.org/10.1542/PEDS.2017-3557.

Salameh, Pascale, Georges Khayat, and Mirna Waked. 2012. "Lower Prevalence of Cigarette and Waterpipe Smoking, But a Higher Risk of Waterpipe Dependence in Lebanese Adult Women than in Men." *Women & Health* 52 (2): 135–150. https://doi.org/10.1080/03630242.2012.656885.

Samet, Jonathan M. 2013. "Tobacco Smoking: The Leading Cause of Preventable Disease Worldwide." *Thoracic Surgery Clinics* 23 (2): 103–112. https://doi.org/10.1016/J.THORSURG.2013.01.009.

Schraufnagel, Dean E., Francesco Blasi, M. Bradley Drummond, David C. L. Lam, Ehsan Latif, Mark J. Rosen, Raul Sansores, Richard van Zyl-Smit, and Forum of International Respiratory Societies. 2014. "Electronic Cigarettes: A Position Statement of the Forum of International Respiratory Societies." *American Journal of Respiratory and Critical Care Medicine* 190 (6): 611–618. https://doi.org/10.1164/RCCM.201407-1198PP.

Schubert, Jens, Frederic D. Müller, Roman Schmidt, Andreas Luch, and Thomas G. Schulz. 2015. "Waterpipe Smoke: Source of Toxic and Carcinogenic VOCs, Phenols and Heavy Metals?" *Archives of Toxicology* 89 (11): 2129–2139. https://doi.org/10.1007/S00204-014-1372-X.

Shihadeh, A. 2003. "Investigation of Mainstream Smoke Aerosol of the Argileh Water Pipe." *Food and Chemical Toxicology* 41 (1): 143–152. https://doi.org/10.1016/S0278-6915(02)00220-X.

Sibai, Abla M., Rania A. Tohme, Mohamad M. Almedawar, Taha Itani, Sara I. Yassine, Eden A. Nohra, and Hussain A. Isma'eel. 2014. "Lifetime Cumulative Exposure to Waterpipe Smoking Is Associated with Coronary Artery Disease." *Atherosclerosis* 234 (2): 454–460. https://doi.org/10.1016/J.ATHEROSCLEROSIS.2014.03.036.

Squeglia, Lindsay M., and Kevin M. Gray. 2016. "Alcohol and Drug Use and the Developing Brain." *Current Psychiatry Reports* 18 (5): 46. https://doi.org/10.1007/S11920-016-0689-Y.

Steven, Sebastian, Katie Frenis, Matthias Oelze, Sanela Kalinovic, Marin Kuntic, Maria Teresa Bayo Jimenez, Ksenija Vujacic-Mirski, et al. 2019. "Vascular Inflammation and Oxidative Stress: Major Triggers for Cardiovascular Disease." *Oxidative Medicine and Cellular Longevity*. https://doi.org/10.1155/2019/7092151.

"Tobacco." N.d. Accessed December 2, 2022. www.who.int/news-room/fact-sheets/detail/tobacco.

Wang, Richard J., Sudhamayi Bhadriraju, and Stanton A. Glantz. 2021. "E-Cigarette Use and Adult Cigarette Smoking Cessation: A Meta-Analysis." *American Journal of Public Health* 111 (2): 230–246. https://doi.org/10.2105/AJPH.2020.305999.

Waziry, Reem, Mohammed Jawad, Rami A. Ballout, Mohammad al Akel, and Elie A. Akl. 2017. "The Effects of Waterpipe Tobacco Smoking on Health Outcomes: An Updated Systematic Review and Meta-Analysis." *International Journal of Epidemiology* 46 (1): 32–43. https://doi.org/10.1093/IJE/DYW021.

Wu, Xianxian, Haiying Zhang, Wei Qi, Ying Zhang, Jiamin Li, Zhange Li, Yuan Lin, et al. 2018. "Nicotine Promotes Atherosclerosis via ROS-NLRP3-Mediated Endothelial Cell Pyroptosis." *Cell Death & Disease* 9 (2). https://doi.org/10.1038/S41419-017-0257-3.

Yatsuya, Hiroshi, and Aaron R. Folsom. 2010. "Risk of Incident Cardiovascular Disease among Users of Smokeless Tobacco in the Atherosclerosis Risk in Communities (ARIC) Study." *American Journal of Epidemiology* 172 (5): 600. https://doi.org/10.1093/AJE/KWQ191.

Yuan, Menglu, Sarah J. Cross, Sandra E. Loughlin, and Frances M. Leslie. 2015. "Nicotine and the Adolescent Brain." *The Journal of Physiology* 593 (16): 3397–3412. https://doi.org/10.1113/JP270492.

Zhang, Yijia, Meghan Angley, Xinran Qi, Liping Lu, Mary E. D'Alton, and Ka Kahe. 2022. "Maternal Electronic Cigarette Exposure in Relation to Offspring Development: A Comprehensive Review." *American Journal of Obstetrics & Gynecology MFM* 4 (5): 100659. https://doi.org/10.1016/J.AJOGMF.2022.100659.

14 The Aryl Hydrocarbon Receptor as an Environmental Sensor and Mediator of Stress and Inflammation

Christoph F. A. Vogel and Thomas Haarmann-Stemmann

ABBREVIATIONS

AD, atopic dermatitis; AHR, aryl hydrocarbon receptor; AIP, AHR-interacting protein; AKR, aldo-keto reductase; ARE, antioxidant response elements; ARNT, aryl hydrocarbon receptor nuclear translocator; BaP, benzo[*a*]pyrene; COPD, chronic obstructive pulmonary disease; COX, cyclooxygenase; CVD, cardiovascular disease; CYP, cytochrome P450; DC, dendritic cells; FICZ, 6-formylindolo[3,2-b]carbazole; HSP90, heat-sock protein 90; KEAP1, Kelch-like ECH-associated protein; LPS, lipopolysaccharide; MCP-1, monocyte chemotactic protein 1; MMP, matrix metalloproteinases; NF-κB, nuclear factor-κB; NKX2.5, NK2 homeobox 5; NOX, NADPH oxidase; NQO1, NADPH:quinone oxidoreductase-1; NRF2, nuclear factor erythroid 2-related factor 2; PAH, polycyclic aromatic hydrocarbon; PCB, polychlorinated biphenyl; PM, particulate matter; POP, persistent organic pollutant; ROS, reactive oxygen species; S100A9, S100 calcium-binding protein A9; SOD, superoxide dismutase; TCDD, 2,3,7,8-tetrachlorodibenzo-*p*-dioxin; TCF21, transcription factor 21; TRAP, traffic-related air pollution; XRE, xenobiotic-responsive element

14.1 INTRODUCTION

The exposure to air pollution and environmental pollutants presents a significant problem by increasing the risk to develop several chronic diseases (Landrigan et al. 2018). Environmental exposure encompasses the inhalation of chemicals and particles through ambient particulate matter (PM) as well as the contact and consumption of harmful chemicals via consumer and food products. Numerous studies have found that the exposure to fine $PM_{2.5}$, i.e. air pollution particles smaller than 2.5 microns in diameter, and other environmental pollutants may cause chronic inflammation in various tissues leading to adverse human health effects, especially lung and cardiovascular diseases (Schraufnagel et al. 2019a; Wu et al. 2018). Furthermore, recent studies reported that PM exposure may also promote the development of neurodegenerative diseases, diabetes, and inflammatory skin disorders (Schraufnagel et al. 2019b). Ambient PM derives from various sources and consist of a complex mixture of chemicals containing numerous compounds including metals, sulfate, nitrate, ammonium, black carbon, polycyclic aromatic hydrocarbons (PAHs), and persistent organic pollutants (POPs) as well as biological substances in the form of endotoxin and other pathogens. Recent studies show that PM that derive from human-made sources like fossil fuel combustion and industrial emissions are more harmful than PM that come from natural sources like soil (Goobie et al. 2022). This study confirms that the composition and the combination of the components of $PM_{2.5}$ are critical factors which determine how harmful they are to cause health effects. Heavy metals (i.e. copper, iron, and lead), PAHs, and

DOI: 10.1201/9781003328100-17

POPs (i.e. dioxins and polychlorinated biphenyls [PCBs]) are harmful chemicals and components of PM generated via combustion and industrial emissions that are responsible to mediate adverse health effects. PAHs result from combustion processes and derive from various sources (e.g. industrial activities, wildfires, and vehicular traffic). Traffic-related air pollution (TRAP) is a major source of ambient PM and PAH exposure. PAHs such as benzo[*a*]pyrene (BaP) have been identified as ligands of the aryl hydrocarbon receptor (AHR), and activation of the AHR signaling pathway via environmental pollutants has been associated with the development of chronic diseases (Vogel et al. 2020). Numerous studies investigated the toxicity of PM focusing on oxidative stress and chronic inflammation as the main biological processes involved in the pathological responses triggered by PM exposure (Wu et al. 2018). In this chapter, we review the role of the AHR signaling pathway and its interaction with other pathways including nuclear factor erythroid 2-related factor 2 (NRF2) and nuclear factor-κB (NF-κB) as relevant mechanisms regarding the generation of oxidative stress and the modulation of immunity in PM-mediated adverse health effects.

14.2 AHR-RELATED GENERATION OF OXIDATIVE STRESS

Exposure to PM and PAHs causes ROS formation, which may result in the oxidative damage of DNA, lipids, and other cellular macromolecules. Probably most prominent pathway by which PAHs induce ROS is the AHR-dependent induction of cytochrome P450 (CYP) enzyme activity. Briefly, PAHs bind to the cytosolic AHR, which subsequently sheds its chaperone molecules, translocated into the nucleus and dimerizes with the AHR nuclear translocator (ARNT) (Figure 14.1) (Rothhammer and Quintana 2019; Vogel et al. 2020). The AHR/ARNT complex binds to so-called xenobiotic-responsive elements (XRE) in the enhancer region of target genes to induce their expression. Prototypical target genes of the canonical AHR pathway encode for CYP1A1, 1A2, and 1B1 (Rothhammer and Quintana 2019; Vogel et al. 2020). These CYP1 isoforms are upregulated by the chemically exposed cell in order to oxidize the invading (AHR-activating) chemical and enable its elimination. Importantly, the AHR/CYP1-related generation of oxidative stress is tightly linked to the catalytic circle of the monooxygenases, more precisely to a phenomenon called "uncoupling" (Albertolle and Guengerich 2018). In particular when compounds that are not oxidizable, such as TCDD, induce the formation of the oxygen-bound CYP complex, the lack of an electron acceptor may cause autooxidation of the CYP enzyme and the subsequent release of superoxide (Albertolle and Guengerich 2018). Superoxide dismutates to hydrogen peroxide and, by undergoing Fe^{2+}-catalyzed Haber-Weiss and Fenton reactions, both superoxide anions and hydrogen peroxide are converted to highly reactive hydroxyl radicals (Wardman and Candeias 1996).

Apart from AHR ligand-triggered events, the PM-induced generation of oxidative stress may involve AHR-independent pathways, which are based on the presence of metals (e.g. iron, nickel, copper), the matrix, or other components of PM. In fact, an AHR-independent ROS production is supported by the direct oxidative potential of PM and the rapid generation of ROS, which can be measured in cell free assays (Ayres et al. 2008). It is important to note that the direct oxidative potential of PM is distinctive from the intracellular production of ROS, which can be mediated via activation of AHR by PAH-rich PM.

In addition, AHR may stimulate the activity of other ROS-generating enzymes in environmentally-stressed cells or tissues. Recently, we reported that AHR-dependent signaling pathways control the expression of several members of the aldo-keto reductase (AKR) family 1 (Vogeley et al. 2022). These cytosolic NADPH-dependent oxidoreductases convert carbonyl groups to their corresponding alcohols and are for instance involved in the metabolism of steroid hormones and eicosanoids (Penning 2014). Some AKR1 isoforms, in particular AKR1A1 and AKR1C1 to 1C4, are capable of oxidizing PAH *trans*-dihydrodiols, intermediates derived from CYP1-mediated oxidation of PAHs, to the respective catechols (Penning 2014). By releasing hydrogen peroxide and superoxide anion

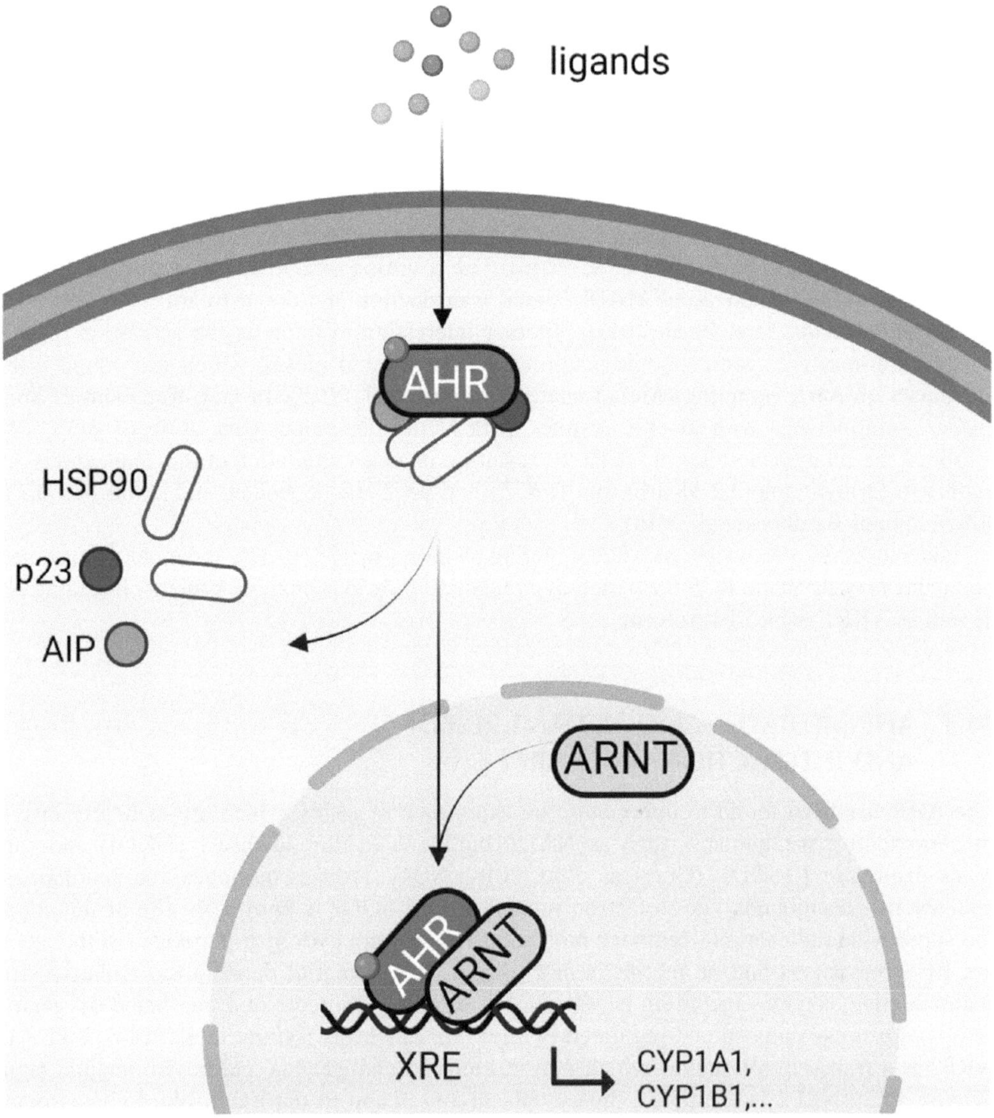

FIGURE 14.1 The AHR signaling pathway. In absence of a ligand, AHR is part of a cytosolic multiprotein complex consisting of heat shock protein 90 (HSP90), AHR-interacting protein (AIP), and p23. Upon binding of a ligand, AHR undergoes conformational changes leading to the dissociation of the multiportion complex and the exhibition of a nuclear localization signal. Subsequently, AHR translocates to the nucleus, where it dimerizes with ARNT and binds to XRE in the enhancer region of target genes, e.g. encoding CYP1A1 and CYP1B1, to induce their expression. The figure was created with BioRender software (www.biorender.com; agreement number PF24R8OWLG).

radicals, the PAH catechol is sequentially oxidized to *o*-semiquinone anion radical and PAH-*o*-quinone. The latter is genotoxic and thus may form DNA adducts or in the presence of NADPH is reduced back to the PAH-catechol (Penning 2014).

Another example is the PM- and/or PAH-stimulated production of ROS by NADPH oxidase (NOX) 2 and NOX4 (Amara et al. 2007; Smith et al. 2019). When inactive, the membrane-associated NOX subunits, three cytoplasmic ($p47^{phox}$, $p67^{phox}$, Rac1) and two intrinsic membrane

($p22^{phox}$, $gp91^{phox}$) components are unassembled (Fisher 2009). Upon activation by triggers such as cytokines, opsonized bacteria, and bacterial lipopolysaccharides (LPS), the complex assembles and the catalytic subunit consisting of $gp91^{phox}$ and $p22^{phox}$ transfers one electron from NADPH to molecular oxygen thereby generating superoxide anions, which then dismutate to hydrogen peroxide (Fisher 2009). As recently summarized (Grishanova and Perepechaeva 2022), several ways of crosstalk between the AHR and the NOX system are known. PM-bound PAHs, for instance, were found to induce a NOX2-mediated generation of ROS in keratinocytes by stimulating the phosphorylation of $p47^{phox}$, a signal for NOX2 complex assembly at the plasma membrane, in an AHR-dependent manner (Lee et al. 2016). The resulting oxidative stress stimulates inflammatory responses by activating MAPK signal transduction and downstream AP-1 and NF-κB transcription factors (Lee et al. 2016). Another interesting example is the activation of NOX4 by the thiol-reactive agents arsenic, cadmium, mercury, and nickel, which was found to have an impact on AHR signaling (Mohammadi-Bardbori et al. 2015). In fact, treatment of human HaCaT keratinocytes with arsenic resulted in a NOX4-dependent generation of ROS, which inhibited the catalytic activity of CYP1A1, resulting in an accumulation of the endogenous AHR agonist 6-formylindolo[3,2-*b*]carbazole (FICZ) and an AHR-dependent induction of CYP1A1 (Mohammadi-Bardbori et al. 2015).

Taken together, a activation of AHR signaling pathways by PM, PAHs, and other environmental pollutants may stimulate ROS formation by upregulating various enzyme systems, including NOX as well as AKR1 and CYP1 isoforms.

14.3 AHR-MEDIATED ANTIOXIDANT RESPONSE AND INTERACTION WITH NRF2

The AHR has been found to upregulate the expression of genes, which are critically involved in antioxidative mechanisms such as NADPH:quinone oxidoreductase-1 (NQO1) and superoxide dismutase 1 (SOD1) (Guerrina et al. 2018). NQO1 reduces quinones and xenobiotics to less-reactive compounds via 2-electron reduction, and SOD1 is known for the dismutation of the superoxide radicals into ordinary molecular oxygen and hydrogen peroxide. Both enzymes are ROS scavengers and protect cells and tissues from potential damage by oxidative stress. Antioxidative enzymes including NQO1, SOD1, catalase, thioredoxin-1 and heme oxygenase 1 (HO-1) are stress-induced and regulated by the NRF2 pathway (Niture et al. 2014). Like AHR, NRF2 is a transcription factor, which may respond to endogenous and environmental signals (Figure 14.2). NRF2 is located in the cytoplasm and bound to the Kelch-like ECH-associated protein (KEAP1). NRF2 is activated by stress signals, like electrophilic molecules or free radicals. After dissociation of KEAP1, NRF2 binds to antioxidant response elements (AREs) in the nucleus inducing the aforementioned antioxidative enzymes (Niture et al. 2014). Interestingly, the promoter regions of NQO1 and SOD1 contain XRE as well as ARE elements and are regulated by both the AHR and NRF2 signaling pathways (Figure 14.2) (Kohle and Bock 2006). Furthermore, a study has found that the induction of SOD1 via AHR ligands requires both regulatory elements, the XRE and ARE element (Park and Rho 2002). The AHR and NRF2 are two separate signaling pathways, which respond to environmental pollutants and stress signals. Interestingly, recent reports demonstrate that both signaling pathways may interact suggesting a tight connection of the AHR and NRF2 pathways as an integrated response to the exposure of environmental chemicals (Kohle and Bock 2006). Of note, the sustained activation of NRF2 in keratinocytes has been associated with a chloracne-like skin phenotype in mice and has been considered as an underlying mechanism in dioxin-induced chloracne (Schafer et al. 2014). Chloracne is a type of acne-like skin disease associated with exposure to certain halogenated aromatic compounds. Noteworthy, chloracne is a very specific biomarker of dioxin exposure in human (Panteleyev and Bickers 2006).

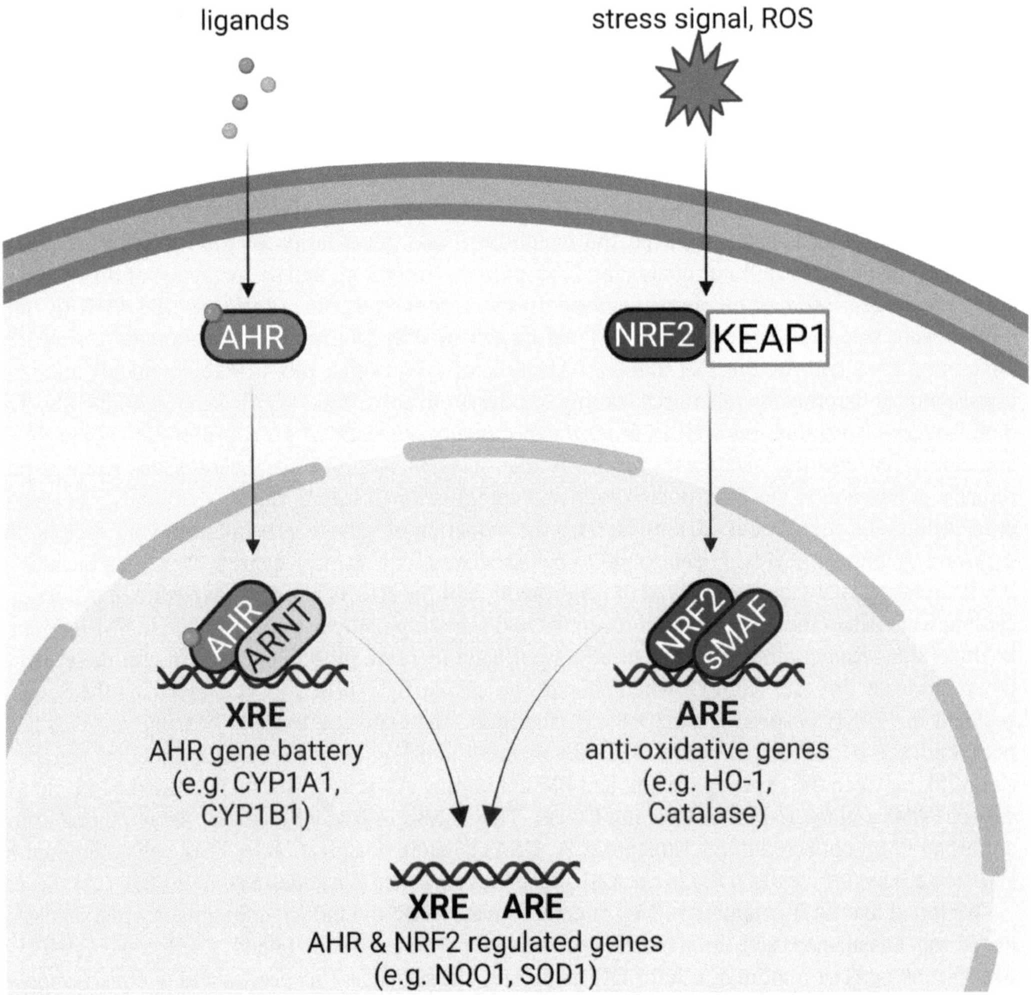

FIGURE 14.2 AHR-mediated and NRF2-mediated gene regulation. Whereas the ligand-activated AHR controls the expression of CYP1 isoforms, NRF2 activated by ROS and other stress signals regulates the expression of antioxidant genes. The expression of several antioxidant enzymes, such as NQO1 and SOD1, is regulated by both signaling pathways by interacting with respective DNA binding sites. The figure was created with BioRender software (www.biorender.com; agreement number UL25K8ZO77).

14.4 ACTIVATION OF AHR VIA PM, PAHS, AND POPS ASSOCIATED WITH ADVERSE HEALTH EFFECTS AND THE IMMUNE SYSTEM

A fine-tuned and orchestrated immune response is the basis of a successful host defense mechanism, however, dysregulated and/or disproportional prolonged inflammatory responses may cause various adverse health effects leading to chronic inflammatory diseases. Early papers have shown that activation of AHR signaling may modify the expression of pro-inflammatory cytokines such as IL-1β (Sutter et al. 1991). Recent studies have shown that the expression of a list of cytokines and chemokines including IL-8, IL-22, IL-33 as well as CXCL and CCL chemokines can be modified via AHR activation, which has been described predominantly in innate immune cells including innate lymphoid cells, macrophages, and dendritic cells (Ishihara et al. 2019; Kado et al. 2017; Quintana et al. 2008; Vogel et al. 2005). Although TCDD has been known to cause immune-suppressive effects, the important role that the AHR is playing in the immune system has been shown only in the last two decades. New findings showed that FICZ is a high-affinity ligand of AHR and may serve as a potential endogenous ligand of AHR. In contrast to the immune-suppressive effects of TCDD, FICZ has been found to trigger the differentiation and function of inflammatory Th17 cells and may exacerbate autoimmunity (Quintana et al. 2008; Veldhoen et al. 2008). Many studies followed up on these initial findings demonstrating that AHR signaling balances the Th17 vs. regulatory T cell (Rothhammer and Quintana 2019). The mechanisms underlying the ligand-specific effects of AHR are still a matter of debate; however, the metabolism and degradation of the ligand, the route of exposure, the dose, the cellular context and the microenvironment, and the presence of other chemicals or pathogens seem to be important aspects directing the outcome of the specific AHR ligand.

Exposure studies indicate that AHR signaling activated by environmental pollutants such as PM containing chemical mixtures of metals, PAHs, dioxins, and other dioxin-like chemicals mediates consistently inflammatory responses leading to adverse health effects (O'Driscoll et al. 2018b). The data from *in vivo* studies, as well as *in vitro* experiments suggest that exposure to PM rich in PAHs and dioxin-like chemicals increases cardiovascular diseases, autoimmune diseases, and worsens pulmonary inflammation via Th17 polarization and upregulation of inflammatory cytokines (O'Driscoll et al. 2018a; O'Driscoll et al. 2018b). Besides the induction of various pro-inflammatory factors, the exposure to ambient PM has been found to be associated with oxidative stress. Whereas the role of AHR signaling is obviously involved in the promotion of the effects after PM exposure, it is not quite clear to what extent the AHR or other receptors and signaling pathways are involved in PM-mediated health effects. Remarkably, TCDD caused a significant increase in the incidence of chronic obstructive pulmonary diseases (COPD) when people were primarily exposed to TCDD during the Seveso-Incident in 1976 (Consonni et al. 2008). Furthermore, the exposure to PAH-bearing particles have been found to be associated with COPD (Brunekreef and Forsberg 2005). These studies suggest a clear link between the exposure to ambient PM containing AHR agonists such as dioxin-like chemicals or PAHs and the risk of developing COPD. This is also in line with studies showing that ambient urban $PM_{2.5}$ collected in Sacramento (CA, USA) triggers polarization of Th17 cells mediated by PM-treated dendritic cells (DC) in an AHR-dependent manner (Castaneda et al. 2018).

We found that PAH-containing $PM_{2.5}$ directly activate AHR and induce inflammatory markers associated with an enhanced activation of DC responsible for differentiation of naive T cells towards a Th17-like phenotype. The important role of DC mediating elevated IL-17 responses and allergic responses has been confirmed in mice after treatment with vehicular ultrafine PM (Xia et al. 2015). The deletion of AHR in $CD11c^+$ cells protected against PM-mediated exacerbation of the allergic responses and PM-mediated IL-17 increase, underlining the key role of DC and AHR. Interestingly, the impact of AHR activation modulating the function of DC during development in mice suggested long-lasting consequences on immune function through environmental exposure in early life (Meyers et al. 2018). The findings of Joshua Mezrich's group revealed that PAHs are a critical component in PM promoting Th17-polarization and secretion of IL-17A from T cells (O'Driscoll et al. 2019). Studies by the same group showed that the PM matrix and chemical components other than PAHs in the organic fraction are also contributing to the observed IL-17 responses *in vivo* and the associated adverse health effects, including autoimmune diseases (O'Driscoll et al. 2019; O'Driscoll et al. 2018a; O'Driscoll et al. 2018b).

Future studies are needed to clearly identify the most critical components of PM and their inflammatory potential and effectiveness to induce chronic inflammatory diseases.

14.5 IMPACT OF ENVIRONMENTAL POLLUTANTS AND PM ON THE CARDIOVASCULAR SYSTEM: ROS AND INFLAMMATION MEDIATED EFFECTS AND POSSIBLE ROLE OF AHR

Over the last two decades numerous studies provided strong evidence for the adverse health effects on the cardiovascular system caused by air pollution and PM exposure. The development of cardiovascular diseases (CVD) promoted by ambient PM exposure includes vascular dysfunction, hypertension, atherosclerosis, acute myocardial infarction, and ischemic heart disease as shown in several studies (Goncharov et al. 2008; Sergeev and Carpenter 2005; Sergeev and Carpenter 2010). Noteworthy, epidemiologic studies also showed that occupational exposure to high doses of dioxin, the prototypical ligand of AHR, is associated with ischemic heart disease (Flesch-Janys et al. 1995). Furthermore, the exposure of U.S. Army veterans during the Vietnam war to Agent Orange, a herbicide contaminated with dioxin, was found to be associated with hypertension, stroke, and coronary heart disease (Kang et al. 2006; Kim et al. 2003). These findings were confirmed in studies including 12 cohorts showing an increased mortality from ischemic heart disease after dioxin exposure (Humblet et al. 2008). These studies strongly suggest that exposure to AHR-activating chemicals, such as dioxins, increases the risk of CVD.

The impact of air pollution on CVD represents a serious public health problem (Mateen and Brook 2011). The development of atherosclerosis and atherosclerotic cardiovascular disease induced by air pollution and the exposure to $PM_{2.5}$ has been found to be associated with an inflammatory condition (Brook et al. 2010; Kunzli et al. 2010; Pope et al. 2004). While there is strong evidence that inflammatory mediators are critically involved in atherosclerosis, it is less clear how the inflammatory mediators and markers of atherogenesis are upregulated through exposure to air pollutants and PM.

PM rich in PAHs and heavy metals (e.g. iron, nickel, copper) have been shown to induce inflammatory responses (Grunig et al. 2014; Ruckerl et al. 2007) and may trigger early atherogenesis (Libby et al. 2002). PM generated by combustion are known to contain significant amounts of PAHs which can bind to and activate the AHR (Ban-Weiss et al. 2010; Phuleria et al. 2006). Moreover, the exposure to TCDD, the prototypical ligand of AHR, led to the progression of atherosclerosis in the apolipoprotein E-deficient (ApoE) mouse model (Wu et al. 2011). These findings confirmed an earlier study by Dalton et al. describing more severe atherosclerotic lesions in descending aortas and an increase in blood pressure and lipids in ApoE mice after sub-chronic exposure to dioxin (Dalton et al. 2001). The development of atherosclerotic lesions induced by TCDD was associated with elevated levels of inflammatory markers and accumulation of foam cells, which is an early indicator for the development of atherosclerotic lesions. Similar findings were described in studies with ApoE mice exposed to $PM_{2.5}$ (Zhang et al. 2020). These reports show that the atherogenic activity of environmental pollutants, such as PM and dioxin-like compounds, contribute to the development of atherosclerosis at least in part through activation of the AHR signaling pathway and the induction of a vascular inflammatory response. Additionally, the exposure to $PM_{2.5}$ may affect the function of endothelial progenitor cells leading to vascular injury (Haberzettl et al. 2018). The authors concluded that oxidative stress induced by $PM_{2.5}$ exposure significantly contributed to the dysfunction of endothelial progenitor cells in the vascular system. Together, these studies suggest that environmental pollutants, such as $PM_{2.5}$ and their chemical components, may induce oxidative stress and inflammatory factors leading to vascular injury and CVD. Beyond its role as a mediator of cardiovascular toxicity, a physiological role of the AHR has been suggested by studies using AHR knockout models showing that the lack of AHR affects the function of cardiomyocytes, endothelial cells, and smooth muscle cells (Yi et al. 2018) (Figure 14.3).

Activation of AHR has been shown to reduce the expression of T-cadherin, which is normally expressed at a high level in cardiac and vascular tissues (Niermann et al. 2003). Further the interaction of AHR with TCF21 resulted in stimulation of pro-inflammatory gene expression in smooth muscle cells (Kim et al. 2017). The effect of PM on human aortic endothelial cells showed that fine and ultrafine PM from urban areas stimulate the expression of inflammatory and atherogenic markers such E-selectin, MCP-1, and COX-2 in an AHR-dependent manner (Aung et al. 2011). Besides endothelial cells and smooth muscle cells, numerous studies have shown the important role of immune cells, such as macrophages, DC, and T lymphocytes in the development of atherosclerosis (Wolf and Ley 2019). Moreover, the AHR plays a critical role in differentiation processes of macrophages, DCs, and T lymphocytes (Rothhammer and Quintana 2019). Consequently, AHR ligands including POPs, dioxins, and PCBs may significantly affect

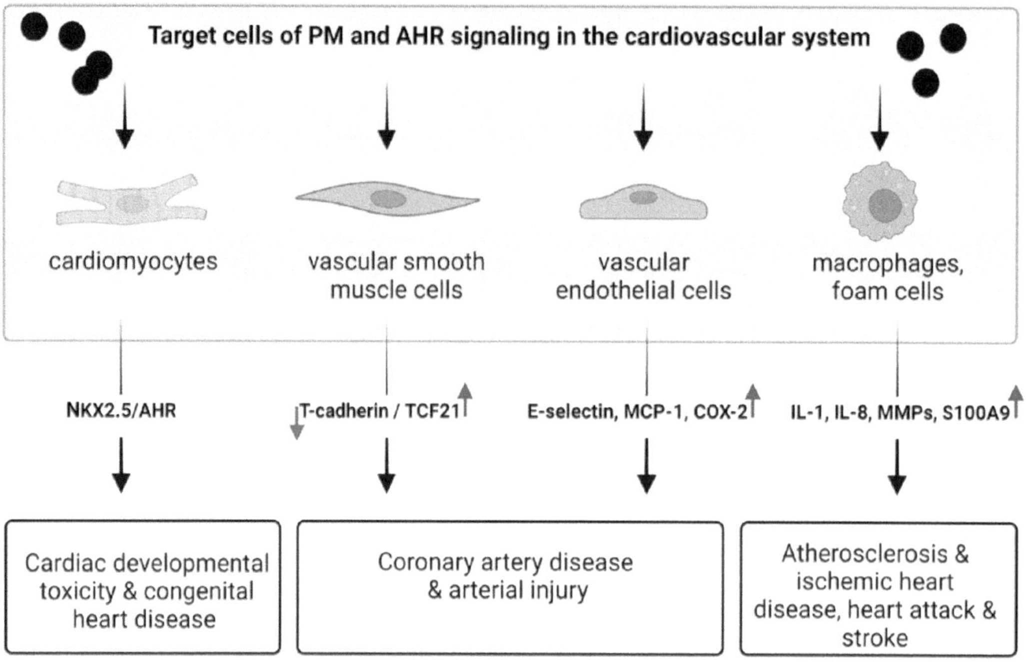

FIGURE 14.3 PM-triggered and AHR-mediated effects on target cells and markers of the cardiovascular system and potential outcomes of CVD. In PM-exposed cardiomyocytes, AHR may interact with the transcription factor NKX2.5, which may be relevant for the development of cardiac developmental toxicity and congenital heart disease. In vascular smooth muscle cells and vascular endothelial cells, the PM-activated AHR may modulate the expression of adhesion proteins and induce proinflammatory mediators, for instance by interacting with the transcription factor TCF21. These events may cause arterial injury and the development of coronary artery disease. In macrophage and foam cells, activation of AHR signaling by PM exposure increases the expression of proinflammatory cytokines and proteins, which may be of pathophysiological relevance for numerous CVD, including atherosclerosis, ischemic heart disease, heart attack, and stroke. The figure was created with BioRender software (www.biorender.com; agreement number CE24QV7E82).

the function of immune cells and modify the expression of cytokines, chemokines, matrix metalloproteinases (MMPs), and S100 calcium-binding protein A9 (S100A9), which are critical factors involved in the development of atherosclerotic lesions and atherosclerotic cardiovascular disease. Developmental effects on the cardiovascular system and the heart after exposure to dioxins and other toxic AHR ligands have been studied in aquarium fish models. TCDD induced pericardial edema in Medaka, which was associated with an increase in COX-2 expression (Dong et al. 2010). In zebrafish TCDD also caused pericardial edema and led to cardiac malformations and reduction of the development of the cardinal vein in an AHR-dependent manner (Bello et al. 2004). Interestingly, the TCDD-mediated induction of cardiac dysfunction has been suggested to involve the interaction of AHR with the cardiomyocyte-specific NK2 homeobox 5 (NKX2.5) gene (Wang et al. 2017) (Figure 14.3). Noteworthy, not only the treatment with TCDD but also the exposure to PM caused developmental toxicity and heart defects mediated via AHR signaling in zebrafish (Ren et al. 2019).

14.6 AHR SIGNALING, ROS, AND INFLAMMATORY SKIN DISEASES

Disturbance of the epidermal barrier integrity and function is a hallmark of common inflammatory skin diseases, including atopic dermatitis (AD) and psoriasis (Schmuth et al. 2015). Exposure of the skin to airborne PM and PAHs may facilitate the development and/or aggravation of cutaneous inflammatory diseases, primarily by inducing ROS generation. Oxidative stress either directly interrupts keratinocyte differentiation and associated epidermal barrier formation or stimulates MAPK signal transduction and downstream transcription factors to induce the expression of proinflammatory cytokines, such as TNFα, IL-4, and IL-13. Upon their release, the Th2 signature cytokines, for instance, repress the expression of loricrin, involucrin, and other proteins involved in keratinocyte differentiation and epidermal barrier formation (Kim et al. 2008). Another cytokine frequently present in chronic inflammatory skin diseases is IL-17, which also decreases filaggrin expression in keratinocytes (Gutowska-Owsiak et al. 2012) and is under transcriptional control of AHR (Quintana et al. 2008).

In fact, several studies provided evidence for an involvement of AHR in the pathogenesis of AD. AD results from a complex interplay between environmental and genetic factors and is characterized by skin barrier dysfunction, overexpression of proinflammatory cytokines, and activation of Th2 cell-driven immune responses. Interestingly, exposure to air pollution, in particular PAH-rich pollutants, such as tobacco smoke and TRAP, is positively associated with the development and worsening of AD in humans (Ahn 2014; Hidaka et al. 2017). In contrast to these findings, exposure to dioxin-like AHR agonists seems to lower the risk of developing AD (Nakamoto et al. 2013). The underlying causes for this difference are quite enigmatic but may involve a ligand-specific modulation of proinflammatory and/or immunosuppressive responses. However, in contrast to an exposure to PAHs, TCDD treatment even accelerates keratinocyte differentiation and cornification in a ROS- and sirtuin-1-dependent manner (Sutter et al. 2019). On the other hand, these halogenated aromatic hydrocarbons also induce immunosuppressive regulatory T cells (Quintana et al. 2008), which may alleviate the contribution of Th2 and Th17 lymphocytes to the pathogenesis of the disease.

14.7 CONCLUSION AND OUTLOOK

There is broad consensus that chronic exposure to air pollution, i.e. PM and associated PAHs, POPs, and heavy metals, significantly contributes to the rising prevalence of several noncommunicable diseases, including allergic, cardiovascular, respiratory, and metabolic diseases as well as cancer in industrialized countries and thus poses a major burden on our healthcare systems. However, the precise molecular and cellular mechanisms by which PM and commonly associated chemicals promote the development of the diseases mentioned earlier are only partially understood.

In the current chapter we highlight the role of AHR as a sensor of PAHs and POPs adsorbed to PM. Much data exist that show how AHR activation by PAHs/POPs and the associated modulation of metabolic reactions, cytokine patterns, and immune responses result in oxidative stress, tissue damage, and proinflammatory conditions in various target tissues, such as the lung, the skin, and the cardiovascular system. However, only recently, studies directly began to reveal the link "PM-PAH-AHR", i.e. showed that PM adsorbed PAHs trigger AHR-dependent cellular changes. This development will give important insights into the differing toxicity of PM whose content of PAHs and other AHR-activating chemicals may change depending on the geographical location or source of PM. Another important consideration regarding predictions of adverse health effects is the complexity of interactions of air pollution with AHR signaling in the target tissue and the potential pathophysiological responses. Both the physicochemical properties of the particles and the context-dependent outcome of AHR signaling in the specific tissues are relevant. An interplay of these variables ultimately defines the toxicodynamic and toxicokinetic characteristics of the respective pollutants. A better understanding of these relationships and mechanisms is necessary for proper risk assessment. Hence, there is an urgent need for further research aiming to unravel the functional relevance of AHR for air pollution-associated adverse health effects and to assess its suitability as a target for molecular prevention.

REFERENCES

Ahn K (2014) The role of air pollutants in atopic dermatitis. *J Allergy Clin Immunol* 134(5):993–9; discussion 1000 doi:10.1016/j.jaci.2014.09.023

Albertolle ME, Guengerich FP (2018) The relationships between cytochromes P450 and H2O2: Production, reaction, and inhibition. *J Inorg Biochem* 186:228–34 doi:10.1016/j.jinorgbio.2018.05.014

Amara N, Bachoual R, Desmard M, et al. (2007) Diesel exhaust particles induce matrix metalloprotease-1 in human lung epithelial cells via a NADP(H) oxidase/NOX4 redox-dependent mechanism. *Am J Physiol Lung Cell Mol Physiol* 293(1):L170–81 doi:10.1152/ajplung.00445.2006

Aung HH, Lame MW, Gohil K, et al. (2011) Comparative gene responses to collected ambient particles in vitro: Endothelial responses. *Physiol Genomics* 43(15):917–29 doi:10.1152/physiolgenomics.00051.2011

Ayres JG, Borm P, Cassee FR, et al. (2008) Evaluating the toxicity of airborne particulate matter and nanoparticles by measuring oxidative stress potential—a workshop report and consensus statement. *Inhal Toxicol* 20(1):75–99 doi:10.1080/08958370701665517

Ban-Weiss GA, Lunden MM, Kirchstetter TW, Harley RA (2010) Size-resolved particle number and volume emission factors for on-road gasoline and diesel motor vehicles. *Journal of Aerosol Science* 41(1):5–12

Bello SM, Heideman W, Peterson RE (2004) 2,3,7,8-Tetrachlorodibenzo-p-dioxin inhibits regression of the common cardinal vein in developing zebrafish. *Toxicol Sci* 78(2):258–66 doi:10.1093/toxsci/kfh065

Brook RD, Rajagopalan S, Pope CA, 3rd, et al. (2010) Particulate matter air pollution and cardiovascular disease: An update to the scientific statement from the American Heart Association. *Circulation* 121(21):2331–78 doi:10.1161/CIR.0b013e3181dbece1

Brunekreef B, Forsberg B (2005) Epidemiological evidence of effects of coarse airborne particles on health. *EurRespirJ* 26(2):309–18

Castaneda AR, Pinkerton KE, Bein KJ, et al. (2018) Ambient particulate matter activates the aryl hydrocarbon receptor in dendritic cells and enhances Th17 polarization. *Toxicol Lett* 29(2):85–96 doi:10.1016/j.toxlet.2018.04.020

Consonni D, Pesatori AC, Zocchetti C, et al. (2008) Mortality in a population exposed to dioxin after the Seveso, Italy, accident in 1976: 25 years of follow-up. *AmJEpidemiol* 167(7):847–58

Dalton TP, Kerzee JK, Wang B, et al. (2001) Dioxin exposure is an environmental risk factor for ischemic heart disease. Cardiovasc Toxicol 1(4):285–98 doi:10.1385/ct:1:4:285

Dong W, Matsumura F, Kullman SW (2010) TCDD induced pericardial edema and relative COX-2 expression in medaka (oryzias latipes) embryos. *Toxicol Sci* 118(1):213–23 doi:10.1093/toxsci/kfq254

Fisher AB (2009) Redox signaling across cell membranes. *Antioxid Redox Signal* 11(6):1349–56 doi:10.1089/ARS.2008.2378

Flesch-Janys D, Berger J, Gurn P, et al. (1995) Exposure to polychlorinated dioxins and furans (PCDD/F) and mortality in a cohort of workers from a herbicide-producing plant in Hamburg, Federal Republic of Germany. *Am J Epidemiol* 142(11):1165–75 doi:10.1093/oxfordjournals.aje.a117575

Goncharov A, Haase RF, Santiago-Rivera A, et al. (2008) High serum PCBs are associated with elevation of serum lipids and cardiovascular disease in a native American population. *Environ Res* 106(2):226–39 doi:10.1016/j.envres.2007.10.006

Goobie GC, Carlsten C, Johannson KA, et al. (2022) Association of particulate matter exposure with lung function and mortality among patients with fibrotic interstitial lung disease. *JAMA Intern Med* 182(12):1248–1259 doi:10.1001/jamainternmed.2022.4696

Grishanova AY, Perepechaeva ML (2022) Aryl hydrocarbon receptor in oxidative stress as a double agent and its biological and therapeutic significance. *Int J Mol Sci* 23(12) doi:10.3390/ijms23126719

Grunig G, Marsh LM, Esmaeil N, et al. (2014) Perspective: Ambient air pollution: Inflammatory response and effects on the lung's vasculature. *Pulm Circ* 4(1):25–35 doi:10.1086/674902

Guerrina N, Traboulsi H, Eidelman DH, Baglole CJ (2018) The aryl hydrocarbon receptor and the maintenance of lung health. *Int J Mol Sci* 19(12) doi:10.3390/ijms19123882

Gutowska-Owsiak D, Schaupp AL, Salimi M, et al. (2012) IL-17 downregulates filaggrin and affects keratinocyte expression of genes associated with cellular adhesion. *Exp Dermatol* 21(2):104–10 doi:10.1111/j.1600-0625.2011.01412.x

Haberzettl P, Conklin DJ, Abplanalp WT, Bhatnagar A, O'Toole TE (2018) Inhalation of fine particulate matter impairs endothelial progenitor cell function via pulmonary oxidative stress. *Arterioscler Thromb Vasc Biol* 38(1):131–42 doi:10.1161/ATVBAHA.117.309971

Hidaka T, Ogawa E, Kobayashi EH, et al. (2017) The aryl hydrocarbon receptor AhR links atopic dermatitis and air pollution via induction of the neurotrophic factor artemin. *Nat Immunol* 18(1):64–73 doi:10.1038/ni.3614

Humblet O, Birnbaum L, Rimm E, Mittleman MA, Hauser R (2008) Dioxins and cardiovascular disease mortality. *Environ Health Perspect* 116(11):1443–8 doi:10.1289/ehp.11579

Ishihara Y, Haarmann-Stemmann T, Kado NY, Vogel CFA (2019) Interleukin 33 expression induced by aryl hydrocarbon receptor in macrophages. *Toxicol Sci* 170(2):404–14 doi:10.1093/toxsci/kfz114

Kado S, Chang WLW, Chi AN, Wolny M, Shepherd DM, Vogel CFA (2017) Aryl hydrocarbon receptor signaling modifies toll-like receptor-regulated responses in human dendritic cells. Arch Toxicol 91(5):2209–21 doi:10.1007/s00204-016-1880-y

Kang HK, Dalager NA, Needham LL, et al. (2006) Health status of army chemical corps Vietnam veterans who sprayed defoliant in Vietnam. *Am J Ind Med* 49(11):875–84 doi:10.1002/ajim.20385

Kim BE, Leung DY, Boguniewicz M, Howell MD (2008) Loricrin and involucrin expression is down-regulated by Th2 cytokines through STAT-6. *Clin Immunol* 126(3):332–7 doi:10.1016/j.clim.2007.11.006

Kim JB, Pjanic M, Nguyen T, et al. (2017) TCF21 and the environmental sensor aryl-hydrocarbon receptor cooperate to activate a pro-inflammatory gene expression program in coronary artery smooth muscle cells. *pLoS Genet* 13(5):e1006750 doi:10.1371/journal.pgen.1006750

Kim JS, Lim HS, Cho SI, Cheong HK, Lim MK (2003) Impact of agent orange exposure among Korean Vietnam veterans. Ind Health 41(3):149–57 doi:10.2486/indhealth.41.149

Kohle C, Bock KW (2006) Activation of coupled Ah receptor and NRF2 gene batteries by dietary phytochemicals in relation to chemoprevention. *Biochem Pharmacol* 72(7):795–805 doi:10.1016/j.bcp.2006.04.017

Kunzli N, Jerrett M, Garcia-Esteban R, et al. (2010) Ambient air pollution and the progression of atherosclerosis in adults. *pLoS One* 5(2):e9096 doi:10.1371/journal.pone.0009096

Landrigan PJ, Fuller R, Acosta NJR, et al. (2018) The lancet commission on pollution and health. *Lancet* 391(10119):462–512 doi:10.1016/S0140-6736(17)32345-0

Lee CW, Lin ZC, Hu SC, et al. (2016) Urban particulate matter down-regulates filaggrin via COX2 expression/PGE2 production leading to skin barrier dysfunction. *Sci Rep* 6:27995 doi:10.1038/srep27995

Libby P, Ridker PM, Maseri A (2002) Inflammation and atherosclerosis. *Circulation* 105(9):1135–43 doi:10.1161/hc0902.104353

Mateen FJ, Brook RD (2011) Air pollution as an emerging global risk factor for stroke. *JAMA* 305(12):1240–1 doi:10.1001/jama.2011.352

Meyers JL, Winans B, Kelsaw E, Murthy A, Gerber S, Lawrence BP (2018) Environmental cues received during development shape dendritic cell responses later in life. *pLoS One* 13(11):e0207007 doi:10.1371/journal.pone.0207007

Mohammadi-Bardbori A, Vikstrom Bergander L, Rannug U, Rannug A (2015) NADPH oxidase-dependent mechanism explains how arsenic and other oxidants can activate aryl hydrocarbon receptor signaling. *Chem Res Toxicol* 28(12):2278–86 doi:10.1021/acs.chemrestox.5b00415

Nakamoto M, Arisawa K, Uemura H, et al. (2013) Association between blood levels of PCDDs/PCDFs/dioxin-like PCBs and history of allergic and other diseases in the Japanese population. *Int Arch Occup Environ Health* 86(8):849–59 doi:10.1007/s00420-012-0819-8

Niermann T, Schmutz S, Erne P, Resink T (2003) Aryl hydrocarbon receptor ligands repress T-cadherin expression in vascular smooth muscle cells. *Biochem Biophys Res Commun* 300(4):943–9 doi:10.1016/s0006-291x(02)02970-4

Niture SK, Khatri R, Jaiswal AK (2014) Regulation of NRF2-an update. *Free Radic Biol Med* 66:36–44 doi:10.1016/j.freeradbiomed.2013.02.008

O'Driscoll CA, Gallo ME, Fechner JH, Schauer JJ, Mezrich JD (2019) Real-world PM extracts differentially enhance Th17 differentiation and activate the aryl hydrocarbon receptor (AHR). *Toxicology* 414:14–26 doi:10.1016/j.tox.2019.01.002

O'Driscoll CA, Gallo ME, Hoffmann EJ, et al. (2018a) Polycyclic aromatic hydrocarbons (PAHs) present in ambient urban dust drive proinflammatory T cell and dendritic cell responses via the aryl hydrocarbon receptor (AHR) in vitro. *pLoS One* 13(12):e0209690 doi:10.1371/journal.pone.0209690

O'Driscoll CA, Owens LA, Gallo ME, et al. (2018b) Differential effects of diesel exhaust particles on T cell differentiation and autoimmune disease. *Part Fibre Toxicol* 15(1):35 doi:10.1186/s12989-018-0271-3

Panteleyev AA, Bickers DR (2006) Dioxin-induced chloracne—reconstructing the cellular and molecular mechanisms of a classic environmental disease. *Exp Dermatol* 15(9):705–30 doi:10.1111/j.1600-0625.2006.00476.x

Park EY, Rho HM (2002) The transcriptional activation of the human copper/zinc superoxide dismutase gene by 2,3,7,8-tetrachlorodibenzo-p-dioxin through two different regulator sites, the antioxidant responsive element and xenobiotic responsive element. *Mol Cell Biochem* 240(1–2):47–55 doi:10.1023/a:1020600509965

Penning TM (2014) Human aldo-keto reductases and the metabolic activation of polycyclic aromatic hydrocarbons. *Chem Res Toxicol* 27(11):1901–17 doi:10.1021/tx500298n [doi]

Phuleria HC, Geller MD, Fine PM, Sioutas C (2006) Size-resolved emissions of organic tracers from light- and heavy-duty vehicles measured in a California roadway tunnel. *Environ Sci Technol* 40(13):4109–18 doi:10.1021/es052186d

Pope CA, 3rd, Burnett RT, Thurston GD, et al. (2004) Cardiovascular mortality and long-term exposure to particulate air pollution: Epidemiological evidence of general pathophysiological pathways of disease. *Circulation* 109(1):71–7 doi:10.1161/01.CIR.0000108927.80044.7F

Quintana FJ, Basso AS, Iglesias AH, et al. (2008) Control of T(reg) and T(H)17 cell differentiation by the aryl hydrocarbon receptor. *Nature* 453(7191):65–71 doi:10.1038/nature06880

Ren F, Ji C, Huang Y, Aniagu S, Jiang Y, Chen T (2019) AHR-mediated ROS production contributes to the cardiac developmental toxicity of PM2.5 in zebrafish embryos. *Sci Total Environ*:135097 doi:10.1016/j.scitotenv.2019.135097

Rothhammer V, Quintana FJ (2019) The aryl hydrocarbon receptor: An environmental sensor integrating immune responses in health and disease. *Nat Rev Immunol* 19(3):184–97 doi:10.1038/s41577-019-0125-8

Ruckerl R, Greven S, Ljungman P, et al. (2007) Air pollution and inflammation (interleukin-6, C-reactive protein, fibrinogen) in myocardial infarction survivors. *Environ Health Perspect* 115(7):1072–80 doi:10.1289/ehp.10021

Schafer M, Willrodt AH, Kurinna S, et al. (2014) Activation of NRF2 in keratinocytes causes chloracne (MADISH)-like skin disease in mice. *EMBO Mol Med* 6(4):442–57 doi:10.1002/emmm.201303281

Schmuth M, Blunder S, Dubrac S, Gruber R, Moosbrugger-Martinz V (2015) Epidermal barrier in hereditary ichthyoses, atopic dermatitis, and psoriasis. *J Dtsch Dermatol Ges* 13(11):1119–23 doi:10.1111/ddg.12827

Schraufnagel DE, Balmes JR, Cowl CT, et al. (2019a) Air pollution and noncommunicable diseases: A review by the forum of international respiratory societies' environmental committee, part 1: The damaging effects of air pollution. *Chest* 155(2):409–16 doi:10.1016/j.chest.2018.10.042

Schraufnagel DE, Balmes JR, Cowl CT, et al. (2019b) Air pollution and noncommunicable diseases: A review by the forum of international respiratory societies' environmental committee, part 2: Air pollution and organ systems. *Chest* 155(2):417–26 doi:10.1016/j.chest.2018.10.041

Sergeev AV, Carpenter DO (2005) Hospitalization rates for coronary heart disease in relation to residence near areas contaminated with persistent organic pollutants and other pollutants. *Environ Health Perspect* 113(6):756–61 doi:10.1289/ehp.7595

Sergeev AV, Carpenter DO (2010) Exposure to persistent organic pollutants increases hospitalization rates for myocardial infarction with comorbid hypertension. *Prim Prev Insights* 2:1–9 doi:10.4137/PPRI.S4332

Smith J, Neupane R, McAmis W, Singh U, Chatterjee S, Raychoudhury S (2019) Toxicity of polycyclic aromatic hydrocarbons involves NOX2 activation. *Toxicol Rep* 6:1176–81 doi:10.1016/j.toxrep.2019.11.006

Sutter CH, Olesen KM, Bhuju J, Guo Z, Sutter TR (2019) AHR regulates metabolic reprogramming to promote SIRT1-dependent keratinocyte differentiation. *J Invest Dermatol* 139(4):818–26 doi:10.1016/j.jid.2018.10.019

Sutter TR, Guzman K, Dold KM, Greenlee WF (1991) Targets for dioxin: Genes for plasminogen activator inhibitor-2 and interleukin-1 beta. *Science* 254(5030):415–18

Veldhoen M, Hirota K, Westendorf AM, et al. (2008) The aryl hydrocarbon receptor links TH17-cell-mediated autoimmunity to environmental toxins. *Nature* 453(7191):106–9 doi:10.1038/nature06881

Vogel CF, Sciullo E, Wong P, Kuzmicky P, Kado N, Matsumura F (2005) Induction of proinflammatory cytokines and C-reactive protein in human macrophage cell line U937 exposed to air pollution particulates. *Environ Health Perspect* 113(11):1536–41 doi:10.1289/ehp.8094

Vogel CFA, Van Winkle LS, Esser C, Haarmann-Stemmann T (2020) The aryl hydrocarbon receptor as a target of environmental stressors—implications for pollution mediated stress and inflammatory responses. *Redox Biol* 34:101530 doi:10.1016/j.redox.2020.101530

Vogeley C, Sondermann NC, Woeste S, et al. (2022) Unraveling the differential impact of PAHs and dioxin-like compounds on AKR1C3 reveals the EGFR extracellular domain as a critical determinant of the AHR response. *Environ Int* 158:106989 doi:10.1016/j.envint.2021.106989

Wang Q, Fan Y, Kurita H, et al. (2017) Aryl hydrocarbon receptor ablation in cardiomyocytes protects male mice from heart dysfunction induced by NKX2.5 haploinsufficiency. *Toxicol Sci* 160(1):74–82 doi:10.1093/toxsci/kfx164

Wardman P, Candeias LP (1996) Fenton chemistry: An introduction. *Radiat Res* 145(5):523–31

Wolf D, Ley K (2019) Immunity and inflammation in atherosclerosis. *Circ Res* 124(2):315–27 doi:10.1161/CIRCRESAHA.118.313591

Wu D, Nishimura N, Kuo V, et al. (2011) Activation of aryl hydrocarbon receptor induces vascular inflammation and promotes atherosclerosis in apolipoprotein E-/- mice. Arterioscler Thromb Vasc Biol 31(6):1260–7 doi:10.1161/ATVBAHA.110.220202

Wu W, Jin Y, Carlsten C (2018) Inflammatory health effects of indoor and outdoor particulate matter. *J Allergy Clin Immunol* 141(3):833–44 doi:10.1016/j.jaci.2017.12.981

Xia M, Viera-Hutchins L, Garcia-Lloret M, et al. (2015) Vehicular exhaust particles promote allergic airway inflammation through an aryl hydrocarbon receptor-notch signaling cascade. *J Allergy Clin Immunol* 136(2):441–53 doi:10.1016/j.jaci.2015.02.014

Yi T, Wang J, Zhu K, et al. (2018) Aryl hydrocarbon receptor: A new player of pathogenesis and therapy in cardiovascular diseases. Biomed Res Int 2018:6058784 doi:10.1155/2018/6058784

Zhang J, Liang S, Ning R, et al. (2020) PM(2.5)-induced inflammation and lipidome alteration associated with the development of atherosclerosis based on a targeted lipidomic analysis. *Environ Int* 136:105444 doi:10.1016/j.envint.2019.105444

Section IV

Other Physical Stressors (Noise, UV, and EMF)

15 Oxidative Stress and Inflammation Contribute to Traffic Noise-Induced Vascular and Cerebral Dysfunction via Uncoupling of Nitric Oxide Synthases

Katie Frenis, Omar Hahad, Thomas Münzel, and Andreas Daiber

15.1 NOISE AS A CARDIOVASCULAR RISK FACTOR

Modernization has brought a myriad of changes to human behavior and environment—more people than ever before are living in cities where environmental noise is omnipresent. These changes in environment are coupled with advances in modern medicine, which shifts the global burden of disease away from communicable toward noncommunicable disease, especially cardiovascular disease and cancer (Collaborators 2017). In this way, human health in the modern age is beset by modern complexities, which includes increasing exposure to traffic noise, recently identified as a cardiovascular and cerebrovascular risk factor. Though traffic noise is a novel risk factor, the evidence is mounting that it plays an important part in the health of modern humans. Noise exposure is associated with hypertension, the most important 'classical' risk factor for cardiovascular and cerebrovascular disease, as well as myocardial infarction and stroke. Additionally, noise is associated with arterial stiffness, a subclinical marker of atherosclerosis (Munzel, Sorensen, and Daiber 2021).

In 2018, to revise the noise guidelines for the European Region, the World Health Organization (WHO) evaluated over 600 studies on the noise-health relationship that were conducted from 2000–2015 (Kempen et al. 2018). The majority of these studies were not of sufficient quality to determine an exposure-response relationship. There were, however, seven longitudinal studies that gave a relative risk (RR) of 1.08 (95% CI: 1.01–1.15) per 10 dB (L_{DEN}) between road traffic noise and the incidence of IHD (Kempen et al. 2018). Since 2018, the field of noise research has expanded considerably and has generated stronger associations between traffic noise exposure and cardiovascular disease. A recent analysis reported that the quality of evidence between road traffic and coronary artery disease incidence is now of high quality, followed by incidence of stroke, depression, and anxiety with moderate quality (Munzel, Sorensen, and Daiber 2021). Traffic noise appears to have strong effects on quality of life: according to the WHO, at least 1.6 million healthy life years are lost annually from traffic-related noise in Western Europe (Fritschi et al. 2011). This is supported by a Swiss study investigating the years of life lost to environmental stressors, which demonstrated that while air pollution was an outsized component of years of life lost due to environmental stressors, noise pollution was primarily responsible for effects on quality of life (Vienneau et al. 2015). A brief outline of modern studies into this topic since the WHO report can be found in Table 15.1.

DOI: 10.1201/9781003328100-19

TABLE 15.1
Translational Noise Studies Since the WHO Report of 2011[a]

Authors/Year	Species	Exposure	Findings
Antunes et al. (2013)	rat	90 dB, low frequency, unknown duration	Significant myocardial fibrosis detected via CAB staining in noise-exposed rats
Arpornchayanon et al. (2013)	guinea pigs	106 dB, unknown type, 30m	TNF-α signaling is activated in the cochlea following noise exposure, causing vessel constriction; improved by etanercept
Gannouni et al. (2013)	rat	70 dB, 80 dB, unknown type, 6h, 90d	Increased corticosterone levels, affected various parameters of the endocrine glands and cardiac function; markers of oxidative stress (catalase, superoxide dismutase and lipid peroxidation) were increased
Gannouni et al. (2014)	rat	70 dBA, unknown type, 6h/d, 3/5m	Structural alterations within the adrenal gland consistent with chronic stress; signs of necrosis and inflammation in myocardium
Said and El-Gohary (2016)	rat	80–100 dB, chronic and intermittent, unknown type, 8h, 20d	Increases in plasma levels of corticosterone, adrenaline, noradrenaline, endothelin-1, nitric oxide and malondialdehyde; decreases in superoxide dismutase
Lyamin et al. (2016)	beluga whale	140–175 dB, unknown type, 2–4h, 60 events	Heart rate acceleration following noise exposure; calves were more susceptible to the effects of noise and did not habituate
Konkle et al. (2017)	rat	87.3 dBA, unknown type, 15m-1h, 21d	Plasma ACTH, adrenal gland weight, IL6, IL1b levels were unchanged following noise exposure; increases in TNFα and CRP were seen
Munzel et al. (2017)	mouse	72 dBA, intermittent aircraft, 4d	Endothelial dysfunction, blood pressure, and redox balance were disturbed following noise exposure
Lousinha et al. (2018)	rat	120 dB, high intensity infrasound, 28 days	Exposed mice had prominent perivascular tissue with notable fibrosis that was mitigated by dexamethasone treatment
Yang and Guthrie (2020)	mouse	105 dB SPL, unknown type, 1/4h	DNA damage response genes appear to fail to respond to noise-induced DNA damage in cochlea, heart, liver, and cortex
Lousinha et al. (2020)	rat	120 dB, high intensity infrasound, 12w	Atrial interstitial fibrosis was increased and connexin 43 was decreased following noise exposure
Kvandova et al. (2020)	mouse	72 dBA, intermittent aircraft, 4d	Oxidative parameters and DNA damage increased following noise exposure with synergetic increases in $Ogg^{-/-}$ mice.
Steven et al. (2020)	mouse	72 dBA, intermittent aircraft, 4d	Preexisting hypertension had an additive effect on endothelial dysfunction, blood pressure, and oxidative stress caused by noise
Gogokhia et al. (2021)	rat	High intensity white noise, 1h, 10d	Male rats show higher anxiety-like response following noise
Bayo Jimenez et al. (2021)	mouse	72 dBA, intermittent aircraft, 4d	Induction of NRF2/HO-1 protected against oxidative damage, normalized blood pressure, and vascular endothelial function
Frenis et al. (2021)	mouse	72 dBA, intermittent aircraft, 4d	Monocytes are effector cells for vascular damage from noise exposure; ablation of monocytes normalizes endothelial function, blood pressure and oxidative parameters; microglia are important actors in the brain
Eckrich et al. (2021)	mouse	72 dBA, intermittent aircraft, 4d	Noise causes leukocyte adherence on the endothelium of microvessels and acute inflammatory changes to the plasma proteome, normalized by genetic deletion of $gp91^{phox}$

a In the decade since the WHO report, the quantity and quality of mechanistic data in animal models exploring noise exposure have increased. Though the field is small, the data is consistent with the noise reaction pathways. However, there is inconsistency in the noise exposure protocols, which may confound associations between noise exposure and cardiovascular disease.

Overall, it has become clear in recent years that transportation noise poses a risk to human health, particularly to those in urban dwellings and with other cardiovascular risk factors. In the present chapter, we will delineate how the effects of noise impart a physiological response and how the physiological response primes the body for cardiovascular and cerebrovascular disease, with focus on the brain, oxidative stress and inflammation.

15.2 TWO PATHWAYS FOR NOISE EXPOSURE

A "noise reaction model" proposing two apparent modes that noise could affect human physiology (Figure 15.1) was put forth by Babisch (2003) and has subsequently been studied in detail. In this model, a 'direct' and an 'indirect' pathway converge at a stress response nexus that perpetuates inflammatory and oxidative signaling, leading to cardiovascular sequelae.

The direct pathway is the physical result of consciously hearing and perceiving noise, which can then damage hearing and impart other auditory effects. This pathway is particularly relevant in the context of occupational noise exposure and hearing loss, a robust field that has been indicated as a significant environmental risk factor in the Global Burden of Disease study (Collaborators 2017). High decibel noise exposure is the most intuitive pathway for causing a hazard to human health and indeed, 16% of disabling hearing loss in adults worldwide can be attributed to occupational noise exposure (Nelson et al. 2005), which is also associated with coronary heart disease, angina pectoris and diastolic hypertension (Gan, Davies, and Demers 2011). The indirect pathway describes a cognitive or emotional response to the noise stimulus (i.e. feeling angry or annoyed) and is particularly important in the mental effects caused by noise: daytime noise annoyance predicted new onset of depressive and anxiety symptoms (Beutel et al. 2020) A "dose dependent" correlation between annoyance of noise of environmental origins and atrial fibrillation was found in the Gutenberg Health Study (Hahad et al. 2018). Sleep disruption is a bridging factor between both the direct and

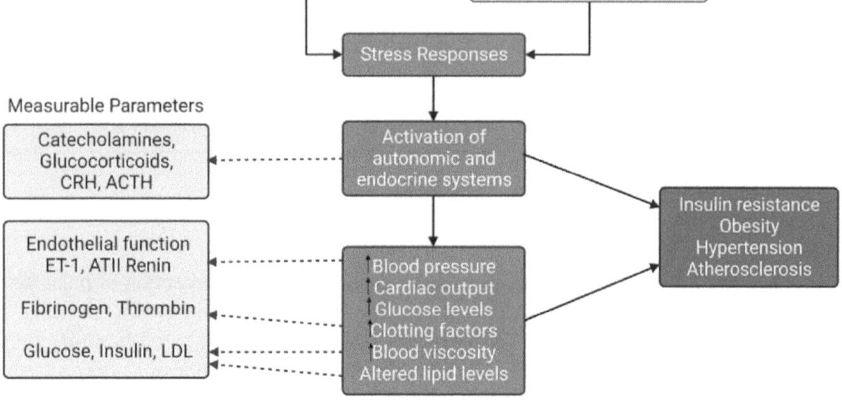

FIGURE 15.1 The direct and indirect pathways of noise exposure. Noise can impart auditory and nonauditory effects on the perceiver, which independently cause sympathoadrenal stress responses. Chronic stress caused mainly by sleep disruption is associated with a myriad of well-established cardiovascular risk factors. In this way, noise influences a variety of risks to prime for the development of CVD.

indirect pathways and is likely to be a core component in propagating the detrimental effects of noise. The physiological effects of sleep disruption include circadian disruption, decreased melatonin production and disruptions in the hypothalamus-pituitary-adrenal (HPA) axis (increased catecholamine, adrenocorticotropic hormone (ACTH) and cortisol secretion). Poor quality sleep is also associated with decreased insulin sensitivity and leptin levels; increases in ghrelin and appetite; upregulation of inflammatory proteins TNF, IL1, IL6 and CRP and increases in oxidative stress (Medic, Wille, and Hemels 2017). While the indirect pathway may appear less detrimental that the direct harmful effects of noise, it plays a crucial role in the emergence of cardiovascular disease (Babisch 2003) by promoting an environment conducive to cardiovascular dysfunction.

Both direct and indirect pathways lead to stress reactions that are risk factors for cardiovascular dysfunction and disease (Figure 15.2). Stress responses activate the sympathetic nervous

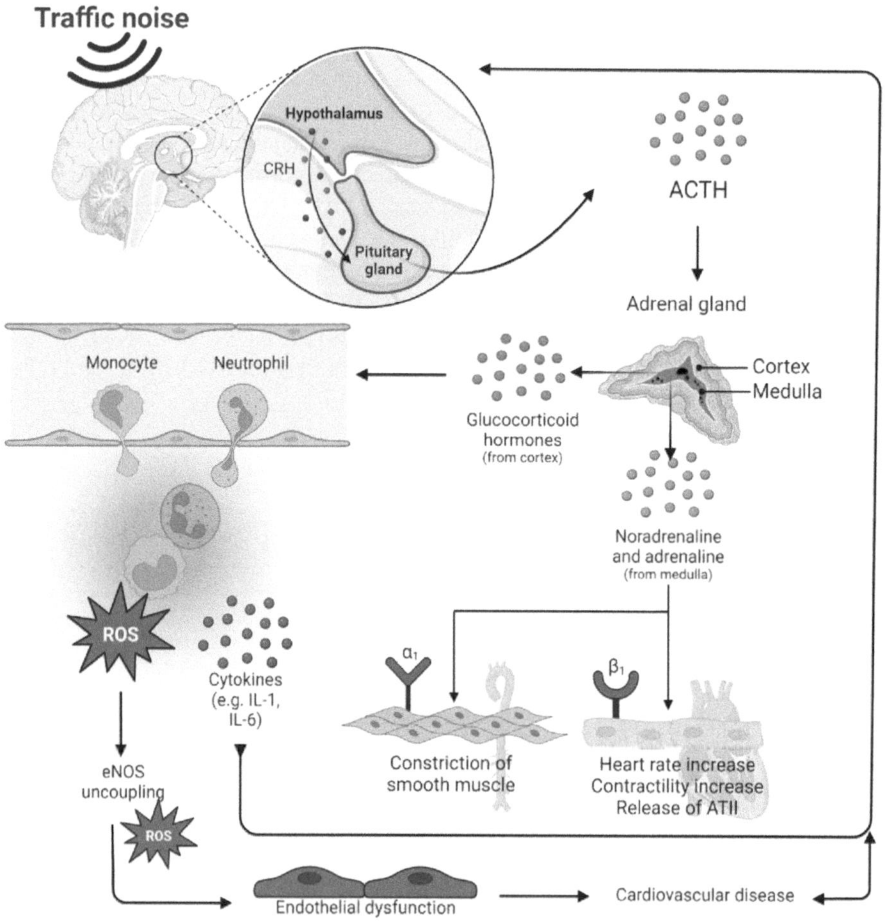

FIGURE 15.2 Noise stress and the onset of endothelial dysfunction. The perception of noise activates the HPA axis in the brain. The hypothalamus releases corticotropin-releasing hormone (CRH), which stimulates the pituitary gland to release adrenocorticotropic hormone (ACTH) crossing the blood-brain-barrier and stimulating the adrenal gland. The adrenal gland is the nexus for the two stress response systems: the HPA axis and the activation of the sympathetic nervous system, producing glucocorticoids and noradrenaline. These substances have a myriad of effects, from direct vasoconstriction to influencing the recruitment of immune cells out of circulation and into the tissue. Once in the tissue, immune cells are activated and produce cytokines, which influence other immune cells and also feedback into the brain, either perpetuating or quenching the immune response. These cells also produce reactive oxygen species (ROS), leading to the uncoupling of eNOS and thus to a worsening of endothelial function and ultimately to cardiovascular disease.

system, HPA axis and endocrine systems to produce responses to stress that, when persistent, lead to increased blood pressure, cardiac output, blood viscosity, glucose and clotting factors as well as alterations in lipid profile, all of which are risk factors for cardiovascular disease (Daiber et al. 2019). The initiating factor, the perception of noise, then activates the HPA axis, whose terminal product is cortisol in the adrenal glands in humans. The adrenal gland also produces adrenaline and noradrenaline (Jedema and Grace 2004), which activates the renin-angiotensin-aldosterone system (RAAS) (Miller and Arnold 2019). Both RAAS (Xiao et al. 2015) and neurotransmitters (Schlaich et al. 2009) activate immune cells, which then produce reactive oxygen species and inflammatory cytokines, promoting a pro-inflammatory and prooxidative milieu. In this way, noise causes sympatho-adrenal activation (Recio et al. 2016) and stress hormone release that perpetuates inflammation, pro-thrombotic pathways and endothelial dysfunction (Munzel et al. 2018) that are hallmarks of acute MI, heart failure, hypertension, arrhythmia and stroke (Babisch 2014).

15.3 ANIMAL MODELS OF NOISE EXPOSURE

Cohort and longitudinal studies have been essential epidemiological data for establishing an exposure-response relationship between transportation noise exposure and cardiovascular disease (CVD) in humans. As previously mentioned, in less than a decade, the quality of this data and the strength of the associations has greatly increased (Munzel, Sorensen, and Daiber 2021). However, gaining mechanistic insight into how noise exposure elicits adverse physiological reactions has relied largely on establishing translational models. The emotional state of mice is difficult to ascertain, and it is not likely that mice can experience "annoyance." Experiments using rodent models are limited only to direct effects on hearing and sleep, somewhat deconvoluting readouts of physiological reactions, which would be difficult to decouple from emotional state in humans.

It is then necessary to consider the conservation of the mechanisms driving human and rodent stress responses. A comparison of the HPA axis between rodents and humans reveals that the essential mechanics are conserved, and similar stimuli elicit activation, though the primary hormonal output is cortisol in humans and corticosterone in rodents (Kolber, Wieczorek, and Muglia 2008). Similarly, though there are differences between human and rodent RAAS, especially regarding the active site of the renin enzyme, and structure of angiotensin and the AT_1 receptor, the overarching function of the system is largely retained between the two species (Hollenberg 2000). Many of the signaling events following noise exposure stem from the activity of the innate immune system, which is structured quite similarly in mice and humans (Mestas and Hughes 2004). Interspecies disparities include a large proportional difference between neutrophils and lymphocytes (50–70% neutrophils in human blood, 10–25% in mouse) and differences in iNOS expression in macrophages (Mestas and Hughes 2004). Additionally, human astrocytes that appear to be more sensitive to oxidative stress are less resilient and more prone to activating inflammatory gene programs than mouse counterparts (Li et al. 2021).

Though there are interspecies differences in mechanisms bearing the stress response, it appears that rodents and humans maintain much similarity and that studies in rodent models of noise exposure are translationally relevant. To that end, translational experiments of acute noise exposure in mice have produced similar effects in both mice and humans: increased adrenaline, reactive oxygen species, endothelial dysfunction, hypertension and pro-thromboinflammatory changes which will be discussed in more detail.

15.4 INFLAMMATION AS A CARDIOVASCULAR RISK

Inflammation is a hallmark in the development of cardiovascular disease, especially through the actions of the innate immune system. The cells of the innate immune system, particularly phagocytes, are tasked with first line defense against invading pathogens and with resolving ongoing inflammatory processes. In some conditions, these cells fail to resolve the insult, which can lead to a persistent inflammatory microenvironment (Mezzaroma, Abbate, and Toldo 2021). For this reason, involvement of innate immune cells in cardiovascular disease is widely reported in atherosclerosis, atrial fibrillation, myocardial infarction and hypertension (Steven et al. 2019). There is a tight association between CVD and inflammation: a genome-wide association study that included 88,192 participants with coronary artery disease and 162,544 participants without revealed 25 SNP-CAD associations in or near genes involved in cellular adhesion, leukocyte migration and atherosclerosis (Howson et al. 2017). Hypertension, the leading cardiovascular risk factor for global mortality and disease burden, has been well associated with an inflammatory component (Harrison et al. 2010).

Inflammation causes endothelial and arterial damage through the chemotactic and phagocytic response: cells move into the subendothelial space to combat an invading pathogen or phagocytose dying cells. They produce reactive oxygen species (ROS) that can disrupt nitric oxide signaling via oxidative mechanisms, e.g. by direct scavenging of ${}^{\bullet}NO$ by $O_2^{\bullet -}$ to form peroxynitrite, eNOS uncoupling via S-glutathionylation, adverse phosphorylation or oxidative tetrahydrobiopterin (BH4) depletion or oxidative inhibition of soluble guanylyl cyclase (Daiber et al. 2017). This can lead to apoptosis, breakdown of the elastic lamella of the media and remodeling of the extracellular matrix (Weber and Noels 2011). Colchicine, a drug with anti-inflammatory properties against several NFκB-promoted genes (Angelidis et al. 2018), has been reported to have significantly reduced major adverse CV events (D'Amario et al. 2021) and similarly, interventions targeting IL6, TNFα, and IL1β, all cytokines promoted by NFκB have benefits for cardiovascular events (Choy 2012). Overall, the evidence is substantial that inflammation is both a product of and a contributing factor to nearly all CVD (summarized in Steven et al. [2019]), and we will detail noise-inflammation interactions in the following section.

15.5 NOISE INDUCES INFLAMMATION OF THE ARTERIAL WALL

There are several studies indicating that inflammation is induced by noise exposure, however, it is important to note that there are many discrepancies in noise exposure protocols: duration, intensity and identity of noise is rarely standardized throughout the field. There are two clear purposes for studying noise exposure. Studies of noise-induced hearing loss (NIHL) typically employ a high-intensity noise exposure, whereas studies investigating the indirect effects of noise generally use noise intensities below the threshold for hearing damage. In both scenarios, inflammation has been reported as an effect of noise exposure. High intensity sound (105 dB) has been reported to cause more damage to mice with diabetes, a pro-inflammatory phenotype, than to wild type mice (Fujita et al. 2012). White noise exposure of 100 dB for only 4 hours per day for 30 days caused an increase blood glucose, markers of inflammation and triglycerides and changes in the microbiome that returned to normal within 14 days of noise cessation (Cui et al. 2016).

Our own data has clearly demonstrated the extravasation of circulating immune cells into the arterial vessel, which is a hallmark of the inflammatory process following noise exposure. We observe overexpression of VCAM-1, a marker of an activated endothelium and crucial facilitator of leukocyte capture and rolling, and monocyte chemokine MCP1, in the aortas of noise-exposed mice (Frenis et al. 2021; Kroller-Schon et al. 2018; Steven et al. 2020). Activation of the endothelium is a process mediated by noradrenaline, which we have also reported to be increased in the plasma of noise-exposed mice (Frenis et al. 2021). This suggests that the stress response initiated by the sympatho-adrenal activation in the brain directly innervates the endothelium or peripheral arteries, stimulating the infiltration of circulating leukocytes. Flow cytometry experiments indicate an influx of $Ly6C^+$ $Ly6G^+$ neutrophils and $Ly6C^{high}$ inflammatory monocytes as well as T cells into the aortas of noise-exposed mice, which is worsened by the presence of preexisting hypertension (Steven et al. 2020). We were also able to directly observe this behavior using the novel dorsal skinfold method of intravital microscopy, which showed a significantly increased presence of adherent leukocytes on the endothelium of subcutaneous microvessels (Eckrich et al. 2021). We aimed to identify the leukocytes primarily responsible for responsible for the vascular damage observed in the form of worsened endothelial function and increased oxidative stress. We used the mouse model LysMCre^{iDTR}, which expresses an inducible diphtheria toxin transgene on LysM^{+} cells, leaving these cells vulnerable to ablation with the toxin. Ablated mice were protected from noise effects: endothelial function and blood pressure were unaffected and oxidative measurements remained at baseline or below. We interpret this result as a clear indication that monocytic infiltration into the vessel is mediating the damage caused therein and that this infiltration spurs the prooxidative milieu within the vessel. Similar results were seen in gp91^{phox}-/y mice lacking NOX2, an oxidative enzyme found on monocytes and neutrophils, indicating that these cells were unable to "spark" the oxidative processes that disturb nitric oxide signaling (Kroller-Schon et al. 2018). Circulating monocytes recruited into a tissue infiltrate and differentiate into M1 macrophages, which is characterized by the expression of NFκB-transcribed genes IL6, IL1β, MCP-1, VCAM-1, gp91^{phox}, HO-1 (Lin et al. 2007), which we routinely report to be increased in the aorta, plasma and brains of noise-exposed mice along with pro-inflammatory changes in the plasma proteome (Eckrich et al. 2021).

Overall, the evidence that inflammation is a critical process in precipitating the detrimental effects of noise is strong, especially within the arterial walls. There has been much progress recently in elucidating the mechanism by which activated immune cells ignite the oxidative blaze that is a hallmark of diseased vessels, but much of this has been focused on the vessels of the periphery, and there is still much to be understood regarding inflammation in other locations.

15.6 NOISE IMPACTS MENTAL HEALTH, COGNITIVE EFFECTS AND NEUROINFLAMMATION

It is evident that the brain plays a pivotal role in initiating the cardiovascular effects of noise exposure and though the effects of noise in the peripheral vasculature are backed by a solid body of evidence, there is much less specifically looking at the role of noise in affecting neuroinflammation. The effects of noise in the brain can be roughly equated to that of chronic mental stress: there is evidence that chronic mental stress produces an inflammatory phenotype, associated with cardiovascular risk (Brotman, Golden, and Wittstein 2007). In fact, a meta-analysis showed that acute mental stress increases circulating proinflammatory cytokines IL-6 and IL-1β (Steptoe, Hamer, and Chida 2007).

Cardiovascular responses to noise seem to originate in the brain, where we observed astrocyte activation (GFAP^{+} staining) was higher in mice exposed to noise while sleeping than their waking phase-exposed counterparts (Kroller-Schon et al. 2018) and a small cumulative effect was evident in noise-exposed mice with hypertension (Steven et al. 2020). Mice exposed to 4 days of noise also have a moderate activation of microglia as determined by flow cytometry of the CD68^{+}CD86^{+}MHCII^{+} state (Frenis et al. 2021). Microglia are the principle immune responders of the brain and react to many of the same physiological cues peripheral macrophages do, including producing superoxide via NOX2. We also observed strong increases in superoxide in the brains of hypertensive mice exposed to noise (Steven et al. 2020), pointing toward an unexplored role for activated microglia. Microglia are not susceptible to ablation in the LysMCre^{iDTR} model and thus, heightened levels of inflammatory markers (NFKb2 and CD40L mRNA and IL6, NLRP3 protein) were still present in the brains of ablated, noise exposed mice (Frenis et al. 2021). However, in experiments exposing gp91$^{phox-/y}$ (NOX2) mice to noise, CD68 levels remained at basal levels in the brains of knockout mice but were elevated in exposed wild type mice (Kroller-Schon et al. 2018), perhaps implying that microglia have a critical role in "noise signaling." Though some progress has been made, it appears that the brain presents an opportunity for new research into how noise stress arises.

15.7 OXIDATIVE STRESS IS A HALLMARK OF NOISE EXPOSURE

Oxidative damage is a hallmark of cardiovascular disease; the evidence of oxidative involvement in perpetuating vascular damage following noise exposure is quite strong and includes both human and rodent studies. Noise-induced oxidative stress is mediated via redox-sensitive kinases (e.g. c-Jun N-terminal kinase) and can be prevented by antioxidant therapies, including antioxidants, synthetic ROS scavengers, cardiovascular drugs, NRF2 activation or mitigated by physical exercise (Daiber et al. 2020). In humans, oxidative damage readouts 3-nitrotyrosine and 8-isoprostane were elevated following exposure to only 46 dB(A) mean sound pressure level (Daiber et al. 2020). Factory workers occupationally exposed to loud noise had higher blood pressure, greater oxidative DNA damage in peripheral blood mononuclear cells and higher GPx-1 compared to office workers with exposure to lower noise levels (40–50 dB[A]) (Daiber et al. 2020).

There are also several direct studies of noise exposure in rodent models. When exposed to high intensity white noise (80–100 dB[A]), rats show oxidative alterations in plasma: higher malondialdehyde, a marker for fatty acid peroxidation, lower superoxide dismutase levels and increased nitrite/nitrate levels along with higher plasma concentration. Similar, white noise exposure at 100 dB(A) for 4 hours per day produced cognitive impairment, more lipid peroxidation and higher corticosterone levels in rats (Manikandan et al. 2006). These detrimental oxidative parameters were accompanied by higher plasma concentrations of corticosterone, catecholamines and endothelin-1(Said and El-Gohary 2016). High intensity noise produces more oxidative DNA damage in rat cardiomyocytes (Lenzi et al. 2003) and more 8-hydroxy-2′deoxyguanosine, a circulating marker of oxidative DNA damage (Zheng and Ariizumi 2007). Notably, the studies detailed earlier use sound pressure levels that are extremely high and would thereby trigger both the auditory and nonauditory noise stress.

We aimed to decouple the auditory and nonauditory effects by performing experiments at a mean sound pressure level of 73 dB(A) for 4 days. Using this standardized model, we observed oxidative stress in the aorta, heart and brain, as indicated through direct measurement via HPLC and read out by oxidative damage markers such as 4-hydroxynonenal-positive proteins, 3NT and MDA. Like other reports, we observed oxidative DNA damage (8-hydroxy-2′deoxyguanosine) in the aorta of aircraft noise-exposed mice (Kvandova et al. 2020).

15.8 ENOS AS A PERPETRATOR OF OXIDATIVE NOISE STRESS

As previously detailed, there is a preponderance of evidence in the form of biomarkers that suggest that noise creates a prooxidative state and implication that NADPH oxidase (Nox2, specifically) is responsible for much of the oxidative stress generated by the immune component of noise reactions. This "first step" in oxidative stress scavenges nitric oxide and sets the stage for a secondary component for ROS generation: uncoupling of nitric oxide synthases (NOS). In highly oxidative environments, NOS becomes "uncoupled" and switches from producing nitric oxide to producing superoxide, effectively switching a beneficial enzyme with inhibitory effects on sympathetic tone (Cunnington and Channon 2010) to a harmful one. eNOS uncoupling can occur when the cofactor for the enzyme, tetrahydrobiopterin (BH4), is depleted or through S-glutathionylation.

Likely due to this overexpression of superoxide-producing enzymes, eNOS in the aorta and nNOS in the brain both become uncoupled following noise exposure. We have demonstrated that the eNOS inhibitor L-NAME prevents oxidative stress levels in stained aortic sections (Kroller-Schon et al. 2018) and nNOS inhibitor ARL-17477 reduced oxidative stress in stained cortical sections (Kroller-Schon et al. 2018). Inhibition of an uncoupled NOS (whose product is superoxide) will lead to a reduction of superoxide, as detectable with a dihydroethidium stain. We observe that in addition to being uncoupled, eNOS and nNOS are transcriptionally overexpressed (Kroller-Schon et al. 2018). These affects appear to be mitigated in gp91^{phox} and exacerbated in preexisting hypertension, suggesting that suppressing the immune cell ignition of the oxidative state is an effective strategy for mitigating the effects of traffic noise. Other reports have suggested that upregulation or induction of heme oxygenase-1, NRF2, or PGC-1α (Bayo Jimenez et al. 2021; Breton et al. 2019) could also have protective effects on eNOS coupling state and prevent noise-induced endothelial dysfunction. Using next generation sequencing we also demonstrated a down-regulation of genes encoding for antioxidant systems (e.g. intracellular superoxide dismutase [*Sod1*], forkhead-box-protein O3 [*Foxo3*]), whereas pro-apoptotic/cell death factors were upregulated (e.g. *caspases, Fas, p38*) (Kroller-Schon et al. 2018).

15.9 THE CIRCADIAN RHYTHM IS DISRUPTED FOLLOWING NOISE EXPOSURE

The circadian rhythm is a powerful mechanism for syncing biological processes with the environment and has important effects on cardiovascular function. About 13% of genes and 8% of proteins in the heart are rhythmically expressed (Thosar, Butler, and Shea 2018), and many of the operant cells in the cardiovascular system (endothelial cells, vascular smooth muscle cells, fibroblasts) oscillate in activity and gene expression to maintain normal cardiovascular homeostasis and core circadian genes *Bmal1* and *Cry1* directly influence the behavior of the vascular endothelium (Takeda et al. 2007). Conversely, polymorphisms in *Bmal1* cause susceptibility of endothelial cells to injury and hypertension (Woon et al. 2007). Finally, mice with *Per2* mutations manifest endothelial dysfunction, produce less nitric oxide and prostaglandins for vasodilation while simultaneously producing more vasoconstrictors (Viswambharan et al. 2007), affecting many aspects of normal vascular control. Importantly, the circadian rhythm is highly dependent on redox reactions (Daiber et al. 2022) that can potentially be disrupted by highly oxidative states and vice versa: Circadian misalignment is commonly seen in shift workers, who have a higher prevalence of obesity, type II diabetes, hypertension, coronary heart disease and ischemic stroke (Koch et al. 2017), as an example.

Animal models of noise exposure are a critical component in understanding the molecular mechanisms that allow noise to affect the organism. A peripheral clock exists in the murine cochlea, implying that sensitivity to noise could be rhythmically controlled (Meltser et al. 2020). Similarly, the neurons in the inferior colliculus, a region for sound processing, also have CLOCK-dependent rhythmic expression that is transcriptionally sensitive to noise exposure (Park et al. 2016). These insights into the basic biology of the circadian rhythm shed light on the effects in noise exposure studies: noise exposure at night has significantly stronger effects than noise exposure during the day, implying that noise exposure exerts its effects through sleep and/or circadian disruption and by shortening the sleep period. When applying noise in the "dark cycle," anxiety-like behavior; memory and learning impairments; reduced brain volume, hippocampal volume and neural density and HPA axis activation was reported (Jafari, Kolb, and Mohajerani 2018). We report dysregulation of multiple circadian genes in the aorta and kidney of mice following exposure of mice to noise in their sleeping phase, which is of similar severity to mice exposed around the clock (Kroller-Schon et al. 2018). Specifically, we found transcriptional reductions via NGS in *Per1* and *Foxo3* and increases of Cry1 and *Arntl* in the aortas of mice exposed to noise around the clock or during their sleeping phase but not during their waking phase, which was mirrored by more severe effects on hypertension, endothelin-1 expression, oxidative stress in the aorta and brain, as well as astrocyte activation (Kroller-Schon et al. 2018). This appeared to be related to impaired *Foxo3* signaling, which was confirmed by the protective actions of bepridil (Kroller-Schon et al. 2018), a calcium antagonist with stimulatory effects on the expression of this transcription factor. Because eNOS also appears to be diurnally regulated via phosphorylation and also through diurnal expression of BH4 (Paschos and FitzGerald 2010), and stress hormones are also under control of the circadian clock and display diurnal rhythm (Wisser and Breuer 1981), we speculate that the oxidative effects of noise exposure may be mediated through this noise disruption of the diurnal hormone-eNOS axis.

15.10 SUMMARY AND FUTURE DIRECTIONS

In the past decade, evidence of a connection between cardiovascular disease and traffic noise exposure has substantially been strengthened, both in epidemiological studies in human populations and translational animal models. Noise operates through a forked pathway of direct auditory effects and indirect nonauditory effects that elicit stress responses and disrupt sleep. In this way, the stress response is propagated affecting the brain, heart, peripheral arteries and microvessels. The innate immune system is responsive to these stress signals, causing an infiltration into tissues and an ignition of an oxidative cascade via their Nox2 isoform, which subsequently creates a prooxidative environment that results in eNOS or nNOS uncoupling. Though much progress has been made in understanding the mechanism and effects on man, there is still much to be understood about the way noise propagates damage within the brain. Similarly, although many epidemiological studies on the effect of noise on human heath have been conducted, there is still only a loose correlation with certain cardiovascular and mental health conditions (hypertension, depression, dementia). Since these are major components of the modern burden of disease, it is important to cultivate an understanding of how any factor could influence their severity or prevalence.

REFERENCES

Angelidis, C., Z. Kotsialou, C. Kossyvakis, A. R. Vrettou, A. Zacharoulis, F. Kolokathis, V. Kekeris, and G. Giannopoulos. 2018. "Colchicine pharmacokinetics and mechanism of action." *Curr Pharm Des* 24 (6): 659–663. https://doi.org/10.2174/1381612824666180123110042. www.ncbi.nlm.nih.gov/pubmed/29359661.

Antunes, E., P. Oliveira, G. Borrecho, M. J. Oliveira, J. Brito, A. Aguas, and S. J. Martins dos. 2013. "Myocardial fibrosis in rats exposed to low frequency noise." *Acta Cardiol* 68 (3): 241–245. https://doi.org/10.1080/ac.68.3.2983417. https://www.ncbi.nlm.nih.gov/pubmed/23882868.

Arpornchayanon, W., M. Canis, F. Ihler, C. Settevendemie, and S. Strieth. 2013."TNF-alpha inhibition using etanercept prevents noise-induced hearing loss by improvement of cochlear blood flow in vivo." *Int J Audiol* 52 (8): 545–552. https://doi.org/10.3109/14992027.2013.790564. https://www.ncbi.nlm.nih.gov/pubmed/23786392.

Babisch, W. 2003. "Stress hormones in the research on cardiovascular effects of noise." *Noise Health* 5 (18): 1–11. https://doi.org/NO_DOI. www.ncbi.nlm.nih.gov/pubmed/12631430.

———. 2014. "Updated exposure-response relationship between road traffic noise and coronary heart diseases: A meta-analysis." *Noise Health* 16 (68): 1–9. https://doi.org/10.4103/1463-1741.127847. www.ncbi.nlm.nih.gov/pubmed/24583674.

Bayo Jimenez, M. T., K. Frenis, S. Kroller-Schon, M. Kuntic, P. Stamm, M. Kvandova, M. Oelze, H. Li, S. Steven, T. Munzel, et al. 2021. "Noise-induced vascular dysfunction, oxidative stress, and inflammation are improved by pharmacological modulation of the NRF2/HO-1 axis." *Antioxidants (Basel)* 10 (4). https://doi.org/10.3390/antiox10040625. www.ncbi.nlm.nih.gov/pubmed/33921821.

Beutel, M. E., E. Brahler, M. Ernst, E. Klein, I. Reiner, J. Wiltink, M. Michal, P. S. Wild, A. Schulz, T. Munzel, et al. 2020. "Noise annoyance predicts symptoms of depression, anxiety and sleep disturbance 5 years later. Findings from the Gutenberg Health Study." *Eur J Public Health* 30 (3): 516–521. https://doi.org/10.1093/eurpub/ckaa015. www.ncbi.nlm.nih.gov/pubmed/32034913.

Breton, C. V., A. Y. Song, J. Xiao, S. J. Kim, H. H. Mehta, J. Wan, K. Yen, C. Sioutas, F. Lurmann, S. Xue, et al. 2019. "Effects of air pollution on mitochondrial function, mitochondrial DNA methylation, and mitochondrial peptide expression." *Mitochondrion* 46: 22–29. https://doi.org/10.1016/j.mito.2019.04.001. www.ncbi.nlm.nih.gov/pubmed/30980914.

Brotman, D. J., S. H. Golden, and I. S. Wittstein. 2007. "The cardiovascular toll of stress." *Lancet* 370 (9592): 1089–1100. https://doi.org/10.1016/S0140-6736(07)61305-1. www.ncbi.nlm.nih.gov/pubmed/17822755.

Choy, E. 2012. "Understanding the dynamics: Pathways involved in the pathogenesis of rheumatoid arthritis." *Rheumatology (Oxford)* 51 Suppl 5: v3–11. https://doi.org/10.1093/rheumatology/kes113. www.ncbi.nlm.nih.gov/pubmed/22718924.

Collaborators, G. B. D. Risk Factors. 2017. "Global, regional, and national comparative risk assessment of 84 behavioural, environmental and occupational, and metabolic risks or clusters of risks, 1990–2016: A systematic analysis for the Global Burden of Disease study 2016." *Lancet* 390 (10100): 1345–1422. https://doi.org/10.1016/S0140-6736(17)32366-8. www.ncbi.nlm.nih.gov/pubmed/28919119.

Cui, B., Z. Gai, X. She, R. Wang, and Z. Xi. 2016. "Effects of chronic noise on glucose metabolism and gut microbiota-host inflammatory homeostasis in rats." *Sci Rep* 6: 36693. https://doi.org/10.1038/srep36693. www.ncbi.nlm.nih.gov/pubmed/27811997.

Cunnington, C., and K. M. Channon. 2010. "Tetrahydrobiopterin: Pleiotropic roles in cardiovascular pathophysiology." *Heart* 96 (23): 1872–1877. https://doi.org/10.1136/hrt.2009.180430. www.ncbi.nlm.nih.gov/pubmed/20837663.

D'Amario, D., D. Cappetta, L. Cappannoli, G. Princi, S. Migliaro, G. Diana, K. Chouchane, J. A. Borovac, A. Restivo, A. Arcudi, et al. 2021. "Colchicine in ischemic heart disease: The good, the bad and the ugly." *Clin Res Cardiol*. https://doi.org/10.1007/s00392-021-01828-9. www.ncbi.nlm.nih.gov/pubmed/33713178.

Daiber, A., K. Frenis, M. Kuntic, H. Li, E. Wolf, A. B. Kilgallen, S. Lecour, L. W. Van Laake, R. Schulz, O. Hahad, and T. Munzel. 2022. "Redox regulatory changes of circadian rhythm by the environmental risk factors traffic noise and air pollution." *Antioxid Redox Signal* 37 (10–12): 679–703. https://doi.org/10.1089/ars.2021.0272. www.ncbi.nlm.nih.gov/pubmed/35088601.

Daiber, A., S. Kroller-Schon, K. Frenis, M. Oelze, S. Kalinovic, K. Vujacic-Mirski, M. Kuntic, M. T. Bayo Jimenez, J. Helmstadter, S. Steven, et al. 2019. "Environmental noise induces the release of stress hormones and inflammatory signaling molecules leading to oxidative stress and vascular dysfunction-signatures of the internal exposome." *Biofactors* 45 (4): 495–506. https://doi.org/10.1002/biof.1506. www.ncbi.nlm.nih.gov/pubmed/30937979.

Daiber, A., S. Kroller-Schon, M. Oelze, O. Hahad, H. Li, R. Schulz, S. Steven, and T. Munzel. 2020. "Oxidative stress and inflammation contribute to traffic noise-induced vascular and cerebral dysfunction via uncoupling of nitric oxide synthases." *Redox Biol* 34: 101506. https://doi.org/10.1016/j.redox.2020.101506. www.ncbi.nlm.nih.gov/pubmed/32371009.

Daiber, A., S. Steven, A. Weber, V. V. Shuvaev, V. R. Muzykantov, I. Laher, H. Li, S. Lamas, and T. Munzel. 2017. "Targeting vascular (endothelial) dysfunction." *Br J Pharmacol* 174 (12): 1591–1619. https://doi.org/10.1111/bph.13517. www.ncbi.nlm.nih.gov/pubmed/27187006.

Eckrich, J., K. Frenis, G. Rodriguez-Blanco, Y. Ruan, S. Jiang, M. T. Bayo Jimenez, M. Kuntic, M. Oelze, O. Hahad, H. Li, et al. 2021. "Aircraft noise exposure drives the activation of white blood cells and induces microvascular dysfunction in mice." *Redox Biol* 46: 102063. https://doi.org/10.1016/j.redox.2021.102063. www.ncbi.nlm.nih.gov/pubmed/34274810.

Frenis, K., J. Helmstadter, Y. Ruan, E. Schramm, S. Kalinovic, S. Kroller-Schon, M. T. Bayo Jimenez, O. Hahad, M. Oelze, S. Jiang, et al. 2021. "Ablation of lysozyme M-positive cells prevents aircraft noise-induced vascular damage without improving cerebral side effects." *Basic Res Cardiol* 116 (1): 31. https://doi.org/10.1007/s00395-021-00869-5. www.ncbi.nlm.nih.gov/pubmed/33929610.

Fritschi, L., A. L. Brown, A. Kim, D. H. Schwela, and S. Kephalopoulos. 2011. "WHO and JRC report: Burden of disease from environmental noise." https://ec.europa.eu/jrc/sites/jrcsh/files/e94888.pdf.

Fujita, T., D. Yamashita, S. Katsunuma, S. Hasegawa, H. Tanimoto, and K. Nibu. 2012. "Increased inner ear susceptibility to noise injury in mice with streptozotocin-induced diabetes." *Diabetes* 61 (11): 2980–2986. https://doi.org/10.2337/db11-1845. www.ncbi.nlm.nih.gov/pubmed/22851574.

Gan, W. Q., H. W. Davies, and P. A. Demers. 2011. "Exposure to occupational noise and cardiovascular disease in the United States: The national health and nutrition examination survey 1999–2004." *Occup Environ Med* 68 (3): 183–190. https://doi.org/10.1136/oem.2010.055269. www.ncbi.nlm.nih.gov/pubmed/20924023.

Gannouni, N., A. Mhamdi, M. El May, O. Tebourbi, and K. B. Rhouma. 2014."Morphological changes of adrenal gland and heart tissue after varying duration of noise exposure in adult rat." *Noise Health* 16 (73): 416–421. https://doi.org/10.4103/1463-1741.144424. https://www.ncbi.nlm.nih.gov/pubmed/25387538.

Gannouni, N., A. Mhamdi, O. Tebourbi, M. El May, M. Sakly, and K. B. Rhouma. 2013."Qualitative and quantitative assessment of noise at moderate intensities on extra-auditory system in adult rats." *Noise Health* 15 (67): 406–411. https://doi.org/10.4103/1463-1741.121236. https://www.ncbi.nlm.nih.gov/pubmed/24231419.

Gogokhia, N., N. Japaridze, Y. Tizabi, L. Pataraya, and M. G. Zhvania. 2021."Gender differences in anxiety response to high intensity white noise in rats." *Neurosci Lett* 742: 135543. https://doi.org/10.1016/j.neulet.2020.135543. https://www.ncbi.nlm.nih.gov/pubmed/33278506.

Hahad, O., M. Beutel, T. Gori, A. Schulz, M. Blettner, N. Pfeiffer, T. Rostock, K. Lackner, M. Sorensen, J. H. Prochaska, P. S. Wild, and T. Munzel. 2018. "Annoyance to different noise sources is associated with atrial fibrillation in the Gutenberg Health Study." *Int J Cardiol* 264: 79–84. https://doi.org/10.1016/j.ijcard.2018.03.126. www.ncbi.nlm.nih.gov/pubmed/29695315.

Harrison, D. G., A. Vinh, H. Lob, and M. S. Madhur. 2010. "Role of the adaptive immune system in hypertension." *Curr Opin Pharmacol* 10 (2): 203–207. https://doi.org/10.1016/j.coph.2010.01.006. www.ncbi.nlm.nih.gov/pubmed/20167535.

Hollenberg, N. K. 2000. "Implications of species difference for clinical investigation: Studies on the renin-angiotensin system." *Hypertension* 35 (1 Pt 2): 150–154. https://doi.org/10.1161/01.hyp.35.1.150. www.ncbi.nlm.nih.gov/pubmed/10642291.

Howson, J. M. M., W. Zhao, D. R. Barnes, W. K. Ho, R. Young, D. S. Paul, L. L. Waite, D. F. Freitag, E. B. Fauman, E. L. Salfati, et al. 2017. "Fifteen new risk loci for coronary artery disease highlight arterial-wall-specific mechanisms." *Nat Genet* 49 (7): 1113–1119. https://doi.org/10.1038/ng.3874. www.ncbi.nlm.nih.gov/pubmed/28530674.

Jafari, Z., B. E. Kolb, and M. H. Mohajerani. 2018. "Chronic traffic noise stress accelerates brain impairment and cognitive decline in mice." *Exp Neurol* 308: 1–12. https://doi.org/10.1016/j.expneurol.2018.06.011. www.ncbi.nlm.nih.gov/pubmed/29936225.

Jedema, H. P., and A. A. Grace. 2004. "Corticotropin-releasing hormone directly activates noradrenergic neurons of the locus ceruleus recorded in vitro." *J Neurosci* 24 (43): 9703–9713. https://doi.org/10.1523/JNEUROSCI.2830-04.2004. www.ncbi.nlm.nih.gov/pubmed/15509759.

Kempen, E. V., M. Casas, G. Pershagen, and M. Foraster. 2018. "WHO environmental noise guidelines for the European Region: A systematic review on environmental noise and cardiovascular and metabolic effects: A summary." *Int J Environ Res Public Health* 15 (2): 379. https://doi.org/10.3390/ijerph15020379. www.ncbi.nlm.nih.gov/pubmed/29470452.

Koch, C. E., B. Leinweber, B. C. Drengberg, C. Blaum, and H. Oster. 2017. "Interaction between circadian rhythms and stress." *Neurobiol Stress* 6: 57–67. https://doi.org/10.1016/j.ynstr.2016.09.001. www.ncbi.nlm.nih.gov/pubmed/28229109.

Kolber, B. J., L. Wieczorek, and L. J. Muglia. 2008. "Hypothalamic-pituitary-adrenal axis dysregulation and behavioral analysis of mouse mutants with altered glucocorticoid or mineralocorticoid receptor function." *Stress* 11 (5): 321–338. https://doi.org/10.1080/10253890701821081. www.ncbi.nlm.nih.gov/pubmed/18609295.

Konkle, A. T. M., S. E. Keith, J. P. McNamee, and D. Michaud. 2017."Chronic noise exposure in the spontaneously hypertensive rat." *Noise Health* 19 (90): 213–221. https://doi.org/10.4103/nah.NAH_15_17. https://www.ncbi.nlm.nih.gov/pubmed/28937015.

Kroller-Schon, S., A. Daiber, S. Steven, M. Oelze, K. Frenis, S. Kalinovic, A. Heimann, F. P. Schmidt, A. Pinto, M. Kvandova, et al. 2018. "Crucial role for Nox2 and sleep deprivation in aircraft noise-induced vascular and cerebral oxidative stress, inflammation, and gene regulation." *Eur Heart J* 39 (38): 3528–3539. https://doi.org/10.1093/eurheartj/ehy333. www.ncbi.nlm.nih.gov/pubmed/29905797.

Kvandova, M., K. Filippou, S. Steven, M. Oelze, S. Kalinovic, P. Stamm, K. Frenis, K. Vujacic-Mirski, K. Sakumi, Y. Nakabeppu, et al. 2020. "Environmental aircraft noise aggravates oxidative DNA damage,

granulocyte oxidative burst and nitrate resistance in Ogg1(-/-) mice." *Free Radic Res* 54 (4): 280–292. https://doi.org/10.1080/10715762.2020.1754410. www.ncbi.nlm.nih.gov/pubmed/32326776.
Lenzi, P., G. Frenzilli, M. Gesi, M. Ferrucci, G. Lazzeri, F. Fornai, and M. Nigro. 2003. "DNA damage associated with ultrastructural alterations in rat myocardium after loud noise exposure." *Environ Health Perspect* 111 (4): 467–471. https://doi.org/10.1289/ehp.5847. www.ncbi.nlm.nih.gov/pubmed/12676600.
Li, J., L. Pan, W. G. Pembroke, J. E. Rexach, M. I. Godoy, M. C. Condro, A. G. Alvarado, M. Harteni, Y. W. Chen, L. Stiles, et al. 2021. "Conservation and divergence of vulnerability and responses to stressors between human and mouse astrocytes." *Nat Commun* 12 (1): 3958. https://doi.org/10.1038/s41467-021-24232-3. www.ncbi.nlm.nih.gov/pubmed/34172753.
Lin, C. C., L. L. Chiang, C. H. Lin, C. H. Shih, Y. T. Liao, M. J. Hsu, and B. C. Chen. 2007. "Transforming growth factor-beta1 stimulates heme oxygenase-1 expression via the PI3K/Akt and NF-kappaB pathways in human lung epithelial cells." *Eur J Pharmacol* 560 (2–3): 101–109. https://doi.org/10.1016/j.ejphar.2007.01.025. www.ncbi.nlm.nih.gov/pubmed/17307160.
Lousinha, A., M. J. R. Oliveira, G. Borrecho, J. Brito, P. Oliveira, A. Oliveira de Carvalho, D. Freitas, A. P. Aguas, and E. Antunes. 2018."Infrasound induces coronary perivascular fibrosis in rats." *Cardiovasc Pathol* 37: 39–44. https://doi.org/10.1016/j.carpath.2018.10.004. https://www.ncbi.nlm.nih.gov/pubmed/30342320.
Lousinha, A., G. Pereira, G. Borrecho, J. Brito, A. Oliveira de Carvalho, D. Freitas, P. Oliveira, M. J. R. Oliveira, and E. Antunes. 2020. "Atrial fibrosis and decreased connexin 43 in rat hearts after exposure to high-intensity infrasound." *Exp Mol Pathol* 114: 104409. https://doi.org/10.1016/j.yexmp.2020.104409. https://www.ncbi.nlm.nih.gov/pubmed/32088192.
Lyamin, O. I., S. M. Korneva, V. V. Rozhnov, and L. M. Mukhametov. 2016."Cardiorespiratory responses to acoustic noise in belugas." *Adv Exp Med Biol* 875: 665–672. https://doi.org/10.1007/978-1-4939-2981-8_80. https://www.ncbi.nlm.nih.gov/pubmed/26611017.
Manikandan, S., M. K. Padma, R. Srikumar, N. Jeya Parthasarathy, A. Muthuvel, and R. Sheela Devi. 2006. "Effects of chronic noise stress on spatial memory of rats in relation to neuronal dendritic alteration and free radical-imbalance in hippocampus and medial prefrontal cortex." *Neurosci Lett* 399 (1–2): 17–22. https://doi.org/10.1016/j.neulet.2006.01.037. www.ncbi.nlm.nih.gov/pubmed/16481110.
Medic, G., M. Wille, and M. E. Hemels. 2017. "Short- and long-term health consequences of sleep disruption." *Nat Sci Sleep* 9: 151–161. https://doi.org/10.2147/NSS.S134864. www.ncbi.nlm.nih.gov/pubmed/28579842.
Meltser, I., C. R. Cederroth, V. Basinou, S. Savelyev, G. S. Lundkvist, and B. Canlon. 2020. "TrkB-mediated protection against circadian sensitivity to noise trauma in the murine cochlea." *Curr Biol* 30 (22): 4547. https://doi.org/10.1016/j.cub.2020.10.090. www.ncbi.nlm.nih.gov/pubmed/33202221.
Mestas, J., and C. C. Hughes. 2004. "Of mice and not men: Differences between mouse and human immunology." *J Immunol* 172 (5): 2731–2738. https://doi.org/10.4049/jimmunol.172.5.2731. www.ncbi.nlm.nih.gov/pubmed/14978070.
Mezzaroma, E., A. Abbate, and S. Toldo. 2021. "NLRP3 inflammasome inhibitors in cardiovascular diseases." *Molecules* 26 (4). https://doi.org/10.3390/molecules26040976. www.ncbi.nlm.nih.gov/pubmed/33673188.
Miller, A. J., and A. C. Arnold. 2019. "The renin-angiotensin system in cardiovascular autonomic control: Recent developments and clinical implications." *Clin Auton Res* 29 (2): 231–243. https://doi.org/10.1007/s10286-018-0572-5. www.ncbi.nlm.nih.gov/pubmed/30413906.
Munzel, T., A. Daiber, S. Steven, L. P. Tran, E. Ullmann, S. Kossmann, F. P. Schmidt, M. Oelze, N. Xia, H. Li, et al. 2017. "Effects of noise on vascular function, oxidative stress, and inflammation: Mechanistic insight from studies in mice." *Eur Heart J* 38 (37): 2838–2849. https://doi.org/10.1093/eurheartj/ehx081. https://www.ncbi.nlm.nih.gov/pubmed/28329261.
Munzel, T., F. P. Schmidt, S. Steven, J. Herzog, A. Daiber, and M. Sorensen. 2018. "Environmental noise and the cardiovascular system." *J Am Coll Cardiol* 71 (6): 688–697. https://doi.org/10.1016/j.jacc.2017.12.015. www.ncbi.nlm.nih.gov/pubmed/29420965.
Munzel, T., M. Sorensen, and A. Daiber. 2021. "Transportation noise pollution and cardiovascular disease." *Nat Rev Cardiol* 18 (9): 619–636. https://doi.org/10.1038/s41569-021-00532-5. www.ncbi.nlm.nih.gov/pubmed/33790462.

Nelson, D. I., R. Y. Nelson, M. Concha-Barrientos, and M. Fingerhut. 2005. "The global burden of occupational noise-induced hearing loss." *Am J Ind Med* 48 (6): 446–458. https://doi.org/10.1002/ajim.20223. www.ncbi.nlm.nih.gov/pubmed/16299704.

Park, J. S., C. R. Cederroth, V. Basinou, I. Meltser, G. Lundkvist, and B. Canlon. 2016. "Identification of a circadian clock in the inferior colliculus and Its dysregulation by noise exposure." *J Neurosci* 36 (20): 5509–5519. https://doi.org/10.1523/JNEUROSCI.3616-15.2016. www.ncbi.nlm.nih.gov/pubmed/27194331.

Paschos, G. K., and G. A. FitzGerald. 2010. "Circadian clocks and vascular function." *Circ Res* 106 (5): 833–841. https://doi.org/10.1161/CIRCRESAHA.109.211706. www.ncbi.nlm.nih.gov/pubmed/20299673.

Recio, A., C. Linares, J. R. Banegas, and J. Diaz. 2016. "Road traffic noise effects on cardiovascular, respiratory, and metabolic health: An integrative model of biological mechanisms." *Environ Res* 146: 359–370. https://doi.org/10.1016/j.envres.2015.12.036. www.ncbi.nlm.nih.gov/pubmed/26803214.

Said, M. A., and O. A. El-Gohary. 2016. "Effect of noise stress on cardiovascular system in adult male albino rat: Implication of stress hormones, endothelial dysfunction and oxidative stress." *Gen Physiol Biophys* 35 (3): 371–377. https://doi.org/10.4149/gpb_2016003. www.ncbi.nlm.nih.gov/pubmed/27174896.

Schlaich, M. P., F. Socratous, S. Hennebry, N. Eikelis, E. A. Lambert, N. Straznicky, M. D. Esler, and G. W. Lambert. 2009. "Sympathetic activation in chronic renal failure." *J Am Soc Nephrol* 20 (5): 933–939. https://doi.org/10.1681/ASN.2008040402. www.ncbi.nlm.nih.gov/pubmed/18799718.

Steptoe, A., M. Hamer, and Y. Chida. 2007. "The effects of acute psychological stress on circulating inflammatory factors in humans: A review and meta-analysis." *Brain Behav Immun* 21 (7): 901–912. https://doi.org/10.1016/j.bbi.2007.03.011. www.ncbi.nlm.nih.gov/pubmed/17475444.

Steven, S., K. Frenis, S. Kalinovic, M. Kvandova, M. Oelze, J. Helmstadter, O. Hahad, K. Filippou, K. Kus, C. Trevisan, et al. 2020. "Exacerbation of adverse cardiovascular effects of aircraft noise in an animal model of arterial hypertension." *Redox Biol* 34: 101515. https://doi.org/10.1016/j.redox.2020.101515. www.ncbi.nlm.nih.gov/pubmed/32345536.

Steven, S., K. Frenis, M. Oelze, S. Kalinovic, M. Kuntic, M. T. Bayo Jimenez, K. Vujacic-Mirski, J. Helmstadter, S. Kroller-Schon, T. Munzel, et al. 2019. "Vascular inflammation and oxidative stress: Major triggers for cardiovascular disease." *Oxid Med Cell Longev* 2019: 7092151. https://doi.org/10.1155/2019/7092151. www.ncbi.nlm.nih.gov/pubmed/31341533.

Takeda, N., K. Maemura, S. Horie, K. Oishi, Y. Imai, T. Harada, T. Saito, T. Shiga, E. Amiya, I. Manabe, N. Ishida, and R. Nagai. 2007. "Thrombomodulin is a clock-controlled gene in vascular endothelial cells." *J Biol Chem* 282 (45): 32561–32567. https://doi.org/10.1074/jbc.M705692200. www.ncbi.nlm.nih.gov/pubmed/17848551.

Thosar, S. S., M. P. Butler, and S. A. Shea. 2018. "Role of the circadian system in cardiovascular disease." *J Clin Invest* 128 (6): 2157–2167. https://doi.org/10.1172/JCI80590. www.ncbi.nlm.nih.gov/pubmed/29856365.

Vienneau, D., L. Perez, C. Schindler, C. Lieb, H. Sommer, N. Probst-Hensch, N. Kunzli, and M. Roosli. 2015. "Years of life lost and morbidity cases attributable to transportation noise and air pollution: A comparative health risk assessment for Switzerland in 2010." *Int J Hyg Environ Health* 218 (6): 514–521. https://doi.org/10.1016/j.ijheh.2015.05.003. www.ncbi.nlm.nih.gov/pubmed/26003939.

Viswambharan, H., J. M. Carvas, V. Antic, A. Marecic, C. Jud, C. E. Zaugg, X. F. Ming, J. P. Montani, U. Albrecht, and Z. Yang. 2007. "Mutation of the circadian clock gene Per2 alters vascular endothelial function." *Circulation* 115 (16): 2188–2195. https://doi.org/10.1161/CIRCULATIONAHA.106.653303. www.ncbi.nlm.nih.gov/pubmed/17404161.

Weber, C., and H. Noels. 2011. "Atherosclerosis: Current pathogenesis and therapeutic options." *Nat Med* 17 (11): 1410–1422. https://doi.org/10.1038/nm.2538. www.ncbi.nlm.nih.gov/pubmed/22064431.

Wisser, H., and H. Breuer. 1981. "Circadian changes of clinical chemical and endocrinological parameters." *J Clin Chem Clin Biochem* 19 (6): 323–337. www.ncbi.nlm.nih.gov/pubmed/7024460.

Woon, P. Y., P. J. Kaisaki, J. Braganca, M. T. Bihoreau, J. C. Levy, M. Farrall, and D. Gauguier. 2007. "Aryl hydrocarbon receptor nuclear translocator-like (BMAL1) is associated with susceptibility to hypertension and type 2 diabetes." *Proc Natl Acad Sci U S A* 104 (36): 14412–14417. https://doi.org/10.1073/pnas.0703247104. www.ncbi.nlm.nih.gov/pubmed/17728404.

Xiao, L., A. Kirabo, J. Wu, M. A. Saleh, L. Zhu, F. Wang, T. Takahashi, R. Loperena, J. D. Foss, R. L. Mernaugh, et al. 2015. "Renal denervation prevents immune cell activation and renal inflammation in Angiotensin II-induced hypertension." *Circ Res* 117 (6): 547–557. https://doi.org/10.1161/CIRCRESAHA.115.306010. www.ncbi.nlm.nih.gov/pubmed/26156232.

Yang, L., and O. W. Guthrie. 2020. "Effects of acute noise exposure on DNA damage response genes in the cochlea, cortex, heart and liver." *Exp Mol Pathol* 114: 104401. https://doi.org/10.1016/j.yexmp.2020.104401. https://www.ncbi.nlm.nih.gov/pubmed/32061943.

Zheng, K. C., and M. Ariizumi. 2007. "Modulations of immune functions and oxidative status induced by noise stress." *J Occup Health* 49 (1): 32–38. https://doi.org/10.1539/joh.49.32. www.ncbi.nlm.nih.gov/pubmed/17314464.

16 Noise Pollution and Neuropsychiatric Disease

Omar Hahad, Thomas Münzel, Donya A. Gilan, Maria Teresa Bayo Jimenez, and Jasmin Ghaemi Kerahrodi

16.1 INTRODUCTION

During the last decades, it has been shown that novel risk factors in the physical environment, such as noise pollution (mostly traffic-related) may also contribute significantly to the burden of disease (Hahad, Kroller-Schon, et al. 2019; Hahad et al. 2021). Moreover, globally, people moving from the countryside to large urban areas where environmental risk factors are an integral part of the physical environment and may independently facilitate the development of noncommunicable (NCD) diseases such as cardiovascular diseases (CVD) or neuronal diseases (Kalsch et al. 2014; Schmidt et al. 2015). The Lancet Commission on pollution and health stated that 16% of global annual deaths are caused by all forms of pollution, representing 15 times more deaths than from all wars and other forms of violence as well as three times more than from AIDS, tuberculosis, and malaria combined (Landrigan et al. 2018). Additionally, the World Health Organization (WHO) estimates that at least 12.6 million premature deaths per year are attributable to unhealthy environments, 8.2 million of which are due to NCD, such as atherosclerosis, metabolic disease, or CVD (including stroke), which is the most significant contributor to the health burden, accounting for nearly 5 million of these deaths ("World Health Organization. Preventing disease through healthy environments: a global assessment of the burden of disease from environmental risks (World Health Organization, 2016). Likewise, the WHO estimated that at least 1.6 million healthy life years are lost every year in Western European countries alone due to environmental noise exposure (*Burden of disease from environmental noise. Quantification of healthy life years lost in Europe* 2011).

While the association between noise pollution and CVD is well-established, emerging evidence also emphasizes the role of chronic exposure to noise in the impairment of the CNS via increased oxidative stress, imbalance in neurotransmitter levels, deterioration of molecular functions, impaired cognition, and genetic modifications and demonstrates its role as a risk factor for mental and cardio-cerebrovascular health (Arjunan and Rajan 2020; Kroller-Schon et al. 2018; Munzel et al. 2017; Munzel, Kroller-Schon, et al. 2020).

16.2 PATHOPHYSIOLOGY OF NOISE-INDUCED DISORDERS

The pathophysiology of health side effects in response to a single exposure to noise has been demonstrated during the last decade. Babisch established the modern noise reaction model demonstrating two pathways through which noise could affect human physiology (Figure 16.1) (Babisch 2002). Noise can exert its effects directly, through synaptic interactions, or indirectly, through sound Feld's emotional and cognitive perception (Babisch 2002). The direct pathway includes all effects that come up through high sound pressure levels; loud noise will damage the hair cells, and sensory receptors of the auditory system and disrupt sleep (Le et al. 2017). Approximately 16% of disabling hearing loss (over 4 million DALYs) is attributed to high noise via this direct pathway,

 DOI: 10.1201/9781003328100-20

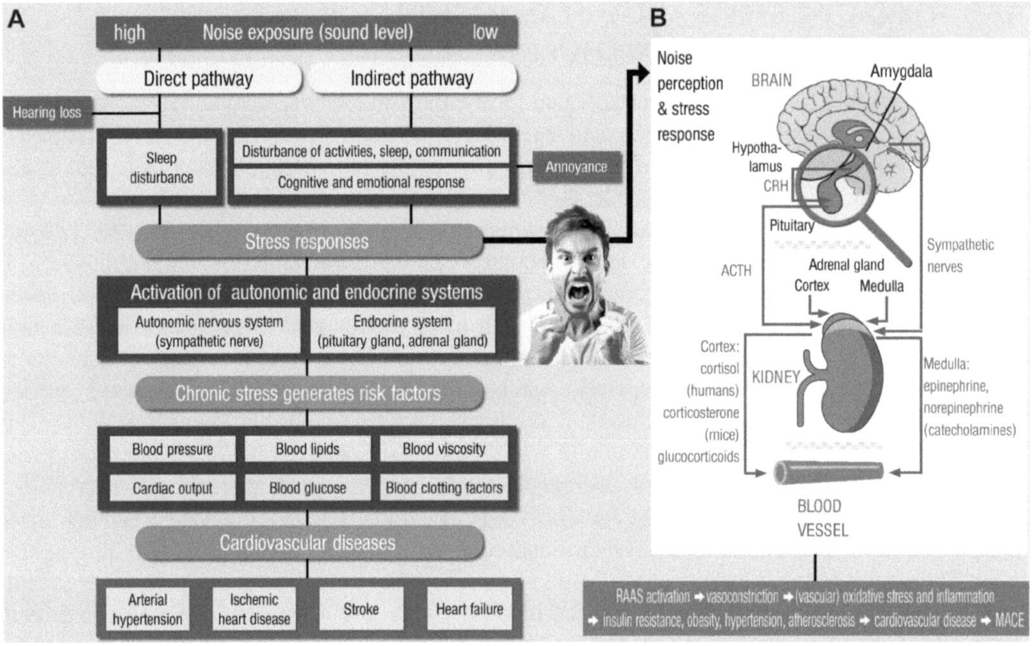

FIGURE 16.1 Noise–stress concept and the adverse health consequences in humans. (A) Noise reaction model and the (B) neuronal activation by noise exposure. Reused from Munzel et al. (2022) with permission.

ranging from 7% to 21% in the various subregions (Nelson et al. 2005). Also, this direct pathway has been linked with sleep disturbances, a well-accepted cardiovascular risk factor (Lechat et al. 2021; Coborn et al. 2019). The physiological effects of sleep disruption include increases in catecholamine, adrenocorticotropic hormone, and cortisol secretion; circadian disruption and decreased melatonin production; decreased insulin sensitivity and leptin levels; increases in ghrelin and appetite; upregulation of inflammatory proteins, as well as increases in oxidative stress in the production of ROS (Medic, Wille, and Hemels 2017). The second indirect pathway involves the cognitive and emotional response to a stimulus (i.e., feeling angry or annoyed) (Beutel et al. 2020; Beutel et al. 2016). Noise stimulates the brain's limbic region to generate an emotional response (Spreng 2000), leading to neuroendocrine arousal that affects the metabolic state (Henry 1992). The altered metabolic state is responsible for many risk factors that are known to be implicated in cardio-cerebrovascular and neurodegenerative disease, such as changes in glucose metabolism (Mottillo et al. 2010; Frisardi et al. 2010), lipid dysregulation, and hemodynamic changes (Wong et al. 2017; Zhao et al. 2015; Sabayan et al. 2012; Zhang, Xu, and Li 2014). The Gutenberg Health Study examined the emotional state of the human experience of noise exposure. They suggested for the first time that annoyance of noise of environmental origins is "dose-dependently" associated with atrial fibrillation (Hahad et al. 2018) because of the activation of the indirect pathway by noise. Both pathways, direct and indirect, lead to three levels of physiological consequences: First, stress indicators (e.g., stress hormones including catecholamine, adrenocorticotropic hormone, and cortisol secretion, or increasing ROS causing oxidative stress); second, enhanced risk factors (e.g., blood pressure, blood lipids, homeostatic factors); and third, manifest diseases (e.g., hypertension, ischemic heart disease) (Hahad, Prochaska, et al. 2019; Daiber et al. 2019; Munzel, Sorensen, and Daiber 2021).

16.3 OXIDATIVE STRESS AND INFLAMMATION INDUCED BY NOISE EXPOSURE

As was introduced earlier, Babisch established for the first-time environmental noise, such as traffic or aircraft, as a recognized cardiovascular risk factor strongly tied to oxidative stress pathways. Moreover, not only does noise promote the initiation of the oxidative stress cascade, but it also causes a cascade of inflammatory and metabolic effects that work in tandem to cause vascular damage and lead to higher cardiovascular and presumably neuropsychiatric risk (Babisch 2002, 2003). Recently this concept was confirmed on a molecular level by positron emission tomography scans in men, demonstrating an association between amygdala activity and arterial inflammation with severe chronic noise exposure, which was associated with more pronounced heightened arterial inflammation and higher cardiovascular risk (more major adverse cardiovascular events) (Osborne et al. 2020) (Figure 16.2). In 2016, another well-established study supported this hypothesis. The effect of acute, chronic intermittent, and chronic continuous exposure to noise of intensity 80–100 dB (8 to 16 kHz) on heart rate and mean systemic arterial blood pressure in rats (8 rats per group) were investigated. Increased levels of stress hormones such as corticosterone, adrenaline, noradrenaline, and vasoconstrictor endothelin-1 were found in the plasma of these rats (Said and El-Gohary 2016), supporting the hypothesis that adverse cardiovascular reactions are induced by daily noise exposure. The effects of stress hormones produced by an emotional state of annoyance are diverse. Still, they are generally associated with increased blood pressure and heart rate, and they can directly contribute to the development of CVD by different mechanisms (Daiber et al. 2019).

Ersoy et al. demonstrated higher levels of ROS in rats in the cerebral cortex and reduced levels of •NO in the cerebellum and brainstem, which was reduced by rosuvastatin, a 3-hydroxy-3-methylglutaryl coenzyme, a reductase inhibitor, and a cholesterol-lowering drug. The drug shows antioxidative properties such as reducing malondialdehyde values in the brain stem (Ersoy et al. 2014). To sum up, many studies provided evidence that noise's effects are not specific to the vasculature but may also have effects in systems other than the cardiovascular. As a result of the adverse effects of this noise stress, which include metabolic and anatomical changes in neurons; a decreased dendritic count; and impaired memory, cognition, and locomotor activity, numerous studies have shown that noise impairs cognition, motor coordination, changes in feeding behavior, fear, and anxiety (Hahad et al. 2022) (Figure 16.3).

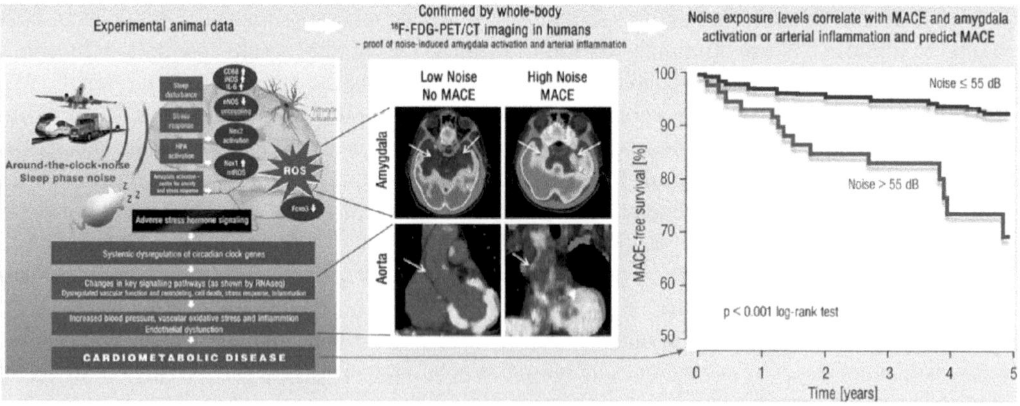

FIGURE 16.2 Animal and human evidence for noise-induced stress response and cardiovascular risk. Noise-induced cerebral (amygdalar) activation and arterial inflammation as critical processes to increase the risk of major cardiovascular events and are likely to increase the risk of neuropsychiatric disease, which is in correspondence with mechanistic data on noise-induced neuroactivation of the endocrine systems, neuroinflammation, and cerebral oxidative stress. Reused from Munzel, Steven, et al. (2020) with permission.

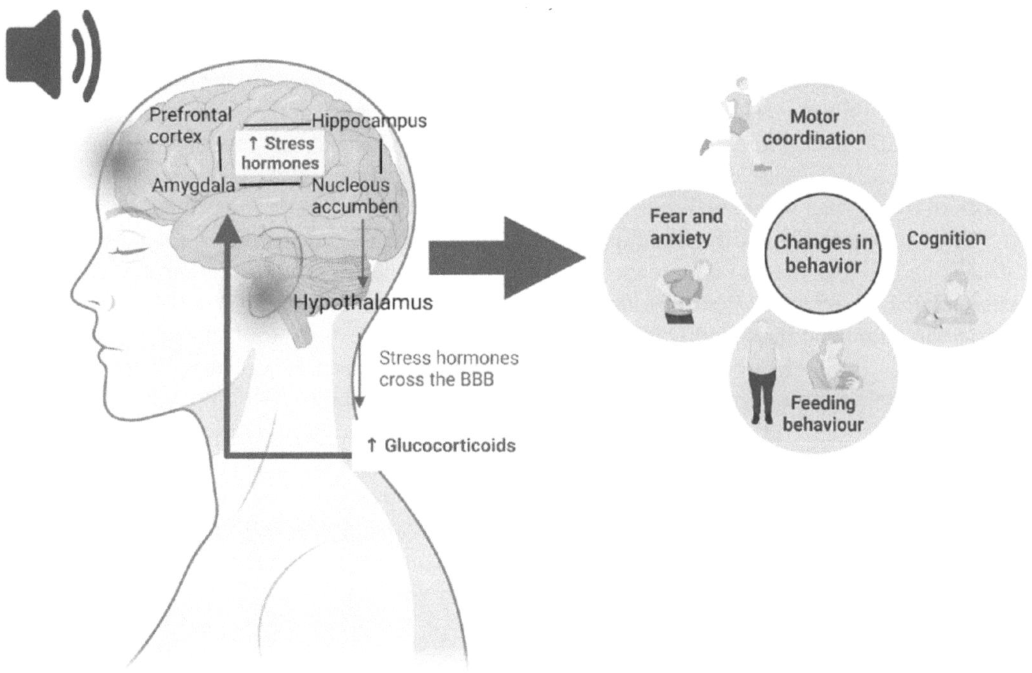

FIGURE 16.3 An illustration of the brain-body reaction to noise-induced stress. On the limbic surfaces of the brain (prefrontal cortex, hippocampus, amygdala, and nucleus accumbens), stressful events alter the release of several neurotransmitters, which promote and modify behavioral processes to deal with the stressor. Dopamine, noradrenaline, acetylcholine, or glutamate are hormones and peptides released from the body and pass across the blood-brain barrier (BBB) to enter the brain when the hypothalamus is activated. Reused from Hahad et al. (2022) with permission.

16.4 EVIDENCE FROM HUMAN OBSERVATIONAL/EPIDEMIOLOGICAL STUDIES ON THE ASSOCIATION BETWEEN NOISE EXPOSURE AND NEUROPSYCHIATRIC DISEASE

16.4.1 Stroke

A large case-control study from Germany (N=1,026,670) examined the stroke risks associated with aircraft, road traffic, and railway noise exposure. An increased risk of stroke was demonstrated in response to aircraft noise (OR 1.07, 95% CI 1.02–1.13 for subjects exposed to < 40 dB of 24-hour continuous aircraft noise ($L_{pAeq,24h}$) and ≥ 6 events of maximum nightly sound pressure levels ≥ 50 dB) (A.L. Seidler et al. 2018). In addition, road traffic and railway noise exposure were positively related to the risk of stroke (OR 1.017, 95% CI 1.003–1.032 and OR 1.018, 95% CI 1.001–1.033, respectively). Herein, the authors discriminated between ischemic and hemorrhagic stroke, demonstrating higher OR for the latter for aircraft and railway noise. The relationship between noise from wind turbines and the risk of incident myocardial infarction and stroke was the subject of a nationwide Danish cohort study from Poulsen et al. from 1982–2013. This field is particularly important since wind turbine noise has been reported as more annoying than traffic noise at similar levels. The authors revealed significant evidence for a relationship between mean 1-year and 5-year nighttime outdoor wind turbine noise exposure groups and risk of stroke (e.g., incidence rate ratio [IRR] 1.10, 95% CI 1.03–1.17 for 30–36 vs. < 24 dB[A]) (Poulsen et al. 2019). In contrast, in a recent comprehensive cohort study of 23,912 Danish nurses, weaker evidence for a relationship between wind turbine noise and stroke incidence was found (Brauner et al. 2019). In a further nationwide Danish study, a 4% (IRR 1.04, 95% CI 1.03–1.05 per 10 dB increase at the most exposed façade) higher risk of incident stroke was found in response to road traffic noise. In contrast, no association was found for railway noise, possibly because railway noise is generally perceived as less annoying than road traffic noise (Sorensen et al. 2021). Interestingly, stroke risk and severity have been associated with residential noise exposure, as revealed in 2,761 patients hospitalized with acute ischemic stroke in Barcelona (Vivanco-Hidalgo et al. 2019).

While higher residential surrounding greenspace was associated with decreased risk of severe stroke, increased residential noise exposure (L_{den}) was associated with a pronounced risk of severe stroke (OR 1.30, 95% CI 1.02–1.65). The influence of exposure to road traffic noise and stroke incidence was examined in 25,660 Danish nurses indicating no substantial effect after adjustment for $PM_{2.5}$ or NO_2 (Cole-Hunter et al. 2021). In contrast, in the recent Danish Nurse Cohort study (n= 22,858), they found more substantial evidence for all-cause mortality (HR 1.06, 95% CI 1.01–1.11, and HR 1.09, 95% CI 1.03–1.15 per 10 dB increase of 5-year and 23-year mean L_{den}, respectively) and stroke mortality (HR 1.10, 95% CI 0.91–1.31 for 5-year mean L_{den}) within the same cohort (Cole-Hunter et al. 2022). The substantial role of traffic noise exposure in the risk of increased stroke mortality was also confirmed in two nationwide studies from Switzerland (Vienneau et al. 2022; Heritier et al. 2017), whereas a small cohort study from Brazil found no significant association between aircraft noise exposure and stroke mortality (Roca-Barcelo et al. 2021).

16.4.2 Mental Illness

Unfortunately, nowadays the WHO evidence on mental disorders could not provide a quantitative estimate of the effect of environmental noise. A recent meta-analysis from Dzhambov and Lercher found road traffic noise exposure to be associated with 4% (OR 1.04, 95% CI 1.03–1.11 per 10 dB [A] increase in L_{den}) higher odds of depression and 12% (OR 1.12, 95% CI 1.04–1.30) of anxiety. However, it is essential to note that most of the included studies were cross-sectional and overall lower quality, but overall indicating an association between road traffic noise and depression and anxiety (Dzhambov and Lercher 2019). Accordingly, Hegewald et al. revealed substantial meta-analytic evidence on the association between traffic noise exposure and depression and anxiety, including 26 studies (Hegewald et al. 2020). Herein, aircraft noise exposure was found to increase depression risk by 12% (effect size 1.12, 95% CI 1.02–1.23 per 10 dB increase in L_{den}), whereas the effect sizes obtained from road traffic and railway noise studies were of smaller magnitude. In the Netherlands Study of Depression and Anxiety (N=2,980), high levels of traffic noise were associated with the presence of depressive and anxiety disorders, and the quality of socioeconomic, physical, and social neighborhoods characteristics were associated with the presence and severity of affective disorders (Generaal et al. 2019). A large case-control study from Germany examined the risk of depression in response to aircraft, road traffic, and railway noise exposure (Seidler et al. 2017). A linear exposure-risk relationship was observed for road traffic noise (OR 1.17, 95% CI 1.10–1.25 for $L_{pAeq,\,24h} \geq 70$ dB). For aircraft noise, the highest risk estimate was observed for exposures at 50–55 dB (OR of 1.23, 95% CI 1.19–1.28) and for railway noise at 60–65 dB (OR 1.15, 95% CI 1.08–1.22). Combining all three exposures (above 50 dB) yielded the highest risk estimate (OR 1.42, 95% CI 1.33–1.52). Baseline data from the UK Biobank study suggested that symptoms of anxiety, tension, depression, and bipolar disorder be positively associated with road traffic noise exposure. Higher road traffic noise exposure was significantly associated with a higher risk of depression symptoms. However, a negative relationship was found in the case of significant depression (Hao et al. 2021). In a cohort of 2,398 men from the United Kingdom, road traffic noise exposure (OR 1.82, 95% CI 1.07–3.07 for 56–60 dB [A]), noise annoyance (OR 2.47, 95% CI 1.00–6.13), and noise sensitivity (OR 1.65, 95% CI 1.09–2.50) were associated with incident psychological ill-health (Stansfeld et al. 2021). In a cohort of 140,456 women from Canada, the relationship between residential noise exposure during pregnancy, especially nighttime noise, and later depression hospitalization were evaluated. The results revealed the strongest association for increased nighttime noise exposure (HR 1.68, 95% CI 1.05–2.67 for 70 vs. 50 dB [A] L_{night}) (He et al. 2019). In conclusion, there are many studies supporting this association. Still, more evidence is needed to explore the underlying molecular mechanisms to finally develop more effective public health control of depression and noise exposure burden.

16.4.3 Dementia and Cognitive Decline

The neurotoxicity of noise might extend to older adults, possibly precipitating cognitive decline and dementia through direct effects on Alzheimer's disease (AD) pathology and inflammatory processes. This is yet a vastly understudied area of research. In 5,227 participants from the Chicago Health and Aging Project (≥ 65 years), residential noise exposure resulted in 36% and 29% higher odds of prevalent mild cognitive impairment (OR 1.36, 95% CI 1.15–1.62 per 10 dB[A] increase) and AD (OR 1.29, 95% CI 1.08–1.55) (Weuve et al. 2021). Additionally, noise exposure was related to lower global cognitive performance (perceptual speed) but not consistent cognitive decline. Likewise, data from the German Heinz Nixdorf Recall study completed on 4,086 participants who were 50–80 years old indicated that traffic-related noise exposure is associated with a lower global cognitive score and a mild cognitive impairment (Tzivian, Dlugaj, Winkler, Weinmayr, et al. 2016). Interestingly, stronger associations were observed in former and current smokers as indicated by a significant interaction, suggesting that lifestyle risk factors may potentiate memory and executive function effects of noise exposure, while there was minor evidence for mutual confounding (Tzivian, Dlugaj, Winkler, Hennig, et al. 2016). Importantly, the authors also demonstrated that air pollution and road traffic noise exposure in 4,086 participants of the population-based Heinz Nixdorf Recall cohort study may act synergistically to negatively influence cognitive function (Tzivian et al. 2017). In 2019, Fuks et al. investigated the association of residential exposure to road traffic noise and annoyance due to road traffic noise with cognitive function in a cohort of 288 elderly women from the German longitudinal study on the influence of air pollution on lung function, inflammation, and aging (SALIA). The results showed that road traffic noise was associated with total impaired cognition and the constructional praxis domain (neuropsychological assessment battery), which remained stable after further adjustment for air pollution exposure. However, this finding should be interpreted with caution since the observations from this study could not control for potential confounding by hearing loss (Fuks et al. 2019). In the Irish Longitudinal Study on Ageing, road traffic noise exposure negatively impacted executive function (Mac Domhnaill et al. 2021). Recently, similar results were found in a study of 1,612 elderly Mexican-American participants from Sacramento in which the authors found suggestive evidence that traffic noise exposure increases the risk of dementia and cognitive impairment (Yu, Mayeda, et al. 2020). Interestingly, a subsequent study demonstrated metabolic dysfunction (hyperglycemia or low HDL cholesterol) to negatively modify the influence of traffic-related air pollution and noise exposure on these outcomes (Yu, Haan, et al. 2020). Linares et al. revealed the short-term association between traffic noise and the risk of dementia-related emergency hospital admissions in Spain (Linares et al. 2017). In contrast, Andersson et al. found no effect of road traffic noise exposure, independently or in combination with traffic-related air pollution, on dementia risk in 1,721 subjects (Andersson et al. 2018).

16.4.4 Neurodevelopmental Disorders

Environmental noise may play a role in the manifestation and severity of attention-deficit/hyperactivity disorder symptoms, but the evidence is limited. Children and adolescents may be vulnerable to environmental influences such as noise, which can affect mental well-being. The meta-analysis of three studies revealed increased odds of hyperactivity/inattention (OR 1.11, 95% CI 1.04–1.19 per 10 dB increase) and total difficulties (OR 1.09, 95% CI 1.02–1.16) in response to road traffic noise exposure in children (Schubert et al. 2019). In contrast, prenatal and childhood road traffic noise exposure were not associated with emotional, aggressive, or attention-deficit/hyperactivity disorder-related symptoms in children obtained from two European birth cohorts (Essers et al. 2022). In agreement, in 1,710 children from the TRAILS study in The Netherlands, road traffic noise exposure was not associated with attention-deficit/hyperactivity disorder symptoms. It may be due to residual confounding or selection bias (Zijlema et al. 2021). However, the meta-analysis of studies investigating the influence of aircraft noise at school on children's reading comprehension and psychology demonstrated that a 1 dB increase in aircraft noise exposure at school led to a −0.007 (95% CI −0.012 to −0.001) decreased reading score and 4% higher odds of scoring well below or below average on the reading test (Clark et al. 2021).

Furthermore, a 1 dB increase in aircraft noise exposure at school was associated with a 0.017 (95% CI 0.007–0.028) increase in hyperactivity score. Lastly, Tangermann et al. studied the response of behavior and behavioral change within 1 year in a cohort of 886 adolescents in Switzerland aged 10–17 years in response to road traffic noise exposure. The results showed a positive relationship in cross-sectional analysis, whereas no association was found in prospective analysis (Tangermann et al. 2022).

16.5 CONCLUSIONS

It is likely assumed that noise pollution can raise the risk of neuropsychiatric diseases based on the literature. It is believed that noise pollution causes brain damage through two main pathophysiological mechanisms: inflammation and oxidative stress. Although the negative health consequences brought on by noise exposure may be minor or mild on an individual level, the overall attributable burden may be significantly higher considering the vast proportion of the population exposed to noise during their lifetime. However, definitive conclusions cannot be drawn because the research on the connection between noise pollution and the CNS is still in its early stages, resulting in inconsistent or partially inconclusive observations within studies. Inconsistencies in the selection of the most appropriate time of exposure for the outcome of interest (acute vs. chronic), assessment of noise exposure (mainly due to exposure measurement error related to spatial misalignment of monitoring data and study participants), measurement and definition of outcome variables, confounding (individual-variables and area-variables), and adjustment for co-exposure are just a few of the reasons why results may differ. Therefore, additional research is needed to clarify these links. Future studies should focus on more precise exposure measurement, clinical and subclinical endpoints, and the combined effects of noise exposure and other stressors. To comprehend why certain people are more resistant to noise-mediated stress reactions, resilience needs also be discussed in greater detail. This may also aid in discovering new targets for pharmaceutical or lifestyle therapies to combat the harmful effects of noise on health. It is crucial to remember that, unlike other risk factors like smoking, inactivity, excessive alcohol use, and poor diets, noise pollution is rarely preventable or ameliorated by changing one's lifestyle. This is why policymakers, who regulate noise pollution to safeguard the public from harmful health effects, are better able to control improvements in environmental quality than patients and doctors. An early step toward achieving this would be recognizing and including environmental stressors, such as noise pollution, as substantial risk factors in official preventative guidelines.

CONFLICT OF INTEREST

None.

REFERENCES

Andersson, J., A. Oudin, A. Sundstrom, B. Forsberg, R. Adolfsson, and M. Nordin. 2018. "Road traffic noise, air pollution, and risk of dementia—results from the Betula project." *Environ Res* 166: 334–339. https://doi.org/10.1016/j.envres.2018.06.008. www.ncbi.nlm.nih.gov/pubmed/29909174.

Arjunan, A., and R. Rajan. 2020. "Noise and brain." *Physiol Behav* 227: 113136. https://doi.org/10.1016/j.physbeh.2020.113136. www.ncbi.nlm.nih.gov/pubmed/32798569.

Babisch, W. 2002. "The noise/stress concept, risk assessment and research needs." *Noise Health* 4 (16): 1–11. www.ncbi.nlm.nih.gov/pubmed/12537836.

———. 2003. "Stress hormones in the research on cardiovascular effects of noise." *Noise Health* 5 (18): 1–11. https://doi.org/NO_DOI. www.ncbi.nlm.nih.gov/pubmed/12631430.

Beutel, M. E., E. Brahler, M. Ernst, E. Klein, I. Reiner, J. Wiltink, M. Michal, P. S. Wild, A. Schulz, T. Munzel, et al. 2020. "Noise annoyance predicts symptoms of depression, anxiety and sleep disturbance 5 years later. Findings from the Gutenberg Health Study." *Eur J Public Health* 30 (3): 516–521. https://doi.org/10.1093/eurpub/ckaa015. www.ncbi.nlm.nih.gov/pubmed/32034913.

Beutel, M. E., C. Junger, E. M. Klein, P. Wild, K. Lackner, M. Blettner, H. Binder, M. Michal, J. Wiltink, E. Brahler, and T. Munzel. 2016. "Noise annoyance is associated with depression and anxiety in the general population- the contribution of aircraft noise." *pLoS One* 11 (5): e0155357. https://doi.org/10.1371/journal.pone.0155357. www.ncbi.nlm.nih.gov/pubmed/27195894.

Brauner, E. V., J. T. Jorgensen, A. K. Duun-Henriksen, C. Backalarz, J. E. Laursen, T. H. Pedersen, M. K. Simonsen, and Z. J. Andersen. 2019. "Association between long-term exposure to wind turbine noise and the risk of stroke: Data from the Danish Nurse Cohort." *J Am Heart Assoc* 8 (14): e013157. https://doi.org/10.1161/JAHA.119.013157. www.ncbi.nlm.nih.gov/pubmed/31310571.

Burden of Disease from Environmental Noise: Quantification of Healthy Life Years Lost in Europe. 2011. World Health Organization. www.euro.who.int/__data/assets/pdf_file/0008/136466/e94888.pdf.

Clark, C., J. Head, M. Haines, I. van Kamp, E. van Kempen, and S. A. Stansfeld. 2021. "A meta-analysis of the association of aircraft noise at school on children's reading comprehension and psychological health for use in health impact assessment." *Journal of Environmental Psychology* 76: 101646. https://doi.org/10.1016/j.jenvp.2021.101646. www.sciencedirect.com/science/article/pii/S0272494421000992.

Coborn, J. E., R. E. Lessie, C. M. Sinton, N. E. Rance, C. E. Perez-Leighton, and J. A. Teske. 2019. "Noise-induced sleep disruption increases weight gain and decreases energy metabolism in female rats." *Int J Obes (Lond)* 43 (9): 1759–1768. https://doi.org/10.1038/s41366-018-0293-9. www.ncbi.nlm.nih.gov/pubmed/30568267.

Cole-Hunter, T., C. Dehlendorff, H. Amini, A. Mehta, Y. H. Lim, J. T. Jorgensen, S. Li, R. So, L. H. Mortensen, R. Westendorp, et al. 2021. "Long-term exposure to road traffic noise and stroke incidence: A Danish Nurse Cohort study." *Environ Health* 20 (1): 115. https://doi.org/10.1186/s12940-021-00802-2. www.ncbi.nlm.nih.gov/pubmed/34740347.

Cole-Hunter, T., R. So, H. Amini, C. Backalarz, J. Brandt, E. V. Brauner, O. Hertel, S. S. Jensen, J. T. Jorgensen, M. Ketzel, et al. 2022. "Long-term exposure to road traffic noise and all-cause and cause-specific mortality: A Danish Nurse Cohort study." *Sci Total Environ*: 153057. https://doi.org/10.1016/j.scitotenv.2022.153057. www.ncbi.nlm.nih.gov/pubmed/35031374.

Daiber, A., S. Kroller-Schon, K. Frenis, M. Oelze, S. Kalinovic, K. Vujacic-Mirski, M. Kuntic, M. T. Bayo Jimenez, J. Helmstadter, S. Steven, et al. 2019. "Environmental noise induces the release of stress hormones and inflammatory signaling molecules leading to oxidative stress and vascular dysfunction-signatures of the internal exposome." *Biofactors* 45 (4): 495–506. https://doi.org/10.1002/biof.1506. www.ncbi.nlm.nih.gov/pubmed/30937979.

Dzhambov, A. M., and P. Lercher. 2019. "Road traffic noise exposure and depression/anxiety: An updated systematic review and meta-analysis." *Int J Environ Res Public Health* 16 (21). https://doi.org/10.3390/ijerph16214134. www.ncbi.nlm.nih.gov/pubmed/31717834.

Ersoy, A., E. R. Koc, S. Sahin, U. Duzgun, B. Acar, and A. Ilhan. 2014. "Possible effects of rosuvastatin on noise-induced oxidative stress in rat brain." *Noise Health* 16 (68): 18–25. https://doi.org/10.4103/1463-1741.127849. www.ncbi.nlm.nih.gov/pubmed/24583676.

Essers, E., L. Perez-Crespo, M. Foraster, A. Ambros, H. Tiemeier, and M. Guxens. 2022. "Environmental noise exposure and emotional, aggressive, and attention-deficit/hyperactivity disorder-related symptoms in children from two European birth cohorts." *Environ Int* 158: 106946. https://doi.org/10.1016/j.envint.2021.106946. www.ncbi.nlm.nih.gov/pubmed/34715431.

Frisardi, V., V. Solfrizzi, D. Seripa, C. Capurso, A. Santamato, D. Sancarlo, G. Vendemiale, A. Pilotto, and F. Panza. 2010. "Metabolic-cognitive syndrome: A cross-talk between metabolic syndrome and Alzheimer's disease." *Ageing Res Rev* 9 (4): 399–417. https://doi.org/10.1016/j.arr.2010.04.007. www.ncbi.nlm.nih.gov/pubmed/20444434.

Fuks, K. B., C. Wigmann, H. Altug, and T. Schikowski. 2019. "Road traffic noise at the residence, annoyance, and cognitive function in elderly women." *Int J Environ Res Public Health* 16 (10). https://doi.org/10.3390/ijerph16101790. www.ncbi.nlm.nih.gov/pubmed/31137595.

Generaal, E., E. J. Timmermans, J. E. C. Dekkers, J. H. Smit, and Bwjh Penninx. 2019. "Not urbanization level but socioeconomic, physical and social neighbourhood characteristics are associated with presence and severity of depressive and anxiety disorders." *Psychol Med* 49 (1): 149–161. https://doi.org/10.1017/S0033291718000612. www.ncbi.nlm.nih.gov/pubmed/29540253.

Hahad, O., J. H. Prochaska, A. Daiber, and T. Muenzel. 2019. "Environmental noise-induced effects on stress hormones, oxidative stress, and vascular dysfunction: Key factors in the relationship between cerebrocardiovascular and psychological disorders." *Oxid Med Cell Longev* 2019: 4623109. https://doi.org/10.1155/2019/4623109. www.ncbi.nlm.nih.gov/pubmed/31814877.

Hahad, O., K. Frenis, M. Kuntic, A. Daiber, and T. Munzel. 2021. "Accelerated aging and age-related diseases (CVD and neurological) due to air pollution and traffic noise exposure." *Int J Mol Sci* 22 (5). https://doi.org/10.3390/ijms22052419. www.ncbi.nlm.nih.gov/pubmed/33670865.

Hahad, O., M. Beutel, T. Gori, A. Schulz, M. Blettner, N. Pfeiffer, T. Rostock, K. Lackner, M. Sorensen, J. H. Prochaska, P. S. Wild, and T. Munzel. 2018. "Annoyance to different noise sources is associated with atrial fibrillation in the Gutenberg Health Study." *Int J Cardiol* 264: 79–84. https://doi.org/10.1016/j.ijcard.2018.03.126. www.ncbi.nlm.nih.gov/pubmed/29695315.

Hahad, O., M. T. Bayo Jimenez, M. Kuntic, K. Frenis, S. Steven, A. Daiber, and T. Munzel. 2022. "Cerebral consequences of environmental noise exposure." *Environ Int* 165: 107306. https://doi.org/10.1016/j.envint.2022.107306. www.ncbi.nlm.nih.gov/pubmed/35635962.

Hahad, O., S. Kroller-Schon, A. Daiber, and T. Munzel. 2019. "The cardiovascular effects of noise." *Dtsch Arztebl Int* 116 (14): 245–250. https://doi.org/10.3238/arztebl.2019.0245. www.ncbi.nlm.nih.gov/pubmed/31092312.

Hao, G., L. Zuo, P. Xiong, L. Chen, X. Liang, and C. Jing. 2021. "Associations of PM2.5 and road traffic noise with mental health: Evidence from UK Biobank." *Environ Res*: 112221. https://doi.org/10.1016/j.envres.2021.112221. www.ncbi.nlm.nih.gov/pubmed/34656633.

He, S., A. Smargiassi, N. Low, M. Bilodeau-Bertrand, A. Ayoub, and N. Auger. 2019. "Residential noise exposure and the longitudinal risk of hospitalization for depression after pregnancy: Postpartum and beyond." *Environ Res* 170: 26–32. https://doi.org/10.1016/j.envres.2018.12.001. www.ncbi.nlm.nih.gov/pubmed/30557689.

Hegewald, J., M. Schubert, A. Freiberg, K. Romero Starke, F. Augustin, S. G. Riedel-Heller, H. Zeeb, and A. Seidler. 2020. "Traffic noise and mental health: A systematic review and meta-analysis." *Int J Environ Res Public Health* 17 (17). https://doi.org/10.3390/ijerph17176175. www.ncbi.nlm.nih.gov/pubmed/32854453.

Henry, J. P. 1992. "Biological basis of the stress response." *Integr Physiol Behav Sci* 27 (1): 66–83. https://doi.org/10.1007/BF02691093. www.ncbi.nlm.nih.gov/pubmed/1576090.

Heritier, H., D. Vienneau, M. Foraster, I. C. Eze, E. Schaffner, L. Thiesse, F. Rudzik, M. Habermacher, M. Kopfli, R. Pieren, et al., and S. N. C. study group. 2017. "Transportation noise exposure and cardiovascular mortality: A nationwide cohort study from Switzerland." *Eur J Epidemiol* 32 (4): 307–315. https://doi.org/10.1007/s10654-017-0234-2. www.ncbi.nlm.nih.gov/pubmed/28280950.

Kalsch, H., F. Hennig, S. Moebus, S. Mohlenkamp, N. Dragano, H. Jakobs, M. Memmesheimer, R. Erbel, K. H. Jockel, B. Hoffmann, and Group Heinz Nixdorf Recall Study Investigative. 2014. "Are air pollution and traffic noise independently associated with atherosclerosis: The Heinz Nixdorf Recall study." *Eur Heart J* 35 (13): 853–860. https://doi.org/10.1093/eurheartj/eht426. www.ncbi.nlm.nih.gov/pubmed/24194529.

Kroller-Schon, S., A. Daiber, S. Steven, M. Oelze, K. Frenis, S. Kalinovic, A. Heimann, F. P. Schmidt, A. Pinto, M. Kvandova, et al. 2018. "Crucial role for Nox2 and sleep deprivation in aircraft noise-induced vascular and cerebral oxidative stress, inflammation, and gene regulation." *Eur Heart J* 39 (38): 3528–3539. https://doi.org/10.1093/eurheartj/ehy333. www.ncbi.nlm.nih.gov/pubmed/29905797.

Landrigan, P. J., R. Fuller, N. J. R. Acosta, O. Adeyi, R. Arnold, N. N. Basu, A. B. Balde, R. Bertollini, S. Bose-O'Reilly, J. I. Boufford, et al. 2018. "The lancet commission on pollution and health." *Lancet* 391 (10119): 462–512. https://doi.org/10.1016/S0140-6736(17)32345-0. www.ncbi.nlm.nih.gov/pubmed/29056410.

Le, T. N., L. V. Straatman, J. Lea, and B. Westerberg. 2017. "Current insights in noise-induced hearing loss: A literature review of the underlying mechanism, pathophysiology, asymmetry, and management options." *J Otolaryngol Head Neck Surg* 46 (1): 41. https://doi.org/10.1186/s40463-017-0219-x. www.ncbi.nlm.nih.gov/pubmed/28535812.

Lechat, B., H. Scott, F. Decup, K. L. Hansen, G. Micic, C. Dunbar, T. Liebich, P. Catcheside, and B. Zajamsek. 2021. "Environmental noise-induced cardiovascular responses during sleep." *Sleep*. https://doi.org/10.1093/sleep/zsab302. www.ncbi.nlm.nih.gov/pubmed/34965303.

Linares, C., D. Culqui, R. Carmona, C. Ortiz, and J. Diaz. 2017. "Short-term association between environmental factors and hospital admissions due to dementia in Madrid." *Environ Res* 152: 214–220. https://doi.org/10.1016/j.envres.2016.10.020. www.ncbi.nlm.nih.gov/pubmed/27810678.

Mac Domhnaill, C., O. Douglas, S. Lyons, E. Murphy, and A. Nolan. 2021. "Road traffic noise and cognitive function in older adults: A cross-sectional investigation of the Irish Longitudinal Study on Ageing." *BMC Public Health* 21 (1): 1814. https://doi.org/10.1186/s12889-021-11853-y. www.ncbi.nlm.nih.gov/pubmed/34625043.

Medic, G., M. Wille, and M. E. Hemels. 2017. "Short- and long-term health consequences of sleep disruption." *Nat Sci Sleep* 9: 151–161. https://doi.org/10.2147/NSS.S134864. www.ncbi.nlm.nih.gov/pubmed/28579842.

Mottillo, S., K. B. Filion, J. Genest, L. Joseph, L. Pilote, P. Poirier, S. Rinfret, E. L. Schiffrin, and M. J. Eisenberg. 2010. "The metabolic syndrome and cardiovascular risk a systematic review and meta-analysis." *J Am Coll Cardiol* 56 (14): 1113–1132. https://doi.org/10.1016/j.jacc.2010.05.034. www.ncbi.nlm.nih.gov/pubmed/20863953.

Munzel, T., A. Daiber, S. Steven, L. P. Tran, E. Ullmann, S. Kossmann, F. P. Schmidt, M. Oelze, N. Xia, H. Li, A. Pinto, P. Wild, K. Pies, E. R. Schmidt, S. Rapp, and S. Kroller-Schon. 2017. "Effects of noise on vascular function, oxidative stress, and inflammation: Mechanistic insight from studies in mice." *Eur Heart J* 38 (37): 2838–2849. https://doi.org/10.1093/eurheartj/ehx081. www.ncbi.nlm.nih.gov/pubmed/28329261.

Munzel, T., M. Sorensen, and A. Daiber. 2021. "Transportation noise pollution and cardiovascular disease." *Nat Rev Cardiol* 18 (9): 619–636. https://doi.org/10.1038/s41569-021-00532-5. www.ncbi.nlm.nih.gov/pubmed/33790462.

Munzel, T., O. Hahad, M. Sorensen, J. Lelieveld, G. D. Duerr, M. Nieuwenhuijsen, and A. Daiber. 2022. "Environmental risk factors and cardiovascular diseases: A comprehensive expert review." *Cardiovasc Res* 118 (14): 2880–2902. https://doi.org/10.1093/cvr/cvab316. www.ncbi.nlm.nih.gov/pubmed/34609502.

Munzel, T., S. Kroller-Schon, M. Oelze, T. Gori, F. P. Schmidt, S. Steven, O. Hahad, M. Roosli, J. M. Wunderli, A. Daiber, and M. Sorensen. 2020. "Adverse cardiovascular effects of traffic noise with a focus on nighttime noise and the new WHO noise guidelines." *Annu Rev Public Health* 41: 309–328. https://doi.org/10.1146/annurev-publhealth-081519-062400. www.ncbi.nlm.nih.gov/pubmed/31922930.

Munzel, T., S. Steven, O. Hahad, and A. Daiber. 2020. "The sixth sense is involved in noise-induced stress responses and vascular inflammation: Evidence for heightened amygdalar activity in response to transport noise in man." *Eur Heart J* 41 (6): 783–785. https://doi.org/10.1093/eurheartj/ehz867. www.ncbi.nlm.nih.gov/pubmed/31859340.

Nelson, D. I., R. Y. Nelson, M. Concha-Barrientos, and M. Fingerhut. 2005. "The global burden of occupational noise-induced hearing loss." *Am J Ind Med* 48 (6): 446–458. https://doi.org/10.1002/ajim.20223. www.ncbi.nlm.nih.gov/pubmed/16299704.

Osborne, M. T., A. Radfar, M. Z. O. Hassan, S. Abohashem, B. Oberfeld, T. Patrich, B. Tung, Y. Wang, A. Ishai, J. A. Scott, L. M. Shin, Z. A. Fayad, K. C. Koenen, S. Rajagopalan, R. K. Pitman, and A. Tawakol. 2020. "A neurobiological mechanism linking transportation noise to cardiovascular disease in humans." *Eur Heart J* 41 (6): 772–782. https://doi.org/10.1093/eurheartj/ehz820. www.ncbi.nlm.nih.gov/pubmed/31769799.

Poulsen, A. H., O. Raaschou-Nielsen, A. Pena, A. N. Hahmann, R. B. Nordsborg, M. Ketzel, J. Brandt, and M. Sorensen. 2019. "Long-term exposure to wind turbine noise and risk for myocardial infarction and stroke: A nationwide cohort study." *Environ Health Perspect* 127 (3): 37004. https://doi.org/10.1289/EHP3340. www.ncbi.nlm.nih.gov/pubmed/30864814.

Roca-Barcelo, A., A. Nardocci, B. S. de Aguiar, A. G. Ribeiro, M. A. Failla, A. L. Hansell, M. R. Cardoso, and F. B. Piel. 2021. "Risk of cardiovascular mortality, stroke and coronary heart mortality associated with aircraft noise around Congonhas airport, Sao Paulo, Brazil: A small-area study." *Environ Health* 20 (1): 59. https://doi.org/10.1186/s12940-021-00746-7. www.ncbi.nlm.nih.gov/pubmed/33985498.

Sabayan, B., S. Jansen, A. M. Oleksik, M. J. van Osch, M. A. van Buchem, P. van Vliet, A. J. de Craen, and R. G. Westendorp. 2012. "Cerebrovascular hemodynamics in Alzheimer's disease and vascular dementia: A meta-analysis of transcranial Doppler studies." *Ageing Res Rev* 11 (2): 271–277. https://doi.org/10.1016/j.arr.2011.12.009. www.ncbi.nlm.nih.gov/pubmed/22226802.

Said, M. A., and O. A. El-Gohary. 2016. "Effect of noise stress on cardiovascular system in adult male albino rat: Implication of stress hormones, endothelial dysfunction and oxidative stress." *Gen Physiol Biophys* 35 (3): 371–377. https://doi.org/10.4149/gpb_2016003. www.ncbi.nlm.nih.gov/pubmed/27174896.

Schmidt, F., K. Kolle, K. Kreuder, B. Schnorbus, P. Wild, M. Hechtner, H. Binder, T. Gori, and T. Munzel. 2015. "Nighttime aircraft noise impairs endothelial function and increases blood pressure in patients with or at high risk for coronary artery disease." *Clin Res Cardiol* 104 (1): 23–30. https://doi.org/10.1007/s00392-014-0751-x. www.ncbi.nlm.nih.gov/pubmed/25145323.

Schubert, M., J. Hegewald, A. Freiberg, K. R. Starke, F. Augustin, S. G. Riedel-Heller, H. Zeeb, and A. Seidler. 2019. "Behavioral and emotional disorders and transportation noise among children and adolescents: A systematic review and meta-analysis." *Int J Environ Res Public Health* 16 (18). https://doi.org/10.3390/ijerph16183336. www.ncbi.nlm.nih.gov/pubmed/31510007.

Seidler, A., J. Hegewald, A. L. Seidler, M. Schubert, M. Wagner, P. Droge, E. Haufe, J. Schmitt, E. Swart, and H. Zeeb. 2017. "Association between aircraft, road and railway traffic noise and depression in a large case-control study based on secondary data." *Environ Res* 152: 263–271. https://doi.org/10.1016/j.envres.2016.10.017. www.ncbi.nlm.nih.gov/pubmed/27816007.

Seidler, A. L., J. Hegewald, M. Schubert, V. M. Weihofen, M. Wagner, P. Droge, E. Swart, H. Zeeb, and A. Seidler. 2018. "The effect of aircraft, road, and railway traffic noise on stroke—results of a case-control study based on secondary data." *Noise Health* 20 (95): 152–161. https://doi.org/10.4103/nah.NAH_7_18. www.ncbi.nlm.nih.gov/pubmed/30136675.

Sorensen, M., A. H. Poulsen, U. A. Hvidtfeldt, T. Munzel, J. D. Thacher, M. Ketzel, J. Brandt, J. H. Christensen, G. Levin, and O. Raaschou-Nielsen. 2021. "Transportation noise and risk of stroke: A nationwide prospective cohort study covering Denmark." *Int J Epidemiol* 50 (4): 1147–1156. https://doi.org/10.1093/ije/dyab024. www.ncbi.nlm.nih.gov/pubmed/33755127.

Spreng, M. 2000. "Central nervous system activation by noise." *Noise Health* 2 (7): 49–58. www.ncbi.nlm.nih.gov/pubmed/12689471.

Stansfeld, S., C. Clark, M. Smuk, J. Gallacher, and W. Babisch. 2021. "Road traffic noise, noise sensitivity, noise annoyance, psychological and physical health and mortality." *Environ Health* 20 (1): 32. https://doi.org/10.1186/s12940-021-00720-3. www.ncbi.nlm.nih.gov/pubmed/33766056.

Tangermann, L., D. Vienneau, J. Hattendorf, A. Saucy, N. Kunzli, B. Schaffer, J. M. Wunderli, and M. Roosli. 2022. "The association of road traffic noise with problem behaviour in adolescents: A cohort study." *Environ Res* 207: 112645. https://doi.org/10.1016/j.envres.2021.112645. www.ncbi.nlm.nih.gov/pubmed/34979122.

Tzivian, L., M. Dlugaj, A. Winkler, F. Hennig, K. Fuks, D. Sugiri, T. Schikowski, H. Jakobs, R. Erbel, K. H. Jockel, et al., and Group Heinz Nixdorf Recall Study Investigative. 2016. "Long-term air pollution and traffic noise exposures and cognitive function: A cross-sectional analysis of the Heinz Nixdorf Recall study." *J Toxicol Environ Health A* 79 (22–23): 1057–1069. https://doi.org/10.1080/15287394.2016.1219570. www.ncbi.nlm.nih.gov/pubmed/27924705.

Tzivian, L., M. Dlugaj, A. Winkler, G. Weinmayr, F. Hennig, K. B. Fuks, M. Vossoughi, T. Schikowski, C. Weimar, R. Erbel, et al. 2016. "Long-term air pollution and traffic noise exposures and mild cognitive Impairment in older adults: A cross-sectional analysis of the Heinz Nixdorf Recall study." *Environ Health Perspect* 124 (9): 1361–1368. https://doi.org/10.1289/ehp.1509824. www.ncbi.nlm.nih.gov/pubmed/26863687.

Tzivian, L., M. Jokisch, A. Winkler, C. Weimar, F. Hennig, D. Sugiri, V. J. Soppa, N. Dragano, R. Erbel, K. H. Jockel, et al. 2017. "Associations of long-term exposure to air pollution and road traffic noise with cognitive function-An analysis of effect measure modification." *Environ Int* 103: 30–38. https://doi.org/10.1016/j.envint.2017.03.018. www.ncbi.nlm.nih.gov/pubmed/28363096.

Vienneau, D., A. Saucy, B. Schaffer, B. Fluckiger, L. Tangermann, M. Stafoggia, J. M. Wunderli, M. Roosli, and S. N. C. study group. 2022. "Transportation noise exposure and cardiovascular mortality: 15-years of follow-up in a nationwide prospective cohort in Switzerland." *Environ Int* 158: 106974. https://doi.org/10.1016/j.envint.2021.106974. www.ncbi.nlm.nih.gov/pubmed/34775186.

Vivanco-Hidalgo, R. M., C. Avellaneda-Gomez, P. Dadvand, M. Cirach, A. Ois, A. Gomez Gonzalez, A. Rodriguez-Campello, P. de Ceballos, X. Basagana, A. Zabalza, et al. 2019. "Association of residential air pollution, noise, and greenspace with initial ischemic stroke severity." *Environ Res* 179 (Pt A): 108725. https://doi.org/10.1016/j.envres.2019.108725. www.ncbi.nlm.nih.gov/pubmed/31539843.

Weuve, J., J. D'Souza, T. Beck, D. A. Evans, J. D. Kaufman, K. B. Rajan, C. F. M. de Leon, and S. D. Adar. 2021. "Long-term community noise exposure in relation to dementia, cognition, and cognitive decline in older adults." *Alzheimers Dement* 17 (3): 525–533. https://doi.org/10.1002/alz.12191. www.ncbi.nlm.nih.gov/pubmed/33084241.

Wong, M. W., N. Braidy, A. Poljak, R. Pickford, M. Thambisetty, and P. S. Sachdev. 2017. "Dysregulation of lipids in Alzheimer's disease and their role as potential biomarkers." *Alzheimers Dement* 13 (7): 810–827. https://doi.org/10.1016/j.jalz.2017.01.008. www.ncbi.nlm.nih.gov/pubmed/28242299.

World Health Organization. "Preventing disease through healthy environments: A global assessment of the burden of disease from environmental risks." Available online: https://apps.who.int/iris/handle/10665/204585 (accessed on 29 April 2020).

Yu, Y., E. R. Mayeda, K. C. Paul, E. Lee, M. Jerrett, J. Su, J. Wu, I. F. Shih, M. Haan, and B. Ritz. 2020. "Traffic-related noise exposure and late-life dementia and cognitive impairment in Mexican-Americans." *Epidemiology* 31 (6): 771–778. https://doi.org/10.1097/EDE.0000000000001249. www.ncbi.nlm.nih.gov/pubmed/33555809.

Yu, Y., M. Haan, K. C. Paul, E. R. Mayeda, M. Jerrett, J. Wu, E. Lee, J. Su, I. F. Shih, K. Inoue, and B. R. Ritz. 2020. "Metabolic dysfunction modifies the influence of traffic-related air pollution and noise exposure on late-life dementia and cognitive impairment: A cohort study of older Mexican-Americans." *Environ Epidemiol* 4 (6): e122. https://doi.org/10.1097/EE9.0000000000000122. www.ncbi.nlm.nih.gov/pubmed/33778355.

Zhang, P. Y., X. Xu, and X. C. Li. 2014. "Cardiovascular diseases: Oxidative damage and antioxidant protection." *Eur Rev Med Pharmacol Sci* 18 (20): 3091–3096. www.ncbi.nlm.nih.gov/pubmed/25392110.

Zhao, Y. Y., H. Miao, X. L. Cheng, and F. Wei. 2015. "Lipidomics: Novel insight into the biochemical mechanism of lipid metabolism and dysregulation-associated disease." *Chem Biol Interact* 240: 220–238. https://doi.org/10.1016/j.cbi.2015.09.005. www.ncbi.nlm.nih.gov/pubmed/26358168.

Zijlema, W. L., Y. de Kluizenaar, I. van Kamp, and C. A. Hartman. 2021. "Associations between road traffic noise exposure at home and school and ADHD in school-aged children: The TRAILS study." *Eur Child Adolesc Psychiatry* 30 (1): 155–167. https://doi.org/10.1007/s00787-020-01521-8. www.ncbi.nlm.nih.gov/pubmed/32246274.

17 Imaging of Metabolic Activity Adaptations to UV Stress and Differentiation at Cellular Resolution in Skin and Skin Equivalents

Implications for Oxidative UV Damage

Florian Gruber and Christopher Kremslehner

17.1 INTRODUCTION

The human skin is a complex organ consisting of an outer epidermal and an inner dermal layer and appendages such as hair follicles or sebaceous glands. The human epidermis is a multi-layered epithelium that consists mainly of keratinocytes (KC); immune cells, most prominently the dendritic epidermal Langerhans cells and melanocytes, the pigment producing cells. The epidermis forms the outmost layer of the skin and is attached to the dermis via the basement membrane. The dermis harbors the skin appendages such as hair follicles and sebaceous glands as well as nerve endings and blood vessels. The dermal fibroblasts that sparsely populate the dermis produce the bulk of the extra cellular matrix that provides the cushioning and elastic properties of this tissue.

A proportion of the keratinocytes, the precursors or the transient amplifying cells, proliferate in the basal layer of the epidermis but start an irreversible transformation process as soon as they detach from the basement membrane to form a multilayered stratified epithelium through the process of differentiation. Once a keratinocyte has left the basal layer its fate is set to progress up and outwards until it becomes a corneocyte and in a final step dismantles its nucleus and nucleic acids at the transition into the stratum corneum. Through this permanent directional process of terminal differentiation the keratinocytes form a continuously renewing barrier to protect the organism from the environment (Eckhart and Zeeuwen, 2018). During this transformation keratinocytes are subjected to massive changes of their organelle composition as well as proteome and lipidome in order to produce the components required for the maintenance of the epidermal permeability barrier (Kezic and Jakasa, 2016; van Smeden and Bouwstra, 2016). Furthermore, the epidermis is protected by a variety of constitutive and adaptive redox regulatory networks that help to counteract the accumulation of reactive oxygen species, mostly superoxide anions and other mitochondrial reactive oxygen species that goes hand in hand with differentiation (Hamanaka et al., 2013; Hornig-Do et al., 2007). Thus, the differentiation state, alongside the cell type and ontology and age, are major factors that influence a skin cell's ability to protect from or adapt to environmental stressors (Schafer et al., 2010; Zhang et al., 2015).

Adaption of the metabolic state and shift in metabolic activity are among the fastest ways an organism can response to external stressors. It was recently shown that keratinocytes and fibroblast

DOI: 10.1201/9781003328100-21

can response to oxidative stress generated by ultraviolet radiation (UV) or hydrogen peroxide by increased glucose flux through the oxidative pentose phosphate pathway (PPP) (Kuehne et al., 2015). UV damage cells are "caught between a rock and a hard place". On the one hand they must cope with excessive reactive oxygen species, and on the other hand they may need to initiate DNA damage repair mechanisms. Rerouting of glucose into the PPP offers multiple benefits in turn of anti-oxidant and anti-genotoxic defense properties. Glucose 6-phosphate dehydrogenase (G6PD), the initial and rate-limiting enzyme of the oxidative PPP, can be rapidly activated. Further, the oxidative PPP yields NADPH that in turn is used to replenish the pool of the reducing equivalent glutathione (Winkler et al., 1986). Last, the end product of the oxidative PPP, Ribulose 5-phosphate, can further be used as the starting point for the production of nucleotide precursors and thus aids DNA damage repair efforts (Ramos-Martinez, 2017; Stanton, 2012). Kuehne et al. (2015) identified this shift in glucose flux as a redox response of skin cells in cultured fibroblasts and keratinocytes so the connection between changes in metabolic activity and stress response was yet to be elucidated with in the (epidermal) tissue context.

As in other tissues glycolysis and oxidative phosphorylation (OXPHOS) are the main energy (ATP) sources in the skin. In the presence of oxygen, the glycolytic end product pyruvate is directed to the citrate cycle and the respiratory chain, while in its absence pyruvate is fermented to ethanol or lactate. However, irrespective of the oxygen concentration, the epidermis produces large amounts of lactate up to the point that the epidermal metabolism is regarded to be functionally anaerobic (Ronquist et al., 2003). On a similar note, mice lacking functional electron transport chain in epidermal progenitor cells do not show impaired differentiation or barrier formation (Baris et al., 2011). As other highly proliferating tissues are known to rely on OXPHOS it is likely that at least keratinocytes within the basal layer show similar metabolic configurations (Chang et al., 2009). Further, electron microscopy studies of the epidermis have shown that only basal keratinocytes had mitochondria capable of OXPHOS (Ronquist et al., 2003). With superoxide produced during OXPHOS and the differentiation-linked accumulation of reactive oxygen species there is a high demand for NADPH generation in the epidermis to fuel the glutathione reductase and thus G6PD's role as a provider of this important antioxidant is well documented in the skin (Shindo et al., 1994).

Although the epidermis consists predominantly of keratinocytes it is far from a homogenous tissue, so insight into the redox-responses gathered from tissue lysates or cultured cells are a good starting point for a hypothesis, but to gain a more holistic picture it is necessary to investigate these reactions within the tissue context and at a single-cell resolution. For our study we used a formazan-based tissue zymographic assay that can localize peak enzyme activity at a cellular level within complex human tissues (Miller et al., 2017). This assay operates under saturated substrate conditions and utilizes tissue native G6PD in non-fixated cryo-sections. In order to allocate enzyme activity to individual cells and specific differentiation states within the epidermis we used a combination of bright-field and immunofluorescence slide scanning microscopy with advanced image analysis and segmentation. To assess the effect of UVB on keratinocytes within the tissue context we further generated three-dimensional organotypic epidermal skin equivalents (Mildner et al., 2010). These were either irradiated one or ten minutes before sample preparation to study immediate redox responses or were created containing a small fraction of pre-irradiated fluorescently labelled keratinocytes to investigate the implications of UVB-caused damage on cell fate during differentiation.

17.2 RESULTS AND DISCUSSION

17.2.1 Epidermal Strata Prediction by Advanced Image Analysis Based on Composite Bright-Field and Immunofluorescence Images

In this study we aimed to build a tool that enables us to correlate tissue enzymatic staining with histological detection of specific cell types and cell-state markers at a high spatial resolution and high throughput. To achieve this, we created a specific analysis profile for the StrataQuest (SQ, Version

5.0.1.336 and above, TissueGnostics GmbH, Vienna, Austria) image and context analysis software. In short, cryo-sections of human tissue or skin equivalents were subjected to the enzymatic activity assay, immunofluorescence detection of keratinocyte differentiation markers (Keratin 10/K10, Keratin 14/K14) and nuclear counterstain with Höchst reagent. Slides were digitized on a TissueFAXS I Plus slide-scanning microscope at 20x magnification. Data sets from bright-field (enzyme assay) and reflected light (K10, K14, nuclei) were imported into SQ stitched to generate a virtual microscopic slide image. Semi-automated detection of the epidermis was achieved by combining a nuclei density map with an inverted and threshold-adjusted bright-field image taking advantage of both the epidermis' higher nuclei count and its different optical properties compared to the dermis. Further we wanted to divide the recognized epidermis into bins of cells that are in similar differentiation states, namely the basal, low suprabasal, high suprabasal strata and stratum corneum (SC) (Figure 17.1A). The discrimination between the basal strata and the stratum corneum was accomplished by measuring the nuclei density inside the epidermal contour. The SC area was grown from the nuclei sparse border to the edge of the nuclei density map, while the basal strata was set to contain the first row of nuclei from the opposite border based on distance. The same distance-based growth step was repeated from the edge of the basal to obtain the low suprabasal strata and the remainder of the nucleated area was attributed to the high suprabasal strata respectively. Within these defined areas the masks for cellular measurements were generated around all detected nuclei (Ecker and Steiner, 2004). Next, we evaluated the correct strata allocation by measuring the staining intensities of differentiation markers within cells belonging to the basal and low suprabasal strata. In undifferentiated keratinocytes, as present in the basal layer, Keratins 5 and 14 (Figure 17.1B and Figure 17.1D) are expressed while differentiating keratinocytes mainly

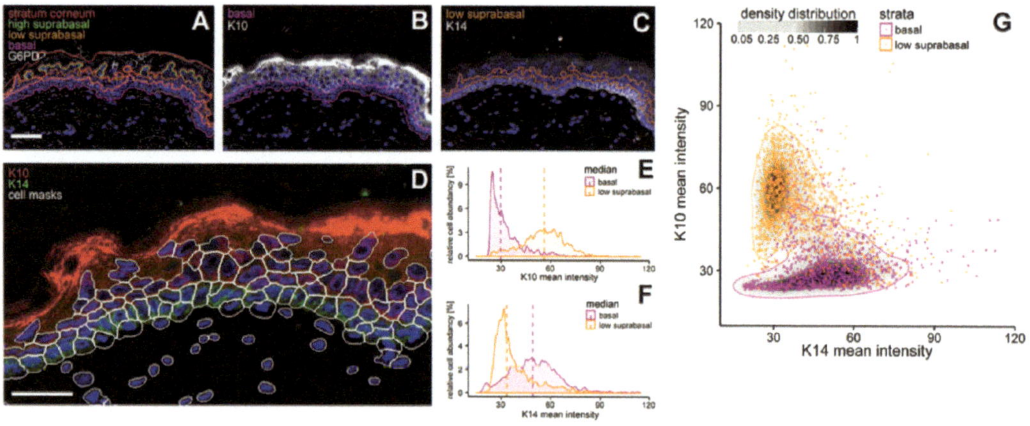

FIGURE 17.1 Automated epidermis recognition and counter-staining based validation of strata prediction. Cryo-sections of human skin biopsies were stained by immunofluorescence for the keratinocyte differentiation markers keratin 10 (K10, only expressed in suprabasal, differentiating KC), keratin 14 (K14, highly expressed in basal KC) and nuclei were stained with Hoechst dye. Whole tissue sections were scanned in fluorescent and brightfield modalities with a TissueFAXS slide scanning microscope at 20-fold magnification and subsequently processed with StrataQuest image analysis software. (A) Overlay of the color-coded predicted strata, G6PD activity staining (white) and nuclei counter stain (blue). (B–D) IF staining for keratin 10 (K10 stained in white in B, red in D) and for keratin 14 (K14 stained white in C, green in D). Histogram presentations of abundance vs. staining intensity of the predicted cells positive for K10 (E) and K14 (F) in the respective predicted strata. The median of the individual populations is indicated by dashed vertical lines. (G) Scatter plot (K10/K14) of predicted cells in basal (pink) and low suprabasal (orange) strata. The density of the different point clouds is indicated by the greyscale color bar and the outlines delimit the 95% density distributions of the respective strata. Size bars: 50 μm. Reused from Kremslehner et al. (2020) with permission.

express Keratins 1 and 10 (Figure 17.1C and Figure 17.1D) among other differentiation markers as they progress from the basal via the spinous and granular layer towards the stratum corneum. Staining intensities of both keratins within cells from the basal and low suprabasal layer showed distinct distributions in the individual histograms (Figure 17.1E and Figure 17.1F) as well as two clearly separated populations in a K10/K14 scatter plot (Figure 17.1G). Expectedly for human skin samples 5 to 7% of the predicted cell masks that were allocated to the basal stratum lacked K14 expression, since on average 8% of the cells present in the basal layer are nucleated non-epithelial cells (Jorizzo et al., 2008). With the technical reliability of our analysis profile confirmed by immunofluorescence detection of basal and suprabasal specific keratinocyte markers, it will be a great asset in further studies within the epidermal microenvironment since it allows for single cell measurements and strata allocation without the need of additional markers besides bright-field and nuclei images.

17.2.2 Spatial Resolution of G6PD Enzyme Activity in the Skin and Skin Equivalents Based on Tissue Zymography and Histocytometry

In this study we adapted a tetrazolium salt-based enzymatic activity assay to assess the distribution of G6PD enzyme activity in cryo-sections of skin and skin equivalent biopsies. Miller et al. (2017) described this assay to work in a variety of complex tissue microenvironments at a wide linear range and high spatial resolution. We measured the enzymatic activity of G6PD, the gatekeeper of glucose flux through the oxidative pentose phosphate pathway. The G6PD reaction yields one NADPH per glucose molecule processed, and products of the PPP can be used for nucleotide biosynthesis. In brief the assay utilizes tissue native enzymes in non-fixated cryo-sections and measures peak enzyme activity at physiological conditions and substrate saturation (Frederiks et al., 2003). In the presence of the electron carrier 1-methoxy phenazinium methylsulfate (PMS) the colorless nitro blue tetrazolium (NBT) is reduced to form a colored, non-soluble formazan precipitate by accepting the electron from NADPH/H^+, the co-factor product of G6PD (Figure 17.2A). Reaction linearity was confirmed in time response curves in cryo-sections from three different donors over a time period of 25 minutes, and background signal was evaluated by addition of the non-competitive G6PD inhibitor dehydroepiandrosterone (DHEA, 1 mM) (Figure 17.1B and 17.1C). Based on the formazan staining intensity, the activity distribution of G6PD in the skin was lowest in the basal layer and reached its maximum in the granular layer, although inhibitable staining intensity was measurable throughout the nucleated epidermis as described before (Pearse et al., 1987). In the next step we applied the assay on samples from artificial reconstituted skin equivalents. In this *in vitro* skin model primary human keratinocytes are seeded on a collagen support matrix containing human dermal fibroblast and differentiate to form a fully stratified epidermis within 7 days of air-liquid interface culture (Mildner et al., 2010; Rendl et al., 2002). It was shown that the skin equivalent highly resembles the epidermis in terms of gene expression, lipid composition, processing of keratinocyte barrier proteins and barrier function (Gschwandtner et al., 2013; Mildner et al., 2010). Further adding to these similarities between the model and native tissue we found comparable G6PD activity distribution in the skin equivalent models (Figure 17.1D). Remarkably the localization of G6PD protein staining as detectable by antibody did neither match the distribution of the enzyme activity in the skin nor in the skin equivalents suggesting that the activity is strongly regulated by differentiation-related factors (Figure 17.1E), probably posttranslational modifications (Frederiks et al., 2003; Stanton, 2012). Further it was shown that differentiation in epidermal keratinocytes leads to the formation of an UVB-protective glutathione barrier (Schafer et al., 2010) that offers added protection to cells in the differentiated layers while it allows for clearance of cells in the basal layer if they received irreparable UV damage. In order to establish this glutathione barrier G6PD activity is required for providing NADPH as a co-factor for GSH reductase to maintain the pool of reduced GSH.

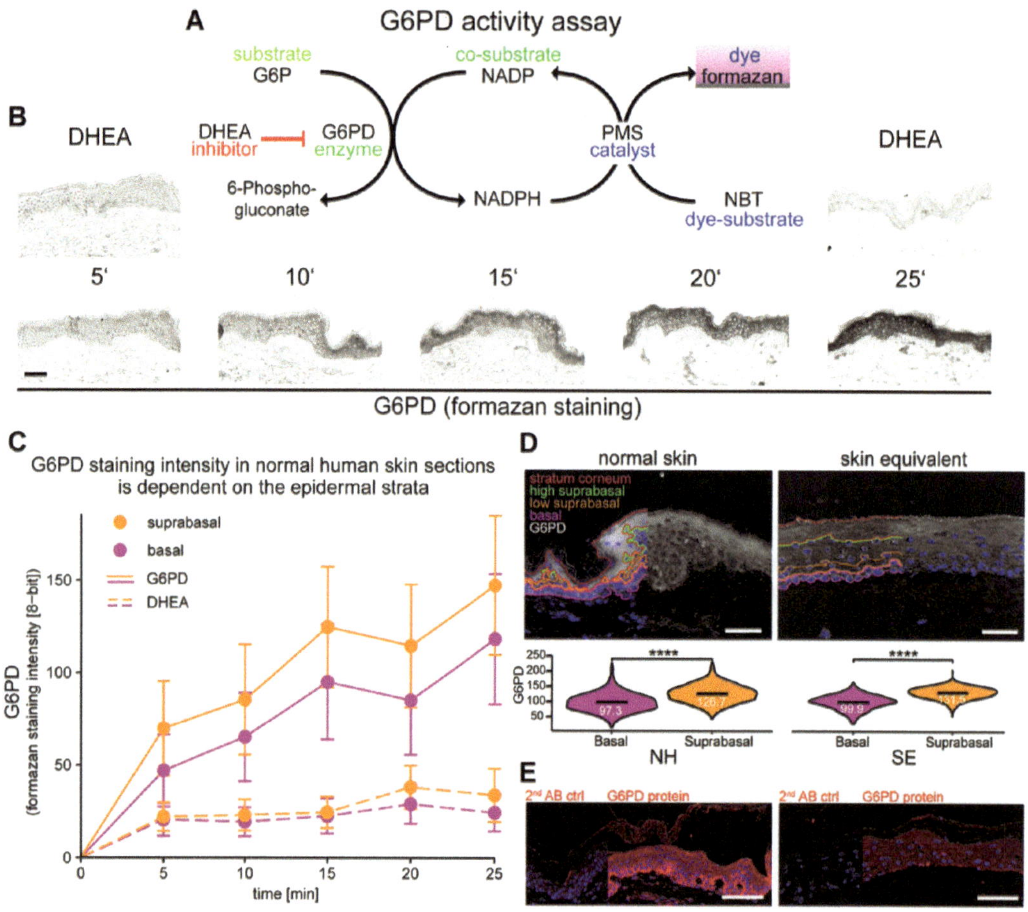

FIGURE 17.2 Formazan-based enzymatic activity assay of G6PD in skin and equivalence of activity distribution between reconstructed epidermal equivalents and normal human skin. Non-fixed cryo-sections of normal human skin were subjected to a formazan-based enzymatic activity assay for glucose 6-phoshate dehydrogenase (G6PD). (A) Schematic mode of action of the enzymatic activity assay. The electron released during the conversion of G6P by the tissue native enzyme G6PD is shuttled via the electron carrier phenazinium methylsulfate (PMS) to nitroblue tetrazolium chloride (NBT) that in turn forms a non-soluble formazan salt. The reaction was stopped by washing at the indicated time points. The non-competitive selective inhibitor of G6PD dehydroepiandrosterone (DHEA, 1 mM) was used as a negative control. (B) Representative brightfield images of cryo-sections subjected to the G6PD activity assay at five indicated time-points. Cryo-sections exposed for the indicated timepoints to which DHEA had been added are shown as negative controls for the start (5′) and end (25′) time points (upper panels). (C) Quantification of the increase in G6PD enzymatic activity staining in basal and suprabasal strata of three consecutive sections from normal human skin biopsies from three donors over time. The dashed line indicates the G6PD activity measured in the sections where the assay reaction was inhibited by addition of DHEA. (D) Top panel: Comparison of G6PD enzymatic activity staining and strata prediction in normal human skin (left) and reconstructed epidermal equivalents (right). Lower panel: Violin plots of cellular G6PD activity distribution in the basal and the suprabasal strata in normal human skin and SE. (E) IF staining for G6PD protein (bottom). Staining control performed with secondary antibody only (narrow sections). For D the mean values are indicated by black bars with subscribed numeric values. Asterisks indicate statistical differences (****-$p < 0.0001$; Student's t-test calculated with R software). Size bars: 50 μm. Reused from Kremslehner et al., (2020) with permission.

17.2.3 G6PD Activity Spikes Immediately after UVB Irradiation in Skin Equivalent Models

Next, we investigated the effect of a single high dose of UVB irradiation on skin equivalent models. Therefore fully stratified skin equivalents were bisected, and one half was used as control while the other half was irradiated with 150 mJ/cm^2 of UVB and cryopreserved either 1 or 10 minutes after the end of irradiation. Single cell measurements of G6PD activity staining in the epidermis revealed an immediate increase of enzyme activity as early as 1 minute after irradiation (Figure 17.3A). The highest relative increase in G6PD activity was measurable in low suprabasal cells already 1 minute after irradiation (Figure 17.3B, middle panel, orange violins). This highly significant rise in activity was

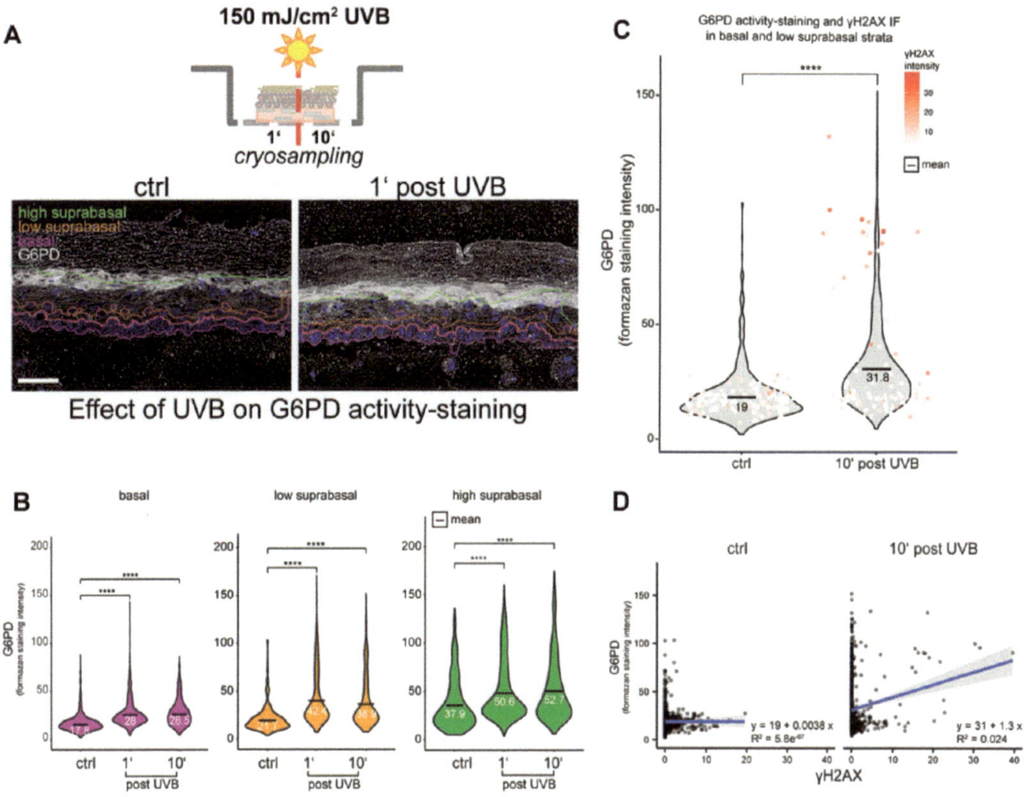

FIGURE 17.3 UVB exposure of human epidermal skin equivalent models lead to immediate increase of G6PD activity staining. Fully stratified human epidermal skin equivalents were irradiated with UVB (150 mJ/cm^2) and cryo-samples were collected 1 and 10 minutes after end of irradiation (A, top scheme). Shown are representative graphs from three independent experiments. (A) Lower panels: Inverted brightfield images of G6PD activity staining in cryo-sections from control skin equivalents and 1 minute post irradiation. Basal and low suprabasal strata indicated by pink and orange contour respectively. (B) Violin plots of G6PD activity staining at the indicated time points within the indicated strata. (C) Violin plots of G6PD activity staining in the combined basal and low suprabasal strata of control samples and samples taken 10 minutes post irradiation to which values of γH2AX positive cells are superimposed, the color gradient indicating relative IF staining intensity. (D) Scatter plots of G6PD activity staining in the combined basal and low suprabasal strata of control samples and samples taken 10 minutes post irradiation versus γH2AX-staining intensity with regression lines (gray: 95% CI). For B and C the mean values are indicated by black bars with subscripted numeric values. Asterisks indicate statistical differences (****- $p < 0.0001$; Student's t-test calculated with R software). Size bars: 50μm. Reused from Kremslehner et al. (2020) with permission.

detectable throughout all the strata of the epidermis independent of the baseline activity in the controls (Figure 17.3B). The significant increase of staining in high suprabasal cells suggest that G6PD activity was increased by a UVB-related mechanism that can further increase enzyme activity on top of the differentiation-related high base activity already found in these cells (Pearse et al., 1987).

Solar radiation is the most prevalent extrinsic damage source to the skin and can directly cause damage to DNA, proteins and lipids or generate photoproducts via the formation of ROS (Bickers and Athar, 2006; Davies, 2003; Schuch et al., 2017). UV light can be the cause of multiple sorts of direct and indirect DNA damage depending on the respective wavelength. Long wave UVA is more likely single strand breaks indirectly by ROS or cause damage via adduct-formation. These issues are resolved by base excision repair (BER), which also leads to elevated glucose consumption and defective BER increases glycolysis via accumulation of ROS and thereby activation of the transcription factor Hif1 alpha (Kamenisch et al., 2016). Short wave UVB, on the other side, generates DNA damage directly in the form of double strand breaks or pyrimidine dimers that are removed via nucleotide excision repair (NER) (Vrouwe et al., 2011). UVB-caused DNA damage is marked for NER by rapid phosphorylation of histone H2AX (γH2AX), and DNA damage response (DDR) can be improved by activation of the PPP or supplementation of its products NADPH and ribose (Marti et al., 2006). Histocytometric evaluation of immunodetected γH2AX in samples harvested 10 minutes after UVB irradiation showed an increase in positive nuclei compared to non-irradiated controls. Predicted cell masks in the basal and low suprabasal strata that contained nuclei with positive γH2AX stain grouped among the cells with higher G6PD activity (Figure 17.3C). Further regression analysis of scatter plots for G6PD activity and γH2AX staining revealed a positive correlation of increased G6PD activity with stronger gH2AX staining (Figure 17.3D). Although this finding matches the idea of DNA damage-induced activation of the PPP, the early timeframe in which we noted the changes suggest activation via mechanism independent of the transcriptionally activation of the PPP through the canonical DDR (Franklin et al., 2016). It was shown that upon UVB-irradiation ATM kinase and Hsp27 are induced and can activate G6PD (Cosentino et al., 2011) making them one of the most likely mechanistic mediators. Although additional studies will be needed to identify the mechanism that link metabolic adaptations to oxidative stress and specific DDR-pathways, here we were the first to show a UVB-inducible immediate increase in G6PD activity within the tissue microenvironment of the skin and were able to show a correlation of this increase with initiation of NER.

17.2.4 UVB Pre-Irradiation of Keratinocytes Affect Their Tissue Localization in Organotypic Skin Models

In terms of UV protection not all keratinocytes are created equal as location within the epidermis has a substantial impact on the present antioxidants. The glutathione barrier that is established in differentiated cells is one of these mechanisms that favors clearance of damaged basal cells of the obtained irreversible damage while it increases survival chances in higher differentiated layers (Schafer et al., 2010). In order to investigate the impact that UVB irradiation has on basal cells and their fate within the tissue context, we adapted our skin equivalent model and spiked them with a fraction of pre-irradiated and fluorescently labelled keratinocytes. Therefor 20% of the primary keratinocytes were either irradiated with 20 mJ/cm^2 UVB or sham-irradiated (cell tracker–ctrl) as internal controls, subsequently labelled with a fluorescent cell tracker, expressed as skin equivalents among non-irradiated, non-labelled keratinocytes from the same donor. After 7 days in air-liquid interface culture skin equivalents were cryo-sampled, and G6PD zymography was performed on cryosections. Of initial interest was the comparison of spatial localization differences between samples containing UVB and sham-irradiated cells. While sham-irradiated cells contributed to all the living layers of the epidermis as well as the stratum corneum equally, cells that were pre-irradiated with UVB were all but absent from the basal and low suprabasal strata as detected by tissue cytometry (Figure 17.4A and 17.4B). When assessing the G6PD activity in labelled and non-labelled cells without discrimination of strata allocation we found no differences in skin equivalents containing sham-irradiated cells (Figure 17.4C, left panel). Interestingly UVB-pretreated cells showed a significant reduction of average G6PD activity compared to non-treated neighboring cells (Figure 17.4C, right panel). Upon further inspection of strata specific differences, we found that the few irradiated cells that were found in the basal layer had no altered G6PD activity compared to non-irradiated cells from the same strata (Figure 17.4D, left pair of violins). Irradiated cells found in the low and high suprabasal strata, on the other hand, showed significantly reduced levels of G6PD activity when compared to their non-irradiated surrounding cells

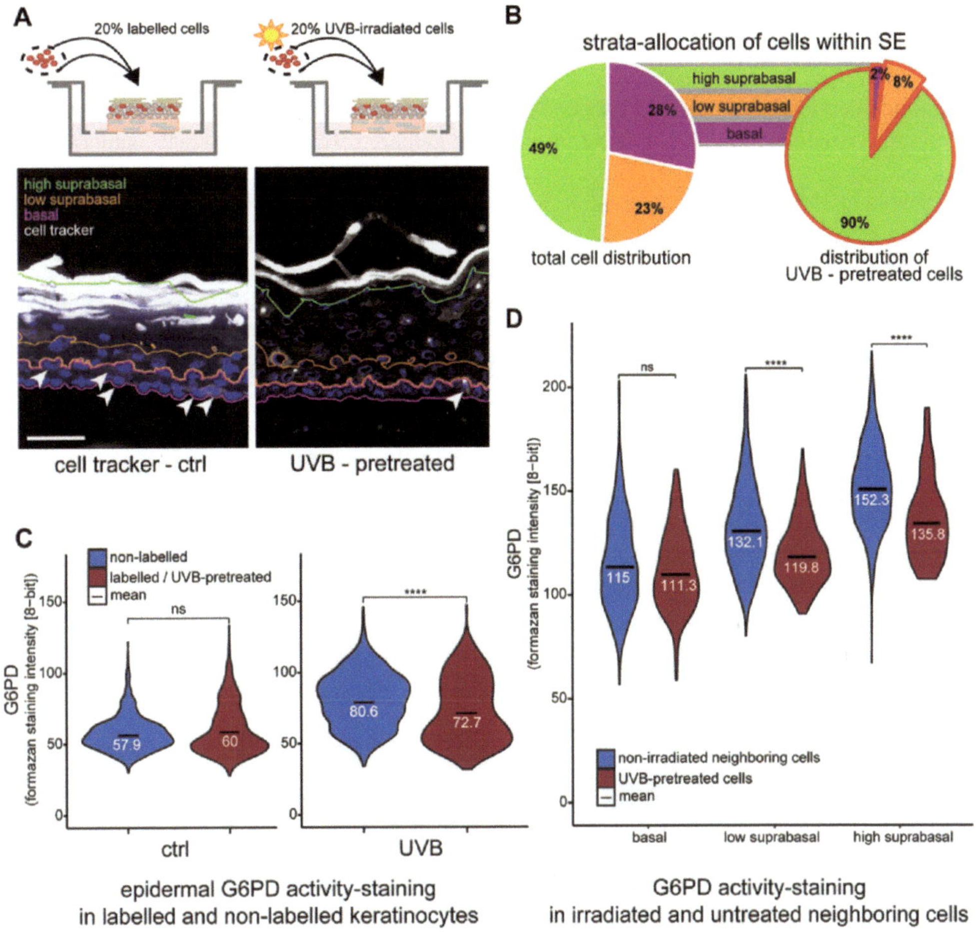

FIGURE 17.4 UVB pre-treatment affects the location of keratinocytes within the skin equivalent model. Skin equivalent models were generated with 20% labelled or UVB pre-treated labelled cells and their location within the different strata of the fully differentiated skin equivalent model as well as their G6PD activity were assessed. (A) IF image of cell tracker (white) labelled keratinocytes within the skin equivalent model, basal, low and high suprabasal strata indicated by pink, orange and green contours respectively. Labelled basal cells are indicated by arrow heads. (B) Pie chart depicting the distribution of all cells to the epidermal strata and the distribution of UVB pre-treated, labelled cells to the strata (red contour pie chart). (C) Violin plots of G6PD activity staining. Left: activity plot of non-labelled cells (blue) in comparison to labelled untreated cells (red) within the same skin equivalent. Right: activity plot of non-labelled cells (blue) and labelled UVB pre-treated cells within the same skin equivalent. (D) Violin plots of G6PD activity staining of UVB pre-treated labelled cells (red) and their non-irradiated neighboring cells (blue) within the different strata of the skin equivalent model. For C and D the mean values are indicated by black bars with subscripted numeric values. Asterisks indicate statistical differences (****$p < 0.0001$; ns-no significant difference; Student's t-test calculated with R). Reused from Kremslehner et al. (2020) with permission.

(Figure 17.4D, center and right pairs of violins). In contrast to the upregulation of G6PD activity that we observed upon UVB irradiation within the tissue context, primary cells that managed to survive irradiation showed reduced enzymatic activity after a week of organotypic culture. The reduction of G6PD activity could be a byproduct of successful DDR or an early form of (photo)aging or senescence (Gruber et al., 2020) comparable to reduced G6PD activity that was observed in keratinocytes isolated from geriatric skin (Kuehne et al., 2017). These cells did not cause any morphological disturbance in the epidermal structure, but their presence was virtually limited to strata of increased differentiation. It still needs further studies to determine whether this altered strata-allocation is due to changed physical properties of the cells or a mechanism to exclude damaged cells by (premature) differentiation.

17.3 CONCLUSION

Metabolic adaptations of cells as reaction to external insults is a novel and highly interesting field of research (Kuehne et al., 2015). Most insights are gathered from cultured cells, supernatants and tissue homogenates measuring cellular oxygen consumption, acidification or key enzyme activities (Schutz et al., 2016) or are provided by proteomic and metabolomic data from tissue extracts (Hosseini et al., 2018). Interesting insights on the dynamics of metabolic adaptations to environmental stress (Bisphenol A exposure) were recently gained by introducing time course metabolomics, and also this stress strongly shifted metabolism toward nucleotide precursor synthesis and oxidative stress responses in a chronologically concerted response (Zhao et al., 2023). Not only chemical stress can lead to alterations in carbohydrate metabolism in the skin, as was demonstrated by a recent study on chronic restraint stress that led to hair loss. There the normal amplitude of glycolytic activity that normally accompanies the hair cycle was lost, as assayed from skin extract metabolomics (Wang et al., 2022). Most studies were lacking *in situ* data and were not able to describe metabolic activity within the tissue microenvironment or at the cellular level. Our study approach is centered around the combination of a highly reliable zymographic assay with fluorescence microscopy and advanced image analysis to allow for detection of metabolic activity at single cell resolution. We investigated changes of G6PD activity in three-dimensional full-thickness organotypic epidermal equivalent models and were, to our knowledge, the first to show a UVB-induced spike of G6PD activity within the tissue context (Figure 17.5). Furthermore we found a

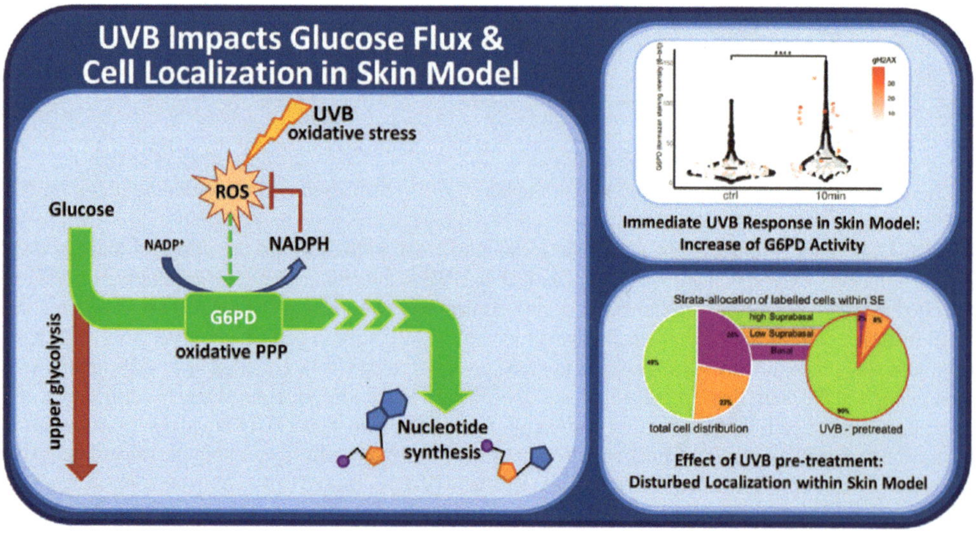

FIGURE 17.5 Summarizing scheme on UVB-mediated effects on G6PD activity, γH2AX expression and long-term damage of the epidermis. Reused from Kremslehner et al. (2020) with permission.

correlation of increased G6PD activity with the presence of γH2AX, an indicator for initiation of DNA damage repair. Subsequently we fine-tuned our skin equivalents to study the long-term ramifications of UVB damage to cells in the basal layer of the epidermis by incorporating a fraction of pre-irradiated fluorescently labelled keratinocytes in the skin equivalents. These UVB-treated cells showed aberrant tissue localization and were all but absent from the basal layer of the model system. Cells that were still present in the differentiated layers showed reduced G6PD activity compared to their non-treated neighboring cells.

ACKNOWLEDGEMENTS

The financial support of the Federal Ministry for Digital and Economic Affairs (BMWFW) of Austria and the National Foundation for Research, Technology, and Development of Austria and of CHANEL Parfums et Beauté to the Christian Doppler Laboratory for Skin Multimodal Imaging of Aging and Senescence (SKINMAGINE) is gratefully acknowledged. This funding body does not issue grant numbers.

COPYRIGHT INFORMATION

REFERENCES

Baris OR, Klose A, Kloepper JE, Weiland D, Neuhaus JF, Schauen M, et al. The mitochondrial electron transport chain is dispensable for proliferation and differentiation of epidermal progenitor cells. Stem Cells 2011;29(9):1459–68.

Bickers DR, Athar M. Oxidative stress in the pathogenesis of skin disease. J Invest Dermatol 2006;126(12):2565–75.

Chang SE, Kim PY, Lee HW, Choi JH, Jeong EJ, Choi SH, et al. Reduced mitochondrial properties in putative progenitor/stem cells of human keratinocytes. Ann Dermatol 2009;21(4):364–8.

Cosentino C, Grieco D, Costanzo V. ATM activates the pentose phosphate pathway promoting anti-oxidant defence and DNA repair. EMBO J 2011;30(3):546–55.

Davies MJ. Singlet oxygen-mediated damage to proteins and its consequences. Biochem Biophys Res Commun 2003;305(3):761–70.

Ecker RC, Steiner GE. Microscopy-based multicolor tissue cytometry at the single-cell level. Cytometry A 2004;59(2):182–90.

Eckhart L, Zeeuwen PLJM. The skin barrier: Epidermis vs environment. Exp Dermatol 2018;27(8):805–6.

Franklin DA, He Y, Leslie PL, Tikunov AP, Fenger N, Macdonald JM, et al. p53 coordinates DNA repair with nucleotide synthesis by suppressing PFKFB3 expression and promoting the pentose phosphate pathway. Sci Rep 2016;6:38067.

Frederiks WM, Bosch KS, De Jong JS, Van Noorden CJ. Post-translational regulation of glucose-6-phosphate dehydrogenase activity in (pre)neoplastic lesions in rat liver. J Histochem Cytochem 2003;51(1):105–12.

Gruber F, Kremslehner C, Eckhart L, Tschachler E. Cell aging and cellular senescence in skin aging—Recent advances in fibroblast and keratinocyte biology. Exp Gerontol 2020;130:110780.

Gschwandtner M, Mildner M, Mlitz V, Gruber F, Eckhart L, Werfel T, et al. Histamine suppresses epidermal keratinocyte differentiation and impairs skin barrier function in a human skin model. Allergy 2013;68(1):37–47.

Hamanaka RB, Glasauer A, Hoover P, Yang S, Blatt H, Mullen AR, et al. Mitochondrial reactive oxygen species promote epidermal differentiation and hair follicle development. Sci Signal 2013;6(261):ra8.

Hornig-Do HT, von Kleist-Retzow JC, Lanz K, Wickenhauser C, Kudin AP, Kunz WS, et al. Human epidermal keratinocytes accumulate superoxide due to low activity of Mn-SOD, leading to mitochondrial functional impairment. J Invest Dermatol 2007;127(5):1084–93.

Hosseini M, Dousset L, Mahfouf W, Serrano-Sanchez M, Redonnet-Vernhet I, Mesli S, et al. Energy metabolism rewiring precedes UVB-induced primary skin tumor formation. Cell Rep 2018;23(12):3621–34.
Jorizzo JL, Bolognia J, Rapini RP. Dermatology 2008.
Kamenisch Y, Baban TSA, Schuller W, von Thaler AK, Sinnberg T, Metzler G, et al. UVA-irradiation induces Melanoma Invasion via the enhanced warburg effect. J Invest Dermatol 2016;136(9):1866–75.
Kezic S, Jakasa I. Filaggrin and skin barrier function. Curr Probl Dermatol 2016;49:1–7.
Kremslehner C, Miller A, Nica R, Nagelreiter IM, Narzt MS, Golabi B, et al. Imaging of metabolic activity adaptations to UV stress, drugs and differentiation at cellular resolution in skin and skin equivalents—Implications for oxidative UV damage. Redox Biol 2020:101583.
Kuehne A, Emmert H, Soehle J, Winnefeld M, Fischer F, Wenck H, et al. Acute activation of oxidative pentose phosphate pathway as first-line response to oxidative stress in human skin cells. Mol Cell 2015;59(3):359–71.
Kuehne A, Hildebrand J, Soehle J, Wenck H, Terstegen L, Gallinat S, et al. An integrative metabolomics and transcriptomics study to identify metabolic alterations in aged skin of humans in vivo. BMC Genomics 2017;18(1):169.
Marti TM, Hefner E, Feeney L, Natale V, Cleaver JE. H2AX phosphorylation within the G1 phase after UV irradiation depends on nucleotide excision repair and not DNA double-strand breaks. Proc Natl Acad Sci U S A 2006;103(26):9891–6.
Mildner M, Jin J, Eckhart L, Kezic S, Gruber F, Barresi C, et al. Knockdown of filaggrin impairs diffusion barrier function and increases UV sensitivity in a human skin model. J Invest Dermatol 2010;130(9):2286–94.
Miller A, Nagy C, Knapp B, Laengle J, Ponweiser E, Groeger M, et al. Exploring metabolic configurations of single cells within complex tissue microenvironments. Cell Metab 2017;26(5):788–800.
Pearse AD, Gaskell SA, Marks R. Epidermal changes in human skin following irradiation with either UVB or UVA. J Invest Dermatol 1987;88(1):83–7.
Ramos-Martinez JI. The regulation of the pentose phosphate pathway: Remember krebs. Arch Biochem Biophys 2017;614:50–2.
Rendl M, Ban J, Mrass P, Mayer C, Lengauer B, Eckhart L, et al. Caspase-14 expression by epidermal keratinocytes is regulated by retinoids in a differentiation-associated manner. J Invest Dermatol 2002;119(5):1150–5.
Ronquist G, Andersson A, Bendsoe N, Falck B. Human epidermal energy metabolism is functionally anaerobic. Exp Dermatol 2003;12(5):572–9.
Schafer M, Dutsch S, auf dem KU, Navid F, Schwarz A, Johnson DA, et al. NRF2 establishes a glutathione-mediated gradient of UVB cytoprotection in the epidermis. Genes Dev 2010;24(10):1045–58.
Schuch AP, Moreno NC, Schuch NJ, Menck CFM, Garcia CCM. Sunlight damage to cellular DNA: Focus on oxidatively generated lesions. Free Radic Biol Med 2017;107:110–24.
Schutz R, Kuratli K, Richard N, Stoll C, Schwager J. Mitochondrial and glycolytic activity of UV-irradiated human keratinocytes and its stimulation by a saccharomyces cerevisiae autolysate. J Photochem Photobiol B 2016;159:142–8.
Shindo Y, Witt E, Han D, Epstein W, Packer L. Enzymic and non-enzymic antioxidants in epidermis and dermis of human skin. J Invest Dermatol 1994;102(1):122–4.
Stanton RC. Glucose-6-phosphate dehydrogenase, NADPH, and cell survival. IUBMB Life 2012;64(5):362–9.
Van SJ, Bouwstra JA. Stratum corneum lipids: Their role for the skin barrier function in healthy subjects and atopic dermatitis patients. Curr Probl Dermatol 2016;49:8–26.
Vrouwe MG, Pines A, Overmeer RM, Hanada K, Mullenders LH. UV-induced photolesions elicit ATR-kinase-dependent signaling in non-cycling cells through nucleotide excision repair-dependent and -independent pathways. J Cell Sci 2011;124(Pt 3):435–46.
Wang X, Cai C, Liang Q, Xia M, Lai L, Wu X, et al. Integrated transcriptomics and metabolomics analyses of stress-induced murine hair follicle growth inhibition. Front Mol Biosci 2022;9:781619.
Winkler BS, DeSantis N, Solomon F. Multiple NADPH-producing pathways control glutathione (GSH) content in retina. Exp Eye Res 1986;43(5):829–47.
Zhang H, Davies KJA, Forman HJ. Oxidative stress response and NRF2 signaling in aging. Free Radic Biol Med 2015;88(Pt B):314–36.
Zhao H. LM, Yang J., Chen Y., Fang M. Time-dependent metabolomics uncover dynamic metabolic adaptions in MCF-7 cells exposed to bisphenol A. Frontiers of Environmental Science and Engineering 2023;17(1).

18 Adverse Health Effects of UVR on Skin and Other Organs

Giuseppe Valacchi and Francesca Ferrara

18.1 INTRODUCTION

Ultraviolet radiations (UVR), also referred as UV light (UV), represent the electromagnetic radiations of the electromagnetic spectra (ES) with a wavelength ranging from 180 nm to 400 nm. These radiations predominantly derive from the sun, but they can also be emitted by man-made sources as UV lamps. UVR are divided in three groups based on their wavelength: UVA (320–400 nm) also known as near UV or black-light and ulteriorly classified in two subgroups: UVA2 (320–340 nm) and UVA1 (340–400 nm) (D'Orazio et al. 2013); UVB (290–320 nm); and UVC (180–290 nm). Despite visible light, UVR are invisible to humans as they are absorbed by the cornea (360 nm) and the internal lens (below 400 nm). As depicted in Figure 18.1, the solar spectrum reaching the earth's surface ranges from 290 to 3000 nm and besides UVR, include visible light (400–760 nm) and IR radiation (760–3000 nm). In particular, the corresponding solar energy of the total solar irradiance reaching the earth is 6.8% in the UV range (0.5% UVB, 6.3% UVA), 38.9% in the visible range and about 54.3% in the IR spectral range. Although UV light represent the lowest part of solar radiations arriving on the earth, they are considered the most dangerous since they can penetrate target tissues, primarily the skin and eyes, inducing oxidative stress reactions as well as DNA damage, and eventually resulting in cancer and skin aging (Zegarska et al. 2017). The ozone (O_3) layer within the atmosphere can absorb part of these dangerous radiation depending on their wavelengths, acting as a natural shield. In particular, it completely absorbs short wavelengths radiation like UVC, that directly interact with the oxygen present within the atmosphere contributing to the formation of O_3 itself. Regarding the other radiations, the UV penetration through the atmosphere increases with higher wavelengths. Therefore, the most predominately and harmful sunlight radiations present in the environment are the longer wavelengths UVA (90–95%) and the fraction of UVB radiations able to pass through the atmosphere (5–10%). In particular, UVA1, the longest radiations, represents 80% of the total solar UV light reaching the earth, and they can interact with the deepest skin layer, the dermis (Tewari et al. 2013). In the past decades the developing urbanization had led to the continuous release of toxic substances into the atmosphere, such as chlorofluorocarbons (CFCs), which able to dissociate in the presence of UV light and deplete the O_3 layer via a catalytic reaction (Montzka et al. 2018; Lickley et al. 2020). The depletion of the ozone layer has been correlated not only with several important climate changes but most importantly to an increase in the onset of different skin pathologies due to the more energized sun radiations.

Indeed even though a short sun exposure (5–15 min) is known to be good for our health due to its involvement in vitamin D production (Nair and Maseeh 2012), a longer exposure, especially during midday, can be extremely dangerous for skin health and cause sunburn, skin cancer, collagen depletion, and premature skin aging (Matsumura and Ananthaswamy 2004). Moreover, UV radiation can affect other organs directly (eyes) or indirectly (nervous system, intestine, and kidneys).

DOI: 10.1201/9781003328100-22

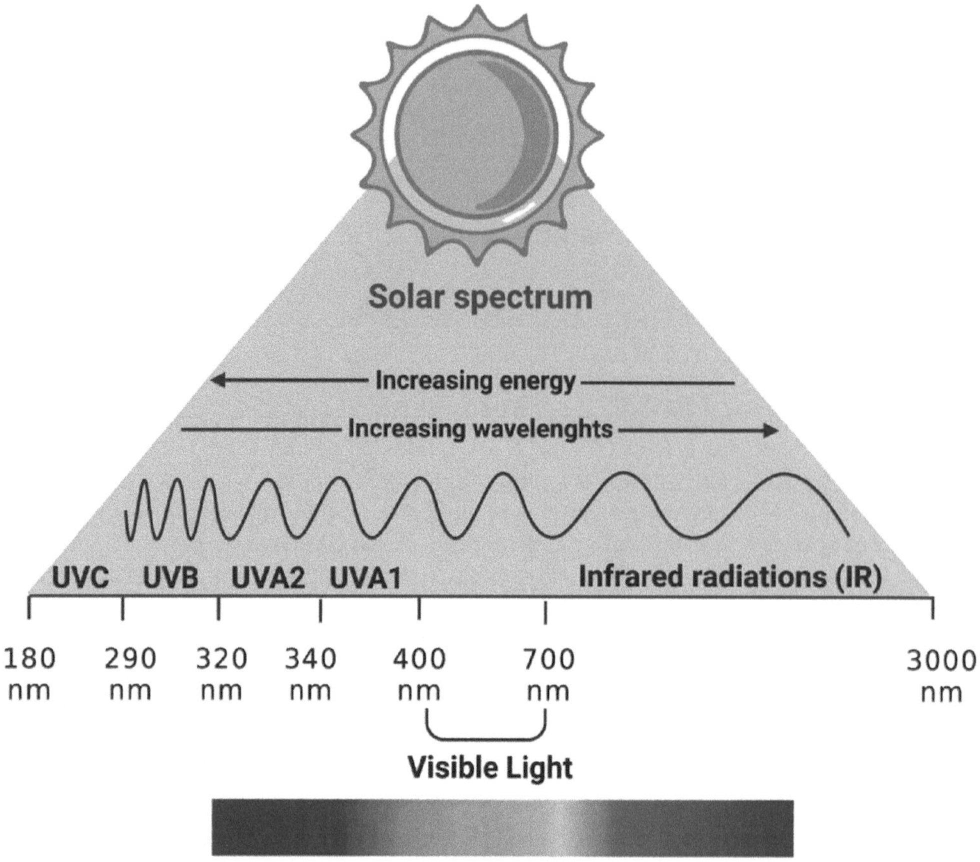

FIGURE 18.1 Solar spectrum. Wavelengths of the solar radiations reaching the earth (UVR, visible light, and infrared radiations). The energy of solar radiations is inversely proportional to their wavelengths, thus making them more dangerous for human skin when having a shorter wavelength. Created with BioRender.com.

18.2 UVR AND ROS: OXINFLAMMATION WITHIN HUMAN SKIN

UVA and UVB radiations are both able to penetrate the skin at different level, inducing DNA damage and reactive oxygen species (ROS) production (Figure 18.2). However, their mechanism of action can differ due to their different wavelengths and energy. UVB radiations, which are normally absorbed by the epidermis, lead to the formation of pyrimidine dimers in DNA and consequent mutations (direct mechanism), whereas UVA, which are able to penetrate more in deep the skin reaching the dermis, lead to the production of ROS able to damage the DNA (indirect mechanism) (Rastogi et al. 2010) and attack other biomolecules such as protein and lipids. Since UVA radiations are lower in energy compared to UVB, their detrimental effect has been underestimated for a long period of time, always considering UVB the most dangerous radiations. However, in the recent years a crescent number of studies demonstrated that UVA are particularly harmful for human skin health contributing to photoaging, carcinogenesis, hyperpigmentation, and immune suppression (Bernerd et al. 2022). In particular, due to their longer wavelength, UVA1 radiations (340 to 400 nm) are able to penetrate deeper into the dermis compared to UVA2 (320 to 340 nm) and UVB, reaching the subcutaneous layer where they can affect different cell types

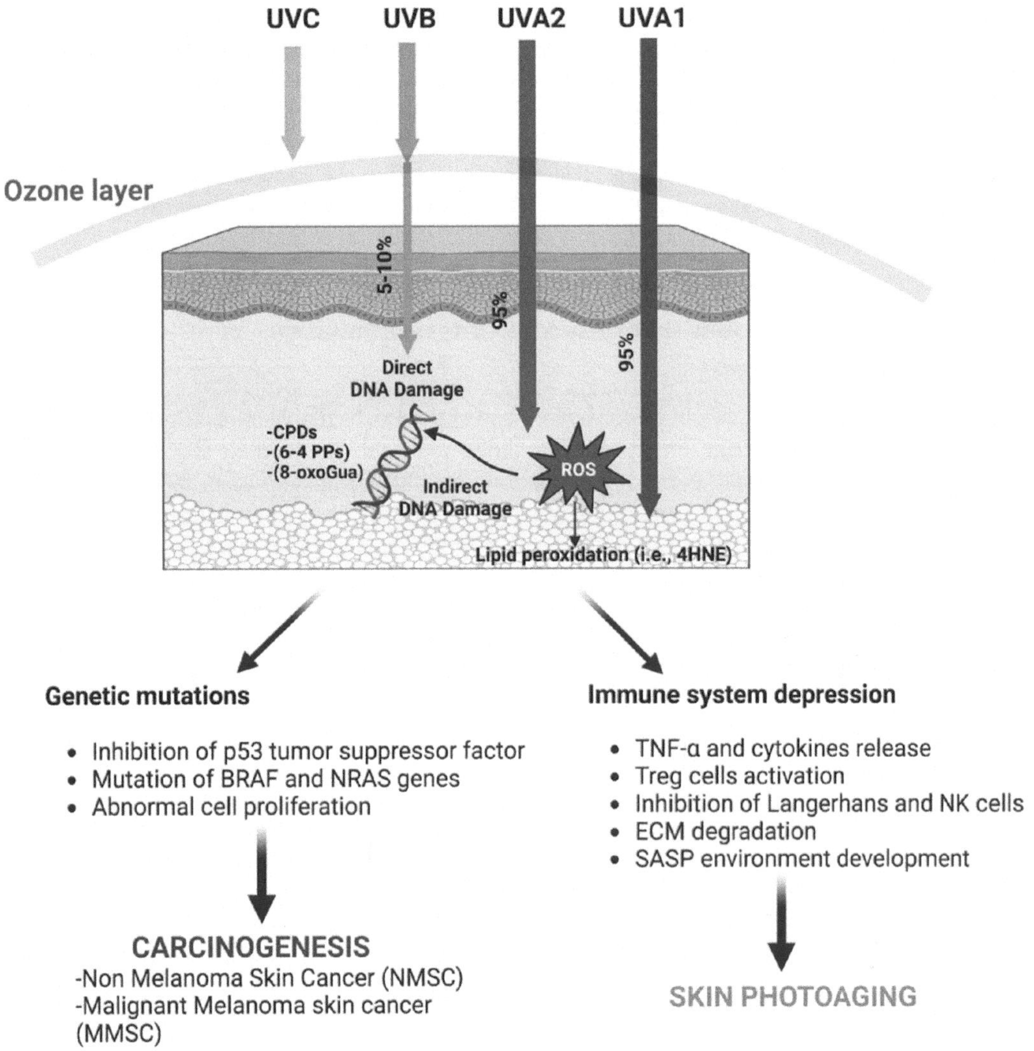

FIGURE 18.2 Skin photoaging and carcinogenesis processes. The portion of UVB radiations reaching the skin (5–10%) and absorbed by the epidermis are the main solar source responsible for direct DNA damage, inducing mutation as CPDs and 6-4 PPs. UVA radiations able to reach the dermis (UVA2) and the subcutaneous tissues (UVA1) can induce oxidative stress reactions leading to ROS production (i.e., H_2O_2) and lipid peroxidation products (i.e., 4HNE) that promote indirect DNA damage (CPDs, 6-4 PPs, 8-oxoGua), as well as immunosuppression. Genetic mutations occurring in p53, BRAF, and NRAS genes are responsible for abnormal cell proliferation and carcinogenesis. Depression of the immune system is responsible for the development of a SASP environment that leads to premature aging of the skin (photoaging). Created with BioRender.com.

including infiltrating T cells and B lymphocytes, fibroblasts, dendritic cells, and immature mast cells. Moreover, UVA1 light has been shown to induce apoptosis of infiltrating T cells therefore modulating the inflammatory response of the skin tissue (Yamauchi et al. 2004). UVA radiations can be absorbed by chromophores present within the skin such as bilirubin, melanin, porphyrins, pterins, B6 vitamins and flavins, vitamin K, and urocanic acid (UCA). When subjected to UVA radiation, chromophores pass to a photoexcited state inducing the production of singlet oxygen ($^1O_2$) and superoxide radical anion ($\bullet O_2^-$) by interacting with molecular oxygen. UVA-induced ROS products can interact with other substrates as lipids, proteins, DNA giving peroxidation reaction products (i.e., reactive aldehydes as 4-hydroxynonenal). UVA radiation can also induce cellular release of free iron and heme that can then participate to the Fenton's reaction. The results of these reactions can either be the formation of superoxide radical anion $\bullet O_2^-$ (type 1 reaction) or of single oxygen $^1O_2$ (type 2 reaction) which is particularly reactive toward biomolecules. $\bullet O_2^-$ can undergo dismutation resulting in H_2O_2 production involved in the Fenton reaction in presence of Fe^{2+}, resulting in hydroxyl radical formation (HO•), that is extremely reactive towards biomolecules as well (Bickers and Athar 2006).

18.2.1 Hydrogen Peroxide (H_2O_2) as Mediators of Cutaneous Induced UV Damage

Among free radicals, H_2O_2 has been considered one of the most suitable molecule for redox signaling in response to UV damage, and it has been shown to be able to modulate several signaling and transcription factors, as well as lipid peroxidation (Sies 2017; Marinho et al. 2014). For instance H_2O_2 has been correlated to the regulation of several transcription factors activity such as nuclear factor erythroid 2-related factor 2 (NRF2), nuclear factor kappa-light-chain-enhancer of activated B cells (NF-κB), activator protein 1 (AP-1), NOTCH, via a redox regulation mechanism, largely and well discussed in several reviews and articles (Marinho et al. 2014; Schreck, Rieberl, and Baeuerle 1991). This redox-regulation mechanism is mediated mainly by the reactivity of H_2O_2 with thiols groups, which are largely present in several biomolecules and that leads to its ability in modulate the synthesis, stability, translocation of these transcription factors, their degradation via the proteasome, and their affinity in binding the DNA. It is now evident that H_2O_2 is able to regulate both the antioxidant and inflammatory response of our organism and therefore be a key mediator for UVR in skin damage. Indeed the oxidative stress status induced by ROS production has been correlated to the exacerbation of several skin pathologies to cancer and skin aging mainly due to skin lipid peroxidation (Bickers and Athar 2006; Kammeyer and Luiten 2015). Inhibition or activation of NF-κB depends on H_2O_2 concentration levels, and therefore H_2O_2 is a fine regulator of pro-inflammatory and anti-inflammatory events within the cells (Jornot, Petersen, and Junod 1997). The dual role of H_2O_2 on NF-κB activity has been longed investigated. NF-κB activation and translocation into nucleus is very sensitive to oxidative events promoted by an oxidative environment whereas a reductive environment is required for its binding with the DNA (Kamata et al. 2002; Jung et al. 2008). Upon ROS productions many other events happen within the cells that serve as propagation of the oxidative damage. Free radicals such as Hydroxyl radicals (HO-radicals), can interact with lipids present within the cell membrane, triggering a free radical chain reaction mechanism able to promote the degradation of lipids. The presence of an intracellular matrix rich in lipids as free fatty acids, triglycerides, cholesterol, and ceramides in the outermost stratum of the skin, makes the skin an easy substrate for lipid peroxidation, with the consequent production of secondary oxidative messengers such as 4-hydroxynoneal (4HNE) and Malondialdehyde (Ayala, Muñoz, and Argüelles 2014). For instance Squalene has been found to be particularly susceptible to solar UVR in human keratinocytes, since its photo-oxidation leads to production of 4HNE and stimulation of inflammatory markers as interleukin-6 (IL-6), Cyclooxygenase-2 (COX-2), and iNOS genes (Kostyuk et al. 2012). 4HNE is able to modulate the expression of several genes, the activation of inflammatory responses such as NF-κB, but also detoxification mechanisms (NRF2) and mitogen-activated protein kinases (MAPKs) as well as important cellular pathways such as apoptosis, necrosis, proteasome, and autophagy, which are all fundamental in the regulation of cell death (Yadav and Ramana 2013; Dinkova-Kostova et al. 2002; Dalleau et al. 2013). It is therefore clear that 4HNE could play an important role in UV-induced damage and new evidences highlight the role of 4HNE as a possible trigger factor in pollution-induced skin OxInflammation (Pecorelli et al. 2019).

18.2.2 UVR, DNA Damage, and Skin Cancer

UV-induced DNA damage is the primary cause of skin carcinoma. UVR exposure can lead to two types of nonmelanoma skin cancers (NMSC) as squamous cell carcinoma (SCC) and basal cell carcinoma (BCC) and to malignant melanomas (MM). The absorption of DNA in the UV range occurs predominantly in the UVB wavelengths of 290–300 nm, resulting in the formation of cyclobutene pyrimidine dimers (CPDs), that induce cell cycle arrest, and pyrimidine (6-4) pyrimidine photoproducts (6-4 PPs) able to then trig apoptosis but less carcinogenic compared to CPDs (Lo et al. 2005). Despite the direct DNA damage induced by UVB, UVA can also induce the formation of CPDs through a mechanism known as triplet-triplet energy transfer (TTET), where UVA energy is absorbed by other molecules and transferred to the DNA that then reach an excited state (Douki, Bérard, and Wack 2014). UVA radiation can also induce oxidation of DNA bases as 8-oxo-7,8-dihydroguanine (8-oxoGua), which is an important hallmark of UVA-induced DNA damage (Dizdaroglu, Kirkali, and Jaruga 2008).

Compared to UVB radiations, UVA are less prone to form CPDs and 6,4 PPs. However, CPDs related to UVA radiations have been shown to be more present in the basal layer of epidermis and therefore linked to carcinomas, whereas UVB induced CPDs are more located in the upper layer of epidermis (Agar et al. 2004). CPDs induced by UVA have been found to be accumulated overtime in human melanocytes and associated to carcinomas (Bernerd et al. 2022). Recently, a Genome-wide nucleotide-resolution analysis of CPDs in melanocytes and fibroblasts has identified "CPD hyper hotspots", pyrimidine-pyrimidine sites that are hundreds of folds more susceptible to the generation of CPDs than the genomic average. These sites called hyper hotspots could be used to predict future skin cancer risk in small skin samples (Garcia-Ruiz and Kornacker 2022). Beside tumor initiation, UVR has been reported to be involved in SCC progression by modulating the expression of miRNA. Indeed an overexpression of miR-7-5p, miR-29a-3p and miR-183-5p have been found either in a primary SCC tumor (Met-1) upon UV irradiation and in the respective un-irradiated metastasis (Met-4) (Chen et al. 2020). UV-induced DNA damage is linked to mutation of several genes involved in cell cycle arrest and apoptosis as the tumor suppressor p53 (Rastrelli et al. 2020). Usually, induction of p53 promotes cell cycle arrest and apoptosis to prevent cell mutations that can lead to carcinogenesis. However, an imbalance between cell cycle arrest and apoptosis due to p53 mutations is an essential key point in UV-induced carcinomas as well as in other human cancers. For instance, more than 90% of SCC and 50% of BCC present p53 mutation due to UVR. As a matter of fact altered levels of proapoptotic Caspases, such as Caspase-3, and antiapoptotic proteins, such as Bcl-2, have been found in many carcinomas including eyelid carcinoma (Koyun et al. 2020). The modulation of p53 has been shown to prevent UV-induced DNA damage. For instance, the β-blocker carvedilol has been found to prevent UV-induced DNA damage in a full-thickness 3D human reconstituted skin, by attenuating UV-mediated epidermal thickening, the number of Ki-67 and p53 positive cells as well as CPD formation (Chen et al. 2020). The immunosuppressant Rapamycin displays protective effects against the UVA damage in human dermal fibroblasts (HDFs) by reducing senescence (reduced levels of SA-β-gal) and the expression of p53, and modulating the levels of metalloproteinases (MPPs) such as MMP-1, MPP-3 as well as collagen type I (Bai et al. 2021). Of note, MPPs modulation has been shown to be an essential event in SCC and BCC development. As extensively summarized by Pilvi Riihilä et al. (Riihilä, Nissinen, and Kähäri 2021), UV-activated fibroblasts promote the production of pro-inflammatory markers as well as proteolytic enzymes as MMPs, able to degrade the extracellular matrix (ECM), favoring the formation of a microenvironment more suitable for tumor progression and metastasis (Hofmann et al. 2005). Different from SCC and BCC, mutations of p53 gene are not the main cause of malignant melanomas (MM) progression. Indeed, melanomas have been associated to the mutation of the two oncogenes neuroblastoma RAS viral oncogene homolog (NRAS) and proto-oncogene B-Raf (BRAF), involved in the regulation of the mitogen activated protein kinase (MAPK) pathway that modulates cells proliferation. While BRAF mutations occur more frequently during acute sun exposure, NRAS mutations are more prone to occur after chronic sun exposure. Besides the most known NRAS and BRAF, other oncogenes have been discovered to be involved in melanoma progression, such as RAC1, PREX2, and ERBB4. The first two can affect the PI3K/AKT pathway, resulting in melanoma cells proliferation upon cross-talking with the MAPK pathway (Kim and Giovannucci 2020).

18.2.3 UVA and Skin Photoaging: Immunosenescence

During aging, the skin starts to display a common pro-inflammatory phenotype called senescence-associated secretory phenotype (SASP), that relies on the secretion of many pro-inflammatory mediators including cytokines, chemokines, and matrix MMP, promoting skin aging (Ghosh and Capell 2016). It is now known that detrimental events as ROS production, DNA damage as well as mitochondrial stress can promote a cutaneous senescent state accompanied to an inflammatory conditions (Inflammaging) (Fulop et al. 2021), resulting in premature skin aging that can even play a role in developing skin conditions. In particular, UVR can promote skin aging (photoaging) by inducing an immunosuppressive effect. Indeed, besides keratinocytes and fibroblasts, human skin presents immune-competent cells as Langerhans cells in epidermis and dendritic cells, T cells, B cells, and natural killer (NK cells) that are fundamental for the skin inflammatory response (Matejuk 2018; Belkaid and Segre 2014). The activation of the cutaneous immune response leads to the production of pro-inflammatory mediators (i.e., TNF-α), cytokines such as interleukins (IL-1β, Il.18, IL-6 etc.) that have been correlated to several skin inflammatory diseases and cancer (Nedoszytko et al. 2014). To counteract the inflammatory damage, senescent cells display an activation of immunosuppressive cells as Treg cells that lead to the production of anti-inflammatory cytokines (i.e., IL-10, IL-4) and TNF-β. These mediators can promote inhibition of cell proliferation as well as ECM degradation by inducing MMPs, favoring cell senescence (Salminen, Kaarniranta, and Kauppinen 2022). Although skin immune system is responsible for the removal of neoplastic cells caused by DNA damage, exposure to UVR can promote skin aging either by directly or indirectly targeting DNA (CPDs formation, ROS production) or by inducing inflammation and immunosuppression. A chronic immunosuppression due to an hyperactivation of cutaneous immune cells as regulatory T (Treg) cells upon UV exposure, and lack in NK and CD8 T cells activity, have been associated to a premature skin aging process called immunosenescence (Ovadya et al. 2018). In this context UVR triggers the development of a SASP phenotype with the concomitant production of pro- and anti-inflammatory mediators (Salminen, Kaarniranta, and Kauppinen 2022). Both UVB and UVA have been shown to promote skin immunosuppression at a different wavelength absorption as 300 nm for UVB and 370 nm for UVA radiation. However, considering that UVA are the most abundant radiations reaching the earth, they are also considered the most dangerous in term of skin aging process, and they promote delayed inflammatory skin response compared to the early and acute cutaneous response to UVB radiations (Damian DL, Matthews YJ, Phan TA, n.d.). UVA radiations, in particular UVA1, can affect cells of the immune system as Langerhans cells, Treg cells, T lymphocytes resulting in alteration of cytokines production, apoptosis and mitigated capability of migration. For instance, under UVA exposure, Langerhans cells and DCs seem to lose their antigen-presenting capacity, and NK cells tend to lose their cytotoxic activity (Achachi et al. 2015). Interesting higher levels of cutaneous DCs induced by UVR have been correlated to an increased induction of Treg cells, promoting an immunosuppressive state (Yamazaki et al. 2018). Excessive production of TNF-β due to UVR exposure has been shown to promote the expression of the Aryl Hydrocarbon Receptor (AhR), which is involved in the maintenance of the cutaneous immunosuppressive state, promoting skin aging and senescence (Vogeley et al. 2019). In addition, UVR can regulate cutaneous immunosuppression by promoting the conversion of the urocanic acid, a product of histidine metabolism within the epidermis, from its *trans* to its *cis* conformation, leading to IL-10 secretion and reduced immune cells contact hypersensitivity (Bernerd et al. 2022). UVR, especially UVA1, can affect the dermis and its functionality by targeting the degradation of structural proteins of the extracellular matrix

(ECM) as collagen, via the induction of MMPs, resulting in premature skin aging (Yamaba et al. 2016). Indeed, UVR have been shown to affect collagen and elastin levels as well as other structural proteins involved in the formation of the skin *stratum corneum*, differentiation, and proliferation processes such as keratins, involucrin, and filaggrin. Downregulation of these proteins can compromise skin barrier functions and promote pro-inflammatory skin conditions.

18.2.4 Sun Exposure and Skin Abnormalities

Beside promoting skin aging and carcinogenesis, UVR exposure has been associated to the development of skin pigmentary disorders, photodermatosis, and elastosis (Table 18.1) (Bernerd et al. 2022). During sun exposure, the melanocytes present in the skin are stimulated to produce the pigment melanin as a defensive mechanism to protect the cutaneous tissue from damage. Melanocytes can be found in the basal layer but also in hair follicles and are involved in keeping skin homeostasis by protecting from oxidative stress induced by UV light and also from microbes injury (Slominski, Paus, and Schadendorf 1993). As described by Joseph Michael Yardman-Frank and colleagues, while UVB radiations are the mainly responsible for melanogenesis and skin tunning, exposure to UVA has been shown to induce persistent pigment darkening (PPD) (Yardman-Frank 2022). Short exposure to UVA radiation (1 J/cm^2) promotes an immediate pigment darkening (IPD) whereas a greater exposure (11.0 J/cm^2) is responsible for persistent skin pigment darkening (PPD). H_2O_2 produced by UVA radiation exposure has been found to oxidize the melanin precursor 5,6-dihyd roxyindole-2-carboxylic acid (DHICA) and its metabolites. These metabolites tend to accumulate outside melanocytes in granulocytes present in the basal and suprabasal layers of the epidermis, favoring an irreversible brownish-black pigmentation of the skin (Maeda and Hatao 2004). On the other hand, UVB radiations can promote delayed skin tanning by inducing melanogenesis via DNA damage and p53 activation. Activation of p53 stimulates keratinocytes to secrete a-melanocyte-stimulating hormone (a-MSH) that promotes melanogenesis (Cui et al. 2007). Differently from skin tanning, darkening of the skin induced by UVA1 radiations does not provide for photoprotection and is linked to inflammation, oxidative stress, and cellular damage. Hyperpigmentation of the skin can promote the formation of lentigines in selected areas of the body or photoaging pathologies as melasma (Passeron and Picardo 2018). Besides skin pigmentation, UVR can promote the development of several skin disorders, such as photodermatosis, depending on the skin type. When exposed to the sun, individuals with a Fitzpatrick skin type 1-IV are most likely subjected to a type of skin dermatosis called polymorphic light eruption (PMLE) that consists in the formation of papules, vesicles, and plaques in specific areas of the skin, accompanied with itching. UVA seem to be more involved in the development of PMLE compared to UVB, by inducing immunosuppression (Das, Lloyd, and Walshaw 2004). Solar Urticaria (SU), another type of photo dermatoses, is also linked to UVA radiations, and it manifests shortly after sun exposure. The detrimental effect of UVR on skin dermis and in particular on fibroblasts, can promote the development of Solar Elastosis (SE), a cutaneous pathology consisting in the accumulation of degraded elastic fibers as elastin, fibrillin, fibronectin, and collagen due to ECM destruction and MPPs induction. As final results, the alteration of the dermal structures results in the formation of dermal solar scars throughout the cutaneous tissue (Bernerd et al. 2022).

18.2.5 UVR Radiations and Eyes

Beside the skin, human eyes are everyday directly exposed to the sun and become an important target for UVR. Since the eyes' function is to convert electromagnetic radiations (the light) into images via phototransduction of the signal in the neural retina, they are very susceptible to sun light damage. Indeed, human eyes can interact with all the portions of the solar electromagnetic spectrum (UVR, visible light, infrared radiations), and based on the wavelengths and the intensity of the radiations, different structures of the eye can be affected. For instance, UVA, visible light, and

IRA radiations (700–1400 nm) can reach the retina, contributing to macular degeneration (Ivanov et al. 2018), while UVB and IRB (400–3000 nm) can impact the crystalline lens, favoring the development of cataract. However, IRC (3000–10,000 n) and in a less extensive way UVC radiations cannot pass through the cornea. Moreover, it has been demonstrated that a part the sun, artificial light as blue light and visible light can contribute to the ocular damage (Zhao et al. 2018). Although UVR do not contribute to visual perception, evidence demonstrates that exposure to these radiations can promote ocular damage, which is a risk factor for the onset of cornea squamous cell carcinoma, photo conjunctivitis, photokeratitis, glaucoma etc. (Table 18.1) (Ivanov et al. 2018; Begaj and Schaal 2018; Delic et al. 2017). A great part of these ocular diseases is due to photo-thermal, photo-mechanical, or photo-chemical reactions happening with the interaction between UVR and the eyes. Visible light and IR are mainly responsible for photo-thermal and photomechanical eye damage, resulting in increased temperature and mechanical compressive forces that promote the denaturation of proteins as well as the deteriorating of the retinal pigment epithelium (Glickman 2002). UVR, instead, plays a pivotal role in photochemical damage via rupture of chemical bonds or via the promotion of oxidative stress reactions. For instance, UVC and UVB radiations with a wavelength below 200 nm can promote chemical bonds breaks, resulting in photons release (photochemical ablation). The use of these radiation is often adopted in eye surgery where the released photons are directed to destroy altered tissue sections (Ivanov et al. 2018). The retinal pigment epithelium (RPE) presents an important number of chromophores (pigments, flavoproteins, melanin, lipofuscin) that can absorb photons and be excited by prolonged exposure to UVA ≥550, resulting in ROS production (Ivanov et al. 2018). As for the skin, H_2O_2 has been demonstrated to be the main reactive molecule mediating the oxidative stress damage within human eye. Water channels aquaporins (AQPs) have been found to allow the passage of H_2O_2 through membranes, favoring its propagation into the RPE. AQP9, as well as AQPs 6–12, are the main channels responsible of H_2O_2 passage, being responsible for ROS accumulation in human eyes leading to glaucoma and other ocular pathologies (Ivanov et al. 2018; Yang, Dibas, and Tyan 2013). Beside the retina, other eye components can also be affected by UVR. For instance cornea can absorb UVB radiations (up to 70%) and a small part of UVA, whereas aqueous humor can absorb radiations with a wavelength inferior to 400 nm (Ivanov et al. 2018). As a matter of fact, protection of aqueous humor by using ultraviolet-blocking contact lenses is becoming a potential strategy to protect from inflammation caused by UVR exposure (Notara et al. 2018). With aging UVR transmission through the eyes decrease since the ocular lens changes in proteins content levels, while children present cloudless lens unable to stop UVA and UVB radiations in reaching the retina, thus being more susceptible to UVR ocular damage. Although the eye is highly susceptible to UVR, it possesses several defensive mechanisms, such as eye lids and eyebrows, in addition to the production of tears, that are able to partially absorb UVR. Moreover, the eye is equipped with molecules such as ascorbic acid and macular pigments able to counteract the oxidative damage (Reddy et al. 1998).

18.2.6 UVR Effect on Other Organs

Although exposure to air pollutants have been demonstrated to affect a variety of tissues (Mousavi et al. 2022), the effect that UVR can exert on other organs is still a developing field. However, few studies demonstrated that the exposure of the skin to UVR can affect other districts as the gut, kidneys, and even the nervous system (Table 18.1). Indeed, the presence of nervous structures and immune cells within the human tissues give to the skin the ability to communicate with multiple organs of our body. In the past decades a crescent number of studies have demonstrated that skin health and its functionality is strictly interconnected with the gut (skin-gut axis) and the brain (skin-gut-brain axis), and many studies demonstrated that a variety of GI tract disorders such as Crohn disease, ulcerative colitis, and celiac disorder were associated to skin conditions (i.e., atopic dermatitis, psoriasis etc.) (Lu et al. 2020). Most of these pathologies were accompanied by an alteration of the skin or gut microbiomes that are essential for the regulation of the tissue immune response. For

TABLE 18.1
Effect of UVR on Target Organs and Related Disorders

Target organ	Disorders	Cause	Effects
Cutaneous tissues	Persistent pigment darkening (PPD)	Oxidation of 5,6-dihydroxyindole-2-carboxylic acid (dHICA)	Irreversible brownish-black pigmentation of the skin
	lentigines Melasma	Increased production of melanin (hyperpigmentation)	Light brown, dark brown and/or blue-gray patches on different skin areas
	Photodermatosis: Potymorphic Light Eruption (PMLE)-Solar urticaria (SU)	Immunosuppression	Formation of papules, vesicles, and plaques
	Solar elastosis	MPPs induction	Formation of dermal scars due to ECM degradation and accumulation of altered elastic fibers, elastin, fibrillin, fibronectin
	Squamous cell carcinoma (SCC) and Basal cell carcinoma (BCC)	P53 mutation. CPDs and 6,4 PPs formation. ROS production MPPs induction, abnormal cell proliferation	Lesion with dark spots, scaly patch with a raised edge, white scar lesions, red nodule
	Malignant melanoma (MM)	BRAF and NRAS genes mutation	Unusual moles with asymmetrical shape that grow overtime
Eyes	Age-related macular degeneration	Oxidative stress reactions occurring in the RPE and cell death	Blindness
	Cataract	Oxidative stress damage of proteins, nucleic acid, lipids and polysaccharides present in the crystallinus	Blurred vision, blindness
	Glaucoma	Oxidation of antioxidant defense genes (AOX) —increase in aqueous humor flow resistance due to OS	Elevated intraocular pressure and blindness
	Keratoconus	Mitochondria DNA damage due to ROS with consequent decrease in number of fibroblasts, keratinocytes and altered lamellae organization	Corneal thinning with a conical shape, blindness
	Photokeratitis Photo conjunctivitis	Sunburn of the cornea Sunburn of the conjunctiva	Eyelid twitching, intense tears, pain, light sensitivity, redness, swelling, seeing halos or temporary loss of vision
	Cornea squamous cell carcinoma	P53 mutation CPDs and 6.4 PPs formation. ROS production, abnormal cell proliferation	Small and hard reddish nodules, verrucous or papilloma, granular surface ulcers
Gut	Gi tracts disorders (Crohn's disease, ulcerative colitis, inflammatory bowel disease	Alteration of skin and gut microbiomes	Irreversible brownish-black pigmentation of the skin
Kidney	Nephritis, kidney inflammation	Infiltration of cutaneous neutrophils within kidneys, foamy urine	High blood pressure, swelling of the legs, ankles, and feet, joint problems, fever, rashes
Central Nervous System	Multiple sclerosis (MP)	Immunosuppression via conversion of the *trans*-UCA in *cis*-UCA	Stimulation of immunosuppressive cells, IL-10 release and inhibition of IFN-Y production

Created with BioRender.com.

instance antimicrobial peptides (AMPs) such as cathelicidins, alarmins etc., play an important role in the first defense of the skin and help in the regulation of the inflammatory response (Boguniewicz and Leung 2011; Schauber and Gallo 2008). Aberrant expression of AMPs has been linked to the development of inflammatory skin diseases (atopic dermatitis, psoriasis) and to increased microbial infections susceptibility.

Although the mechanism underneath the regulation of the skin-gut interplay is still not completely known, the immune and endocrine systems seem to play a crucial role (Ellis et al. 2019). Altered activity of gut bacteria or migration of bacteria from the gut to the systemic circulation can promote systemic inflammation that can affect human skin resulting in the development of skin conditions such as psoriasis (Ellis et al. 2019). Some studies in the recent years also demonstrated that skin is able to promote gut microbiome alteration. A recent study showed that skin exposure to Narrow Band UVB light can affect gut microbiota and that vitamin D intake could prevent the induction of many commensal bacteria as *Lachnospiracheae*, *Rikenellaceae*, *Desulfobacteraceae*, *Ruminococcus*, etc. (Bosman et al. 2019). A similar result was obtained in mice exposed to UVR and supplemented with Vitamin D (Ghaly et al. 2018). Moreover patients subjected to metastatic melanomas have been found to experience generalized gastrointestinal symptoms such as abdominal pain or constipation (Syed, Shekar, and Aravantagi 2021). The skin is equipped with nervous structures belonging to the serotoninergic system that make the cutaneous tissue directly in contact with the brain report noxious external stimuli (Slominski et al. 2012). Indeed, sensory nerves present within the skin can express serotonin receptors that when activated by environmental insults can communicate with the brain. The cutaneous serotoninergic receptor 5-HT2A has been found involved in the conversion of the *trans*-UCA in *cis*-UCA, favoring skin immunosuppression (Walterscheid et al. 2006). As a matter of fact, *cis*-UCA induced by UVR has been found to be involved in multiple sclerosis development (MS) (Correale and Farez 2013), an immune-mediated demyelinating disease of the nervous system. Beside the gut and the nervous system, exposure of the skin to UVR seems to affect kidneys most likely via migration of the immune cells from the cutaneous tissue. For instance, a recent study showed that patients affected by lupus nephritis present blood neutrophils with a phenotypes similar to neutrophiles observed within the skin upon UVR exposure, suggesting that these cells have migrated from the cutaneous tissues to the kidneys inducing inflammation and nephritis (Skopelja-Gardner et al. 2021). Considering the interconnection between skin and other organs via microorganisms (microbiome) and infiltrating immune cells regulating the cutaneous inflammatory status, it might be plausible that alteration of the skin homeostasis induced by environmental stressors can indirectly affect other organs, via OxInflammatory reactions, which makes it even more important for us to be protected from these radiations.

18.2.7 UVR, Vitamin D, and Photoprotection

Beside the noxious effects that prolonged sun exposure can exert on human skin, UVR are essential to produce Vitamin D, a fat soluble secosteroid responsible for calcium maintenance and bone homeostasis. Vitamin D is present in some foods as salmon or liver oil, and it can be assumed as oral supplementation. The recommended daily intake of Vitamin D for adults is 8000IU/mL (200mcg/mL), however, the main source of Vitamin D is due to sun exposure by cutaneous absorption of UVB radiations When absorbed, UVB radiations cause the excitation of 7-dehydrocholesterol, which is then transformed in pre-vitamin D_3. Pre-vitamin D_3 easily isomerizes in Vitamin D_3 that is metabolized in the liver into 25(OH)D, the inactive form of Vitamin D circulating in the blood and then activated in 1,25$(OH)_2$D by the 25-hydroxtvitamin D-1-α-hydroxylase (CYP27B1) within kidneys. The synthesis of cutaneous Vitamin D is regulated by the presence of the pigment melanin that absorbs a great part or UVB radiations. For instance, individuals with a skin phototype of 5 or 6 produce a very low amount of Vitamin D due to the high presence of melanin. On the contrary, skin phototypes of 1–2 is equipped with low levels of melanin that on one side allow a greater production of Vitamin D but on the other side makes the skin more prone to sunburn. Vitamin D production is strictly related to the season, latitude, and time of the day. During winter UVB radiations are almost completely absorbed by the earth's atmosphere, and early in the morning or in the late afternoon sun radiations are more oblique to the earth, which make easier for the ozone layer to absorb them (Kim and Giovannucci 2020).

Vitamin D has been shown to regulate cell growth by inhibiting angiogenesis thus preventing the proliferation of malignant cells (Wu et al. 2014) and protecting against UV-induced DNA damage as well as oxidative stress. For instance, individuals exposed to low-dose summer sunlight present levels of Vitamin D sufficient to protect against CPDs formation as well as 8-oHdG production (Felton et al. 2016). The mechanism of action behind Vitamin D photoprotection is mostly related to its ability to activate antioxidant pathways, such as NRF2-KEAP1, thus inducing the production of the radical scavenger metallothionein. A study demonstrated that 25(OH)2D3 can delay skin aging in mice by upregulating NRF2 pathway, resulting in decreased oxidative stress, p53 activity and a reduced SASP senescence profile (Chen et al. 2019). Due to the regulation of many physiological processes as cell proliferation, differentiation, and apoptosis, the use of Vitamin D as pharmacological treatment has been approved not only for many skin pathologies (i.e., psoriasis, atopic dermatitis) but also in other conditions, such as colitis, kidneys injury, and pulmonary diseases as COVID-19 (Umar et al. 2018). As a matter of fact more than 1000 studies have been conducted in the past 3 years proving the efficiency of Vitamin D intake against COVID-19 progression (Singh et al. 2020). Phototherapy has been used for decades to treat dermatological disorders by inducing the production of Vitamin D able to regulate and prevent the activation of inflammatory pathways associated with skin pathologies. UVA1 therapy has been shown to ameliorate skin conditions linked to an increased immune response, such as atopic dermatitis, urticaria, and lupus erythematosus. Indeed, due to immunosuppressive capacity, UVA can promote the inhibition of inflammatory cytokines as IL-5 and IL-13 (Vangipuram and Feldman 2016).

Thus, although excessive exposure to sun radiations can promote the development of skin cancers, a correct exposure to sun can promote Vitamin D production that protects from photo carcinogenesis, skin aging, and many other conditions.

However, the lack of information concerning an adequate exposure to the sun is leading to Vitamin D deficiency in great part of the world population, resulting in several health consequences as osteoporosis, malabsorption and gastrointestinal disorders, renal failures, and pulmonary conditions as chronic obstructive pulmonary disease (COPD) and asthma (Kim and Giovannucci 2020). Therefore, learning how to protect ourselves from UVR by being exposed to the sun in the correct period of the day (early morning or late afternoon rather than in the worst hours, 12–4 pm) while using sun lotions, can be beneficial for our health and prevent from skin carcinogenesis, sunburn, and DNA damage.

18.3 CONCLUSIONS

Overall, the present chapter has given the main knowledge we should have regarding sun exposure and UVR. Indeed, the incorrect exposure to UVR as well as total lack of sun exposure can be deleterious for our health. Understanding all the biological and clinical effects that UVR can exert on our body is of greatest importance to learn how we can benefit from sun exposure and avoid its damaging effects.

REFERENCES

Achachi, Amine, Marc Vocanson, Philippe Bastien, Josette Péguet-Navarro, Sophie Grande, Catherine Goujon, Lionel Breton, Isabelle Castiel-Higounenc, Jean François Nicolas, and Audrey Gueniche. 2015. "UV Radiation Induces the Epidermal Recruitment of Dendritic Cells That Compensate for the Depletion of Langerhans Cells in Human Skin." *Journal of Investigative Dermatology* 135 (8): 2058–2067. https://doi.org/10.1038/jid.2015.118.

Agar, Nita S., Gary M. Halliday, Ross St C. Barnetson, Honnavara N. Ananthaswamy, Mark Wheeler, and Alexandra M. Jones. 2004. "The Basal Layer in Human Squamous Tumors Harbors More UVA Than UVB Fingerprint Mutations: A Role for UVA in Human Skin Carcinogenesis." *Proceedings of the National Academy of Sciences of the United States of America* 101 (14): 4954–4959. https://doi.org/10.1073/pnas.0401141101.

Ayala, Antonio, Mario F. Muñoz, and Sandro Argüelles. 2014. "Lipid Peroxidation: Production, Metabolism, and Signaling Mechanisms of Malondialdehyde and 4-Hydroxy-2-Nonenal." *Oxidative Medicine and Cellular Longevity*. https://doi.org/10.1155/2014/360438.

Bai, Gen Long, Ping Wang, Xin Huang, Zi Yue Wang, Di Cao, Chuan Liu, Yi Yi Liu, Ruo Lin Li, and Ai Jun Chen. 2021. "Rapamycin Protects Skin Fibroblasts from UVA-Induced Photoaging by Inhibition of P53 and Phosphorylated HSP27." *Frontiers in Cell and Developmental Biology* 9 (February): 1–11. https://doi.org/10.3389/fcell.2021.633331.

Begaj, Tedi, and Shlomit Schaal. 2018. "Sunlight and Ultraviolet Radiation: Pertinent Retinal Implications and Current Management." *Survey of Ophthalmology* 63 (2): 174–192. https://doi.org/10.1016/j.survophthal.2017.09.002.

Belkaid, Yasmine, and Julia A. Segre. 2014. "Dialogue between Skin Microbiota and Immunity." *Science* 346 (6212): 954–959. https://doi.org/10.1126/science.1260144.

Bernerd, Françoise, Thierry Passeron, Isabelle Castiel, and Claire Marionnet. 2022. "The Damaging Effects of Long UVA (UVA1) Rays: A Major Challenge to Preserve Skin Health and Integrity." *International Journal of Molecular Sciences* 23 (15): 1–33. https://doi.org/10.3390/ijms23158243.

Bickers, David R., and Mohammad Athar. 2006. "Oxidative Stress in the Pathogenesis of Skin Disease." *Journal of Investigative Dermatology* 126 (12): 2565–2575. https://doi.org/10.1038/sj.jid.5700340.

Boguniewicz, Mark, and Donald Y. M. Leung. 2011. "Atopic Dermatitis: A Disease of Altered Skin Barrier and Immune Dysregulation." *Immunological Reviews* 242 (1): 233–246. https://doi.org/10.1111/j.1600-065X.2011.01027.x.

Bosman, Else S., Arianne Y. Albert, Harvey Lui, Jan P. Dutz, and Bruce A. Vallance. 2019. "Skin Exposure to Narrow Band Ultraviolet (Uvb) Light Modulates the Human Intestinal Microbiome." *Frontiers in Microbiology* 10 (October): 1–11. https://doi.org/10.3389/fmicb.2019.02410.

Chen, I. Peng, Marc Bender, Ivelina Spassova, Stefan Henning, Linda Kubat, Kaiji Fan, Sarah Degenhardt, et al. 2020. "UV-Type Specific Alteration of MiRNA Expression and Its Association with Tumor Progression and Metastasis in SCC Cell Lines." *Journal of Cancer Research and Clinical Oncology* 146 (12): 3215–3231. https://doi.org/10.1007/s00432-020-03358-9.

Chen, Lulu, Renlei Yang, Wanxin Qiao, Wei Zhang, Jie Chen, Li Mao, David Goltzman, and Dengshun Miao. 2019. "1,25-Dihydroxyvitamin D Exerts an Antiaging Role by Activation of NRF2-Antioxidant Signaling and Inactivation of P16/P53-Senescence Signaling." *Aging Cell* 18 (3): 1–18. https://doi.org/10.1111/acel.12951.

Chen, Mengbing, Sherry Liang, Ayaz Shahid, Bradley T. Andresen, and Ying Huang. 2020. "The β-Blocker Carvedilol Prevented Ultraviolet-Mediated Damage of Murine Epidermal Cells and 3D Human Reconstructed Skin." *International Journal of Molecular Sciences* 21 (3). https://doi.org/10.3390/ijms21030798.

Correale, Jorge, and Mauricio F. Farez. 2013. "Modulation of Multiple Sclerosis by Sunlight Exposure: Role of Cis-Urocanic Acid." *Ournal of Neuroimmunology* 26 (1–2): 134–140. https://doi.org/10.1016/j.jneuroim.2013.05.014.

Cui, Rutao, Hans R. Widlund, Erez Feige, Jennifer Y. Lin, Dara L. Wilensky, Viven E. Igras, John D'Orazio, et al. 2007. "Central Role of P53 in the Suntan Response and Pathologic Hyperpigmentation." *Cell* 128 (5): 853–864. https://doi.org/10.1016/j.cell.2006.12.045.

Das, S., Lloyd, J. J., Walshaw, D., and Farr, P. M. 2004. "Provocation Testing in Polymorphic Light Eruption Using Fluorescent Ultraviolet (UV) A and UVB Lamps." *Br J Dermatol* 151 (5): 1066–1070. https://doi.org/10.1111/j.1365-2133.2004.06235.x.

Dalleau, S., M. Baradat, F. Guéraud, and L. Huc. 2013. "Cell Death and Diseases Related to Oxidative Stress:4-Hydroxynonenal (HNE) in the Balance." *Cell Death and Differentiation* 20 (12): 1615–1630. https://doi.org/10.1038/cdd.2013.138.

Damian, D. L., Y.J. Matthews, T. A. Phan, G. M. Halliday. n.d. "An Action Spectrum for Ultraviolet Radiation-Induced Immunosuppression in Humans." *Br J Dermatol* 164 (3): 657–659. https://doi.org/10.1111/j.1365-2133.2010.10161.x.

Delic, Naomi C., J. Guy Lyons, Nick Di Girolamo, and Gary M. Halliday. 2017. "Damaging Effects of Ultraviolet Radiation on the Cornea." *Photochemistry and Photobiology* 93 (4): 920–929. https://doi.org/10.1111/php.12686.

Dinkova-Kostova, Albena T., W. David Holtzclaw, Robert N. Cole, Ken Itoh, Nobunao Wakabayashi, Yasutake Katoh, Masayuki Yamamoto, and Paul Talalay. 2002. "Direct Evidence That Sulfhydryl Groups of Keap1 Are the Sensors Regulating Induction of Phase 2 Enzymes That Protect against Carcinogens and Oxidants." *Proceedings of the National Academy of Sciences of the United States of America* 99 (18): 11908–11913. https://doi.org/10.1073/pnas.172398899.

Dizdaroglu, Miral, Güldal Kirkali, and Pawel Jaruga. 2008. "Formamidopyrimidines in DNA: Mechanisms of Formation, Repair, and Biological Effects." *Free Radical Biology and Medicine* 45 (12): 1610–1621. https://doi.org/10.1016/j.freeradbiomed.2008.07.004.

D'Orazio, John, Stuart Jarrett, Alexandra Amaro-Ortiz, and Timothy Scott. 2013. "UV Radiation and the Skin." *International Journal of Molecular Sciences* 14 (6): 12222–12248. https://doi.org/10.3390/ijms140612222.

Douki, T., I. Bérard, A. Wack, and S. Andrä. 2014. "Contribution of Cytosine-Containing Cyclobutane Dimers to DNA Damage Produced by Photosensitized Triplet-Triplet Energy Transfer." *Chemistry* 20 (19): 5787–5794. https://doi.org/10.1002/chem.201303905.

Ellis, Samantha R., Mimi Nguyen, Alexandra R. Vaughn, Manisha Notay, Waqas A. Burney, Simran Sandhu, and Raja K. Sivamani. 2019. "The Skin and Gut Microbiome and Its Role in Common Dermatologic Conditions." *Microorganisms* 7 (11): 1–19. https://doi.org/10.3390/microorganisms7110550.

Felton, S. J., M. S. Cooke, R. Kift, J. L. Berry, A. R. Webb, P. M. W. Lam, F. R. de Gruijl, A. Vail, and L. E. Rhodes. 2016. "Concurrent Beneficial (Vitamin D Production) and Hazardous (Cutaneous DNA Damage) Impact of Repeated Low-Level Summer Sunlight Exposures." *British Journal of Dermatology* 175 (6): 1320–1328. https://doi.org/10.1111/bjd.14863.

Fulop, T., A. Larbi, G. Pawelec, A. Khalil, A. A. Cohen, K. Hirokawa, J. M. Witkowski, and C. Franceschi. 2021. "Immunology of Aging: The Birth of Inflammaging." *Clinical Reviews in Allergy and Immunology* 18: 1–14. https://doi.org/10.1007/s12016-021-08899-6.

Garcia-Ruiz, A., K. Kornacker, and D. E. Brash. 2022. "Cyclobutane Pyrimidine Dimer Hyperhotspots as Sensitive Indicators of Keratinocyte UV Exposure." *Photochem Photobiol* 98 (5): 987–997. https://doi.org/doi: 10.1111/php.13683.

Ghaly, Simon, Nadeem O. Kaakoush, Frances Lloyd, Lavinia Gordon, Cynthia Forest, Ian C. Lawrance, and Prue H. Hart. 2018. "Ultraviolet Irradiation of Skin Alters the Faecal Microbiome Independently of Vitamin D in Mice." *Nutrients* 10 (8): 1–15. https://doi.org/10.3390/nu10081069.

Ghosh, Kanad, and Brian C. Capell. 2016. "The Senescence-Associated Secretory Phenotype: Critical Effector in Skin Cancer and Aging." *J Invest Dermatol* 136 (11): 2133–2139. https://doi.org/10.1016/j.jid.2016.06.621.

Glickman, Randolph D. 2002. "Phototoxicity to the Retina: Mechanisms of Damage." *International Journal of Toxicology* 21 (6): 473–490. https://doi.org/10.1080/10915810290169909.

Hofmann, Uta B., Roland Houben, Eva-b Bröcker, and Jürgen C Becker. 2005. "Role of Matrix Metalloproteinases in Melanoma Cell Invasion." *Biochimie* 87: 307–314. https://doi.org/10.1016/j.biochi.2005.01.013.

Ivanov, Iliya V., Timo Mappes, Patrick Schaupp, Christian Lappe, and Siegfried Wahl. 2018. "Ultraviolet Radiation Oxidative Stress Affects Eye Health." *Journal of Biophotonics* 11 (7): 1–13. https://doi.org/10.1002/jbio.201700377.

Jornot, Lan, Hilke Petersen, and Alain F. Junod. 1997. "Modulation of the DNA Binding Activity of Transcription Factors CREP, NFκB and HSF by H2O2 and TNFα: Differences between in Vivo and in Vitro Effects." *FEBS Letters* 416 (3): 381–386. https://doi.org/10.1016/S0014-5793(97)01244-1.

Jung, Yuyeon, Hojin Kim, Hee Min Sun, Goo Rhee Sue, and Woojin Jeong. 2008. "Dynein Light Chain LC8 Negatively Regulates NF-KB through the Redox-Dependent Interaction with IκBα." *Journal of Biological Chemistry* 283 (35): 23863–23871. https://doi.org/10.1074/jbc.M803072200.

Kamata, Hideaki, Tomoyuki Manabe, Shin ichi Oka, Keiko Kamata, and Hajime Hirata. 2002. "Hydrogen Peroxide Activates IκB Kinases through Phosphorylation of Serine Residues in the Activation Loops." *FEBS Letters* 519 (1–3): 231–237. https://doi.org/10.1016/S0014-5793(02)02712-6.

Kammeyer, A., and R. M. Luiten. 2015. "Oxidation Events and Skin Aging." *Ageing Research Reviews* 21: 16–29. https://doi.org/10.1016/j.arr.2015.01.001.

Kim, Hanseul, and Edward Giovannucci. 2020. "Vitamin d Status and Cancer Incidence, Survival, and Mortality". *Advances in Experimental Medicine and Biology* 1268. https://doi.org/10.1007/978-3-030-46227-7_3.

Kostyuk, Vladimir, Alla Potapovich, Andrea Stancato, Chiara de Luca, Daniela Lulli, Saveria Pastore, and Liudmila Korkina. 2012. "Photo-Oxidation Products of Skin Surface Squalene Mediate Metabolic and Inflammatory Responses to Solar UV in Human Keratinocytes." *pLoS One* 7 (8): 1–11. https://doi.org/10.1371/journal.pone.0044472.

Koyun, Efe, Remzi Karadag, Seyma Ozkanli, Serpil Oguztuzun, Arzu Kaya Kocdogan, and Isilay Ozsoy. 2020. "Caspase-3, P53 and Bcl-2 Expression in Basal Cell Carcinoma of the Eyelid." *Postepy Dermatologii i Alergologii* 37 (4): 535–539. https://doi.org/10.5114/ada.2020.98285.

Lickley, Megan, Susan Solomon, Sarah Fletcher, Guus J. M. Velders, John Daniel, Matthew Rigby, Stephen A. Montzka, Lambert J. M. Kuijpers, and Kane Stone. 2020. "Quantifying Contributions of Chlorofluorocarbon Banks to Emissions and Impacts on the Ozone Layer and Climate." *Nature Communications* 11 (1). https://doi.org/10.1038/s41467-020-15162-7.

Lo, Hsin Lung, Satoshi Nakajima, Lisa Ma, Barbara Walter, Akira Yasui, Douglas Ethell, and Laurie B. Owen. 2005. "Differential Biologic Effects of CPD and 6-4PP UV-Induced DNA Damage on the Induction of Apoptosis and Cell-Cycle Arrest." *BMC Cancer* 5: 1–9. https://doi.org/10.1186/1471-2407-5-135.

Lu, C. Y., M. S. Hsieh, K. C. Wei, M. Ezmerli, C. H. Kuo, and W. Chen. 2020. "Gastrointestinal Involvement of Primary Skin Diseases." *J Eur Acad Dermatol Venereol* 34 (12): 2766–2774. https://doi.org/10.1111/jdv.16676.

Maeda, Kazuhisa, and Masato Hatao. 2004. "Involvement of Photooxidation of Melanogenic Precursors in Prolonged Pigmentation Induced by Ultraviolet A." *Journal of Investigative Dermatology* 122 (2): 503–509. https://doi.org/10.1046/j.0022-202X.2004.22223.x.

Marinho, H. Susana, Carla Real, Luísa Cyrne, Helena Soares, and Fernando Antunes. 2014. "Hydrogen Peroxide Sensing, Signaling and Regulation of Transcription Factors." *Redox Biology* 2 (1): 535–562. https://doi.org/10.1016/j.redox.2014.02.006.

Matejuk, Agata. 2018. "Skin Immunity." *Archivum Immunologiae et Therapiae Experimentalis* 66 (1): 45–54. https://doi.org/10.1007/s00005-017-0477-3.

Matsumura, Yasuhiro, and Honnavara N. Ananthaswamy. 2004. "Toxic Effects of Ultraviolet Radiation on the Skin." *Toxicology and Applied Pharmacology* 195 (3): 298–308. https://doi.org/10.1016/j.taap.2003.08.019.

Stephen A. Montzka, Geoff S. Dutton, Pengfei Yu, Eric Ray, Robert W. Portmann, John S. Daniel, Lambert Kuijpers, Brad D. Hall, Debra Mondeel, Carolina Siso, J. David Nance, Matt Rigby, Alistair J. Manning, Lei Hu, Fred Moore, Ben R. Miller & James W. Elkins. 2018. "An Unexpected and Persistent Increase in Global Emissions of Ozone-Depleting CFC-11." *Nature* 557 (7705): 413–417. doi: 10.1038/s41586-018-0106-2.

Mousavi, Sayed Esmaeil, Juana Maria Delgado-Saborit, Anna Adivi, Sara Pauwels, and Lode Godderis. 2022. "Air Pollution and Endocrine Disruptors Induce Human Microbiome Imbalances: A Systematic Review of Recent Evidence and Possible Biological Mechanisms." *Science of the Total Environment* 816: 151654. https://doi.org/10.1016/j.scitotenv.2021.151654.

Nair, Rathish, and Arun Maseeh. 2012. "Vitamin D: The Sunshine Vitamin." *Journal of Pharmacology and Pharmacotherapeutics* 3 (2): 118–126. https://doi.org/10.4103/0976-500X.95506.

Nedoszytko, Bogusław, Małgorzata Sokołowska-Wojdyło, Katarzyna Ruckemann-Dziurdzińska, Jadwiga Roszkiewicz, and Roman J. Nowicki. 2014. "Chemokines and Cytokines Network in the Pathogenesis of the Inflammatory Skin Diseases: Atopic Dermatitis, Psoriasis and Skin Mastocytosis." *Postepy Dermatologii i Alergologii* 31 (2): 84–91. https://doi.org/10.5114/pdia.2014.40920.

Notara, M., S. Behboudifard, M. A. Kluth, C. Maßlo, C. Ganss, M. H. Frank, B. Schumacher, and C. Cursiefen. 2018. "UV Light-Blocking Contact Lenses Protect against Short-Term UVB-Induced Limbal Stem Cell Niche Damage and Inflammation." *Scientific Reports* 8 (1): 1–12. https://doi.org/10.1038/s41598-018-30021-8.

Ovadya, Yossi, Tomer Landsberger, Hanna Leins, Ezra Vadai, Hilah Gal, Anat Biran, Reut Yosef, et al. 2018. "Impaired Immune Surveillance Accelerates Accumulation of Senescent Cells and Aging." *Nature Communications* 9 (1). https://doi.org/10.1038/s41467-018-07825-3.

Passeron, Thierry, and Mauro Picardo. 2018. "Melasma, a Photoaging Disorder." *Pigment Cell and Melanoma Research* 31 (4): 461–465. https://doi.org/10.1111/pcmr.12684.

Pecorelli, Alessandra, Brittany Woodby, Roxane Prieux, and Giuseppe Valacchi. 2019. "Involvement of 4-Hydroxy-2-Nonenal in Pollution-Induced Skin Damage." *BioFactors* 45 (4): 536–547. https://doi.org/10.1002/biof.1513.

Rastogi, Rajesh P., Richa, Ashok Kumar, Madhu B. Tyagi, and Rajeshwar P. Sinha. 2010. “Molecular Mechanisms of Ultraviolet Radiation-Induced DNA Damage and Repair.” *Journal of Nucleic Acids* 2010. https://doi.org/10.4061/2010/592980.

Rastrelli, Marco, Saveria Tropea, Carlo Riccardo Rossi, and Mauro Alaibac. 2020. “Melanoma: Epidemiology, Risk Factors, Pathogenesis, Pathogenesis, Diagnosis and Classification.” *In Vivo* 28 (6): 1005–1011. https://doi.org/10.32388/7xj0gw.

Reddy, Venkat N., Frank J. Giblin, Li Ren Lin, and Bhargavan Chakrapani. 1998. “The Effect of Aqueous Humor Ascorbate on Ultraviolet-B-Induced DNA Damage in Lens Epithelium.” *Investigative Ophthalmology and Visual Science* 39 (2): 344–350.

Riihilä, Pilvi, Liisa Nissinen, and Veli Matti Kähäri. 2021. “Matrix Metalloproteinases in Keratinocyte Carcinomas.” *Experimental Dermatology* 30 (1): 50–61. https://doi.org/10.1111/exd.14183.

Salminen, Antero, Kai Kaarniranta, and Anu Kauppinen. 2022. “Photoaging: UV Radiation-Induced Inflammation and Immunosuppression Accelerate the Aging Process in the Skin.” *Inflammation Research* 71 (7–8): 817–831. https://doi.org/10.1007/s00011-022-01598-8.

Schauber, Jürgen, and Richard L. Gallo. 2008. “Antimicrobial Peptides and the Skin Immune Defense System.” *Journal of Allergy and Clinical Immunology*. https://doi.org/10.1016/j.jaci.2008.03.027.

Schreck, Ralf, Peter Rieberl, and Patrick A. Baeuerle. 1991. “Reactive Oxygen Intermediates as Apparently Widely Used Messengers in the Activation of the NF-XB Transcription Factor and HIV-1.” *The EMBO Journal* 10 (8): 2247–2258. https://doi.org/10.1016/j.bbrc.2010.02.147.

Sies, Helmut. 2017. “Hydrogen Peroxide as a Central Redox Signaling Molecule in Physiological Oxidative Stress: Oxidative Eustress.” *Redox Biology* 11 (November 2016): 613–619. https://doi.org/10.1016/j.redox.2016.12.035.

Singh, Shipra K., Rujul Jain, Shipra K. Singh, Cristian Arvinte, Maharaj Singh, Paul E. Marik, Chuen Wen Tan M. R. C. P., et al. 2020. “Role of Vitamin D in Preventing of COVID-19 Infection, Progression and Severity.” *Journal Of Infection and Public Health* 74 (January): 9. https://doi.org/10.1016/j.lfs.2020.117808%0Ahttps://doi.org/10/1016/j/numecd.2020.05.020.

Skopelja-Gardner, Sladjana, Joyce Tai, Xizhang Sun, Lena Tanaka, James A. Kuchenbecker, Jessica M. Snyder, Paul Kubes, Tomas Mustelin, and Keith B. Elkon. 2021. “Acute Skin Exposure to Ultraviolet Light Triggers Neutrophil-Mediated Kidney Inflammation.” *Proceedings of the National Academy of Sciences of the United States of America* 118 (3): 1–12. https://doi.org/10.1073/pnas.2019097118.

Slominski, Andrzej, Ralf Paus, and Dirk Schadendorf. 1993. “Melanocytes as ‘Sensory’ and Regulatory Cells in the Epidermis.” *Journal of Theoretical Biology*. https://doi.org/10.1006/jtbi.1993.1142.

Slominski, Andrzej T., Michal A. Zmijewski, Cezary Skobowiat, Blazej Zbytek, Radomir M. Slominski, and Jeffery D. Steketee. 2012. “Sensing the Environment: Regulation of Localand Global Homeostasis by the Skin’s Neuroendocrine System.” *Advances in Anatomy Embryology and Cell Biology* 212. https://doi.org/10.1007/978-3-642-19683-6.

Syed, Hassnain R., Suman Shekar, and Avinash Aravantagi. 2021. “Melanoma and the Gastrointestinal (GI) Tract: Maintaining a High Index of Suspicion.” *Cureus* 13 (2): 3–7. https://doi.org/10.7759/cureus.13408.

Tewari, Angela, Mette M. L. Grage, Graham I. Harrison, Robert Sarkany, and Antony R. Young. 2013. “UVA1 Is Skin Deep: Molecular and Clinical Implications.” *Photochemical and Photobiological Sciences* 12 (1): 95–103. https://doi.org/10.1039/c2pp25323b.

Umar, Meenakshi, Konduru S. Sastry, Fatima Al Ali, Moza Al-Khulaifi, Ena Wang, and Aouatef I. Chouchane. 2018. “Vitamin D and the Pathophysiology of Inflammatory Skin Diseases.” *Skin Pharmacology and Physiology* 31 (2): 74–86. https://doi.org/10.1159/000485132.

Vangipuram, R., and S. R. Feldman. 2016. “Ultraviolet Phototherapy for Cutaneous Diseases: A Concise Review.” *Oral Diseases* 22 (4): 253–259. https://doi.org/10.1111/odi.12366.

Vogeley, Christian, Charlotte Esser, Thomas Tüting, Jean Krutmann, and Thomas Haarmann-Stemmann. 2019. “Role of the Aryl Hydrocarbon Receptor in Environmentally Induced Skin Aging and Skin Carcinogenesis.” *International Journal of Molecular Sciences* 20 (23): 6005. https://doi.org/10.3390/ijms20236005.

Walterscheid, Jeffrey P., Dat X. Nghiem, Nasser Kazimi, Leta K. Nutt, David J. Mcconkey, Mary Norval, and Stephen E. Ullrich. 2006. “Immunosuppressive Factor, Activates Immune Suppression via the 5-HT 2A Receptor.” *Proceedings of the National Academy of Sciences* 103 (46): 17420–17425.

Wu, Xiayu, Tao Zhou, Neng Cao, Juan Ni, and Xu Wang. 2014. "Role of Vitamin D Metabolism and Activity on Carcinogenesis." *Oncology Research* 22 (3): 129–137. https://doi.org/10.3727/096504015X14267282610894.

Yadav, Umesh C. S., and Kota V. Ramana. 2013. "Regulation of NF-KB-Induced Inflammatory Signaling by Lipid Peroxidation-Derived Aldehydes." *Oxidative Medicine and Cellular Longevity* 2013. https://doi.org/10.1155/2013/690545.

Yamaba, Hiroyuki, Manami Haba, Mayumi Kunita, Tsutomu Sakaida, Hiroshi Tanaka, Youichi Yashiro, and Satoru Nakata. 2016. "Morphological Change of Skin Fibroblasts Induced by UV Irradiation Is Involved in Photoaging." *Experimental Dermatology* 25 (May): 45–51. https://doi.org/10.1111/exd.13084.

Yamauchi, Ritsuko, Akimichi Morita, Yoko Yasuda, Susanne Grether-Beck, Lars Oliver Klotz, Takuo Tsuji, and Jean Krutmann. 2004. "Different Susceptibility of Malignant versus Nonmalignant Human T Cells Toward Ultraviolet A-1 Radiation-Induced Apoptosis." *Journal of Investigative Dermatology* 122 (2): 477–483. https://doi.org/10.1046/j.0022-202X.2003.22106.x.

Yamazaki, Sayuri, Mizuyu Odanaka, Akiko Nishioka, Saori Kasuya, Hiroaki Shime, Hiroaki Hemmi, Masaki Imai, et al. 2018. "Ultraviolet B: Induced Maturation of CD11b-Type Langerin – Dendritic Cells Controls the Expansion of Foxp3 + Regulatory T Cells in the Skin." *The Journal of Immunology* 200 (1): 119–129. https://doi.org/10.4049/jimmunol.1701056.

Yang, Ming Hui, Adnan Dibas, and Yu Chang Tyan. 2013. "Changes in Retinal Aquaporin-9 (AQP9) Expression in Glaucoma." *Bioscience Reports* 33 (2): 379–385. https://doi.org/10.1042/BSR20130005.

Yardman-Frank, Joseph Michael, and David E. Fisher. 2022. "Skin Pigmentation and Its Control: From Ultraviolet Radiation to Stem Cells." *Physiology & Behavior* 4 (30): 560–571. https://doi.org/10.1111/exd.14260.

Zegarska, Barbara, Katarzyna Pietkun, Wojciech Zegarski, Paulina Bolibok, Marek Wiśniewski, Katarzyna Roszek, Joanna Czarnecka, and Maciej Nowacki. 2017. "Air Pollution, UV Irradiation and Skin Carcinogenesis : What We Know, Where We Stand and What Is Likely to Happen in the Future ?" *Advances in Dermatology and Allergology*: 6–14.

Zhao, Zhi Chun, Ying Zhou, Gang Tan, and Juan Li. 2018. "Research Progress about the Effect and Prevention of Blue Light on Eyes." *International Journal of Ophthalmology* 11 (12): 1999–2003. https://doi.org/10.18240/ijo.2018.12.20.

19 Electromagnetic Fields Redox Signaling Modulation in Brain

Giorgia Innamorati, Barbara Benassi, and Claudia Consales

19.1 INTRODUCTION

Electromagnetic fields (EMFs) are generated by moving electric charges and are categorized as non-ionizing radiation since they are unable to ionize the matter due to their insufficient energy. EMFs derive from natural and artificial sources. The first one includes solar radiation (photosynthetically active radiation, UV, and IR rays), atmospheric electric discharge (lightning), and the geomagnetic field; examples of artificial sources are power lines, radars, radio-television systems, and radio base stations for cellular telephony.

Electromagnetic fields are mainly characterized by frequency and wavelength, two parameters inversely related: the higher the frequency, the shorter the wavelength and vice versa. Based on their frequency, EMFs can be classified as extremely low frequency electromagnetic fields (ELF-MFs < 300 Hz), intermediate frequency (300 Hz–10 MHz), and radiofrequency (RF-EMFs 10 MHz–300 GHz) (Figure 19.1).

Over the past 20 years, the enormous spread of both ELF and RF sources among the population has raised doubts and concerns about the effect of this non-ionizing radiation on biological systems. For this reason, several experiments have been performed in the presence of both low and high frequency electromagnetic fields, which has led the International Commission on Non-Ionizing Radiation Protection (ICNIRP) to establish guidelines to different exposure limits for workers and people from the public. The necessity of applying this difference lies in the fact that, while workers are aware of being exposed to sources of electromagnetic fields, the rest of the population is not, so they do not take the necessary measures to protect themselves. The limits for whole body exposure to 50 Hz, the frequency most present in household and commonly used appliances, is 1 mT for workers and 200 μT for the public ("Guidelines for Limiting Exposure to Time-Varying Electric and Magnetic Fields [1 Hz TO 100 KHz]" 2010). About RF, considering mobile frequency, the whole-body limit of the specific absorption rate (SAR, which is a measure of the rate at which energy is absorbed per unit mass by a human body when exposed to RF) for workers is 0,4 W/kg, while for public is 0,08 W/Kg ("Guidelines for Limiting Exposure to Electromagnetic Fields [100 KHz to 300 GHz]" 2020). In 2021, an integrated exposure model was developed to estimate the daily RF dose to the brain and the whole body. The study was conducted in four European countries (France, Spain, Switzerland, and the Netherlands) considering several parameters, involving the individual (age, sex, body mass) and the RF source (output power, type of device, the way devices are being used, etc.). The results indicate a median whole-body and whole-brain dose of 183.7 mJ/kg/day and 204.4 mJ/kg/day, respectively (van Wel et al. 2021).

Our attention here is focused on the possible modulation of redox balance operated by EMFs exposure. The results achieved so far on this topic have not yet described a definitive role of EMFs on these molecular aspects; indeed beneficial and adverse effects are both described resulting in controversial opinions.

The hypothetic mechanism majorly consolidated through which the EMFs could cause the imbalance of redox signaling is the radical pair. EMFs, through the magnetic component of their field, can stabilize the radical pair, allowing their dispersion and preventing their return to ground

DOI: 10.1201/9781003328100-23

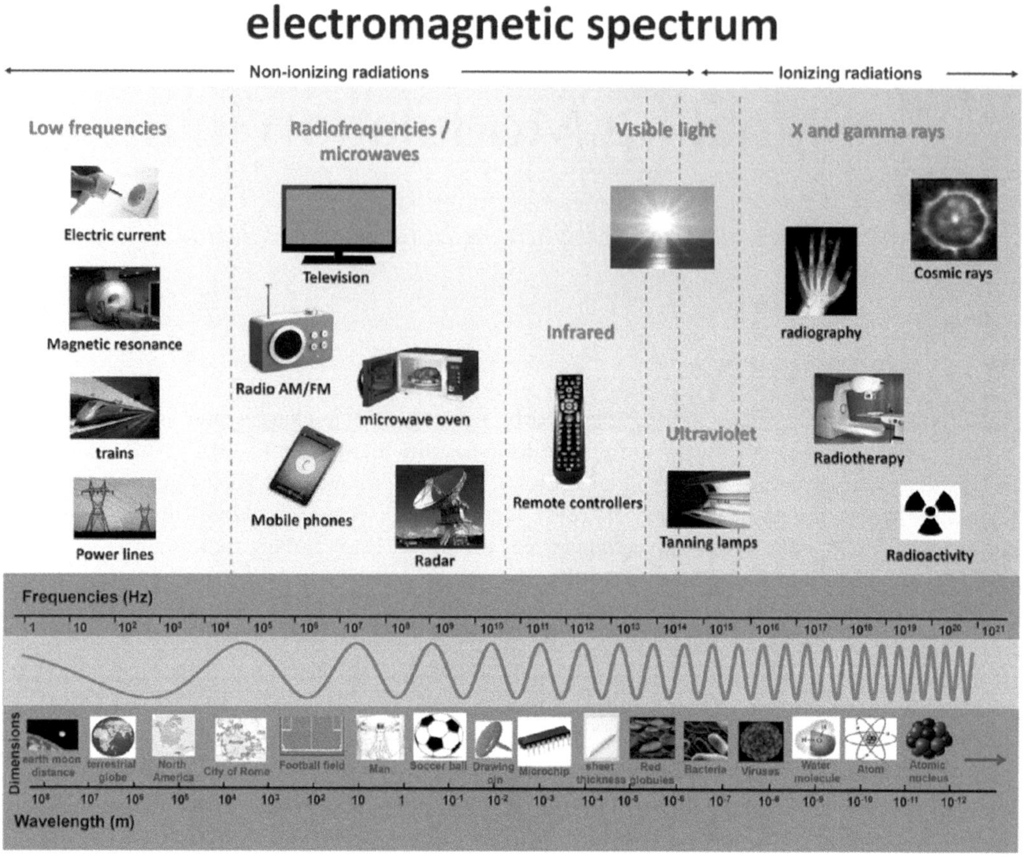

FIGURE 19.1 Electromagnetic spectrum and electronic devices emission ranges.

state. This mechanism also explains how a substantially weak interaction, such as that of magnetic fields with organisms, can result in a clearly visible effect (Zadeh-Haghighi and Simon 2022).

The brain is the organ most sensitive to changes in the redox balance. Redox homeostasis, indeed, is involved in all aspects of central nervous system (CNS), such as development, function, aging, and disease, and can be considered an integrated process among all the different cell populations, like astrocytes and glia, that compose it. The brain is a high energy consumer, which makes it highly vulnerable to oxidative damage (Lee et al. 2021).

Here we will explore how the EMF can impact the redox homeostasis of CNS; we will report results from epidemiology, *in vivo*, and *in vitro* studies.

19.2 EXTREMELY LOW FREQUENCY MAGNETIC FIELDS AND REDOX HOMEOSTASIS

19.2.1 ELF-MF Epidemiological Studies

The possible association of exposure to ELF-MFs and the onset of juvenile acute leukemia has been the subject of studies for about 40 years, ever since Wertheimer and Leeper published a retrospective study in 1979 in which they highlighted an excess of electrical wiring in California, suggesting a high current-flow in the proximity of the homes of children who had fallen ill with leukemia

(Wertheimer and Leeper Wertheimer 1979). These data were only partially confirmed by other epidemiological results, which led, in 2002, the IARC to classify the ELF-MFs as 2B, so possibly carcinogenic to humans (IARC Monographs on the Evaluation of carcinogenic risks to humans Non-Ionizing Radiation, Part 1: Static and Extremely Low-Frequency (ELF) Electric and Magnetic Fields 2001).

Besides leukemia, the possible effect of occupational or residential exposure on the onset of brain disorders, both tumors and neurodegenerative diseases, was investigated.

The first epidemiological work that dealt with analyzing an association between ELF-MFs and Alzheimer's disease (AD) dates back to 1995. The authors, analyzing three cohorts of patients from Koskela hospital and neurology department both in Helsinki and the California Alzheimer's disease center, demonstrated an association between patients exposed to medium and high intensity ELF-MFs during their working life and the disease (Sobel et al. 1995). The effect of ELF-MFs on the onset of AD was confirmed by the same authors in a work in which a larger patient population was considered (Davanipour et al. 2007).

However, further studies have not completely confirmed this data, therefore a direct association between exposure to ELF and AD has not been yet 100% proven (Gunnarsson and Bodin 2019).

As for AD, epidemiological results about the influence of ELF-MFs exposure on amyotrophic lateral sclerosis pathogenesis are conflicting, even though a slight, but significant association seems to be present, as opposed to Parkinson's disease, where the results of the epidemiological works agree that there is no association (Turner et al. 2017).

Several epidemiological studies have taken into consideration that ELF-MFs exposure could be associated with brain tumors, both in adult patients exposed for occupational and in children exposed for residential reasons. The tumors considered are mainly gliomas and meningiomas and, as for the neurodegenerative diseases, no firm conclusion was reached (Turner et al. 2017; Kheifets et al. 2010).

From the aforementioned, there is no definitive confirmation of a possible effect of exposure to ELF on the onset of brain diseases to date. This is due to the intrinsic difficulty of this type of study to trace the real exposure in terms of time, intensity, distance from the source of individuals during their life, and to the lack of a full knowledge of the molecular and cellular pathways affected by ELF-MF exposure.

19.2.2 ELF-MF *In Vivo* Studies

As described before, the possible link among the effect of ELF-MFs on neurodegenerative diseases could be a possible influence of magnetic fields on redox homeostasis imbalance.

Klimek et al. (2022) described that repeated exposure to ELF-MF (50 Hz 1 mT and 7 mT) causes, in rat prefrontal cortex, the induction of oxidative stress, demonstrated by an increase of protein carbonyl (CP) groups and 8-isoprostanes levels and a reduction of total antioxidant capacity, in an intensity and number of exposure dependent manner. A dose effect response was also observed in rats' brains exposed to 50 Hz, 50 and 100 μT for 90 days (Manikonda et al. 2014) and to 100 and 500 μT, 2h/day for 10 months (Akdag et al. 2010), where the higher dose induced a bigger impairment of oxidative stress measured through the increase of reactive oxygen species (ROS) production and of lipid peroxidation (LPO) and a reduction of antioxidant content as determined by the evaluation of reduced (GSH) and oxidized (GSSG) ratio.

Even though it did not induce cell apoptosis, interestingly, the oxidative response induced by ELF-MF was found to be correlated to the age of animals exposed. 50 Hz, 0.1 mT magnetic field applied for 10 days enhanced the activity of superoxide dismutase (SOD), catalase (CAT), glutathione peroxidase (GPX), and glutathione S transferase (GST), which constitute the anti-oxidative enzymatic defense, in young female rats; while, in aged animals, not only was this increase not visible, but a significant decrease in the major antioxidant enzymatic activities was shown (Falone et al. 2008).

Differently from the previous papers, a restoring of antioxidant status of cerebrum induced by 7 days exposure to 60 Hz, 2.5 mT field, was shown by Quesnel-Galván et al. (2021). The results were confirmed in a rat model for spinal cord injury, where a 50 Hz and 17.96 μT field applied just after the injury had beneficial effects on antioxidant capacity by improving SOD and catalase activities and reducing cell apoptosis (Ambalayam and Mathur 2022).

Furthermore, a chronic exposure (2 hours/day for 27 days) to 50 hz, 1 mT field increased the neurogenesis of hippocampal neurons of mice presenting focal cerebral ischemia (Gao et al. 2021). In a mouse model of Huntington disease, the exposure to 60 Hz 0.7 mT for 21 days reversed the neuronal death and the oxidative stress, also favoring the release of neurotrophic factors and the improvement of animals' behavior (Tasset et al. 2012). The effect of stimulation on oxidative stress was NRF2 mediated since a significant increase of its expression was found in the cytoplasm and nucleus of neurons of exposed mice (Tasset et al. 2013).

On the same line several studies have analyzed the influence of this field on specific hallmarks of AD (memory, Aβ amyloid plaque, neurofibrillary tangles, and inflammation).

One of the first papers devoted to understanding the effects of ELF-MFs on spatial memory dates back to 1998, where the authors analyzed the effect of acute exposure to 50 Hz field between 7.5 mT and 7.5 mT for 45 min before the test. Mice exposed just before the test showed an impairment of the performance, which was dependent on the intensity of the magnetic field applied, even though their overall learning ability was not affected. If the test was executed 45 minutes after the ELF-MFs exposure, no difference in the performance was observed among the two groups of animals, control and exposed (Sienkiewicz et al. 1998). An impairment of memory was also demonstrated by Fu and coauthors (2008). Exposing mice to 25 and 50 Hz 140 mT for 7 and 25 days, they observed an effect on mice subjected to the higher frequency and for the longer period. These data haven't been confirmed by Zhang et al. (2015), who demonstrated that 12 weeks exposure to 50 Hz 400 μT did not alter rats' memory, nor did it affect the process of Aβ deposition.

ELF-MFs effects were also investigated in different AD animal model. The chronic (12 weeks) exposure of AD mice, obtained by aluminum administration, to 50 Hz 100 μT did not affect the AD pathogenesis, either in terms of memory worsening or in terms of deposition of Aβ deposition (Zhang et al. 2013). Several papers, which share the same frequency investigated, 50 Hz, but at different intensities and exposure time, show a beneficial effect induced by magnetic fields in terms of spatial memory improvement, increasing of synaptic plasticity and hippocampal neurogenesis, and reduction of β-amyloid plaques (Liu et al. 2015; Hu et al. 2016; Sakhaie et al. 2017; Bobkova et al. 2018; Akbarnejad et al. 2018; Zuo et al. 2018). Investigating the possible pathway involved in this effect, Zuo and coauthors (2018) identified the nuclear factor kB (NF-kB) pathway as involved in the response to ELF-MFs. NF-kB has been demonstrated to play a central role in AD pathogenesis, being able both to carry out a neuroprotective and neurodegenerative response. In their paper, Zuo et al. found that in the hippocampus of AD rat model, 50 Hz, 400 μT field applied continuously for 1 month could upregulate the expression of Raf kinase inhibitor protein (RKIP), of phosphorylated IkB kinase (IKK), an important mediator of NF-kB function, and of the phosphorylated NF-kB that were all downregulated in AD rats (Zuo et al. 2018).

19.2.3 ELF-MF *In Vitro* Studies

To support *in vivo* studies, the effect of ELF-MFs exposure has been extensively investigated in cultured neuronal cells, both of tumor origin and primary cultures. The results of these studies are summarized in Figure 19.2.

Exposing the neuroblastoma cell line SH-SY5Y to 50 Hz 1 mT from 24 up to 72 hours, an increase in ROS production, a decrease of antioxidant content, and an increase of protein carbonylation was observed, even though without any influence on cell death (Benassi et al. 2016). The impact on redox homeostasis was confirmed in Reale et al. (2014), where they observed an increase of iNOS expression and a modulation of CYP-450 and catalase activity.

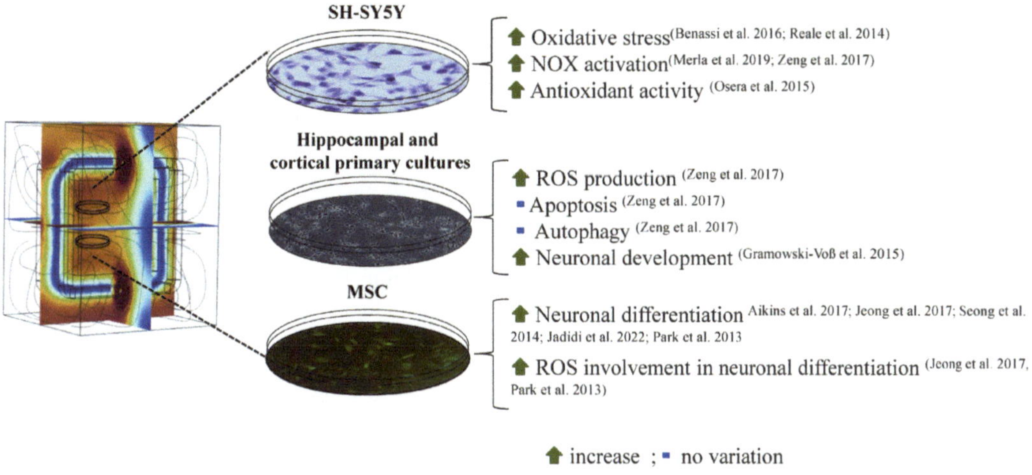

FIGURE 19.2 Schematic representation of *in vitro* ELF-MF exposure results. The magnetic field (also termed "B") distribution within the exposure setup as shown in Benassi et al., 2016, and Merla et al., 2019. It is here reported as an example of ELF-MF device for *in vitro* exposure.

The increase of cytoplasmic ROS was demonstrated, in SH-SY5Y, to be partially due to the activation of nicotinamide adenine dinucleotide phosphate (NADPH oxidase) (Merla et al. 2019). The involvement of NADPH oxidase was shown to be responsible for ROS increase also in hippocampal primary cultures. These cultures were exposed at 50-Hz 2-mT ELF-MF once for 30 min, 8 h, or 24 h from DIV 1–7 or DIV 7–14 (days *in vitro*). ROS increase was observable in cells at DIV 7 only when the field was applied for 8 hours and became significant in cells at DIV 14 with both the time exposures. Analyzing the expression of NOX2, which is the most abundant in the brain, the authors found that its expression was enhanced after 30 minutes and 8 hours repeated exposures (Zeng et al. 2017). Also in this paper, cells did not show, other than ROS modulation, any effect on apoptosis, autophagy, or DNA damage.

Different from previous results, applying a field at 75 Hz and 2 mT 10 min/week for 4 weeks, SH-SY5Y cells did not show any increase of ROS, but, when they were treated with H_2O_2, they showed a lower amount of ROS than sham exposed cells through the activation of mnSOD (Osera et al. 2015).

Really interesting are the results related to the ability of ELF-MFs to influence neuronal differentiation, also by impacting the ROS signaling.

The exposure of primary cortical cell cultures to 30 min of a modulating frequency of 10 Hz followed by 30 min at 16 Hz twice/week for 24 days, an enhancement of GABAergic neurons was observed, together with an increase of the cortical activity (Gramowski-Voß et al. 2015).

Mesenchymal stem cells (MSCs) exposed to the same field 50 Hz, 1 mT but for different days (12 days [Aikins et al. 2017; Jeong et al. 2017]; 8 days [Seong et al. 2014]), 75 Hz, 400 μT 1h/week or 7h/day (Jadidi et al. 2022), differentiated toward a neuronal phenotype. Several pathways have been investigated in these studies, and Jeong et al. showed that the neuronal transformation of MSCs was mediated by increasing ROS that upregulated SIRT1, with a consequent modulation of its downstream factors HES1 and MASH1 (Jeong et al. 2017).

The involvement of ROS in MSCs neuronal differentiation was also demonstrated by Park and coauthors (Park et al. 2013). By exposing cells to 50 or 100 Hz, 1 mT, the authors observed an increase of ROS and, subsequently, an up-regulation of the expression of MAP2, NF-L, and Neurod1 6 days after the end of the exposure. Pretreating the cells with scavenging agents just before the ELF-MFs exposure, neither ROS, nor neuronal markers increase was visible.

19.3 RADIOFREQUENCY AND REDOX HOMEOSTASIS

19.3.1 RF Epidemiological Studies

In the last 20 years, radiofrequency exposure has sharply increased, mainly as a result of the spread and daily use of mobile phones (Jalilian et al. 2019). As described before, this has raised concerns about the possible toxic effects of these EMFs, especially on the nervous system.

In May 2011, IARC classified radiofrequencies as possibly carcinogenic to humans (Group 2B), based on the increased risk of developing glioma and acoustic neuroma (IARC Monographs on the Evaluation of carcinogenic risks to humans Non-Ionizing Radiation, Part 2: Radiofrequency Electromagnetic Fields, 2011).

In a study published in 2002 and conducted on individuals of both sexes who received diagnosis of brain cancer between 1 January 1997 and 30 June 2000, it highlighted a correlation between mobile phones use and an increased risk in developing brain cancer, especially astrocytoma (Hardell et al. 2002). The same authors, through a meta-analysis and a case study published in 2013, reiterated the close correlation between mobile phone use and development of brain tumors (Hardell et al. 2013a, 2013b). In 2011, another epidemiological study conducted on patients with brain cancer from Australia, Canada, France, Israel, and New Zealand showed an increased risk in developing localize glioma following long-term use of mobile phone (Cardis et al. 2011).

In contrast to the works just mentioned, there are several epidemiological studies that have not identified correlations between mobile phones and neurological pathologies. A recent MOBI-Kids case-control study, conducted in several countries on youth (average age 16.5–16.6) showed no significant brain tumors increases due to long-term use of mobile phone (Castaño-Vinyals et al. 2022). Moreover, studies focused on specific countries showed no significant variation in terms of cancer development (Bodewein et al. 2022).

As for ELF-MFs exposure, epidemiological studies do not reach a common result about its possible harmful effects; that's why *in vivo* and *in vitro* studies were performed to settle this issue.

19.3.2 RF *In Vivo* Studies

Variations in the level of enzymes involved in redox homeostasis were detected in studies conducted on male Fisher-344 rats exposed to several frequencies (900, 1800, and 2450 MHz) for 2h/day, 5 days/week, for a total of 60 days (average whole-body SAR 0.6 W/kg). The exposure induced in the hippocampus an increase in the levels of malondialdehyde (MDA), protein carbonyl (PCO), and CAT, with reduced GSH and SOD levels (Megha et al. 2015).

A statistically significant increase in lipid peroxidation was observed in the brain of male Wistar rats exposed to 2.45 GHz for 2h/day for 35 days, in presence of tissue remodeling with reduced pyramidal cell density, and degeneration of neurons in the cerebral cortex (Chauhan et al. 2017). In the same animal model, a similar increase in LPO, associated with reduced SOD and CAT activity, was detected following exposure at 900 MHz for 1h, 2h and 4h/day for 90 days. There was also a considerable reduction in glutathione content and an inhibited activity of GPX, glutathione reductase (GR), GST, and glucose-6-phosphate dehydrogenase (G-6-PDH) compared to control group (Sharma and Shukla 2020). Similar results were obtained in two other studies, one with the same animal model exposed to 2100 MHz for 3 months (4h/day, 5 days/week) (Sharma et al. 2022) and the other with Wistar Albino rats exposed for 30 days (30 min/day, 5 days/week) at 2600 MHz (Delen et al. 2021).

In addition, several experiments have been performed to assess the effect of RF-EMFs on pregnancy and offspring.

Experiments on dams and offspring (BALB/c mice), exposed at 900 and 1800 MHz for 2h/day on days 0–19 of pregnancy, showed in different tissues statistically significant increase in MDA levels and reduction of total thiol groups, CAT, and SOD activity (Bahreyni Toossi et al. 2018).

Focusing on the offspring, a study evaluating the RF effect (2450 MHz) on prenatal (exposure in utero and from day 21 until puberty) and postnatal (exposure from day 21 until puberty) pups, showed a significant increase in total oxidant status and oxidative stress index in both groups compared to the control (Sangun et al. 2015).

In contrast to the earlier described pro-oxidant effect, some studies highlight an absence of effect in terms of redox balance following the RF exposure.

For instance, exposure to 1950 MHz for 8 months (2h/day, 5 days/week) did not induce significant changes in the oxidative state and lipid peroxidation levels in female C57BL/6J mice (Jeong et al. 2018). The oxidative state, already altered in a rat model of Alzheimer's disease, has not been further increased by exposure to 900 MHz, regardless of the different brain average SAR and the exposure time (1.5 W/kg for 15 min, 6 W/kg for 15 and 45 min) (Bouji et al. 2020).

19.3.3 RF *In Vitro* Studies

To support *in vivo* studies, the effect of RF-EMFs exposure has been investigated in several *in vitro* studies conducted on tumoral and primary cell lines. The results of these studies are summarized in Figure 19.3.

The hypothesis of a prooxidative effect following exposure to RF-EMFs has been demonstrated on human SH-SY5Y neuroblastoma cells exposed at 872 MHz for 1h, following treatment with menadione. The post-treatment production of ROS was further increased, in a statistically significant way, after exposure to CW RF, at time point 30 and 60 minutes (Luukkonen et al. 2009).

The production of ROS was increased for 24h also in cortical neurons from newborn Sprague Dawley rats, exposed intermittently for 24h at 1800 MHz (Xu et al. 2010). A similar stimulation protocol (24h intermittent exposure to 1800 MHz GSM), evaluating a SAR-related effect (2W/kg and 4 W/kg), showed again an increase in ROS levels in spiral ganglion neurons from neonatal Sprague Dawley rats. This variation was observed with a SAR of 4 W/kg, with and without LPS treatment (Zuo et al. 2015).

SAR-dependent redox balance alterations have also been studied in HT22 mouse hippocampal neurons cells, exposed to an RF signal of 1950 MHz for 2h with a SAR of 2 W/kg, 4 W/kg, and 6 W/kg. After the exposure, cells were treated with glutamate, and the results were evaluated at several time

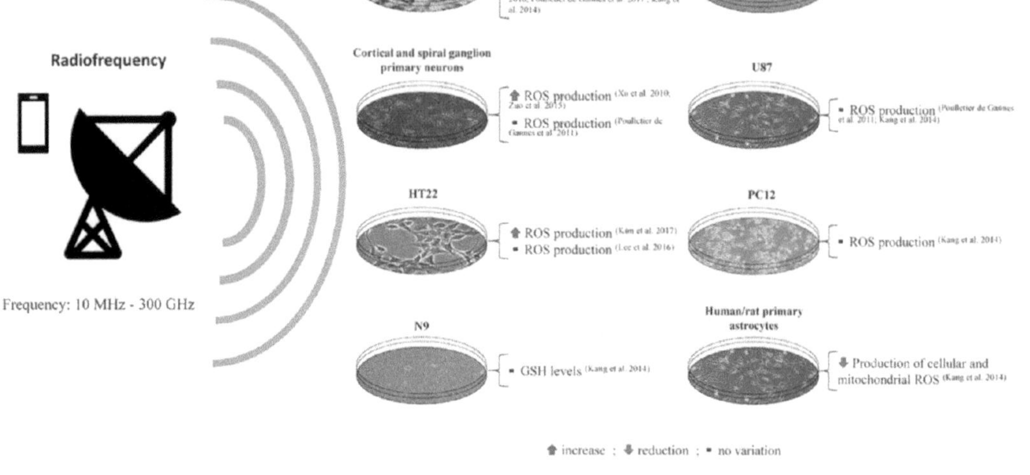

FIGURE 19.3 Schematic representation of *in vitro* RF exposure results.

points: in the early stages (1–6h), when the effect of glutamate is minimal, RF enhances ROS, and this was even more accentuated after 12h, when the glutamate-induced ROS was maximum. In this study, cells were also treated with H_2O_2, resulting in an up-regulation of ROS production following the exposure (Kim et al. 2017).

The prooxidative effect depends, not only on the stimulation protocol, but also on the cell line used, and this was demonstrated in a study conducted on human SH-SY5Y and murine N9 microglial cells, exposed at the same signal (GSM 935 MHz carrier frequency intermittent signal with an average SAR of 4 W/kg +/– 0.16 SD). After 6h from the stimulation, in SH-SY5Y there was a statistically significant reduction in GSH levels that did not occur in N9. No effect was observed in the hydrogen peroxide levels (Zielinski et al. 2020).

In contrast to the afore-described prooxidant effect, some studies highlight an absence of effect or even an antioxidant one following the RF exposure.

Human cells SH-SY5Y, stimulated with 872 MHz CW or GSM, have not reported alterations in the ROS production, not even after treatments with ferrous chloride (Luukkonen et al. 2010). A study conducted on several cell lines (SH-SY5Y, U87 human astrocytoma cells, CHME5 human microglial cells, and cortical neuron from Sprague-Dawley rat fetuses) exposed at 1800 MHz for 1h and 24h with SAR 2 W/kg and 10 W/kg, has again shown no variation in relation to the oxidative stress compared to the sham (Poulletier de Gannes et al. 2011).

Even the use of combined radiofrequency radiation (CDMA signal of 837 MHz and a WCDMA signal of 1950 MHz) did not affect the ROS levels in U87, PC12 (rat pheochromocytoma cells), and SH-SY5Y (Kang et al. 2014). Similar results were obtained on HT22 mouse hippocampal neuronal cells treated with Aβ: the increase in ROS levels induced by the treatment was not affected by the exposure (Lee et al. 2016). On human and rat primary astrocytes, stimulation with RF-EMF (frequency of 918 MHz at 10 W peak power pulsed amplitude modulation with a pulse width of 0.580 ms and a repetition rate of 217 Hz) attenuated the production of cellular and mitochondrial ROS Aβ42- and H_2O_2-induced (Tsoy et al. 2019). This last study pointed out a possible therapeutic approach for EMF in the treatment of Alzheimer's disease.

Some issues related to exposure conditions settings are often observed in RF experiments. Unfortunately, parameters, such as dosimetry and temperature monitoring, are not always controlled, which makes results inconsistent and hard to compare among them.

19.4 CONCLUSIONS

From what exposed earlier, it appears that the results from EMFs biological studies are somehow inconsistent and, sometimes, non-comparable. This relies almost on two aspects: 1) the weakness of the interaction of EMFs with the biological matter, which makes difficult to consistently find the molecular pathways activated by this interaction; 2) the numerous different experimental parameters (frequency, flux density, and duration of EMF exposure) and subjects (cell lines, species, strains, sex, and age) that are applied in all the studies, which sometimes makes it difficult to reproduce the results. Together with this, too often the studies show methodological uncertainties or weaknesses or are not very comprehensive regarding the parameters related to the exposure. Since an interaction between a physical agent and a biological system is investigated, it is mandatory that the whole system is fully controlled. Each analysis should provide the exposure set-up dosimetry, both numerically and experimentally, the uncertainty evaluation, both for measurements and numerical calculations, and the monitoring of local and environmental temperature, which is widely known to be affected by EMF exposure.

Despite the mixing and sometimes contrasting results, the ability to EMF exposure to modulate neuronal cells redox homeostasis appears clear. Indeed, results show that this field can induce an increase in ROS production, eventually by also affecting the cell antioxidant defensive system, acting on NADPH oxidase enzyme. Usually, this ROS enhancement is not accompanied by an induction of cell apoptosis unless another stressor intervenes (Benassi et al. 2016). Results from RF exposure are still too controversial to properly establish a direct effect of RF on redox homeostasis.

ROS are molecules derived from oxygen and formed during redox reaction or by electron excitation. In the last decades, it has been demonstrated that, together with their damaging activities, in physiological response they act as fundamental signaling molecules in a process defined oxidative eustress (Sinenko et al. 2021). This could explain why the increase in ROS usually observed after exposure to electromagnetic fields, particularly after ELF-MFs, does not have a detrimental effect but often has a beneficial one to the point of being widely used in brain disorders as therapeutical approaches (Guidetti et al. 2022).

In conclusion, in the light of what has been said so far, it is evident that it is absolutely necessary to continue to investigate and apply increasingly refined study designs.

REFERENCES

Aikins, Anastasia Rosebud, Sung Won Hong, Hyun Jung Kim, Cheol Ho Yoon, Joo Hee Chung, Mi Jung Kim, and Chan Wha Kim. 2017. "Extremely Low-Frequency Electromagnetic Field Induces Neural Differentiation of HBM-MSCs through Regulation of (Zn)-Metallothionein-3." *Bioelectromagnetics* 38 (5). Wiley-Liss Inc.: 364–373. doi:10.1002/bem.22046.

Akbarnejad, Zeinab, Khadijeh Esmaeilpour, Mohammad Shabani, Majid Asadi-Shekaari, Monavvar Saeedi goraghani, and Meysam Ahmadi-Zeidabadi. 2018. "Spatial Memory Recovery in Alzheimer's Rat Model by Electromagnetic Field Exposure." *International Journal of Neuroscience* 128 (8). Taylor and Francis Ltd: 691–696. doi:10.1080/00207454.2017.1411353.

Akdag, Mehmet Zulkuf, Suleyman Dasdag, Engin Ulukaya, Ali Kemal Uzunlar, Mustafa Ayberk Kurt, and Abdullah Taşkın. 2010. "Effects of Extremely Low-Frequency Magnetic Field on Caspase Activities and Oxidative Stress Values in Rat Brain." *Biological Trace Element Research* 138 (1–3). Humana Press Inc.: 238–249. doi:10.1007/s12011-010-8615-3.

Ambalayam, Sajeev, and Rashmi Mathur. 2022. "Extremely Low Frequency Magnetic Exposure Attenuates Oxidative Stress and Apoptotic Cell Death in Injured Spinal Cord of Rats." *Indian Journal of Experimental Biology* 60: 248–257.

Bahreyni Toossi, Mohammad Hossein, Hamid Reza Sadeghnia, Maryam Mohammad Mahdizadeh Feyzabadi, Mahmoud Hosseini, Mahdiyeh Hedayati, Razieh Mosallanejad, Farimah Beheshti, and Zeynab Alizadeh Rahvar. 2018. "Exposure to Mobile Phone (900–1800 MHz) during Pregnancy: Tissue Oxidative Stress after Childbirth." *Journal of Maternal-Fetal and Neonatal Medicine* 31 (10). Taylor and Francis Ltd: 1298–1303. doi:10.1080/14767058.2017.1315657.

Benassi, Barbara, Giuseppe Filomeni, Costanza Montagna, Caterina Merla, Vanni Lopresto, Rosanna Pinto, Carmela Marino, and Claudia Consales. 2016. "Extremely Low Frequency Magnetic Field (ELF-MF) Exposure Sensitizes SH-SY5Y Cells to the Pro-Parkinson's Disease Toxin MPP+." *Molecular Neurobiology* 53 (6). Humana Press Inc.: 4247–4260. doi:10.1007/s12035-015-9354-4.

Bobkova, Natalia v., Vadim v. Novikov, Natalia I. Medvinskaya, Irina Y. Aleksandrova, Inna v. Nesterova, and Eugenii E. Fesenko. 2018. "Effect of Weak Combined Static and Extremely Low-Frequency Alternating Magnetic Fields on Spatial Memory and Brain Amyloid-β in Two Animal Models of Alzheimer's Disease." *Electromagnetic Biology and Medicine* 37 (3). Taylor and Francis Ltd: 127–137. doi:10.1080/15368378.2018.1471700.

Bodewein, Lambert, Dagmar Dechent, David Graefrath, Thomas Kraus, Tobias Krause, and Sarah Driessen. 2022. "Systematic Review of the Physiological and Health-Related Effects of Radiofrequency Electromagnetic Field Exposure from Wireless Communication Devices on Children and Adolescents in Experimental and Epidemiological Human Studies." *PLoS One* 17 (6 June). Public Library of Science. doi:10.1371/journal.pone.0268641.

Bouji, Marc, Anthony Lecomte, Christelle Gamez, Kelly Blazy, and Anne-Sophie Villégier. 2020. "Impact of Cerebral Radiofrequency Exposures on Oxidative Stress and Corticosterone in a Rat Model of Alzheimer's Disease." *Journal of Alzheimer's Disease*. doi:10.3233/JAD-190593.

Cardis, E., B. K. Armstrong, J. D. Bowman, G. G. Giles, M. Hours, D. Krewski, M. McBride, et al. 2011. "Risk of Brain Tumours in Relation to Estimated RF Dose from Mobile Phones: Results from Five Interphone Countries." *Occupational and Environmental Medicine* 68 (9): 631–640. doi:10.1136/oemed-2011-100155.

Castaño-Vinyals, G., S. Sadetzki, R. Vermeulen, F. Momoli, M. Kundi, F. Merletti, M. Maslanyj, et al. 2022. "Wireless Phone Use in Childhood and Adolescence and Neuroepithelial Brain Tumours: Results from the International MOBI-Kids Study." *Environment International* 160 (February). Elsevier Ltd. doi:10.1016/j.envint.2021.107069.

Chauhan, Parul, H. N. Verma, Rashmi Sisodia, and Kavindra Kumar Kesari. 2017. "Microwave Radiation (2.45 GHz)-Induced Oxidative Stress: Whole-Body Exposure Effect on Histopathology of Wistar Rats." *Electromagnetic Biology and Medicine* 36 (1). Taylor and Francis Ltd: 20–30. doi:10.3109/15368378.2016.1144063.

Davanipour, Zoreh, Chiu Chen Tseng, Pey Jiuan Lee, and Eugene Sobel. 2007. "A Case-Control Study of Occupational Magnetic Field Exposure and Alzheimer's Disease: Results from the California Alzheimer's Disease Diagnosis and Treatment Centers." *BMC Neurology* 7 (June). doi:10.1186/1471-2377-7-13.

Delen, Kevser, Bahriye Sırav, Sinem Oruç, Cemile M. Seymen, Dilek Kuzay, Korkut Yeğin, and Gülnur Take Kaplanoğlu. 2021. "Effects of 2600 MHz Radiofrequency Radiation in Brain Tissue of Male Wistar Rats and Neuroprotective Effects of Melatonin." *Bioelectromagnetics* 42 (2). Wiley-Liss Inc.: 159–172. doi:10.1002/bem.22318.

Falone, Stefano, Alessandro Mirabilio, Maria Cristina Carbone, Vincenzo Zimmitti, Silvia di Loreto, Maria Addolorata Mariggiò, Rosa Mancinelli, Carmine di Ilio, and Fernanda Amicarelli. 2008. "Chronic Exposure to 50 Hz Magnetic Fields Causes a Significant Weakening of Antioxidant Defence Systems in Aged Rat Brain." *International Journal of Biochemistry and Cell Biology* 40 (12): 2762–2770. doi:10.1016/j.biocel.2008.05.022.

Fu, Yu, Cangkai Wang, Jianhong Wang, Yanlin Lei, and Yuanye Ma. 2008. "Long-Term Exposure to Extremely Low-Frequency Magnetic Fields Impairs Spatial Recognition Memory in Mice." *Clinical and Experimental Pharmacology and Physiology* 35 (7): 797–800. doi:10.1111/j.1440-1681.2008.04922.x.

Gao, Qiang, Aaron Leung, Yong Hong Yang, Benson Lau, Qian Wang, Ling Yi Liao, Yun Juan Xie, and Cheng Qi He. 2021. "Extremely Low Frequency Electromagnetic Fields Promote Cognitive Function and Hippocampal Neurogenesis of Rats with Cerebral Ischemia." *Neural Regeneration Research* 16 (7). Wolters Kluwer Medknow Publications: 1252–1257. doi:10.4103/1673-5374.301020.

Gramowski-Voß, Alexandra, Hans Joachim Schwertle, Anna Maria Pielka, Luise Schultz, Anne Steder, Konstantin Jügelt, Jürgen Axmann, and Wolfgang Pries. 2015. "Enhancement of Cortical Network Activity in Vitro and Promotion of GABAergic Neurogenesis by Stimulation with an Electromagnetic Field with a 150 MHz Carrier Wave Pulsed with an Alternating 10 and 16 Hz Modulation." *Frontiers in Neurology* 6 (June). Frontiers Media S.A. doi:10.3389/fneur.2015.00158.

Guidetti, Matteo, Alessandro Bertini, Francesco Pirone, Gessica Sala, Paola Signorelli, Carlo Ferrarese, Alberto Priori, and Tommaso Bocci. 2022. "Neuroprotection and Non-Invasive Brain Stimulation: Facts or Fiction?" *International Journal of Molecular Sciences* 23 (22). MDPI AG: 13775. doi:10.3390/ijms232213775.

Gunnarsson, Lars Gunnar, and Lennart Bodin. 2019. "Occupational Exposures and Neurodegenerative Diseases: A Systematic Literature Review and Meta-Analyses." *International Journal of Environmental Research and Public Health.* MDPI AG. doi:10.3390/ijerph16030337.

Hardell, L., K. H. Mild, and M. Carlberg. 2002. "Case-Control Study on the Use of Cellular and Cordless Phones and the Risk for Malignant Brain Tumours." *Int. j. Radiat. Biol* 78 (10): 931–936. doi:10.1080/0955300021015803.

Hardell, Lennart, Michael Carlberg, and Kjell Hansson Mild. 2013a. "Use of Mobile Phones and Cordless Phones Is Associated with Increased Risk for Glioma and Acoustic Neuroma." *Pathophysiology* 20 (2): 85–110. doi:10.1016/j.pathophys.2012.11.001.

Hardell, Lennart, Michael Carlberg, Fredrik Söderqvist, and Kjell Hansson Mild. 2013b. "Case-Control Study of the Association between Malignant Brain Tumours Diagnosed between 2007 and 2009 and Mobile and Cordless Phone Use." *International Journal of Oncology* 43 (6): 1833–1845. doi:10.3892/ijo.2013.2111.

Hu, Yu, Jinsheng Lai, Baoquan Wan, Xingfa Liu, Yemao Zhang, Jiangong Zhang, Dongsheng Sun, et al. 2016. "Long-Term Exposure to ELF-MF Ameliorates Cognitive Deficits and Attenuates Tau Hyperphosphorylation in 3xTg AD Mice." *NeuroToxicology* 53 (March). Elsevier B.V.: 290–300. doi:10.1016/j.neuro.2016.02.012.

IARC Monographs on the Evaluation of Carcinogenic Risks to Humans Non-Ionizing Radiation, Part 1: Static and Extremely Low-Frequency (ELF) Electric and Magnetic Fields. 2001. *World Health Organization International Agency for Research on Cancer.* Vol. 80. http://monographs.iarc.fr/.

IARC Monographs on the Evaluation of Carcinogenic Risks to Humans Non-Ionizing Radiation, Part 2: Radiofrequency Electromagnetic Fields. 2011. Vol. 102. https://monographs.iarc.who.int/wp-content/uploads/2018/06/mono102.pd.

ICNIRP. 2020“Guidelines for Limiting Exposure to Electromagnetic Fields (100 KHz to 300 GHz).” *Health Physics* 118. Lippincott Williams and Wilkins. doi:10.1097/HP.0000000000001210.

ICNIRP. 2010“Guidelines for Limiting Exposure to Time-Varying Electric and Magnetic Fields (1 Hz TO 100 KHz).” *Health Physics* 99. doi:10.1097/HP.0b013e3181f06c86.

Jadidi, Taha, Nader Asadian, Majid Jadidi, and Abbas Ali Vafaei. 2022. “EMF Promote BMSCs Differentiation and Functional Recovery in Hemiparkinsonian Rats.” *Neuroscience Letters*. doi:10.1016/j.neulet.2022.136765.

Jalilian, Hamed, Marloes Eeftens, Mansour Ziaei, and Martin Röösli. 2019. “Public Exposure to Radiofrequency Electromagnetic Fields in Everyday Microenvironments: An Updated Systematic Review for Europe.” *Environmental Research*. Academic Press Inc. doi:10.1016/j.envres.2019.05.048.

Jeong, Won Yong, Jun Beom Kim, Hyun Jung Kim, and Chan Wha Kim. 2017. “Extremely Low-Frequency Electromagnetic Field Promotes Astrocytic Differentiation of Human Bone Marrow Mesenchymal Stem Cells by Modulating SIRT1 Expression.” *Bioscience, Biotechnology and Biochemistry* 81 (7). Japan Society for Bioscience Biotechnology and Agrochemistry: 1356–1362. doi:10.1080/09168451.2017.1308243.

Jeong, Ye Ji, Yeonghoon Son, Na Kyung Han, Hyung do Choi, Jeong Ki Pack, Nam Kim, Yun Sil Lee, and Hae June Lee. 2018. “Impact of Long-Term RF-EMF on Oxidative Stress and Neuroinflammation in Aging Brains of C57BL/6 Mice.” *International Journal of Molecular Sciences* 19 (7). MDPI AG. doi:10.3390/ijms19072103.

Kang, Kyoung Ah, Hyung Chul Lee, Je Jung Lee, Mi Na Hong, Myung Jin Park, Yun Sil Lee, Hyung do Choi, Nam Kim, Young Gyu Ko, and Jae Seon Lee. 2014. “Effects of Combined Radiofrequency Radiation Exposure on Levels of Reactive Oxygen Species in Neuronal Cells.” *Journal of Radiation Research* 55 (2). Oxford University Press: 265–276. doi:10.1093/jrr/rrt116.

Kheifets, L., A. Ahlbom, C. M. Crespi, M. Feychting, C. Johansen, J. Monroe, M. F. G. Murphy, et al. 2010. “A Pooled Analysis of Extremely Low-Frequency Magnetic Fields and Childhood Brain Tumors.” *American Journal of Epidemiology* 172 (7). Oxford University Press: 752–761. doi:10.1093/aje/kwq181.

Kim, Jeong Yub, Hee Jin Kim, Nam Kim, Jong Hwa Kwon, and Myung Jin Park. 2017. “Effects of Radiofrequency Field Exposure on Glutamate-Induced Oxidative Stress in Mouse Hippocampal HT22 Cells.” *International Journal of Radiation Biology* 93 (2). Taylor and Francis Ltd: 249–256. doi:10.1080/09553002.2017.1237058.

Klimek, Angelika, Anna Nowakowska, Hanna Kletkiewicz, Joanna Wyszkowska, Justyna Maliszewska, Milena Jankowska, Lukasz Peplowski, and Justyna Rogalska. 2022. “Bidirectional Effect of Repeated Exposure to Extremely Low-Frequency Electromagnetic Field (50 Hz) of 1 and 7 MT on Oxidative/Antioxidative Status in Rat’s Brain: The Prediction for the Vulnerability to Diseases.” *Oxidative Medicine and Cellular Longevity* 2022. Hindawi Limited. doi:10.1155/2022/1031211.

Lee, Jong Sun, Jeong Yub Kim, Hee Jin Kim, Jeong Cheol Kim, Jae Seon Lee, Nam Kim, and Myung Jin Park. 2016. “Effects of Combined Radiofrequency Field Exposure on Amyloid-Beta-Induced Cytotoxicity in HT22 Mouse Hippocampal Neurones.” *Journal of Radiation Research* 57 (6). Oxford University Press: 620–626. doi:10.1093/jrr/rrw040.

Lee, Kyung Hee, Myeounghoon Cha, and Bae Hwan Lee. 2021. “Crosstalk between Neuron and Glial Cells in Oxidative Injury and Neuroprotection.” *International Journal of Molecular Sciences* 22 (24). MDPI. doi:10.3390/ijms222413315.

Liu, Xiao, Hongyan Zuo, Dewen Wang, Ruiyun Peng, Tao Song, Shuiming Wang, Xinping Xu, et al. 2015. “Improvement of Spatial Memory Disorder and Hippocampal Damage by Exposure to Electromagnetic Fields in an Alzheimer’s Disease Rat Model.” *PLoS One* 10 (5). Public Library of Science. doi:10.1371/journal.pone.0126963.

Luukkonen, Jukka, Pasi Hakulinen, Jorma Mäki-Paakkanen, Jukka Juutilainen, and Jonne Naarala. 2009. “Enhancement of Chemically Induced Reactive Oxygen Species Production and DNA Damage in Human SH-SY5Y Neuroblastoma Cells by 872 MHz Radiofrequency Radiation.” *Mutation Research: Fundamental and Molecular Mechanisms of Mutagenesis* 662 (1–2): 54–58. doi:10.1016/j.mrfmmm.2008.12.005.

Luukkonen, Jukka, Jukka Juutilainen, and Jonne Naarala. 2010. “Combined Effects of 872 MHz Radiofrequency Radiation and Ferrous Chloride on Reactive Oxygen Species Production and DNA Damage in Human SH-SY5Y Neuroblastoma Cells.” *Bioelectromagnetics* 31 (6): 417–424. doi:10.1002/bem.20580.

Manikonda, Pavan K., Pilankatta Rajendra, D. Devendranath, B. Gunasekaran, Channakeshava, Shivakumara R. S. Aradhya, Rao B. Sashidhar, and Chivukula Subramanyam. 2014. “Extremely Low Frequency Magnetic Fields Induce Oxidative Stress in Rat Brain.” *General Physiology and Biophysics* 33 (1): 81–90. doi:10.4149/gpb_2013059.

Megha, Kanu, Pravin Suryakantrao Deshmukh, Basu Dev Banerjee, Ashok Kumar Tripathi, Rafat Ahmed, and Mahesh Pandurang Abegaonkar. 2015. "Low Intensity Microwave Radiation Induced Oxidative Stress, Inflammatory Response and DNA Damage in Rat Brain." *NeuroToxicology* 51 (December). Elsevier B.V.: 158–165. doi:10.1016/j.neuro.2015.10.009.

Merla, Caterina, Micaela Liberti, Claudia Consales, Agnese Denzi, Francesca Apollonio, Carmela Marino, and Barbara Benassi. 2019. "Evidences of Plasma Membrane-Mediated ROS Generation upon ELF Exposure in Neuroblastoma Cells Supported by a Computational Multiscale Approach." *Biochimica et Biophysica Acta: Biomembranes* 1861 (8). Elsevier B.V.: 1446–1457. doi:10.1016/j.bbamem.2019.06.005.

Osera, Cecilia, Marialaura Amadio, Stefano Falone, Lorenzo Fassina, Giovanni Magenes, Fernanda Amicarelli, Giovanni Ricevuti, Stefano Govoni, and Alessia Pascale. 2015. "Pre-Exposure of Neuroblastoma Cell Line to Pulsed Electromagnetic Field Prevents H2O2-Induced ROS Production by Increasing MnSOD Activity." *Bioelectromagnetics* 36 (3). Wiley-Liss Inc.: 219–232. doi:10.1002/bem.21900.

Park, Jeong Eun, Young Kwon Seo, Hee Hoon Yoon, Chan Wha Kim, Jung Keug Park, and Songhee Jeon. 2013. "Electromagnetic Fields Induce Neural Differentiation of Human Bone Marrow Derived Mesenchymal Stem Cells via ROS Mediated EGFR Activation." *Neurochemistry International* 62 (4): 418–424. doi:10.1016/j.neuint.2013.02.002.

Poulletier de Gannes, F., E. Haro, A. Hurtier, M. Taxile, G. Ruffié, B. Billaudel, B. Veyret, and I. Lagroye. 2011. "Effect of Exposure to the Edge Signal on Oxidative Stress in Brain Cell Models." *Radiation Research* 175 (2): 225–230. doi:10.1667/RR2320.1.

Quesnel-Galván, Leticia R., Patricia v. Torres-Durán, David Elías-Viñas, and Leticia Verdugo-Díaz. 2021. "Effect of Extremely Low Frequency Magnetic Fields on Oxidative Balance in Rat Brains Subjected to an Experimental Model of Chronic Unpredictable Mild Stress." *BMC Neuroscience* 22 (1). BioMed Central Ltd. doi:10.1186/s12868-021-00656-x.

Reale, Marcella, Mohammad A. Kamal, Antonia Patruno, Erica Costantini, Chiara D'Angelo, Miko Pesce, and Nigel H. Greig. 2014. "Neuronal Cellular Responses to Extremely Low Frequency Electromagnetic Field Exposure: Implications Regarding Oxidative Stress and Neurodegeneration." *PLoS One* 9 (8). Public Library of Science. doi:10.1371/journal.pone.0104973.

Sakhaie, Mohammad Hassan, Mansoureh Soleimani, Bagher Pourheydar, Zahra Majd, Pezhman Atefimanesh, Sara Soleimani Asl, and Mehdi Mehdizadeh. 2017. "Effects of Extremely Low-Frequency Electromagnetic Fields on Neurogenesis and Cognitive Behavior in an Experimental Model of Hippocampal Injury." *Behavioural Neurology* 2017. Hindawi Limited. doi:10.1155/2017/9194261.

Sangun, Ozlem, Bumin Dundar, Hakan Darici, Selcuk Comlekci, Duygu Kumbul Doguc, and Suheyla Celik. 2015. "The Effects of Long-Term Exposure to a 2450 MHz Electromagnetic Field on Growth and Pubertal Development in Female Wistar Rats." *Electromagnetic Biology and Medicine* 34 (1). Informa Healthcare: 63–71. doi:10.3109/15368378.2013.871619.

Seong, Yeju, Jihye Moon, and Jongpil Kim. 2014. "Egr1 Mediated the Neuronal Differentiation Induced by Extremely Low-Frequency Electromagnetic Fields." *Life Sciences*, 16–27. doi:10.1016/j.lfs.2014.02.022.

Sharma, Anjali, Sadhana Shrivastava, Asha Singh, Shamli S. Gupte, Arti Rathour, Mohd Salim Reshi, and Sangeeta Shukla. 2022. "Evidences of the Radiofrequency Exposure on the Antioxidant Status, Potentially Contributing to the Inflammatory Response and Demyelination in Rat Brain." *Environmental Toxicology and Pharmacology* 94 (August). Elsevier B.V. doi:10.1016/j.etap.2022.103903.

Sharma, Samta, and Sangeeta Shukla. 2020. "Effect of Electromagnetic Radiation on Redox Status, Acetylcholine Esterase Activity and Cellular Damage Contributing to the Diminution of the Brain Working Memory in Rats." *Journal of Chemical Neuroanatomy* 106 (July). Elsevier B.V. doi:10.1016/j.jchemneu.2020.101784.

Sienkiewicz, Zenon J., Richard G. E. Haylock, Rachel Bartrum, and Richard D. Saunders. 1998. "50 Hz Magnetic Field Effects on the Performance of a Spatial Learning Task by Mice." *Bioelectromagnetics* 19 (8). Wiley-Liss Inc.: 486–493. doi:10.1002/(SICI)1521-186X(1998)19:8<486::AID-BEM6>3.0.CO;2-K.

Sinenko, Sergey A., Tatiana Yu Starkova, Andrey A. Kuzmin, and Alexey N. Tomilin. 2021. "Physiological Signaling Functions of Reactive Oxygen Species in Stem Cells: From Flies to Man." *Frontiers in Cell and Developmental Biology*. Frontiers Media S.A. doi:10.3389/fcell.2021.714370.

Sobel, Eugene, Zoreh Davanipour, Raimo Sulkava, Timo Erkinjuntti, Juhani Wikstrom, Victor W Henderson, Galen Buckwalter, Joseph D Bowman, and Pey-Jiuan Lee. 1995. "Occupations with Exposure to Electromagnetic Fields: A Possible Risk Factor for Alzheimer's Disease." *American Journal of Epidemiology* 142 (5). doi:https://doi.org/10.1186/1471-2377-7-13.

Tasset, I., F. J. Medina, I. Jimena, E. Agüera, F. Gascón, M. Feijóo, F. Sánchez-López, et al. 2012. "Neuroprotective Effects of Extremely Low-Frequency Electromagnetic Fields on a Huntington's Disease Rat Model: Effects on Neurotrophic Factors and Neuronal Density." *Neuroscience* 209 (May): 54–63. doi:10.1016/j.neuroscience.2012.02.034.

Tasset, I., Aleyda Pérez-Herrera, Francisco J. Medina, Óscar Arias-Carrión, René Drucker-Colín, and Isaac Túnez. 2013. "Extremely Low-Frequency Electromagnetic Fields Activate the Antioxidant Pathway NRF2 in a Huntington's Disease-Like Rat Model." *Brain Stimulation* 6 (1): 84–86. doi:10.1016/j.brs.2012.03.015.

Tsoy, Andrey, Timur Saliev, Elvira Abzhanova, Anel Turgambayeva, Aiym Kaiyrlykyzy, Mars Akishev, Samat Saparbayev, Bauyrzhan Umbayev, and Sholpan Askarova. 2019. "The Effects of Mobile Phone Radiofrequency Electromagnetic Fields on β-Amyloid-Induced Oxidative Stress in Human and Rat Primary Astrocytes." *Neuroscience* 408 (June). Elsevier Ltd: 46–57. doi:10.1016/j.neuroscience.2019.03.058.

Turner, Michelle C., Geza Benke, Joseph D. Bowman, Jordi Figuerola, Sarah Fleming, Martine Hours, Laurel Kincl, et al. 2017. "Interactions between Occupational Exposure to Extremely Low Frequency Magnetic Fields and Chemicals for Brain Tumour Risk in the INTEROCC Study." *Occupational and Environmental Medicine* 74 (11). BMJ Publishing Group: 802–809. doi:10.1136/oemed-2016-104080.

van Wel, Luuk, Ilaria Liorni, Anke Huss, Arno Thielens, Joe Wiart, Wout Joseph, Martin Röösli, et al. 2021. "Radio-Frequency Electromagnetic Field Exposure and Contribution of Sources in the General Population: An Organ-Specific Integrative Exposure Assessment." *Journal of Exposure Science and Environmental Epidemiology* 31 (6). Springer Nature: 999–1007. doi:10.1038/s41370-021-00287-8.

Wertheimer, Nancy, and E D Leeper Wertheimer. 1979. "Original Contributions Electrical Wiring Configurations and Childhood Cancer." *American Journal of Epidemiology* 109 (3). doi:10.1093/oxfordjournals.aje.a112681.

Xu, Shangcheng, Zhou Zhou, Lei Zhang, Zhengping Yu, Wei Zhang, Yuan Wang, Xubu Wang, et al. 2010. "Exposure to 1800 MHz Radiofrequency Radiation Induces Oxidative Damage to Mitochondrial DNA in Primary Cultured Neurons." *Brain Research* 1311 (January): 189–196. doi:10.1016/j.brainres.2009.10.062.

Zadeh-Haghighi, Hadi, and Christoph Simon. 2022. "Magnetic Field Effects in Biology from the Perspective of the Radical Pair Mechanism." *Journal of the Royal Society Interface*. Royal Society Publishing. doi:10.1098/rsif.2022.0325.

Zeng, Ying, Yunyun Shen, Ling Hong, Yanfeng Chen, Xiaofang Shi, Qunli Zeng, and Peilin Yu. 2017. "Effects of Single and Repeated Exposure to a 50-Hz 2-MT Electromagnetic Field on Primary Cultured Hippocampal Neurons." *Neuroscience Bulletin* 33 (3). Science Press: 299–306. doi:10.1007/s12264-017-0113-6.

Zhang, Cheng, Yue Li, Chao Wang, Ruili Lv, and Tao Song. 2013. "Extremely Low-Frequency Magnetic Exposure Appears to Have No Effect on Pathogenesis of Alzheimer's Disease in Aluminum-Overloaded Rat." *PLoS One* 8 (8). doi:10.1371/journal.pone.0071087.

Zhang, Yemao, Xingfa Liu, Jiangong Zhang, and Ni Li. 2015. "Short-Term Effects of Extremely Low Frequency Electromagnetic Fields Exposure on Alzheimer's Disease in Rats." *International Journal of Radiation Biology* 91 (1). Informa Healthcare: 28–34. doi:10.3109/09553002.2014.954058.

Zielinski, Jana, Angélique D. Ducray, Anja M. Moeller, Manuel Murbach, Niels Kuster, and Meike Mevissen. 2020. "Effects of Pulse-Modulated Radiofrequency Magnetic Field (RF-EMF) Exposure on Apoptosis, Autophagy, Oxidative Stress and Electron Chain Transport Function in Human Neuroblastoma and Murine Microglial Cells." *Toxicology in Vitro* 68 (October). Elsevier Ltd. doi:10.1016/j.tiv.2020.104963.

Zuo, Hongyan, Xiao Liu, Dewen Wang, Yang Li, Xinping Xu, Ruiyun Peng, and Tao Song. 2018. "RKIP-Mediated NF-KB Signaling Is Involved in ELF-MF-Mediated Improvement in AD Rat." *International Journal of Medical Sciences* 15 (14). Ivyspring International Publisher: 1658–1666. doi:10.7150/ijms.28411.

Zuo, Wen Qi, Yu Juan Hu, Yang Yang, Xue Yan Zhao, Yuan Yuan Zhang, Wen Kong, and Wei Jia Kong. 2015. "Sensitivity of Spiral Ganglion Neurons to Damage Caused by Mobile Phone Electromagnetic Radiation Will Increase in Lipopolysaccharide-Induced Inflammation in Vitro Model." *Journal of Neuroinflammation* 12 (1). BioMed Central Ltd. doi:10.1186/s12974-015-0300-1.

Section V

Mental Stress

20 How Adverse Childhood Experiences Increase Inflammation and Oxidative Stress

Jasmin Ghaemi Kerahrodi and Matthias Michal

20.1 INTRODUCTION

It is a commonly acknowledged fact that mental stress has negative consequences for health (Meyer and Wirtz 2018a). Mental stress-related disorders have been linked to a higher risk of cardiometabolic diseases and a notable decrease in life expectancy and quality of life. The correlation remains stable even after adjustment for other risk factors like cardiovascular risk factors. The molecular mechanisms mediating mental stress and somatic disease are manifold and most likely interact with each other. These pathways involve a complex interplay of social, psychological, neural, immunologic, (epi)genetic, and genomic processes. In addition, oxidative stress as well as endothelial dysfunction are key players when talking about negative consequences of mental stress on the cardiovascular system (Broadley et al. 2005; Gottdiener et al. 2003; Lind, Johansson, and Hall 2002). In this chapter, we will focus on the impact of adverse childhood experiences (ACEx) on mental stress and physical disease in later life and discuss possible molecular pathways linking early stress in life with inflammation and increased oxidative stress in later life.

In this chapter, we will begin by discussing potential sources of mental stress, including emotions, which constitute the primary innate motivational and regulatory system in humans. Next, we will examine the various pathways involved in the progression from amygdala activation to autonomic dysfunction and oxidative stress, while also providing an overview of the fear-defense system. We will then highlight the crucial roles of the immune system and the autonomic nervous system (ANS) in mediating the relationship between mental stress and somatic disease, as well as their interaction with psychological and behavioral factors that may contribute to increased risk for health issues via oxidative stress. Finally, we will explore the clinical and social implications of these findings.

20.2 ADVERSE CHILDHOOD EXPERIENCES, AN OVERVIEW

ACExs have been subjects of research and health care since 1995. The Center for Disease Control and Prevention and Kaiser Permantente developed the ACEx scale. Kaiser Permanente is an integrated managed care consortium based in Oakland, California, United States. It was founded in 1945 by industrialist Henry J. Kaiser and physician Sidney Garfield as a prepaid health insurance program. ACExs are traumatic childhood events effecting biopsychosocial health across the lifespan (Bryan 2019). ACExs are some of the most intense sources of stress children can be exposed to. In this context, childhood bullying is among the top 20 second level risk factors for global deaths and disability-adjusted life years (DALYs) (Murray et al. 2020). Figure 20.1 summarizes the categories of ACExs and the consequences on the individual. In Europe, a prevalence of 23.5% for one ACEx

DOI: 10.1201/9781003328100-25

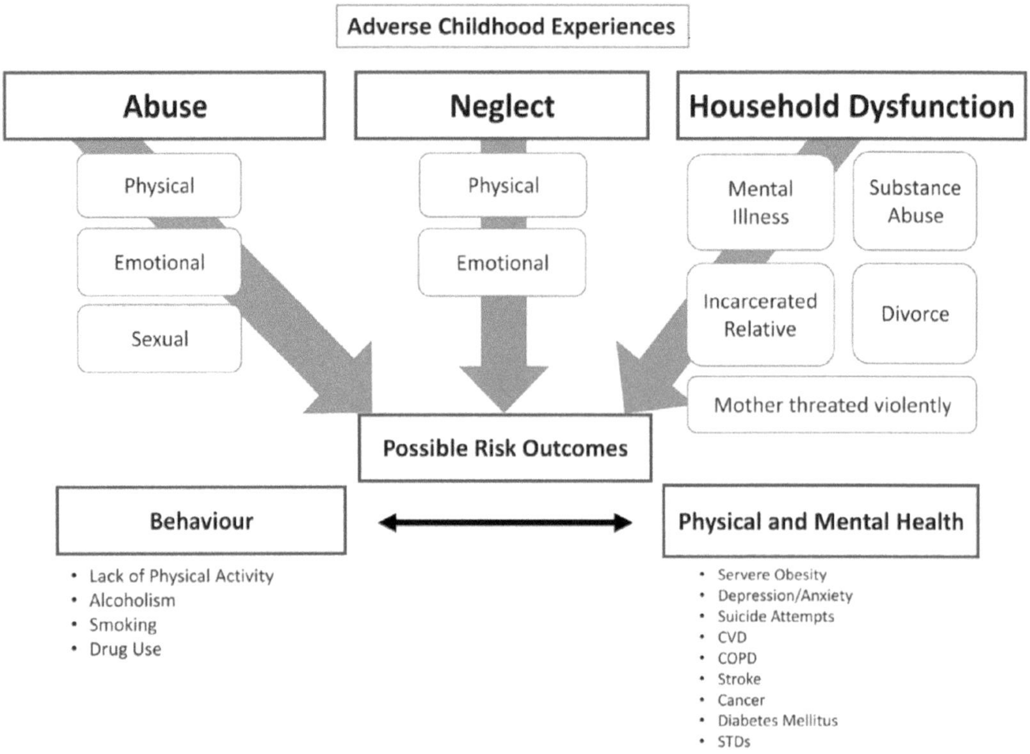

FIGURE 20.1 Overview of ten adverse childhood experiences (ACExs). Some publications also included racial or cultural discrimination and bullying. Abbreviations: CVD, cardiovascular disease; COPD, chronic obstructive pulmonary disease; STD, sexually transmitted diseases.

and 18.7% for two or more ACExs is estimated (Bellis et al. 2019). The prevalence of ACExs differs regarding the specific countries and cultures. For example, in a study of 8629 adults in an American population, 67.3% of participants reported experiencing at least one ACEx, with the most common ACExs being substance abuse in the household (28.2%), physical abuse (26.4%), domestic violence (24.1%), sexual abuse (21.0%), and mental illness in the household (20.3%) (Deighton et al. 2018). ACExs include emotional or physical child maltreatment; the experience of violence, abuse, neglect, and parental substance use. Figure 20.1 offers an overview of recognized ACExs. Childhood exposure to abuse, neglect, and household dysfunction can initiate a chain of events that may result in the development of maladaptive affective states such as depression. One central physiological explanation is the long-term effect of activation of the fear-defense system. Additionally, studies have found that exposure to one ACEx significantly increases the incidence of having additional ACExs (Deighton et al. 2018). An increasing number of studies are identifying associations between ACEx and mental and physical health throughout life (Bellis et al. 2019). ACExs are linked to higher risks of maladaptive behaviors, including smoking, abuse of alcohol, and other drug use, as well as high-risk HIV behavior, obesity, diabetes, myocardial infarction, coronary heart disease, and stroke, taken together, leading to premature mortality (Hughes et al. 2017; Bellis et al. 2014; Campbell, Walker, and Egede 2016; Bellis et al. 2015). Exposure to ACExs is also linked to an increased risk for mental diseases like depression and personality disorders (Riccardi et al. 2020). The effect of ACExs on health-harming behaviors or diseases can be additive or multiplicative. Those with multiple types of ACEx are therefore at a substantially increased risk (Hughes et al. 2017). The increased maladaptive, self-harming behavior in individuals with ACExs is one mechanism linking

ACExs to the development of physical disease (Su et al. 2015). On the other hand, evidence suggests that ACExs also directly affect neurological, hormonal, and immunological processes (Danese and McEwen 2012). For example, ACExs are associated with increased biomarkers for inflammation and shortened telomeres (Holman et al. 2016). Importantly, traumatic events can have transgenerational effects via epigenetic processes (Yehuda and Lehrner 2018).

20.3 THE SOURCES OF MENTAL STRESS AND THE ROLE OF EMOTIONS

In human development, emotions are the primary motivational system. Infants, unable to speak, communicate by expressing their emotions, and the caregivers respond. Usually, these interactions are successful, and infants develop secure attachments, a healthy sense of self, trust in others, and functional capacities to regulate emotions and social relationships. However, early adverse childhood experiences can profoundly affect these neurodevelopmental emotional learning processes. Adverse childhood experiences are characterized by violations of the infant's developmental needs. Adverse childhood experiences comprise early losses of attachment figures, emotional abuse, physical and sexual abuse, and neglect of the infant's emotional, physical, and social needs. Early adverse childhood experiences have lasting impacts on multiple outcomes such as social achievement (education, income, etc.), health, and life expectancy (Merrick et al. 2018).

Human brain development is prolonged compared to other species, allowing optimal adaptation by acquisition of complex behaviors. Early attachment experiences with parents are crucial for cognitive and affective learning (Tottenham et al. 2019). Attachment trauma, i.e., disruptions of the bond between caregivers and the infant by abuse or neglect of the emotional needs of the child (Fonagy and Target 2008), is a highly potent stressor for neurodevelopment (Tottenham et al. 2019). The detriment effects of early adverse childhood experiences are mainly processed by amygdala and the medial prefrontal cortex (mPFC) (Tottenham et al. 2019). The mPFC is a brain structure important for social cognition and regulating emotions and behavior. During human development, the amygdala and mPFC form rich interconnections (Tottenham et al. 2019). The quality of attachment correlates with the neurodevelopment of the amygdala and the prefrontal cortex. Infants who have secure attachment experiences tend to show more adaptive maturation of the amygdala-mPFC connectivity and smaller amygdala volumes compared to those with insecure attachment experiences (Callaghan et al. 2019; Tottenham et al. 2019). Adverse childhood experiences and attachment trauma predict higher amygdala volume in adulthood with the consequence of increased amygdala reactivity to salient stimuli (Teicher et al. 2016). The amygdala hyperactivation means increased autonomic dysregulation and threat-related hypervigilance (Tottenham et al. 2019; Laricchiuta et al. 2023).

In this context, it is important to recognize that the attachment relationship between infants and parents is regulated by basic emotions such as joy, sadness, anger, disgust, surprise, and fear (Nadel and Camaioni 2017; Pace-Schott et al. 2019). Emotions are the primary motivational system of humans. Since infants cannot yet speak, they communicate with their parents by expressing their emotions. Emotions are intrapersonal and interpersonal regulators (Panksepp and Biven 2012; Pace-Schott et al. 2019). Anger, for example, initiates self-assertive behaviors by the infant toward the parent. Adaptive parental responses to the angry infant increase the infant's self-efficacy and trust in the caregiver. However, the parent's response can also cause fear conditioning of the emotion "anger". Imagine the parent reacting with anxiety or terrifying the infant by becoming aggressive or turning away, then the emotion anger becomes a threat signal for the infant. Because of the immense dependence on the caregiver, the infant's primary motive is to is to maintain the bond with the parent and to avoid any behavior that might jeopardize the bond with the parent (Allen 2018). Mental disorders result from such learned affect phobias and the avoidance or defense against such emotions later in life (Abbass and Schubiner 2018; McCullough et al. 2021). Since emotions are the basic motivational system of humans, the conditioned fear of emotions and the defense against one's own emotions has a profound impact on the development of identity, self-regulation, and interpersonal skills. Once fear conditioning has happened, the fear-defense system will be activated as soon as certain emotions are mobilized, with the result that these emotions automatically trigger fear and defensive behavior (e.g., fight, flight, freezing, etc.). This process has the consequence that emotions become conflictual and activate the fear-defense system (Abbass and Schubiner 2018; McCullough et al. 2021; Schore 2019).

20.4 ADVERSE CHILDHOOD EXPERIENCES ON A MOLECULAR LEVEL

The link between chronic psychosocial stress, like ACExs, and somatic as well as physical diseases are extensively documented. The cellular mechanisms leading from stressful life experiences and their neuroendocrine mediators to biological damage remain still poorly understood (Picard, Juster, and McEwen 2014). Indeed, psychosocial stressors have the potential to disrupt a physiological stress response (Franke 2014). Figure 20.2 offers an overview of the complex interplay between trauma, like ACExs, and further variables that affect early mortality.

Acute stress is usually a trigger that activates the fear-defense system, which then activates the hypothalamic-pituitary-adrenal (HPA) axis and/or the sympathetic nervous system (SNS) in order to enable the organism to adapt to a threat by making physiological adjustments (Black 2003; Glaser and Kiecolt-Glaser 2005), also known as allostasis (Cohen, Tyrrell, and Smith 1991; Glaser and Kiecolt-Glaser 2005). Maladaptive stress, on the other hand, increases allostatic load resulting in

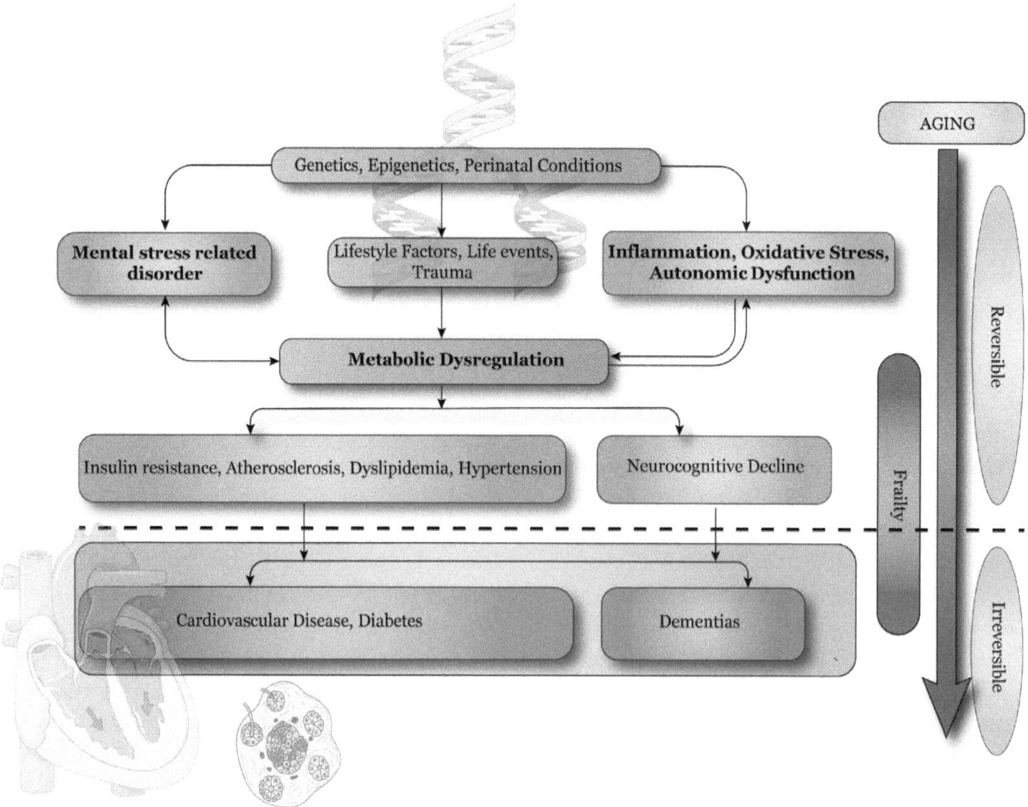

FIGURE 20.2 Summarized mechanisms and interactions that lead to cardiovascular disease, dementias, and aging.

physiological dysfunction (Franco et al. 2009) like enhanced oxidative stress and mitochondrial dysfunction (Picard, Juster, and McEwen 2014).

In the following section, we will describe different molecular pathways that can be activated through maladaptive stress and ACExs: the fear-defense system and the function of the amygdala, the impact of mental stress on oxidative stress and gamma-aminobutyric acid (GABA), the relationship between mental stress and mitochondrial dysfunction, and the role of the immune system in mediating the effects of mental stress.

20.5 THE AMYGDALA AND THE FEAR-DEFENSE SYSTEM

Situated in the temporal lobe, the amygdala is a complex of nuclei that has an almond-like shape. It is a complex structure and comprises approximately 13 nuclei. Further, it can be subdivided into extensive intranuclear and internuclear connections (Rajmohan and Mohandas 2007).

Activation of the ANS and hypothalamic-pituitary-adrenal (HPA) axis are prominent features of the stress response, which in turn affects the function of other body organs. In 1988, Sterling and Eyer forwarded the concept of "allostasis", referring to a collection of internal mechanisms for reestablishing homeostasis during challenging conditions (McEwen 1998).

It is known to be engaged in the acquisition, retrieval, and expression of aversive memories and plays a crucial role in the activation of the fear-defense system (LeDoux 2000; Pape and Pare 2010;

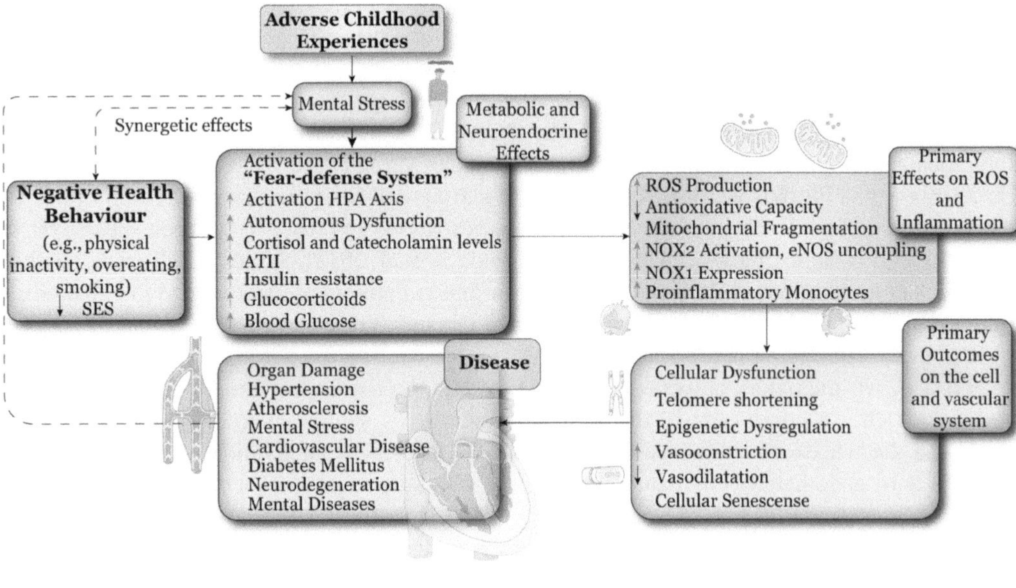

FIGURE 20.3 Adverse childhood experiences (ACExs) are a form of (chronic) mental stress. ACExs can lead to chronic activation of the amygdala. This leads to activation of the hypothalamic-pituitary-adrenal axis as well as autonomous dysfunction. Activation of the sympathetic nervous system (SNS) leads to renin secretion and release of angiotensin II (ATII). ATII activates NOX2 (NADPH oxidase 2) in endothelial cells resulting in oxidative stress. This can result in uncoupling of endothelial nitric oxide synthase (eNOS). Oxidative stress in endothelial cells activates NF-kB (nuclear factor k-light-chain-enhancer of activated B cells), leading to the induction of adhesion molecules leading to vascular inflammation. When stimulated, the HPA axis rapidly releases glucocorticoids (GC). GC enhances NOX1 (NADPH oxidase 1) expression in vascular smooth muscle cells. GC and NA (noradrenalin) can both lead to decreased insulin sensitivity. Chronic mental stress can also lead to mitochondrial allostatic load by perturbing adaptive GC signaling that disrupts mitochondrial structure and function, leading to enhanced reactive oxygen species (ROS) formation and cellular damage.

Sah et al. 2003). The fear-defense system is a built-in mechanism that coordinates automatic defensive reactions to threats, which aid in promoting survival (Gu et al. 2020; Kozlowska et al. 2015). Arousal is the first step of the activation of the fear-defense system. Before conscious awareness, the amygdala initiates the fear response (Kozlowska et al. 2015; Cecala 2016). The fear-defense system triggers a physiological response that leads to the feeling of anxiety, characterized by muscle tension, sweating, increased heart rate, hyperventilation, and cold hands due to vasospasm, which is consciously perceived by the individual. The patterns of defensive responses can be categorized as follows (Kozlowska et al. 2015): When faced with a threat, humans may respond with active behaviors such as becoming angry and speaking up or becoming submissive, which are characteristic of the fight or flight response. Freezing is a state of attentive immobility, enabling the mammal to scan the environment and to prepare fight or flight reactions (e.g., state of enhanced vigilance with activated, tense body). When faced with an inescapable threat, mammals often exhibit tonic immobility, a terminal defense mechanism that deactivates the predator's killing reflex when the mammal is captured. In humans, this defense mechanism is associated with feelings of numbness, fear, and perceptual distortions such as derealization, depersonalization, and hopelessness. A comparable defense response is the state of collapsed immobility, which can result in fear-induced fainting due to cerebral hypoxia in humans. Another response is quiescent immobility, which occurs after periods of acute stress when the mammal has returned to a safe environment and facilitates recovery. This defensive response is also the underlying brain mechanism of clinical conditions such as chronic pain syndromes or prolonged exhaustion.

20.6 FEAR-DEFENSE SYSTEM AND OXIDATIVE STRESS

Long-lasting consequences such as inflammation, atherosclerosis, changes in insulin sensitivity, and cardiovascular disease can result from the maladaptive activation of the fear-defense system due to chronic stress (Kerahrodi and Michal 2020). One link between mental stress and somatic disease is oxidative stress. Oxidative stress is typically described as an imbalance between prooxidant factors, such as reactive oxygen species (ROS) and reactive nitrogen species (RNS), and antioxidants (Lichtenberg and Pinchuk 2015b). The integrity of various cell components can be compromised by the presence of high levels of reactive oxygen species (ROS) and reactive nitrogen species (RNS) along with a low antioxidative capacity (de M. Bandeira et al. 2013). Severe cell distress, resulting from high concentrations of ROS and RNS and low antioxidative capacity, can lead to impaired cell function and ultimately cell death (Lichtenberg and Pinchuk 2015a).

The hypothalamic-pituitary-adrenal (HPA) axis as well as the autonomous nervous system (ANS) are two mediators activated by the amygdala. Activation of the sympathetic nervous system, which is a component of the autonomic nervous system, can result in the secretion of renin and the release of angiotensin II (ATII). ATII triggers the activation of NADPH oxidase 2 (NOX2) in endothelial cells, causing oxidative stress through the uncoupling of endothelial nitric oxide synthase (eNOS). The activation of ROS signaling pathways can lead to increased expression of nuclear factor k-light-chain-enhancer of activated B cells (NF-κB), which in turn induces the expression of adhesion molecules and promotes inflammation within blood vessels (Li et al. 2020). Various stimuli, such as acute alcohol exposure, cigarette smoke, physiological stress, and mental stress, can lead to the transient activation of NF-κB, resulting in neuroinflammatory responses in mice (Ahn and Aggarwal 2005; Yakovleva et al. 2011). Therefore, it can be considered as a "stress-sensor". NF-κB has a wide range of regulatory functions and controls the expression of approximately 500 genes, including cytokines, tumor necrosis factor (TNF), and enzymes such as inducible NO synthase (iNOS) (Shih, Wang, and Yang 2015). The HPA axis, which is mediated by corticotropin-releasing factor (CRF), adrenocorticotropic hormone (ACTH), and corticosteroids, also plays a role in this process. When activated, the HPA axis rapidly releases glucocorticoids (GC), which increase the expression of NADPH oxidase 1 (NOX1) in vascular smooth muscle cells. Both GC and noradrenaline (NA) can reduce insulin sensitivity.

Additionally, ROS signaling can activate transcription factors that regulate genes involved in tumor suppression and antioxidative action.

20.7 FEAR-DEFENSE SYSTEM, GABA, AND ROS

Activation of the amygdala can be mediated due to GABAergic neurotransmission that inhibits the amygdala and prevents the activation of the fear-defense system (Jie et al. 2018). Stress, whether acute or chronic, may result in a reduction of the GABAergic neuronal network, which is responsible for inhibition. Reactive oxygen species (ROS) might play a role in the regulation of the GABAergic network. ROS are well known to be involved in cellular metabolism but also play a role in synaptic and non-synaptic signaling to regulate nervous system function (Rice 2011), eventually impacting synaptic plasticity (Giniatullin, Petrov, and Giniatullin 2019). In the last years, the crucial role of ROS in the physiological regulation of inhibitory neurotransmission mediated by GABA has been recognized increasingly (Beltrán González, López Pazos, and Calvo 2020). In the nervous system, hydrogen peroxide (H_2O_2), as well as superoxide anion, are relevant endogenous ROS. The production of ROS primarily occurs in the mitochondria of neurons and glial cells. These ROS can affect GABA-mediated neurotransmission through both presynaptic and postsynaptic mechanisms, leading to alterations in the activity of phasic and tonic $GABA_A$ receptors as well as GABA release from presynaptic terminals (Beltrán González, López Pazos, and Calvo 2020). The effects of ROS on $GABA_B$ receptor-mediated neuromodulation are less documented.

20.8 THE ROLE OF MITOCHONDRIAL DYSFUNCTION IN CHRONIC MENTAL STRESS

Chronic psychosocial stress can lead to metabolic stress. ACExs have been associated with obesity and metabolic syndrome. Besides hyperactivation of the HPA axis activity leading to enhanced basal and stress-induced cortisol levels, the allostatic load of the mitochondrion might play a crucial role. Metabolic stress, like diabetes mellitus, leads to oxidative stress, inflammation, and telomere shortening resulting in a reduced lifespan of 5–20 years (Olshansky et al. 2005). Metabolic stress can also lead to reduced cognitive function, underlining its ability to harm the CNS, ANS, and other organs. The enhanced telomere shortening that can be observed in diabetes mellitus (Testa et al. 2011) is also a feature of individuals under chronic psychosocial stress (Epel et al. 2004). These observations indicate a possible common underlying mechanism in which the mitochondria might play an important role (Picard, Juster, and McEwen 2014; Manoli et al. 2007). During acute stress, GC receptors can translocate from the cytoplasm into the mitochondria in presence of their ligand (Picard and Turnbull 2013). The exact effect of GC on the mitochondrial (dys)function is not yet fully understood. GC seem to increase binding of the GC receptor to mtDNA, leading to enhanced mitochondrial gene expression (Psarra and Sekeris 2011). The increased mtDNA gene expression augments the capacity of energy production in the mitochondria. GC are also able to increase expression of autosomal genes (in the cell nucleus) that are required when producing new mitochondria (Psarra and Sekeris 2011). Taken together, the process increases the total number of mitochondria per cell, leading to a higher total cellular energy capacity. On the other hand, chronic high levels of GC can lead to maladaptive reactions, like a higher apoptotic sensitivity in nerve cells (Du et al. 2009), as well as leading to increased production of ROS, leading to molecular damage like a depressed activity of the mitochondrial respiratory chain activity (Tang et al. 2013). Besides the direct effects of GC on the mitochondria, the hyperglycemic effect of GC might also play a role. High levels of glucose lead to fragmentation of the mitochondria causing increased production of ROS, leading to accumulation of mtDNA damage (Yu, Robotham, and Yoon 2006). Mitochondrial fragmentation activates proapoptotic pathways leading to programmed cell death (Picard et al. 2012; Lee et al. 2004). On the other hand, mitochondrial fusion promotes the survival of the cell (Gomes, Di Benedetto, and Scorrano 2011). Collectively, data demonstrate that mitochondria readily sense both GC and glucose as primary stress mediators. In response, mitochondria undergo dynamic morphological and functional changes associated with deleterious effects on cell function and senescence. Importantly, oxidative stress (ROS and RNS) is able to activate inflammatory pathways as well as recruitment of immune cells by multiple redox mechanisms, e.g., oxidative modification of cytokines, chemokines, and inflammatory transcription factors via tyrosine nitration and thiol oxidation/nitrosation (Wenzel et al. 2017). This provides an important link between stress response pathways and down-stream activation of oxidative stress as well as inflammation.

20.9 THE ROLE OF MENTAL STRESS, THE CENTRAL AND AUTONOMOUS NERVOUS SYSTEM, AND ITS INTERPLAY WITH THE IMMUNE SYSTEM

The immune system's aim is to protect the body from foreign pathogens. One significant breakthrough in the past decade is the recognition that mental stress can impact the immune system, which is distinct from traditional immune system models indicating that inflammation is primarily governed by internal interactions without the participation of the brain and ANS. The interaction between social-environmental processes, including emotional stress, and inflammatory activity is widely recognized. Psychoneuroimmunology (PNI) is currently exploring questions about the communication and interaction between the immune system and the brain, as well as how this interaction can shape and be influenced by mental stress (Maier 2003).

The relationship between inflammatory processes and mental stress has been investigated since the 1^{9t}h century (Steinberg and Himmerich 2013). Shortly after the discovery of the first cytokine, INF-α (Isaacs and Lindenmann 1957), it was clarified that it is able to influence immunological response in the brain, even when administered peripherally (Hitchcock and Porterfjeld 1961; Himmerich et al. 2019). Immune defense in a nonspecific manner begins rapidly after the entry of infectious microorganisms leading to activation of immune cells. The activated immune cells are able to kill microorganisms, e.g., by ROS, that can lyse pathogens, as well as nitric oxide, which interferes with mitochondrial respiration leading to slower replication (Maier 2003). Importantly, the ROS formation by immune cells, which is called oxidative burst in leukocytes, closes the vicious circle within the adverse interplay of oxidative stress and inflammation by increasing oxidative damage at sites of infiltrated/activated immune cells (Daiber et al. 2020).

Activated immune cells (including monocytes/macrophages, dendritic cells, and neutrophils) synthesize and release cytokines. Cytokines, which are synthesized and released by activated immune cells such as monocytes/macrophages, dendritic cells, and neutrophils, can also be produced by neurons, astrocytes, and microglia in the brain. Despite their relatively large size (6–70 kDa), cytokines produced peripherally can also have an impact on the brain. Multiple pathways have been described on how cytokines can reach the brain, including humoral, neural, and cellular pathways (Capuron and Miller 2011; Haroon, Raison, and Miller 2012). For example, cytokines can penetrate the BBB by active transport, passive passage through "leaky" regions in the BBB and transmission along afferent vagal pathways (the 'neural route') (Haroon, Raison, and Miller 2012).

Cytokines have the ability to recruit more immune cells to the site of infection and trigger the peripheral acute phase response. They are also believed to act as messenger molecules, similar to hormones and neurotransmitters, by transmitting physiological responses through receptor-ligand interactions, and they can have autocrine, paracrine, and endocrine effects (Jain and Mills 2007; Slavich and Irwin 2014).

Activation of the sympathetic nervous system (SNS) can have a significant impact on the immune system. It has been found that SNS activation can increase the production of proinflammatory monocytes through monocytopoiesis in the bone marrow. Chronic inflammation can further alter the hematopoietic topography by shifting it from the bone marrow to the spleen. As a result, the migration of hematopoietic progenitor cells to the periphery can contribute to the production of more leukocytes. Recent evidence suggests that psychosocial stress and an unhealthy lifestyle can trigger the release of hematopoietic stem cells and progenitor cells from the bone marrow to the periphery (Nahrendorf and Swirski 2015). In addition, heightened sympathetic nervous system activity can decrease the expression of C-X-C motif chemokine 12 (CXCL12) in the hematopoietic stem cell niche, leading to increased production of neutrophils and monocytes in mice exposed to stress. This can cause a substantial release of inflammatory leukocytes into circulation and promote plaque inflammation (Heidt et al. 2014).

Interestingly, lymphocytes by itself can synthesize hormones (e.g., ACTH, prolactin, growth hormones) (Weigent and Blalock 1997; Glaser and Kiecolt-Glaser 2005). It is postulated that "lymphocyte-derived" hormones might have a role in modulating cell function within the environment of lymphoid organs. Also, there seem to be a connection between the SNS and lymphoid organs via nerve fibers in the spleen and thymus (Lorton, Lubahn, and Bellinger 2001). Hundreds of cytokines have been identified, with some being extensively studied while others remain mostly unexplored (Slavich 2020). Cytokine research in the field of PNI and mental stress have to deal with difficulties. Results are somehow conflicting, e.g., a high variance of cytokine values within samples. It is currently discussed if, and how, results from serum or plasma cytokine levels can be translated to the state of the brain or (dys)function of ANS (Capuron and Miller 2011). The heterogeneity of the results between studies is large (Köhler et al. 2018), and the results are conflicting. For example, studies showed elevated levels of TNF-α in the serum of depressed individuals, but such positive results have not been obtained by all studies (Himmerich et al. 2019).

An example of the bidirectional relationship between the brain and immune system is the relation between IL-1 and corticotropin-releasing hormone (CRH). IL-1 interact with production of CRH by the hypothalamus. CRH can affect the HPA axis followed by an increase in stress hormone levels resulting in dysregulation of immune function.

Another recognized cytokine in mental stress is IL-6. Production of IL-6 has shown to be increased in states of mental stress (anxiety, depression) (Penninx et al. 2003; Glaser et al. 2003; Maes et al. 2001; Glaser and Kiecolt-Glaser 2005). One possible explanation for the link between cytokines and the main components of the HPA axis is that IL-6 may stimulate the release of cortisol from the adrenal cortex by tyrosine phosphorylation of the transcription factor STAT3. This process involves mitochondrial electron transfer reactions that produce ATP through oxidative respiration and adrenocortical steroid biosynthesis. The involvement of STAT3 in oxidative respiration and steroidogenesis suggests that it integrates both nuclear and mitochondrial actions, which helps to maintain the main steps of glucocorticoid biosynthesis in the adrenal gland during periods of psychosocial stress (Meyer and Wirtz 2018b).

20.10 CLINICAL IMPLICATIONS

The violation of the basic needs of human beings triggers the overactivation of the fear-defense system and results in increased inflammation and oxidative stress with harmful consequences for health and life expectancy.

At a population level, promoting healthy environments and reducing social inequalities can potentially reduce anxiety, inflammation, and oxidative stress. Legislative actions can also play a crucial role in promoting antioxidant-rich diets and physical activity, leading to a healthier population (Kivimäki and Steptoe 2018; Nahrendorf and Swirski 2015). At an individual level, pharmacological interventions could potentially be beneficial. Selective serotonin reuptake inhibitors (SSRIs) have been shown to reduce amygdala reactivity, and some studies have found that they can also reduce inflammation markers (Pasquereau et al. 2021) and oxidative stress (Gałecki, Mossakowska-Wójcik, and Talarowska 2018). In major depressive disorder patients, long-term treatment with antidepressant drugs had positive effects on oxidative damage and inflammatory profile as well as antioxidant enzyme activities (Hernandez et al. 2013). Beta-blockers attenuate the stress-induced catecholamine responses centrally (including in the amygdala) and peripherally (Dar et al. 2019). Statins have anti-inflammatory and antioxidant effects and might be beneficial for the course of mental disorders (Dar et al. 2019; Kim et al. 2019). Significantly, adopting a healthy lifestyle and engaging in exercise can have various positive effects, including enhancing mitochondrial content and function. Furthermore, these lifestyle changes can potentially reduce the effects of prolonged activation of the fear-defense system and assist in preserving telomere length (Puterman et al. 2010). Further, effective psychotherapy models have shown that they can normalize the synaptic and metabolic activity in limbic and prefrontal regions and reduce the hyperactivation of the fear-defense system (Abbass et al. 2014). Psychotherapy has the potential to modify oxidative stress levels in individuals diagnosed with major depression. In fact, this treatment approach has been shown to decrease serum levels of NO in these patients, bringing them closer to levels observed in healthy individuals (Kaufmann et al. 2015).

Last but not least, we believe that a holistic biopsychosocial approach to health comprising the individual and its social relationships, the society, the healthcare system, and the environment is best qualified to reduce inflammation and oxidative stress.

REFERENCES

Abbass, A.A., S.J. Nowoweiski, D. Bernier, R. Tarzwell, and M.E. Beutel. 2014. "Review of psychodynamic psychotherapy neuroimaging studies." *Psychother Psychosom* 83 (3):142–147. doi: 10.1159/000358841.

Abbass, A., and H. Schubiner. 2018. *Hidden from View: A Clinician's Guide to Psychophysiologic Disorders*: Mind Body Publishing.

AHN, KWANG SEOK, and B.B. AGGARWAL. 2005. "Transcription factor NF-κB: A sensor for smoke and stress signals." *Annals of the New York Academy of Sciences* 1056 (1):218–233. doi: https://doi.org/10.1196/annals.1352.026.

Allen, J.G. 2018. *Mentalizing in the Development and Treatment of Attachment Trauma*: Taylor & Francis.

Bellis, M.A., K. Hughes, N. Leckenby, K.A. Hardcastle, C. Perkins, and H. Lowey. 2015. "Measuring mortality and the burden of adult disease associated with adverse childhood experiences in England: A national survey." *Journal of Public Health* 37 (3):445–454.

Bellis, M.A., K. Hughes, K. Ford, G. Ramos Rodriguez, D. Sethi, and J. Passmore. 2019. "Life course health consequences and associated annual costs of adverse childhood experiences across Europe and North America: A systematic review and meta-analysis." *The Lancet Public Health* 4 (10):e517–e528.

Bellis, M.A., K. Hughes, N. Leckenby, C. Perkins, and H. Lowey. 2014. "National household survey of adverse childhood experiences and their relationship with resilience to health-harming behaviors in England." *BMC Medicine* 12 (1):72. doi: 10.1186/1741-7015-12-72.

Beltrán González, A.N., M.I. López Pazos, and D.J. Calvo. 2020. "Reactive oxygen species in the regulation of the GABA mediated inhibitory neurotransmission." *Neuroscience* 439:137–145. doi: 10.1016/j.neuroscience.2019.05.064.

Black, P.H. 2003. "The inflammatory response is an integral part of the stress response: Implications for atherosclerosis, insulin resistance, type II diabetes and metabolic syndrome X." *Brain, Behavior, and Immunity* 17 (5):350–364.

Broadley, A.J.M., A. Korszun, E. Abdelaal, V. Moskvina, C.J.H. Jones, G.B. Nash, C. Ray, J. Deanfield, and M.P. Frenneaux. 2005. "Inhibition of cortisol production with metyrapone prevents mental stress-induced endothelial dysfunction and baroreflex impairment." *Journal of the American College of Cardiology* 46 (2):344–350.

Bryan, R.H. 2019. "Getting to why: Adverse childhood experiences' impact on adult health." *The Journal for Nurse Practitioners* 15 (2):153–157.e1.

Callaghan, B.L., D.G. Gee, L. Gabard-Durnam, E.H. Telzer, K.L. Humphreys, B. Goff, M. Shapiro, J. Flannery, D.S. Lumian, D.S. Fareri, C. Caldera, and N. Tottenham. 2019. "Decreased Amygdala reactivity to parent cues protects against anxiety following early adversity: An examination across 3 years." *Biological Psychiatry: Cognitive Neuroscience and Neuroimaging* 4 (7):664–671. https://doi.org/10.1016/j.bpsc.2019.02.001.

Campbell, J.A., R.J. Walker, and L.E. Egede. 2016. "Associations between adverse childhood experiences, high-risk behaviors, and morbidity in adulthood." *American Journal of Preventive Medicine* 50 (3):344–352. https://doi.org/10.1016/j.amepre.2015.07.022.

Capuron, L., and A.H. Miller. 2011. "Immune system to brain signaling: Neuropsychopharmacological implications." *Pharmacology & Therapeutics* 130 (2):226–238.

Cecala, A.L. 2016. "Anxious: Using the brain to understand and treat fear and anxiety." *Journal of Undergraduate Neuroscience Education* 14 (2):R22–R23.

Cohen, S., D.A.J. Tyrrell, and A.P. Smith. 1991. "Psychological stress and susceptibility to the common cold." *New England Journal of Medicine* 325 (9):606–612.

Daiber, A., S. Steven, K. Vujacic-Mirski, S. Kalinovic, M. Oelze, F. Di Lisa, and T. Münzel. 2020. "Regulation of vascular function and inflammation via cross talk of reactive oxygen and nitrogen species from Mitochondria or NADPH oxidase-implications for diabetes progression." *International Journal of Molecular Sciences* 21 (10). doi: 10.3390/ijms21103405.

Danese, A., and B.S. McEwen. 2012. "Adverse childhood experiences, allostasis, allostatic load, and age-related disease." *Physiology & Behavior* 106 (1):29–39.

Dar, T., A. Radfar, S. Abohashem, R.K. Pitman, A. Tawakol, and M.T. Osborne. 2019. "Psychosocial stress and cardiovascular disease." *Current Treatment Options in Cardiovascular Medicine* 21 (5):1–17.

de, M. Bandeira S., L.J. da Fonseca, S. Guedes G. da, L.A. Rabelo, M.O. Goulart, and S.M. Vasconcelos. 2013. "Oxidative stress as an underlying contributor in the development of chronic complications in diabetes mellitus." *International Journal of Molecular Sciences* 14 (2):3265–3284. doi: 10.3390/ijms14023265.

Deighton, S., A. Neville, D. Pusch, and K. Dobson. 2018. "Biomarkers of adverse childhood experiences: A scoping review." *Psychiatry Research* 269:719–732.

Du, J., Y. Wang, R. Hunter, Y. Wei, R. Blumenthal, C. Falke, R. Khairova, R. Zhou, P. Yuan, R. Machado-Vieira, B.S. McEwen, and H.K. Manji. 2009. "Dynamic regulation of mitochondrial function by glucocorticoids." *Proceedings of the National Academy of Sciences of the United States of America* 106 (9):3543–3548. doi: 10.1073/pnas.0812671106.

Epel, E.S., E.H. Blackburn, J. Lin, F.S. Dhabhar, N.E. Adler, J.D. Morrow, and R.M. Cawthon. 2004. "Accelerated telomere shortening in response to life stress." *Proceedings of the National Academy of Sciences* 101 (49):17312–17315. doi: 10.1073/pnas.0407162101.

Fonagy, P., and M. Target. 2008. "Attachment, trauma, and psychoanalysis: Where psychoanalysis meets neuroscience." In *Mind to Mind: Infant Research, Neuroscience, and Psychoanalysis*, 15–49. New York, NY: Other Press.

Franco, O.H., K. Karnik, G. Osborne, J.M. Ordovas, M. Catt, and F. van der Ouderaa. 2009. "Changing course in ageing research: The healthy ageing phenotype." *Maturitas* 63 (1):13–19. doi: 10.1016/j.maturitas.2009.02.006.

Franke, H.A. 2014. "Toxic stress: Effects, prevention and treatment." *Children* 1 (3):390–402.

Gałecki, P., J. Mossakowska-Wójcik, and M. Talarowska. 2018. "The anti-inflammatory mechanism of antidepressants—SSRIs, SNRIs." *Progress in Neuro-Psychopharmacology and Biological Psychiatry* 80:291–294.

Giniatullin, A., A. Petrov, and R. Giniatullin. 2019. "Action of hydrogen peroxide on synaptic transmission at the mouse neuromuscular junction." *Neuroscience* 399:135–145. doi: 10.1016/j.neuroscience.2018.12.027.

Glaser, R., and J.K. Kiecolt-Glaser. 2005. "Stress-induced immune dysfunction: Implications for health." *Nature Reviews Immunology* 5 (3):243–251. doi: 10.1038/nri1571.

Glaser, R., T.F. Robles, J. Sheridan, W.B. Malarkey, and J.K. Kiecolt-Glaser. 2003. "Mild depressive symptoms are associated with amplified and prolonged inflammatory responses after influenza virus vaccination in older adults." *Archives of General Psychiatry* 60 (10):1009–1014.

Gomes, L.C., G. Di Benedetto, and L. Scorrano. 2011. "During autophagy mitochondria elongate, are spared from degradation and sustain cell viability." *Nature Cell Biology* 13 (5):589–598. doi: 10.1038/ncb2220.

Gottdiener, J.S., W.J. Kop, E. Hausner, M.K. McCeney, D. Herrington, and D.S. Krantz. 2003. "Effects of mental stress on flow-mediated brachial arterial dilation and influence of behavioral factors and hypercholesterolemia in subjects without cardiovascular disease." *The American Journal of Cardiology* 92 (6):687–691.

Gu, Y., W.T. Piper, L.A. Branigan, E.M. Vazey, G. Aston-Jones, L. Lin, J.E. LeDoux, and R.M. Sears. 2020. "A brainstem-central amygdala circuit underlies defensive responses to learned threats." *Molecular Psychiatry* 25 (3):640–654. doi: 10.1038/s41380-019-0599-6.

Haroon, E., C.L. Raison, and A.H. Miller. 2012. "Psychoneuroimmunology meets neuropsychopharmacology: Translational implications of the impact of inflammation on behavior." *Neuropsychopharmacology* 37 (1):137–162. doi: 10.1038/npp.2011.205.

Heidt, T., H.B. Sager, G. Courties, P. Dutta, Y. Iwamoto, A. Zaltsman, C. von zur Muhlen, C. Bode, G.L. Fricchione, J. Denninger, et al. 2014. "Chronic variable stress activates hematopoietic stem cells." *Nature Medicine* 20 (7):754–758. doi: 10.1038/nm.3589.

Hernandez, M.E., D. Mendieta, M. Pérez-Tapia, R. Bojalil, I. Estrada-Garcia, S. Estrada-Parra, and L. Pavón. 2013. "Effect of selective serotonin reuptake inhibitors and immunomodulator on cytokines levels: An alternative therapy for patients with major depressive disorder." *Clinical and Developmental Immunology* 2013.

Himmerich, H., O. Patsalos, N. Lichtblau, M.A.A. Ibrahim, and B. Dalton. 2019. "Cytokine research in depression: Principles, challenges, and open questions." *Frontiers in Psychiatry* 10. doi: 10.3389/fpsyt.2019.00030.

Hitchcock, G., and J.S. Porterfjeld. 1961. "The production of interferon in brains of mice infected with an arthropod-borne virus." *Virology* 13 (3):363–365.

Holman, D.M., K.A. Ports, N.D. Buchanan, N.A. Hawkins, M.T. Merrick, M. Metzler, and K.F. Trivers. 2016. "The association between adverse childhood experiences and risk of cancer in adulthood: A systematic review of the literature." *Pediatrics* 138 (Supplement_1):S81–S91.

Hughes, K., M.A. Bellis, K.A. Hardcastle, D. Sethi, A. Butchart, C. Mikton, L. Jones, and M.P. Dunne. 2017. "The effect of multiple adverse childhood experiences on health: A systematic review and meta-analysis." *The Lancet Public Health* 2 (8):e356–e366.

Isaacs, A., and J. Lindenmann. 1957. "Virus interference. I. The interferon." *Proceedings of the Royal Society of London. Series B-Biological Sciences* 147 (927):258–267.

Jain, S., and P.J. Mills. 2007. "Cytokines, chronic stress, and fatigue."

Jie, F., G. Yin, Wei Yang, M. Yang, S. Gao, J. Lv, and B. Li. 2018. "Stress in regulation of GABA amygdala system and relevance to neuropsychiatric diseases." *Frontiers in Neuroscience* 12. doi: 10.3389/fnins.2018.00562.

Kaufmann, F.N., M. Gazal, T.C. Mondin, T.A. Cardoso, L.Á. Quevedo, L.D.M. Souza, K. Jansen, E. Braganhol, J.P. Oses, and R.T. Pinheiro. 2015. "Cognitive psychotherapy treatment decreases peripheral oxidative stress parameters associated with major depression disorder." *Biological Psychology* 110:175–181.

Kerahrodi, Jasmin Ghaemi, and Matthias Michal. 2020. "The fear-defense system, emotions, and oxidative stress." *Redox Biology* 37:101588.

Kim, Sung-Wan, Hee-Ju Kang, Min Jhon, Ju-Wan Kim, Ju-Yeon Lee, Adam J Walker, Bruno Agustini, Jae-Min Kim, and Michael Berk. 2019. "Statins and inflammation: New therapeutic opportunities in psychiatry." *Frontiers in psychiatry* 10:103.

Kivimäki, Mika, and Andrew Steptoe. 2018. "Effects of stress on the development and progression of cardiovascular disease." *Nature Reviews Cardiology* 15 (4):215–229.

Köhler, Cristiano A., Thiago H. Freitas, Brendon Stubbs, Michael Maes, Marco Solmi, Nicola Veronese, Nayanna Q. de Andrade, Gerwyn Morris, Brisa S. Fernandes, and André R. Brunoni. 2018. "Peripheral alterations in cytokine and chemokine levels after antidepressant drug treatment for major depressive disorder: Systematic review and meta-analysis." *Molecular Neurobiology* 55 (5):4195–4206.

Kozlowska, K., P. Walker, L. McLean, and P. Carrive. 2015. "Fear and the Defense Cascade: Clinical Implications and Management." *Harvard Review of Psychiatry* 23 (4):263–287. doi: 10.1097/hrp.0000000000000065.

Laricchiuta, D., A. Panuccio, E. Picerni, D. Biondo, B. Genovesi, and L. Petrosini. 2023. "The body keeps the score: The neurobiological profile of traumatized adolescents." *Neuroscience & Biobehavioral Reviews* 145:105033. doi: 10.1016/j.neubiorev.2023.105033.

LeDoux, Joseph E. 2000. "Emotion circuits in the brain."

Lee, Y.J., S.Y. Jeong, M. Karbowski, C.L. Smith, and R.J. Youle. 2004. "Roles of the mammalian mitochondrial fission and fusion mediators Fis1, Drp1, and Opa1 in apoptosis." *Molecular Biology of the Cell* 15 (11):5001–5011. doi: 10.1091/mbc.e04-04-0294.

Li, H., A.B. Kilgallen, T. Münzel, E. Wolf, S. Lecour, R. Schulz, A. Daiber, and L.W. Van Laake. 2020. "Influence of mental stress and environmental toxins on circadian clocks: Implications for redox regulation of the heart and cardioprotection." *British Journal of Pharmacology* 177 (23):5393–5412. doi: 10.1111/bph.14949.

Lichtenberg, D., and I. Pinchuk. 2015a. "Oxidative stress, the term and the concept." *Biochemical and Biophysical Research Communications* 461 (3):441–444. doi: 10.1016/j.bbrc.2015.04.062.

Lichtenberg, D., and I. Pinchuk. 2015b. "Oxidative stress, the term and the concept." *Biochemical and Biophysical Research Communications* 461 (3):441–444. doi: https://doi.org/10.1016/j.bbrc.2015.04.062.

Lind, Lars, Kristina Johansson, and Jan Hall. 2002. "The effects of mental stress and the cold pressure test on flow-mediated vasodilation." *Blood Pressure* 11 (1):22–27.

Lorton, D., C. Lubahn, and D. Bellinger. 2001. "Introduction to biological signaling in psychoneuroimmunology." *Psychoneuroimmunology* 3.

Maes, Michael, Willem Ombelet, Raf De Jongh, Gunter Kenis, and Eugene Bosmans. 2001. "The inflammatory response following delivery is amplified in women who previously suffered from major depression, suggesting that major depression is accompanied by a sensitization of the inflammatory response system." *Journal of Affective Disorders* 63 (1–3):85–92.

Maier, Steven F. 2003. "Bi-directional immune—brain communication: Implications for understanding stress, pain, and cognition." *Brain, Behavior, and Immunity* 17 (2):69–85.

Manoli, I., S. Alesci, M. R. Blackman, Y. A. Su, O. M. Rennert, and G. P. Chrousos. 2007. "Mitochondria as key components of the stress response." *Trends in Endocrinology & Metabolism* 18 (5):190–198. doi: 10.1016/j.tem.2007.04.004.

McCullough, L., N. Kuhn, S. Andrews, A. Kaplan, J. Wolf, and C.L. Hurley. 2021. *Treating Affect Phobia: A Manual for Short-Term Dynamic Psychotherapy*: Guilford Publications.

McEwen, B.S. 1998. "Stress, adaptation, and disease: Allostasis and allostatic load." *Annals of the New York Academy of Sciences* 840 (1):33–44.

Merrick, M.T., D.C. Ford, K.A. Ports, and A.S. Guinn. 2018. "Prevalence of adverse childhood experiences from the 2011–2014 behavioral risk factor surveillance system in 23 states." *JAMA Pediatrics* 172 (11):1038–1044.

Meyer, T., and P.H. Wirtz. 2018a. "Mechanisms of Mitochondrial Redox Signaling in Psychosocial Stress-Responsive Systems: New Insights into an Old Story." *Antioxid Redox Signal* 28 (9):760–772. doi: 10.1089/ars.2017.7186.

Meyer, T., and P.H. Wirtz. 2018b. "Mechanisms of mitochondrial redox signaling in psychosocial stress-responsive systems: new insights into an old story." *Antioxidants & Redox Signaling* 28 (9):760–772.

Murray, C.J.L., A.Y. Aravkin, P. Zheng, C. Abbafati, K.M. Abbas, M. Abbasi-Kangevari, and Lim, S.S. (2020). Global burden of 87 risk factors in 204 countries and territories, 1990–2019: A systematic analysis for the Global Burden of Disease Study 2019." *Lancet (London, England)* 396 (10258):1223–49.

Nadel, J., and L. Camaioni. 2017. *New Perspectives in Early Communicative Development.* Vol. 8: Routledge.

Nahrendorf, M., and F.K. Swirski. 2015. "Lifestyle effects on hematopoiesis and atherosclerosis." *Circulation Research* 116 (5):884–894.

Olshansky, S.J., D.J. Passaro, R.C. Hershow, J. Layden, B.A. Carnes, J. Brody, L. Hayflick, R.N. Butler, D.B. Allison, and D.S. Ludwig. 2005. "A potential decline in life expectancy in the United States in the 2[1s]t century." *New England Journal of Medicine* 352 (11):1138–1145. doi: 10.1056/NEJMsr 043743.

Pace-Schott, E.F., M.C. Amole, T. Aue, M. Balconi, L.M. Bylsma, H. Critchley, H.A. Demaree, B.H. Friedman, A.E.K. Gooding, O. Gosseries, et al. 2019. "Physiological feelings." *Neuroscience & Biobehavioral Reviews* 103:267–304. doi: 10.1016/j.neubiorev.2019.05.002.

Panksepp, J., and L. Biven. 2012. "A meditation on the affective neuroscientific view of human and animalian MindBrains." *From the Couch to the Lab: Trends in Psychodynamic Neuroscience*: 145–175.

Pape, Hans-Christian, and Denis Pare. 2010. "Plastic synaptic networks of the amygdala for the acquisition, expression, and extinction of conditioned fear." *Physiological Reviews* 90 (2):419–463.

Pasquereau, B., G. Drui, Y. Saga, A. Richard, M. Millot, E. Météreau, V. Sgambato, P.N. Tobler, and L. Tremblay. 2021. "Selective serotonin reuptake inhibitor treatment retunes emotional valence in primate ventral striatum." *Neuropsychopharmacology* 46 (12):2073–2082.

Penninx, B.W.J.H., S.B. Kritchevsky, K. Yaffe, A.B. Newman, E.M. Simonsick, S. Rubin, L. Ferrucci, T. Harris, and M. Pahor. 2003. "Inflammatory markers and depressed mood in older persons: Results from the health, aging and body composition study." *Biological Psychiatry* 54 (5):566–572.

Picard, M., B. Jung, F. Liang, I. Azuelos, S. Hussain, P. Goldberg, R. Godin, G. Danialou, R. Chaturvedi, K. Rygiel, et al. 2012. "Mitochondrial dysfunction and lipid accumulation in the human diaphragm during mechanical ventilation." *American Journal of Respiratory and Critical Care Medicine* 186 (11):1140–1149. doi: 10.1164/rccm.201206-0982OC.

Picard, M., and D.M. Turnbull. 2013. "Linking the metabolic state and mitochondrial DNA in chronic disease, health, and aging." *Diabetes* 62 (3):672–678. doi: 10.2337/db12-1203.

Picard, M., R.-P. Juster, and B.S. McEwen. 2014. "Mitochondrial allostatic load puts the'gluc'back in glucocorticoids." *Nature Reviews Endocrinology* 10 (5):303–310.

Psarra, A.M., and C.E. Sekeris. 2011. "Glucocorticoids induce mitochondrial gene transcription in HepG2 cells: Role of the mitochondrial glucocorticoid receptor." *Biochimica et Biophysica Acta (BBA) – Molecular Cell Research* 1813 (10):1814–1821. doi: 10.1016/j.bbamcr.2011.05.014.

Puterman, Eli, Jue Lin, Elizabeth Blackburn, Aoife O'donovan, Nancy Adler, and Elissa Epel. 2010. "The power of exercise: buffering the effect of chronic stress on telomere length." *PloS one* 5 (5):e10837.

Rajmohan, V., and E. Mohandas. 2007. "The limbic system." *Indian Journal of Psychiatry* 49 (2):132–139. doi: 10.4103/0019-5545.33264.

Riccardi, I., E. Bilotta, L. Leone, G. Nicolò, M. Procacci, A. Semerari, and A. Carcione. 2020. "Adverse experiences in childhood: Association with metacognition, personality disorders and distress." *Journal of Psychopathology* 26:46–53.

Rice, M.E. 2011. "H2O2: A dynamic neuromodulator." *The Neuroscientist* 17 (4):389–406. doi: 10.1177/1073858411404531.

Sah, P., E.S. Lopez Faber, M. Lopez de Armentia, and J.M.J.P.R. Power. 2003. "The amygdaloid complex: Anatomy and physiology." *Physiological Reviews* 83 (3):803–834.

Schore, A.N. 2019. *The Development of the Unconscious Mind (Norton Series on Interpersonal Neurobiology)*: W. W. Norton.

Shih, R.-H., C.-Y. Wang, and C.-M. Yang. 2015. "NF-kappaB signaling pathways in neurological inflammation: A mini review." *Frontiers in Molecular Neuroscience* 8:77.

Slavich, G.M. 2020. "Psychoneuroimmunology of stress and mental health." In *The Oxford Handbook of Stress and Mental Health*, 519–546.

Slavich, G.M., and M.R. Irwin. 2014. "From stress to inflammation and major depressive disorder: A social signal transduction theory of depression." *Psychological Bulletin* 140 (3):774.

Steinberg, H., and H. Himmerich. 2013. "Emil Kraepelin's habilitation and his thesis: A pioneer work for modern systematic reviews, psychoimmunological research and categories of psychiatric diseases." *The World Journal of Biological Psychiatry* 14 (4):248–257.

Su, S., M.P. Jimenez, C.T.F. Roberts, and E.B. Loucks. 2015. "The role of adverse childhood experiences in cardiovascular disease risk: A review with emphasis on plausible mechanisms." *Current Cardiology Reports* 17 (10):1–10.

Tang, V.M., A.H. Young, H. Tan, C. Beasley, and J.F. Wang. 2013. "Glucocorticoids increase protein carbonylation and mitochondrial dysfunction." *Hormone and Metabolic Research* 45 (10):709–715. doi: 10.1055/s-0033-1345119.

Teicher, M.H., J.A. Samson, C.M. Anderson, and K. Ohashi. 2016. "The effects of childhood maltreatment on brain structure, function and connectivity." *Nature Reviews Neuroscience* 17 (10):652–666.

Testa, R., F. Olivieri, C. Sirolla, L. Spazzafumo, M.R. Rippo, M. Marra, A.R. Bonfigli, A. Ceriello, R. Antonicelli, C. Franceschi, et al. 2011. "Leukocyte telomere length is associated with complications of type 2 diabetes mellitus." *Diabetic Medicine* 28 (11):1388–1394. doi: 10.1111/j.1464-5491.2011.03370.x.

Tottenham, N., M. Shapiro, J. Flannery, C. Caldera, and R.M. Sullivan. 2019. "Parental presence switches avoidance to attraction learning in children." *Nature Human Behaviour* 3 (10):1070–1077.

Weigent, D.A., and J.E. Blalock. 1997. "Production of Peptide Hormones and." *Neuroimmunoendocrinology* 69:1–30.

Wenzel, P., S. Kossmann, T. Münzel, and A. Daiber. 2017. "Redox regulation of cardiovascular inflammation—Immunomodulatory function of mitochondrial and Nox-derived reactive oxygen and nitrogen species." *Free Radical Biology and Medicine* 109:48–60. doi: 10.1016/j.freeradbiomed.2017.01.027.

Yakovleva, T., I. Bazov, H. Watanabe, K.F. Hauser, and G. Bakalkin. 2011. "Transcriptional control of maladaptive and protective responses in alcoholics: A role of the NF-κB system." *Brain, Behavior, and Immunity* 25:S29–S38. https://doi.org/10.1016/j.bbi.2010.12.019.

Yehuda, R., and A. Lehrner. 2018. "Intergenerational transmission of trauma effects: Putative role of epigenetic mechanisms." *World Psychiatry* 17 (3):243–257.

Yu, T., J.L. Robotham, and Y. Yoon. 2006. "Increased production of reactive oxygen species in hyperglycemic conditions requires dynamic change of mitochondrial morphology." *Proceedings of the National Academy of Sciences* 103 (8):2653–2658. doi: 10.1073/pnas.0511154103.

21 The Role of Oxidative Stress in Cardiovascular Disease Caused by Social Isolation and Loneliness

Huige Li and Ning Xia

21.1 INTRODUCTION

Emerging evidence indicates that psychosocial stress is the third most important modifiable risk factor for coronary heart disease and myocardial infarction, ranking only behind lipids and smoking (Rosengren et al. 2004). Accounting for about one-third of the attributable risk of coronary heart disease, psychosocial stress represents a strong cardiovascular risk factor (Rosengren et al. 2004).

Mental stress in modern human societies occurs largely during social interactions. In addition to work-related stress, loneliness and social isolation represent major sources of chronic stress in humans (Steptoe and Kivimaki 2013). Moreover, social and demographic changes in modern society have put an increasing number of individuals at risk for loneliness (Xia and Li 2018; Li and Xia 2020).

21.2 LONELINESS AND SOCIAL ISOLATION CAUSE CARDIOVASCULAR DISEASE IN HUMANS

There is clear evidence that strong social relationships increase the likelihood of better survival (Holt-Lunstad, Smith, and Layton 2010) whereas loneliness and social isolation are associated with higher mortality (Steptoe et al. 2013; Holt-Lunstad et al. 2015). The increased mortality risk is as high as light smoking or alcohol consumption and larger than that caused by obesity and hypertension (Holt-Lunstad, Smith, and Layton 2010; Holt-Lunstad et al. 2015). Loneliness and social isolation are independent risk factors for cardiovascular disease (CVD) (Holt-Lunstad and Smith 2016). Lonely individuals have greater total peripheral vascular resistance and higher risk to develop hypertension (Hawkley et al. 2010). Consistently, loneliness and social isolation have been shown to increase the risks of coronary heart disease and stroke (Valtorta et al. 2016).

21.3 SOCIAL ISOLATION PROMOTES CARDIOVASCULAR DISEASE IN ANIMAL EXPERIMENTS

Social isolation of experimental animals exacerbates atherogenesis. This has been shown in different atherosclerosis models. Individual housing of cynomolgus monkeys increases atherosclerosis development in the coronary artery (Shively, Clarkson, and Kaplan 1989). Likewise, increased atherosclerotic lesion has been found in socially isolated Watanabe heritable hyperlipidemic rabbits (McCabe et al. 2002; Paredes et al. 2006) and apolipoprotein E-knockout mice (Bernberg et al. 2008). Interestingly, the enhanced atherosclerosis in animals exposed to social isolation is associated with physical inactivity, increased activity of the sympathetic nervous system (SNS), as well

DOI: 10.1201/9781003328100-26

as enhanced inflammation and vascular oxidative stress (McCabe et al. 2002; Nation et al. 2008; Nation et al. 2010).

21.4 LONELINESS AND SOCIAL ISOLATION LEAD TO ACTIVATION OF THE HPA AXIS

There is clear evidence that loneliness leads to activation of the hypothalamic-pituitary-adrenocortical (HPA) axis in humans (Hawkley et al. 2012; Cacioppo, Cacioppo, Capitanio, et al. 2015). The HPA axis is the main producer of glucocorticoids, including cortisol in humans and corticosterone in rodents. The cortisol production follows the circadian rhythm with higher levels in the morning and lower levels in the evening. Lonely individuals have greater morning cortisol increases (Adam et al. 2006), elevated circulating cortisol concentrations (Doane and Adam 2010; Cole, Capitanio, et al. 2015), and impaired glucocorticoid receptor (GR) sensitivity (Cole et al. 2007; Cole 2008), suggesting that loneliness causes overaction of the HPA axis (Cacioppo, Cacioppo, Cole, et al. 2015).

Activation of the HPA axis has also been observed in animal models of social isolation. Monogamous prairie voles build pair bonds with mating partners and are used as an animal model to study the consequences of pair bonding or partner loss (Bosch et al. 2009). Separation of pair-bonded prairie voles from the partner leads to an elevation of circulating corticosterone concentrations (Bosch et al. 2009; McNeal et al. 2014; Sun et al. 2014). In contrast, separation from a same-sex sibling has no effect on corticosterone levels (Bosch et al. 2009). The separation of pair-bonded prairie voles is associated with an increase of corticotrophin-releasing hormone and adrenocorticotropic hormone (Bosch et al. 2009; McNeal et al. 2014; Sun et al. 2014), indicating an activation of the entire HPA axis.

In the vascular system, glucocorticoids reduce endothelial nitric oxide (NO) production and enhance oxidative stress (Xia and Li 2018; Li and Xia 2020). Because endothelial NO is a key factor preventing the development of hypertension and atherosclerosis, the reduced endothelial NO production caused by glucocorticoid is likely to be involved in the blood pressure-elevating and atherogenic effects of social isolation.

21.5 DYSREGULATION OF THE AUTONOMIC NERVOUS SYSTEM

The findings on SNS activation in loneliness and social isolation are less consistent than the activation of the HPA axis (Cacioppo, Cacioppo, Capitanio, et al. 2015; Hawkley et al. 2012). Chronic social isolation in humans and macaques is associated with increased urinary levels of norepinephrine metabolites but not epinephrine (Cole, Capitanio, et al. 2015) (Table 21.1). The effect of social isolation on local catecholamine concentrations in SNS-innervated tissues seems to be greater than that on systemic catecholamine levels (Li and Xia 2020).

In adult rats, chronic social isolation has been found to increase the plasma concentrations of both epinephrine and norepinephrine (Gavrilovic, Spasojevic, and Dronjak 2010). Social isolation of male prairie voles from the bonded partner results in increased heart rate, heart rhythm dysregulation, and autonomic imbalance characterized by increased sympathetic and decreased parasympathetic drive to the heart (McNeal et al. 2014).

A well balanced vagal system is important to prevent the detrimental effects of SNS overactivation. The vagally-mediated parasympathetic activity can be monitored with the high frequency heart rate variability (HF-HRV), which denotes heart rate variations associated with respiration. The vagal outflow is inhibited by the cardiovascular center during inhalation and restored during exhalation. Although the heart is innervated both by the SNS and the vagal system, the effect of the SNS is too slow to modify the beat-to-beat changes. As a result, the HF-HRV can be considered a direct measure of the parasympathetic control of the heart. A lower HF-HRV is correlated with depression, reduced cognitive function, CVD, and all-cause mortality (Thayer, Yamamoto, and Brosschot 2010).

TABLE 21.1
Dysregulation of the Autonomic Nervous System by Social Isolation

Species	Model	Effects	References
Human	Perceived social isolation	Elevated urinary levels of norepinephrine metabolites	(Cole, Capitanio, et al. 2015)
Macaques	Perceived social isolation	Elevated urinary levels of norepinephrine metabolites	(Cole, Capitanio, et al. 2015)
Ovarian cancer patients	Poor social support	Higher norepinephrine levels in tumor tissues	(Lutgendorf et al. 2011, 2009)
Rats	Chronic social isolation	Increased plasma epinephrine & norepinephrine	(Gavrilovic, Spasojevic, and Dronjak 2010)
Prairie voles	Social isolation	Increased sympathetic and decreased parasympathetic drive to the heart	(McNeal et al. 2014)
Prairie voles	Social isolation	Lower HF-HRV	(Grippo et al. 2007, 2009)
Human	Lower social integration	Lower HF-HRV	(Hemingway et al. 2005; Gouin, Zhou, and Fitzpatrick 2015)

It has been reported that social environment has an impact on HF-HRV (Gouin, Zhou, and Fitzpatrick 2015). High HF-HRV is observed in married individuals (Randall, Bhattacharyya, and Steptoe 2009) and even higher HF-HRV in people with happy marriage life (Smith et al. 2011). In the Whitehall cohort in the United Kingdom, a smaller HF-HRV has been found to be associated with a lower social integration (Hemingway et al. 2005) (Table 21.1). A similar link between social integration and the autonomic nervous system has been shown for students moving to other countries to study. The poor social integration in the initial time period in a foreign county is associated with higher heart rate and lower HF-HRV (Gouin, Zhou, and Fitzpatrick 2015).

Regulation of HF-HRV by social environment has also been shown in animal studies. Compared to socially paired prairie voles, those suffering from social isolation have lower HF-HRV, both at baseline and in response to stress (Grippo et al. 2007, 2009). Interestingly, the poor HF-HRV in socially isolated voles can be normalized by a treatment with exogenous oxytocin (Grippo et al. 2009).

21.6 PROINFLAMMATORY RESPONSE OF THE IMMUNE SYSTEM

Loneliness and social isolation in humans cause differentiated gene expression in circulating leukocytes leading to the so-called conserved transcriptional response to adversity (CTRA). The CTRA is characterized by up-regulation of proinflammatory genes and down-regulation of antiviral immunity-related genes resulting in enhanced inflammation and an impaired antiviral response (Powell et al. 2013; Cole, Levine, et al. 2015). Loneliness leads to a selective expansion of the immature classical monocyte subset without changing the number of total circulating leukocytes (Cole, Capitanio, et al. 2015; Cole et al. 2011). The increased circulating monocyte frequencies and percentages are the main source of the proinflammatory function of the CTRA (Powell et al. 2013). The CTRA has been observed in both humans and macaques with high perceived social isolation (Cole et al. 2011; Cole, Capitanio, et al. 2015).

The proinflammatory CTRA observed in individuals exposed to loneliness and social isolation is likely to be caused by the SNS activation rather than the HPA axis (Powell et al. 2013). Loneliness increases the levels of norepinephrine metabolites (but not epinephrine) in urine (Cole, Capitanio, et al. 2015).

In mouse models of repeated social defeat (Powell et al. 2013) and chronic variable stress (Heidt et al. 2014), SNS activation has been shown to enhance myelopoiesis leading to an increased output of neutrophils and inflammatory monocytes. The effects can be prevented by pharmacological antagonism or genetic disruption of the β3-adrenoreceptor (Powell et al. 2013; Heidt et al. 2014), indicating a causal role of these receptors. However, this mechanism (SNS-mediated myelopoiesis and CTRA) has not been verified in the mouse model of social isolation, so far.

Remarkably, the CTRA-induced proinflammatory monocytes can traffic into the brain, cause "sickness behaviors", and augment loneliness, creating a vicious cycle (Cole 2014; Cole, Capitanio, et al. 2015). The CTRA and proinflammatory response are likely to represent crucial mechanisms mediating the harmful health effects. Recent epidemiologic studies have associated CTRA with increased risk of cardiovascular, metabolic, and neoplastic diseases.

Moreover, chronic psychosocial stress has been associated with elevated levels of circulating proinflammatory cytokines, particularly of IL-6. Socially less well connected individuals have higher IL-6 levels (Meyer and Wirtz 2018).

21.7 OXIDATIVE STRESS IN THE HYPOTHALAMUS IS REQUIRED FOR HPA ACTIVATION

Oxidative stress is a key molecular mechanism linking chronic psychosocial stress to cardiovascular disease. In socially isolated animals, oxidative stress has been observed both in the brain and in peripheral tissues. In the brain, oxidative stress is required for the social isolation-induced HPA activation (see later). In the vascular tissue, oxidative stress is likely to be the result of social isolation-induced activation of HPA, SNS, and proinflammatory immune response. A recent clinical study demonstrating a dose-dependent association between of HPA activation, SNS activation, and inflammation and oxidative damage supports this concept (Black et al. 2017).

In the rat model of social isolation rearing, the animals are housed individually starting at the age of 21 days. Interestingly, the induction of oxidative stress is an early event that precedes the activation of the HPA axis (Colaianna et al. 2013). Oxidative stress markers in the hypothalamus are increased as early as 2 weeks after social isolation (Colaianna et al. 2013). This is followed by elevated protein levels of corticotropin-releasing factor in the hypothalamus and of adrenocorticotropic hormone in the plasma after 4 weeks of isolation, whereas the increased corticosterone levels in plasma and saliva can be observed after 7 weeks (Colaianna et al. 2013).

The enhanced oxidative stress in the brain of socially isolated rats is attributable to an induction of NADPH oxidase 2 (NOX2), with no changes in the mRNA expression of NOX1, NOX3, or NOX4 (Schiavone et al. 2009; Colaianna et al. 2013). Treatment with the NOX inhibitor apocynin prevents social isolation-induced HPA activation (Schiavone et al. 2009; Colaianna et al. 2013). In addition, rats with a loss-of-function mutation in $p47^{phox}$, which is an essential component of the NOX2 complex, are protected from social isolation-induced oxidative stress, HPA activation, and behavior changes (Colaianna et al. 2013). These results indicate that NOX2-mediated oxidative stress is an early trigger of HPA activation and is causally involved in social isolation-induced pathology. Moreover, the NOX2 induction by social isolation in the hypothalamus is prevented by apocynin treatment and absent in rats with $p47^{phox}$ mutation (Colaianna et al. 2013), suggesting a positive feedback mechanism of social isolation-induced oxidative stress in the brain (Figure 21.1).

A second positive feedback mechanism may be the HPA axis itself. Although glucocorticoids have been shown to inhibit NOX-mediated ROS production in some cell types (Schiavone et al. 2013), dexamethasone has been shown to up-regulate NOX expression in hippocampal neurons (You et al. 2009). Moreover, the dexamethasone-induced ROS production in hippocampal neurons can be prevented by GC receptor blockade, indicating a specific effect of glucocorticoids. Nevertheless, this experiment was performed using organotypic hippocampal slice cultures (You et al. 2009). It is still unclear whether it applies to *in vivo* situations in social isolation animals.

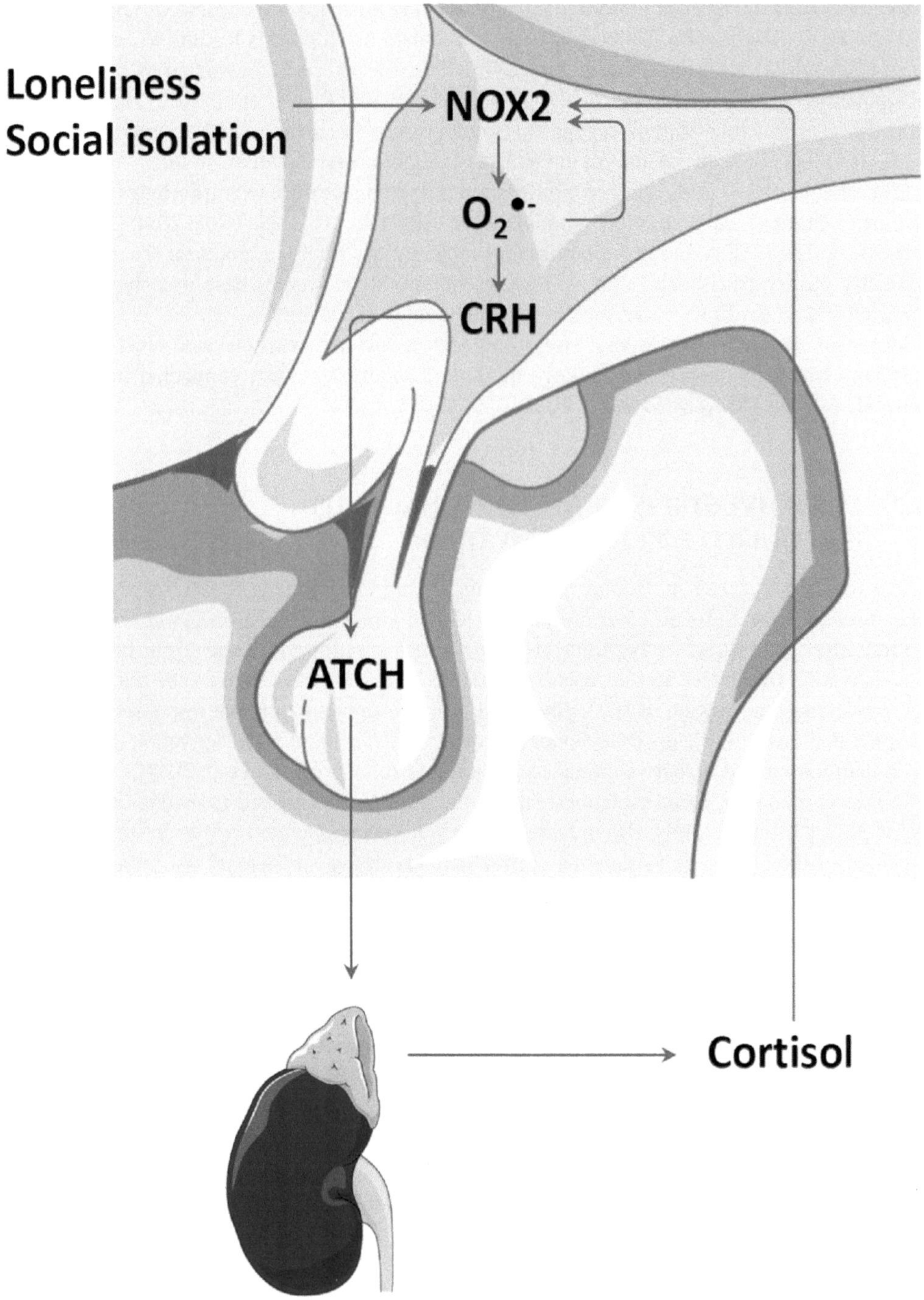

FIGURE 21.1 Role of oxidative stress in social isolation-induced HPA activation. Social isolation increases the expression of NOX2 and its components in the hypothalamus before the elevation of corticotropin-releasing hormone (CRH) in the hypothalamus and the concentration of adrenocorticotropic hormone (ACTH) in the plasma. Inhibition of NOX2 activity prevents the activation of the hypothalamic-pituitary-adrenocortical (HPA) axis and the NOX2 up-regulation, indicating a positive feedback mechanism. NOX2 expression in the hypothalamus may be further potentiated by glucocorticoids such as cortisol produced by the adrenal gland. Reproduced from Li and Xia (2020), which is an open access article distributed under the terms of the Creative Commons CC BY license.

21.8 OXIDATIVE STRESS IN THE BRAIN ENHANCES SNS ACTIVITY

ROS stimulates central and peripheral SNS activity. The sympathetic nervous control center is located in the rostral ventrolateral medulla (RVLM). Redox status in the RVLM is crucial in regulating the sympathetic outflows and blood pressure. Induction of oxidative stress in the RVLM increases sympathetic outflow and elevates blood pressure (Wang et al. 2018). On the contrary, reducing ROS levels in the RVLM inhibits SNS activity and lowers blood pressure (Wang et al. 2018).

The sympathoexcitation induced by oxidative stress in the RVLM has been shown to play a crucial role in causing blood pressure elevation in a number of hypertension models, including obesity-induced hypertension, neurogenic hypertension, and angiotensin II-induced hypertension, as well as in the spontaneously hypertensive rats (Li and Xia 2020).

NADPH oxidases seems to play a key role in mediating RVLM oxidative stress (Figure 21.2). Bilateral microinjection of angiotensin II into the RVLM leads to oxidative stress and blood pressure

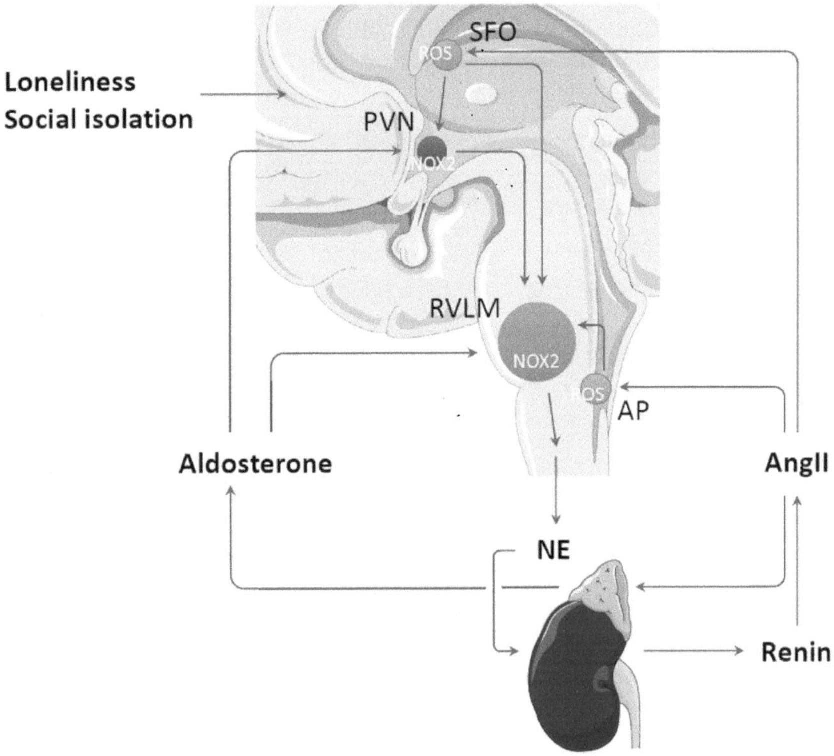

FIGURE 21.2 Role of oxidative stress in regulating sympathetic nerve activity. NOX2-mediated oxidative stress in the rostral ventrolateral medulla (RVLM) is crucial in regulating the sympathetic outflows and blood pressure. This has been shown in numerous hypertension models but not in models of social isolation so far. Social isolation leads to up-regulation of NOX2 gene expression in several brain regions; it is yet still unknown whether this also occurs in RVLM. Theoretically, SNS stimulation leads to activation of the renin-angiotensin system resulting elevated levels of circulating angiotensin II (AngII) and aldosterone. Aldosterone can penetrate the blood-brain barrier and induce NOX2 expression in the paraventricular nucleus (PVN), perhaps also in RVLM. AngII, on the other hand, can cause oxidative stress in blood-brain barrier lacking circumventricular organs, such as area postrema (AP) and subfornical organ (SFO). Oxidative stress in such structures may spread to the RVLM via direct diffusion of oxidants or indirect mechanisms. SFO oxidative stress can also cause PVN activation. PVN neurons, in turn, enhance SNS activity by stimulating sympathetic preganglionic neurons in the spinal cell column and by activating cells in the RVLM. Modified from Li and Xia (2020), which is an open access article distributed under the terms of the Creative Commons CC BY license.

elevation. These effects are mediated by the angiotensin receptor subtype 1 and can be prevented by inhibiting NADPH oxidase. The main hypertension-related RVLM isoform is NOX2 (Tanaka and Laurindo 2018).

21.9 OXIDATIVE STRESS IN THE VASCULATURE

Social isolation leads to oxidative stress in the vascular tissue (Figure 21.3). Higher NADPH oxidase activity is observed in the aortic arch of individually caged Watanabe heritable hyperlipidemic rabbits (Nation et al. 2008). NADPH oxidases represent the major ROS producers in blood vessels and the NADPH oxidase-mediated oxidative stress promotes atherosclerosis (Li and Xia 2020). Indeed, increased atherosclerotic lesions have been found in socially isolated Watanabe heritable hyperlipidemic rabbits (McCabe et al. 2002) as well as apolipoprotein E-knockout mice (Bernberg et al. 2008). In cell culture experiments, treatment with oxytocin reduces NADPH oxidase activity in endothelial cells, smooth muscle cells, monocytes and macrophages, and endothelial cells (Szeto et al. 2008). Oxytocin also inhibits proinflammatory cytokine secretion from endothelial cells and macrophages. Chronic *in vivo* administration with oxytocin attenuates atherosclerosis lesion in socially isolated animals, both in Watanabe heritable hyperlipidemic rabbits (Szeto et al.

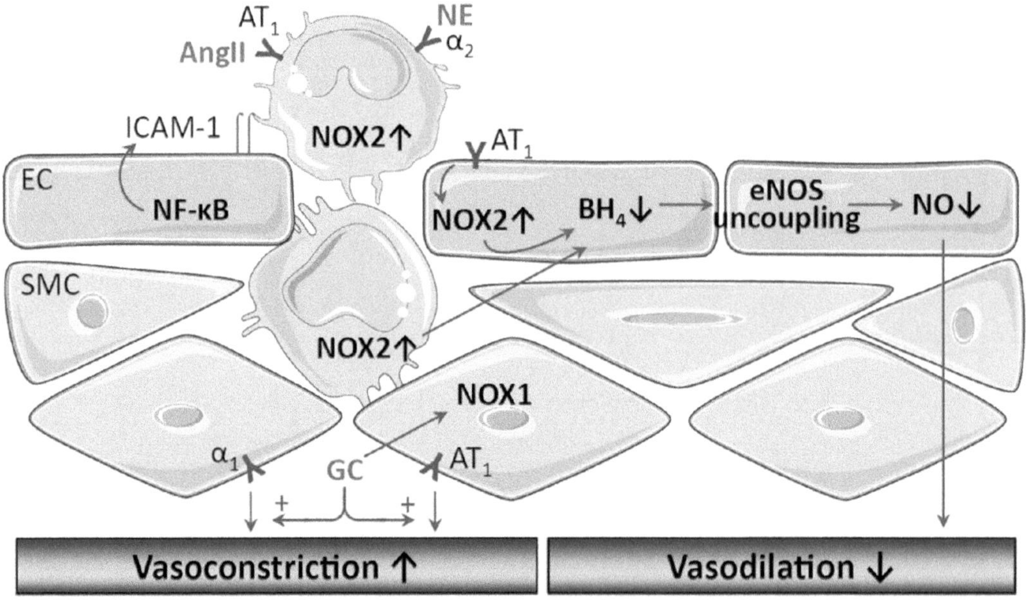

FIGURE 21.3 Social isolation-induced inflammation and oxidative stress in the vasculature. Social isolation leads to activation of the HPA axis and the SNS. SNS stimulates renin secretion and the production of angiotensin II (AngII). AngII activates NOX2 in endothelial cells (EC) causing oxidative stress, which may induce uncoupling of the endothelial nitric oxide synthase (eNOS) through tetrahydrobiopterin (BH4) oxidation. Oxidative stress in EC also leads to activation of NF-κB and the induction of adhesion molecules resulting in vascular inflammation. AngII and the SNS neurotransmitter norepinephrine (NE) stimulate NOX2-medaited superoxide production via AT1 and α2 receptors, respectively, and promote the infiltration of immune cells into the vascular wall. Glucocorticoids (GC) enhance NOX1 expression in vascular smooth muscle cells (SMC) and potentiate vasoconstriction by elevated NE (via α1 receptor) or AngII (AT1 receptors). Moreover, GCs also reduce eNOS gene expression and serine 1177 phosphorylation resulting in decreased NO production and impaired vasodilation. Reproduced from Li and Xia (2020), which is an open access article distributed under the terms of the Creative Commons CC BY license.

2013) and in socially isolated apolipoprotein E-knockout mice (Nation et al. 2010). Although it is not clear to what extent the anti-atherosclerotic effect of oxytocin is attributable to the inhibition of NADPH oxidase activity, it is conceivable that the reduction of oxidative stress contributes to the vasoprotective effect of the compound.

Mechanistically, social isolation-induced vascular oxidative stress can be mediated by HPA axis, SNS activation, and inflammatory cells (Figure 21.3). The effects of glucocorticoids on NADPH oxidase are likely to be cell type-dependent and context-dependent (Schiavone et al. 2013). Treatment with dexamethasone has been shown to enhance the expression of NOX1 in cultured vascular smooth muscle cells as well as in vascular tissues *in vivo* (Siuda et al. 2014).

SNS activation may also contribute to social isolation-induced vascular oxidative stress. Enhanced systemic oxidative stress has been observed in rats treated *in vivo* with the sympathetic neurotransmitter norepinephrine. Treatment of isolated human peripheral blood mononuclear cells (PBMCs) with norepinephrine leads to up-regulation of NOX2 and its component proteins $p22^{phox}$ and $p67^{phox}$ resulting in enhanced superoxide production (Deo et al. 2013). The norepinephrine-stimulated superoxide production is mainly mediated by the α2-receptor and involves protein kinase C activity (Deo et al. 2013). In addition, treatment of monocytes with norepinephrine enhances the adhesion capacity with endothelial cells, indicating a proinflammatory phenotype (Deo et al. 2013). This is consistent with the finding that SNS activation in lonely individuals elevates the number of proinflammatory monocytes (Cole, Capitanio, et al. 2015).

21.10 MITOCHONDRIAL ROS

In addition to NADPH oxidase, increased ROS production by mitochondria also contributes to social isolation-induced oxidative stress (Schiavone et al. 2013). The mitochondrial respiratory chain is a major source of cellular ROS that is counterbalanced by glutathione and endogenous antioxidant systems.

Chronic social isolation inhibits the activities of antioxidant enzymes catalase, glutathione peroxidase, and superoxide dismutase in the rat hippocampus (Shao et al. 2015). Juvenile social isolation stress induces mitochondrial dysfunction in adulthood, evidenced as impaired respiratory chain complex resulting in mitochondrial ROS formation, oxidative damage, and ATP reduction in both brain and heart (Sonei et al. 2017). Chronic social isolation in rats inhibits mitochondrial oxidative metabolism by ROS-dependent inhibition of citric acid cycle enzymes containing redox-sensitive active sites (Zhuravliova et al. 2009). Mitochondrial dysfunction and oxidative stress have also been observed in hippocampus of mice subjected to early social isolation stress (Amiri et al. 2015). In female rhesus macaques, social interactions have been shown to influence mitochondrial DNA copy number in immune cells (Debray et al. 2019).

The exact mechanisms of how mitochondrial ROS participate in social isolation-induced pathology are incompletely understood. It has been shown that mitochondrial ROS promote the production of proinflammatory cytokines such as IL-6 and TNF-α. Moreover, there exists a cross-talk between mitochondrial ROS and NADPH oxidase-derived ROS, resulting in an amplification mechanism for cellular oxidative stress (Daiber 2010). The involvement of such mechanisms in social isolation-induced cardiovascular disease, however, remains elusive.

21.11 SUMMARY AND FUTURE DIRECTIONS

There is clear evidence that loneliness and social isolation impairs cardiovascular health. The underlying mechanisms may include HPA activation, SNS hyperactivity, parasympathetic dysfunction, and a proinflammatory immune response. Oxidative stress in the brain is required for the social isolation-induced HPA activation and is probably also involved in SNS activation. In the vasculature, oxidative stress impairs endothelial function and promotes atherosclerosis. Both in the central nervous system and in the peripheral system, NOX2 seems to be a key superoxide producer mediating social isolation-induced oxidative stress. This, however, remains to be validated in future studies using transgenic animal models. Another point is that some mechanisms presented in the present chapter are puzzled together from studies using different models. Whether these mechanisms apply to social isolation remains elusive.

ACKNOWLEDGEMENTS

Grants from the Deutsche Forschungsgemeinschaft (DFG; LI-1042/1-1, LI-1042/3-1, LI-1042/5-1 and XI 139/2-1) and intramural fund (Stufe I) of the Johannes Gutenberg University Medical Center Mainz supported the original work from the authors' laboratory included in this review.

AUTHOR DISCLOSURE STATEMENT

The authors declare that they have no conflicts of interest.

REFERENCES

Adam, E. K., L. C. Hawkley, B. M. Kudielka, and J. T. Cacioppo. 2006. "Day-to-day dynamics of experience: Cortisol associations in a population-based sample of older adults." *Proc Natl Acad Sci U S A* 103 (45): 17058–17063. https://doi.org/10.1073/pnas.0605053103. www.ncbi.nlm.nih.gov/pubmed/17075058.

Amiri, S., H. Amini-Khoei, A. Haj-Mirzaian, M. Rahimi-Balaei, P. Naserzadeh, A. Dehpour, S. E. Mehr, and M. J. Hosseini. 2015. "Tropisetron attenuated the anxiogenic effects of social isolation by modulating nitrergic system and mitochondrial function." *Biochim Biophys Acta* 1850 (12): 2464–2475. https://doi.org/10.1016/j.bbagen.2015.09.009. www.ncbi.nlm.nih.gov/pubmed/26367080

Bernberg, E., I. J. Andersson, L. M. Gan, A. S. Naylor, M. E. Johansson, and G. Bergstrom. 2008. "Effects of social isolation and environmental enrichment on atherosclerosis in ApoE-/- mice." *Stress* 11 (5): 381–389. https://doi.org/10.1080/10253890701824051. www.ncbi.nlm.nih.gov/pubmed/18609305.

Black, C. N., M. Bot, D. Revesz, P. G. Scheffer, and B. Penninx. 2017. "The association between three major physiological stress systems and oxidative DNA and lipid damage." *Psychoneuroendocrinology* 80: 56–66. https://doi.org/10.1016/j.psyneuen.2017.03.003. www.ncbi.nlm.nih.gov/pubmed/28319849.

Bosch, O. J., H. P. Nair, T. H. Ahern, I. D. Neumann, and L. J. Young. 2009. "The CRF system mediates increased passive stress-coping behavior following the loss of a bonded partner in a monogamous rodent." *Neuropsychopharmacology* 34 (6): 1406–1415. https://doi.org/10.1038/npp.2008.154. www.ncbi.nlm.nih.gov/pubmed/18923404

Cacioppo, J. T., S. Cacioppo, J. P. Capitanio, and S. W. Cole. 2015. "The neuroendocrinology of social isolation." *Annu Rev Psychol* 66: 733–767. https://doi.org/10.1146/annurev-psych-010814-015240. www.ncbi.nlm.nih.gov/pubmed/25148851

Cacioppo, J. T., S. Cacioppo, S. W. Cole, J. P. Capitanio, L. Goossens, and D. I. Boomsma. 2015. "Loneliness across phylogeny and a call for comparative studies and animal models." *Perspect Psychol Sci* 10 (2): 202–212. https://doi.org/10.1177/1745691614564876. www.ncbi.nlm.nih.gov/pubmed/25910390

Colaianna, M., S. Schiavone, M. Zotti, P. Tucci, M. G. Morgese, L. Backdahl, R. Holmdahl, K. H. Krause, V. Cuomo, and L. Trabace. 2013. "Neuroendocrine profile in a rat model of psychosocial stress: Relation to oxidative stress." *Antioxid Redox Signal* 18 (12): 1385–1399. https://doi.org/10.1089/ars.2012.4569. www.ncbi.nlm.nih.gov/pubmed/23320850

Cole, S. W. 2008. "Social regulation of leukocyte homeostasis: The role of glucocorticoid sensitivity." *Brain Behav Immun* 22 (7): 1049–1055. https://doi.org/10.1016/j.bbi.2008.02.006. www.ncbi.nlm.nih.gov/pubmed/18394861

Cole, S. W. 2014. "Human social genomics." *PLoS Genet* 10 (8): e1004601. https://doi.org/10.1371/journal.pgen.1004601.

Cole, S. W., J. P. Capitanio, K. Chun, J. M. Arevalo, J. Ma, and J. T. Cacioppo. 2015. "Myeloid differentiation architecture of leukocyte transcriptome dynamics in perceived social isolation." *Proc Natl Acad Sci U S A* 112 (49): 15142–15147. https://doi.org/10.1073/pnas.1514249112. www.ncbi.nlm.nih.gov/pubmed/26598672

Cole, S. W., L. C. Hawkley, J. M. Arevalo, and J. T. Cacioppo. 2011. "Transcript origin analysis identifies antigen-presenting cells as primary targets of socially regulated gene expression in leukocytes." *Proc Natl Acad Sci U S A* 108 (7): 3080–3085. https://doi.org/10.1073/pnas.1014218108. www.ncbi.nlm.nih.gov/pubmed/21300872

Cole, S. W., L. C. Hawkley, J. M. Arevalo, C. Y. Sung, R. M. Rose, and J. T. Cacioppo. 2007. "Social regulation of gene expression in human leukocytes." *Genome Biol* 8 (9): R189. https://doi.org/10.1186/gb-2007-8-9-r189. www.ncbi.nlm.nih.gov/pubmed/17854483.

Cole, S. W., M. E. Levine, J. M. Arevalo, J. Ma, D. R. Weir, and E. M. Crimmins. 2015. "Loneliness, eudaimonia, and the human conserved transcriptional response to adversity." *Psychoneuroendocrinology* 62: 11–7. https://doi.org/10.1016/j.psyneuen.2015.07.001. www.ncbi.nlm.nih.gov/pubmed/26246388

Daiber, A. 2010. "Redox signaling (cross-talk) from and to mitochondria involves mitochondrial pores and reactive oxygen species." *Biochim Biophys Acta* 1797 (6–7): 897–906. https://doi.org/10.1016/j.bbabio.2010.01.032. www.ncbi.nlm.nih.gov/pubmed/20122895.

Debray, R., N. Snyder-Mackler, J. N. Kohn, M. E. Wilson, L. B. Barreiro, and J. Tung. 2019. "Social affiliation predicts mitochondrial DNA copy number in female rhesus macaques." *Biol Lett* 15 (1): 20180643. https://doi.org/10.1098/rsbl.2018.0643. www.ncbi.nlm.nih.gov/pubmed/30958211

Deo, S. H., N. T. Jenkins, J. Padilla, A. R. Parrish, and P. J. Fadel. 2013. "Norepinephrine increases NADPH oxidase-derived superoxide in human peripheral blood mononuclear cells via alpha-adrenergic receptors." *Am J Physiol Regul Integr Comp Physiol* 305 (10): R1124–R1132. https://doi.org/10.1152/ajpregu.00347.2013. www.ncbi.nlm.nih.gov/pubmed/24068047

Doane, L. D., and E. K. Adam. 2010. "Loneliness and cortisol: Momentary, day-to-day, and trait associations." *Psychoneuroendocrinology* 35 (3): 430–441. https://doi.org/10.1016/j.psyneuen.2009.08.005. www.ncbi.nlm.nih.gov/pubmed/19744794.

Gavrilovic, L., N. Spasojevic, and S. Dronjak. 2010. "Chronic individual housing-induced stress decreased expression of catecholamine biosynthetic enzyme genes and proteins in spleen of adult rats." *Neuroimmunomodulation* 17 (4): 265–269. https://doi.org/10.1159/000290042. www.ncbi.nlm.nih.gov/pubmed/20203532

Gouin, J. P., B. Zhou, and S. Fitzpatrick. 2015. "Social integration prospectively predicts changes in heart rate variability among individuals undergoing migration stress." *Ann Behav Med* 49 (2): 230–238. https://doi.org/10.1007/s12160-014-9650-7. www.ncbi.nlm.nih.gov/pubmed/25212509

Grippo, A. J., D. G. Lamb, C. S. Carter, and S. W. Porges. 2007. "Social isolation disrupts autonomic regulation of the heart and influences negative affective behaviors." *Biol Psychiatry* 62 (10): 1162–1170. https://doi.org/10.1016/j.biopsych.2007.04.011. www.ncbi.nlm.nih.gov/pubmed/17658486

Grippo, A. J., D. M. Trahanas, R. R. Zimmerman, 2nd, S. W. Porges, and C. S. Carter. 2009. "Oxytocin protects against negative behavioral and autonomic consequences of long-term social isolation." *Psychoneuroendocrinology* 34 (10): 1542–1553. https://doi.org/10.1016/j.psyneuen.2009.05.017. www.ncbi.nlm.nih.gov/pubmed/19553027.

Hawkley, L. C., S. W. Cole, J. P. Capitanio, G. J. Norman, and J. T. Cacioppo. 2012. "Effects of social isolation on glucocorticoid regulation in social mammals." *Horm Behav* 62 (3): 314–323. https://doi.org/10.1016/j.yhbeh.2012.05.011. www.ncbi.nlm.nih.gov/pubmed/22663934.

Hawkley, L. C., R. A. Thisted, C. M. Masi, and J. T. Cacioppo. 2010. "Loneliness predicts increased blood pressure: 5-year cross-lagged analyses in middle-aged and older adults." *Psychol Aging* 25 (1): 132–141. https://doi.org/10.1037/a0017805. www.ncbi.nlm.nih.gov/pubmed/20230134

Heidt, T., H. B. Sager, G. Courties, P. Dutta, Y. Iwamoto, A. Zaltsman, C. von Zur Muhlen, C. Bode, G. L. Fricchione, J. Denninger, et al. 2014. "Chronic variable stress activates hematopoietic stem cells." *Nat Med* 20 (7): 754–758. https://doi.org/10.1038/nm.3589. www.ncbi.nlm.nih.gov/pubmed/24952646

Hemingway, H., M. Shipley, E. Brunner, A. Britton, M. Malik, and M. Marmot. 2005. "Does autonomic function link social position to coronary risk? The Whitehall II study." *Circulation* 111 (23): 3071–3077. https://doi.org/10.1161/CIRCULATIONAHA.104.497347. www.ncbi.nlm.nih.gov/pubmed/15939818.

Holt-Lunstad, J., and T. B. Smith. 2016. "Loneliness and social isolation as risk factors for CVD: Implications for evidence-based patient care and scientific inquiry." *Heart* 102 (13): 987–989. https://doi.org/10.1136/heartjnl-2015-309242. www.ncbi.nlm.nih.gov/pubmed/27091845

Holt-Lunstad, J., T. B. Smith, M. Baker, T. Harris, and D. Stephenson. 2015. "Loneliness and social isolation as risk factors for mortality: A meta-analytic review." *Perspect Psychol Sci* 10 (2): 227–237. https://doi.org/10.1177/1745691614568352. www.ncbi.nlm.nih.gov/pubmed/25910392.

Holt-Lunstad, J., T. B. Smith, and J. B. Layton. 2010. "Social relationships and mortality risk: A meta-analytic review." *PLoS Med* 7 (7): e1000316. https://doi.org/10.1371/journal.pmed.1000316. www.ncbi.nlm.nih.gov/pubmed/20668659

Li, H., and N. Xia. 2020. "The role of oxidative stress in cardiovascular disease caused by social isolation and loneliness." *Redox Biol* 37: 101585. https://doi.org/10.1016/j.redox.2020.101585. www.ncbi.nlm.nih.gov/pubmed/32709420

Lutgendorf, S. K., K. DeGeest, L. Dahmoush, D. Farley, F. Penedo, D. Bender, M. Goodheart, T. E. Buekers, L. Mendez, G. Krueger, L. Clevenger, D. M. Lubaroff, A. K. Sood, and S. W. Cole. 2011. "Social isolation is associated with elevated tumor norepinephrine in ovarian carcinoma patients." *Brain Behav Immun* 25 (2): 250–255. https://doi.org/10.1016/j.bbi.2010.10.012. www.ncbi.nlm.nih.gov/pubmed/20955777

Lutgendorf, S. K., K. DeGeest, C. Y. Sung, J. M. Arevalo, F. Penedo, J. Lucci, 3rd, M. Goodheart, D. Lubaroff, D. M. Farley, A. K. Sood, and S. W. Cole. 2009. "Depression, social support, and beta-adrenergic transcription control in human ovarian cancer." *Brain Behav Immun* 23 (2): 176–183. https://doi.org/10.1016/j.bbi.2008.04.155. www.ncbi.nlm.nih.gov/pubmed/18550328

McCabe, P. M., J. A. Gonzales, J. Zaias, A. Szeto, M. Kumar, A. J. Herron, and N. Schneiderman. 2002. "Social environment influences the progression of atherosclerosis in the watanabe heritable hyperlipidemic rabbit." *Circulation* 105 (3): 354–359. www.ncbi.nlm.nih.gov/pubmed/11804992

McNeal, N., M. A. Scotti, J. Wardwell, D. L. Chandler, S. L. Bates, M. Larocca, D. M. Trahanas, and A. J. Grippo. 2014. "Disruption of social bonds induces behavioral and physiological dysregulation in male and female prairie voles." *Auton Neurosci* 180: 9–16. https://doi.org/10.1016/j.autneu.2013.10.001. www.ncbi.nlm.nih.gov/pubmed/24161576

Meyer, T., and P. H. Wirtz. 2018. "Mechanisms of mitochondrial redox signaling in psychosocial stress-responsive systems: New insights into an old story." *Antioxid Redox Signal* 28 (9): 760–772. https://doi.org/10.1089/ars.2017.7186. www.ncbi.nlm.nih.gov/pubmed/28558479.

Nation, D. A., J. A. Gonzales, A. J. Mendez, J. Zaias, A. Szeto, L. G. Brooks, J. Paredes, A. D'Angola, N. Schneiderman, and P. M. McCabe. 2008. "The effect of social environment on markers of vascular oxidative stress and inflammation in the Watanabe heritable hyperlipidemic rabbit." *Psychosom Med* 70 (3): 269–275. https://doi.org/10.1097/PSY.0b013e3181646753. www.ncbi.nlm.nih.gov/pubmed/18256340.

Nation, D. A., A. Szeto, A. J. Mendez, L. G. Brooks, J. Zaias, E. E. Herderick, J. Gonzales, C. M. Noller, N. Schneiderman, and P. M. McCabe. 2010. "Oxytocin attenuates atherosclerosis and adipose tissue inflammation in socially isolated ApoE-/- mice." *Psychosom Med* 72 (4): 376–382. https://doi.org/10.1097/PSY.0b013e3181d74c48. www.ncbi.nlm.nih.gov/pubmed/20368478 www.ncbi.nlm.nih.gov/pmc/articles/PMC4784697/pdf/nihms764058.pdf.

Paredes, J., A. Szeto, J. E. Levine, J. Zaias, J. A. Gonzales, A. J. Mendez, M. M. Llabre, N. Schneiderman, and P. M. McCabe. 2006. "Social experience influences hypothalamic oxytocin in the WHHL rabbit." *Psychoneuroendocrinology* 31 (9): 1062–1075. https://doi.org/10.1016/j.psyneuen.2006.06.007. www.ncbi.nlm.nih.gov/pubmed/16963189

Powell, N. D., E. K. Sloan, M. T. Bailey, J. M. Arevalo, G. E. Miller, E. Chen, M. S. Kobor, B. F. Reader, J. F. Sheridan, and S. W. Cole. 2013. "Social stress up-regulates inflammatory gene expression in the leukocyte transcriptome via beta-adrenergic induction of myelopoiesis." *Proc Natl Acad Sci U S A* 110 (41): 16574–16579. https://doi.org/10.1073/pnas.1310655110. www.ncbi.nlm.nih.gov/pubmed/24062448

Randall, G., M. R. Bhattacharyya, and A. Steptoe. 2009. "Marital status and heart rate variability in patients with suspected coronary artery disease." *Ann Behav Med* 38 (2): 115–123. https://doi.org/10.1007/s12160-009-9137-0. www.ncbi.nlm.nih.gov/pubmed/19806415.

Rosengren, A., S. Hawken, S. Ounpuu, K. Sliwa, M. Zubaid, W. A. Almahmeed, K. N. Blackett, C. Sitthi-amorn, H. Sato, S. Yusuf, and Interheart investigators. 2004. "Association of psychosocial risk factors with risk of acute myocardial infarction in 11119 cases and 13648 controls from 52 countries (the INTERHEART study): Case-control study." *Lancet* 364 (9438): 953–962. https://doi.org/10.1016/S0140-6736(04)17019-0. www.ncbi.nlm.nih.gov/pubmed/15364186

Schiavone, S., V. Jaquet, L. Trabace, and K. H. Krause. 2013. "Severe life stress and oxidative stress in the brain: From animal models to human pathology." *Antioxid Redox Signal* 18 (12): 1475–1490. https://doi.org/10.1089/ars.2012.4720. www.ncbi.nlm.nih.gov/pubmed/22746161

Schiavone, S., S. Sorce, M. Dubois-Dauphin, V. Jaquet, M. Colaianna, M. Zotti, V. Cuomo, L. Trabace, and K. H. Krause. 2009. "Involvement of NOX2 in the development of behavioral and pathologic alterations in isolated rats." *Biol Psychiatry* 66 (4): 384–392. https://doi.org/10.1016/j.biopsych.2009.04.033. www.ncbi.nlm.nih.gov/pubmed/19559404

Shao, Y., G. Yan, Y. Xuan, H. Peng, Q. J. Huang, R. Wu, and H. Xu. 2015. "Chronic social isolation decreases glutamate and glutamine levels and induces oxidative stress in the rat hippocampus." *Behav Brain Res* 282: 201–208. https://doi.org/10.1016/j.bbr.2015.01.005. www.ncbi.nlm.nih.gov/pubmed/25591473

Shively, C. A., T. B. Clarkson, and J. R. Kaplan. 1989. "Social deprivation and coronary artery atherosclerosis in female cynomolgus monkeys." *Atherosclerosis* 77 (1): 69–76. www.ncbi.nlm.nih.gov/pubmed/2719764.

Siuda, D., S. Tobias, A. Rus, N. Xia, U. Forstermann, and H. Li. 2014. "Dexamethasone upregulates Nox1 expression in vascular smooth muscle cells." *Pharmacology* 94 (1–2): 13–20. https://doi.org/10.1159/000365932. www.ncbi.nlm.nih.gov/pubmed/25171190 www.karger.com/Article/Abstract/365932.

Smith, T. W., M. R. Cribbet, J. B. Nealey-Moore, B. N. Uchino, P. G. Williams, J. Mackenzie, and J. F. Thayer. 2011. "Matters of the variable heart: Respiratory sinus arrhythmia response to marital interaction and associations with marital quality." *J Pers Soc Psychol* 100 (1): 103–119. https://doi.org/10.1037/a0021136. www.ncbi.nlm.nih.gov/pubmed/20954783.

Sonei, N., S. Amiri, I. Jafarian, M. Anoush, M. Rahimi-Balaei, H. Bergen, A. Haj-Mirzaian, and M. J. Hosseini. 2017. "Mitochondrial dysfunction bridges negative affective disorders and cardiomyopathy in socially isolated rats: Pros and cons of fluoxetine." *World J Biol Psychiatry* 18 (1): 39–53. https://doi.org/10.3109/15622975.2016.1149218. www.ncbi.nlm.nih.gov/pubmed/27031288

Steptoe, A., and M. Kivimaki. 2013. "Stress and cardiovascular disease: An update on current knowledge." *Annu Rev Public Health* 34: 337–354. https://doi.org/10.1146/annurev-publhealth-031912-114452. www.ncbi.nlm.nih.gov/pubmed/23297662.

Steptoe, A., A. Shankar, P. Demakakos, and J. Wardle. 2013. "Social isolation, loneliness, and all-cause mortality in older men and women." *Proc Natl Acad Sci U S A* 110 (15): 5797–5801. https://doi.org/10.1073/pnas.1219686110. www.ncbi.nlm.nih.gov/pubmed/23530191

Sun, P., A. S. Smith, K. Lei, Y. Liu, and Z. Wang. 2014. "Breaking bonds in male prairie vole: Long-term effects on emotional and social behavior, physiology, and neurochemistry." *Behav Brain Res* 265: 22–31. https://doi.org/10.1016/j.bbr.2014.02.016. www.ncbi.nlm.nih.gov/pubmed/24561258

Szeto, A., D. A. Nation, A. J. Mendez, J. Dominguez-Bendala, L. G. Brooks, N. Schneiderman, and P. M. McCabe. 2008. "Oxytocin attenuates NADPH-dependent superoxide activity and IL-6 secretion in macrophages and vascular cells." *Am J Physiol Endocrinol Metab* 295 (6): E1495–E1501. https://doi.org/10.1152/ajpendo.90718.2008. www.ncbi.nlm.nih.gov/pubmed/18940936.

Szeto, A., M. A. Rossetti, A. J. Mendez, C. M. Noller, E. E. Herderick, J. A. Gonzales, N. Schneiderman, and P. M. McCabe. 2013. "Oxytocin administration attenuates atherosclerosis and inflammation in Watanabe Heritable Hyperlipidemic rabbits." *Psychoneuroendocrinology* 38 (5): 685–693. https://doi.org/10.1016/j.psyneuen.2012.08.009. www.ncbi.nlm.nih.gov/pubmed/22998949

Tanaka, L. Y., and F. R. M. Laurindo. 2018. "The eye of the needle: Redox mechanisms of acupuncture effects in hypertension." *Hypertension* 71 (2): 224–226. https://doi.org/10.1161/HYPERTENSIONAHA.117.09821. www.ncbi.nlm.nih.gov/pubmed/29229747

Thayer, J. F., S. S. Yamamoto, and J. F. Brosschot. 2010. "The relationship of autonomic imbalance, heart rate variability and cardiovascular disease risk factors." *Int J Cardiol* 141 (2): 122–131. https://doi.org/10.1016/j.ijcard.2009.09.543. www.ncbi.nlm.nih.gov/pubmed/19910061.

Valtorta, N. K., M. Kanaan, S. Gilbody, S. Ronzi, and B. Hanratty. 2016. "Loneliness and social isolation as risk factors for coronary heart disease and stroke: Systematic review and meta-analysis of longitudinal observational studies." *Heart* 102 (13): 1009–1016. https://doi.org/10.1136/heartjnl-2015-308790. www.ncbi.nlm.nih.gov/pubmed/27091846

Wang, X. R., J. W. Yang, C. S. Ji, X. H. Zeng, G. X. Shi, M. Fisher, and C. Z. Liu. 2018. "Inhibition of NADPH oxidase-dependent oxidative stress in the rostral ventrolateral medulla mediates the antihypertensive effects of acupuncture in spontaneously hypertensive rats." *Hypertension* 71 (2): 356–365. https://doi.org/10.1161/HYPERTENSIONAHA.117.09759. www.ncbi.nlm.nih.gov/pubmed/29229746

Xia, N., and H. Li. 2018. "Loneliness, social isolation, and cardiovascular health." *Antioxid Redox Signal* 28 (9): 837–851. https://doi.org/10.1089/ars.2017.7312. www.ncbi.nlm.nih.gov/pubmed/28903579

You, J. M., S. J. Yun, K. N. Nam, C. Kang, R. Won, and E. H. Lee. 2009. "Mechanism of glucocorticoid-induced oxidative stress in rat hippocampal slice cultures." *Can J Physiol Pharmacol* 87 (6): 440–447. https://doi.org/10.1139/y09-027. www.ncbi.nlm.nih.gov/pubmed/19526038

Zhuravliova, E., T. Barbakadze, E. Zaalishvili, M. Chipashvili, N. Koshoridze, and D. Mikeladze. 2009. "Social isolation in rats inhibits oxidative metabolism, decreases the content of mitochondrial K-Ras and activates mitochondrial hexokinase." *Behav Brain Res* 205 (2): 377–383. https://doi.org/10.1016/j.bbr.2009.07.009. www.ncbi.nlm.nih.gov/pubmed/19616040

Index

Note: Page numbers in *italics* indicate figures.

M

N

R

S

T

U

V

W

X

Z